# Ecology and Management of Inland Waters

## A Californian Perspective with Global Applications

# Ecology and Management of Inland Waters

## A Californian Perspective with Global Applications

**Marc Los Huertos**

Elsevier
Radarweg 29, PO Box 211, 1000 AE Amsterdam, Netherlands
The Boulevard, Langford Lane, Kidlington, Oxford OX5 1GB, United Kingdom
50 Hampshire Street, 5th Floor, Cambridge, MA 02139, United States



**Library of Congress Cataloging-in-Publication Data**
A catalog record for this book is available from the Library of Congress

**British Library Cataloguing-in-Publication Data**
A catalogue record for this book is available from the British Library

ISBN: 978-0-12-814266-0



*Publisher:* Candice Janco
*Acquisitions Editor:* Louisa Murno
*Editorial Project Manager:* Kelsey Connors
*Production Project Manager:* Kiruthika Govindaraju
*Designer:* Mark Rogers

Typeset by VTeX

*Dedicated to those who appreciate walking in streams and my son, Dylan, and daughters, Nicole and Kaia*

*Enlightenment is like the moon reflected on the water. The moon does not get wet, nor is the water broken.*

**Dogen Zenji (1200–1253)**

# Contents

# List of Figures



## List of Figures – Part 1

## List of Figures – Part 2

## List of Figures – Part 3

# List of Tables

# Preface

## MOTIVATIONS FOR THE PROJECT

This book is born as a curiosity as I was teaching 'Freshwater Ecology.' In spite of wonderful examples and explanations of aquatic ecosystems in many textbooks, they seem to miss the mark when I introduced students to inland waters of California. In fact, California is an odd place for limnologists, a word largely forgotten; the standard textbooks seem to describe standard systems, e.g., dimictic, midwest lakes, that are foreign bodies to most of California's residents, whereas familiar, iconic waters are missing (Fig. I).

While doing field work in ephemeral streams, alpine lakes, vernal pools, ephemeral streams, and regulated rivers in a Mediterranean-type climate, I began to wonder if a textbook would be substantially different if a Californian lens were used. And would residents of California see their inland waters differently if limnology was taught using California examples? I think the answer is yes. But this text is only a step toward that goal.

Of course, California and Mediterranean-type climate ecosystems have fascinating characteristics, but the patterns and processes that occur in these inland waters adhere to the same physical, chemical, and biological constraints as any aquatic system. But as I started supplementing popular texts with examples from California or other Mediterranean-climate type regions, I found divergent intellectual

**FIGURE I**

Yosemite National Park Yosemite Falls winter 2010 (Source: Wiki Commons).

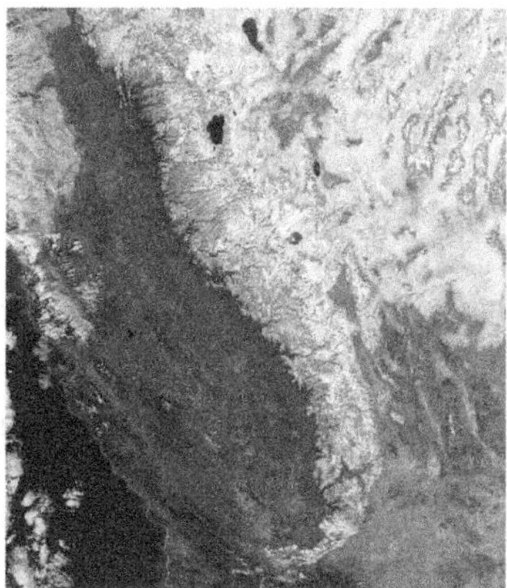

**FIGURE II**

The natural-color images of the Sierra Nevada between Feb. 15, 2018 (left), and Feb. 11, 2019 (right) (Source: NASA Earth Observatory). Much of the drinking water of the state is stored in winter snow, but questions about the impact of climate change on snow pack and the future of inland waters in the state requires a basic understanding of Climate Science and drivers of long-term and short-term variability.

trajectories from mainstream limnology narratives. In fact, Californian ecologists have had a long, impressive history that deserve a review of their intellectual legacy in the state and beyond and it seems to me that Californian's deserve to know how their ecosystems work from practitioners in their own state.

## ENVIRONMENTAL SCIENCE OF INLAND WATERS

### WHAT IS ENVIRONMENTAL SCIENCE OF INLAND WATERS?

For this book, inland waters are waters found in the terrestrial landscape, including the tidal freshwaters. To include estuaries would be ideal, especially since these waters depend so much on terrestrial waters, but was not included to maintain manageability and coherence, but I hope to include these waters in a future edition.

Environmental Science is an integrative science and relies on principles of Physics, Chemistry, Geology, and Biology. The relevance of each of these depend on the questions being asked about inland waters. For example, specialized fields such as Hydrology, Fluid Mechanics, Geomorphology, Applied Mathematics, Systematics, and Ecology can provide critical insight to manage inland waters. Ecologists working in inland waters might rely on specific sub-disciplines, such as Population Biology, Community Ecology, Ecophysiology, and Ecosystem Science to evaluate specific issues in Aquatic Systems. In addition, the management of inland waters requires the use of other integrative sciences, such as Climate Science (Fig. II), Biogeochemistry, Ecotoxicology, and Sustainability Science. With a goal to demonstrate the value of these topics, I use primary literature to demonstrate a variety tools that need to be applied to inland waters and their management.

Environmental Science is also a normative science, i.e., it has value-based goals. Thus as an applied field of study, Environmental Science values its subject matter, i.e., the environment for the sake of humans and our co-inhabitants. In this context, an Environmental Scientist will need to appreciate aspects of Environmental History, Environmental and Resource Economics, Political Ecology, and Public Policy. Finally, Environmental Scientists must appreciate the social stratification in society and locate Environmental Science within environmental justice and social movements; science studies; and cultural studies.

Writing a book that accomplished all of these is an ideal that I am moving towards—but remains an aspiration. Seen in this context, I expect the reader to ask questions about the ethical and economic concerns that influence what and how we think about our inland waters. For example, topics such as bioassessment and sustainable use of water involve both scientific and ethical questions and deserve to be treated from within and across a range of disciplines.

## SUBJECTIVITY IN PLACE AND TIME

I was trained in Wetland Ecology and Biogeochemistry, so I approach this topic from particular position. By integrating an Ecosystem Ecology perspective with evolutionary perspective, I find limited value in viewing the world's biota in terms of functional groups. Although functional groups can provide a useful lens considering the complexity of Aquatic Systems, it is only one lens. In response, I frame ecological processes in inland waters within the context of the geologic history of the planet and diversification of extinct and extant species. To accomplish this I became a student of the widest range of topics in my career, such as Tectonics, Phylogeny, Comparative Physiology. The greatest challenge has been to appreciate the complexities of these topics and then turn around to write and explain how these topics inform the Environmental Science of inland waters. As such, this text is also a manifesto to help frame the future of what I believe to be Environmental Science.

## TARGETED REGION AND AUDIENCE

The textbook focuses on examples from California's Mediterranean-type of climate, but by no means are the examples exclusive to California. The summer drought implies that making a living in surface waters can be a challenge that may be a relatively recent climate type. For California the climate has only existed in a relatively short geologic and evolutionary timeframe, perhaps 15 million years (Fig. III). With the topographic barriers, California's biota have experienced strong directional selection as the climate has become cooler and dryer. And with anthropogenic climate change, further selection will have a profound impact on inland waters. I believe that we can only manage our inland waters by acknowledging the linkages between tectonics, evolution diversification, and climate and to accomplish this I rely on concrete examples. I have not written this text to exclude other regions, but to manage the complexity of these linkages with a place-based approach.

This book is far from a complete picture of the complexities of Environmental Science of inland waters. However, I have attempted to fill a specific void in Environmental Science texts that provide readers the context, knowledge, and skills to address a range of issues that face our inland waters. The target audience include undergraduate and graduate students of Applied Ecology, perhaps as part of a semester course on the ecology of inland waters.

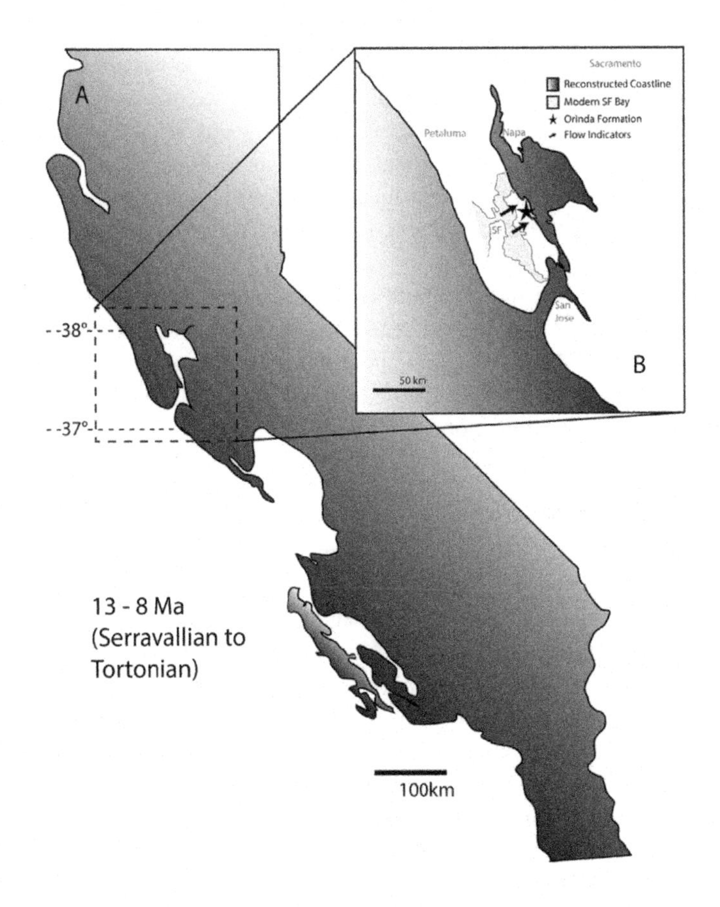


**FIGURE III**

Palaeogeographic reconstruction of California during the late Miocene (redrawn from Hall, 2002).

## TEXTBOOK GOALS AND CHAPTER SUMMARIES

As a former faculty member in the California State University system, of which I am a big fan, I use learning outcomes or objectives to constrain and organize the text. As I understand the pedagogical value of these, learning outcomes provide additional transparency for students and accountability for teachers. The alignment with learning outcomes and content is a laborious process and subject to constant revision, thus as a part of that process I developed tacit objectives for the text and each chapter. Students and instructors should use these to their advantage, taking note of sections that provide clear and compelling linkages between the outcomes and the Environmental Science of inland waters.

After a complete and careful perusal of the text in its entirety, readers should be able to:

1. Describe how *geological history* and *fossil record* inform our understanding of inland water ecology.

2. Summarize how *physical properties of water* select for morphological, physiological, and behavioral characters.
3. Explain why *water chemistry* varies across and within Aquatic Ecosystems and how this variation influences aquatic organisms.
4. Synthesize *biotic and abiotic factors* that control community structure and ecosystem function in streams, rivers, wetlands, lakes, reservoirs, and estuaries.
5. Appreciate the *diversity of aquatic taxa* and describe their role in inland waters.
6. Assess how *human activities* alter community structure and ecosystem function of Aquatic Systems.

## Part I: Coevolving Aquascapes

An evolutionary view of California's Aquatic System begins with a basic understanding of the relevant geologic history. Given the complexity of California's geology, I have attempted to write these chapters accessible for readers with limited Earth Science background. These chapters (1) provide succinct, accurate explanations of relevant geological principles and present essential elements of Geologic History, (2) demonstrate the ecological implications of this history for inland waters, and (3) frame the human development and the anthropocene as part of a trajectory easily explained by social processes.

### 1. A Coevolving Planet

This chapter explores how various planetary processes have selected for a wide range of species to invade inland waters while placing this evolution expansion within the context of major extinctions. Crossing disciplines to understand how ecosystems work and how they evolved is something the Environmental Scientists do on regular basis. In this case, we will focus on the geologic history of the earth to frame the development of inland waters with a long history.

Learning Outcomes:

1. Describe the evolution of the Earth's atmosphere and the implications for life on Earth;
2. Recognize the limitations of the fossil record and summarize what they tell us about evolution of inland waters;
3. Summarize the causes of major extinction events and describe why the current extinction process is unique in the Earth's history.

### 2. Archeology of Inland Waters

The human condition depends on our planet's inland waters. An effective use of water can impart success, whereas an ineffective use can lead to disaster. If nothing else, like all life forms on Earth, humans require water—water that is both reliably available and free of contaminants and pathogens. Meeting these goals is, perhaps, one of the greatest challenges of humankind. In this chapter, we introduce and analyze case studies to expose the evolving nature of our relationship to water.

Learning Outcomes:

1. Describe how water use has been central to the development of human societies;
2. Explain how the geography of a place provides specific opportunities and constraints in human-water relations;

3. Describe unintended consequences of water development projects;
4. Summarize examples aimed to improve the sustainable use of water.

## Part II: The Ecological Theater and the Evolutionary Play

A basic knowledge of the natural history of aquatic taxa is essential to an understanding of the ecology of inland waters—and I believe no existing Limnology text treats this topic effectively. I appreciate perspectives that combine both Systematics and Ecology in the teaching of this natural history. Using this integrated approach, these chapters integrate ecological principles within a context of evolutionary perceptive of inland taxa. My goal is to describe the evolutionary trends within taxonomic groups and highlight important ecological roles of particular aquatic groups, drawing on selected narratives about California's aquatic taxa.

The matrix of the water column is a foreign environment for many terrestrial organisms, including humans. By writing a rather long chapter on the physical and chemical of aspects of life in the water, my intention is to help readers appreciate the conditions that differentiate an aquatic existence from a terrestrial one. In some academic settings, the readership for this relatively technical part of the book may be limited to upper-division Biology majors and graduate students. If used carefully, however, this chapter may also serve as the basis for an introduction of Physics and Chemistry to readers who are interested in Ecology, but lack a strong background in Chemistry or Physics.

### 3. Evolving Aquatic Species

This chapter summarizes major divergences in taxa—with a specific interest in characteristics that may confer adaptive value for life in inland waters. For Environmental Scientists, the Tree of Life allows us to organize the evolutionary relationships among taxa and appreciate how species lineages radiate through geologic time, thus generating biodiversity. Furthermore, with a better understanding of the biodiversity, we can identify the ecological foundation in the patterns and processes of Aquatic Systems.

In addition, we summarize some characteristics of species introductions and how these processes are unique to inland waters on time scales that we can best appreciate.

Learning Outcomes:

1. Explore the evidence that demonstrates that species diversity lies at the microscopic level;
2. Summarize major branches of evolutionary divergence that has created the Earth's biodiversity;
3. Describe how hypothesized ancestral relationships help us appreciate the complex divergences in the taxa that have invaded inland waters;
4. Synthesize the tension between native and introduced species, while appreciating the history of ecological patterns of invasions, range expansions, relative to the current pace of introductions.

### 4. Population Growth and Competition

As groups of interbreeding organisms, populations level is where natural selection is often observed and provides analytical tools to help us predict how Aquatic Systems are structured. As a way to quantify survivorship, growth, and reproduction, population models are central in how we understand Ecological Systems.

Learning Outcomes:

1. Use density-dependent and density-independent equations to predict populations sizes;
2. Evaluate how intra- and interspecific competition can influence population growth;
3. Understand the role of competition within ecological and evolutionary time scales, and
4. Summarize the limitations of competition.

## 5. Beyond Competition

Whereas competition may influence population dynamics, exclude weaker competitors, and promote niche differentiation, we also recognize numerous other types of species interactions occur in streams, lakes, wetlands, etc. Predation, cannibalism, disease, and parasitism are commonly observed factors that affect populations dynamics.

Learning Outcomes:

1. Use Predator-Prey Equations to evaluate the role of trophic interactions to predict population dynamics;
2. Summarize how sexual selection can influence population dynamics, e.g., predator/prey and infection vector and host;
3. Describe succession and how communities change with time;
4. Evaluate various Lotic and Lentic Systems using energy transfer and nutrient flux in ecosystems.

## 6. Physical and Chemical Properties

Understanding the physical properties of water relies on a wide range of topics, such as Hydrology, Geomorphology, Fluid Dynamics, Optics, Temperature, and Chemistry. Far from the deep-end of the pool, we will wade in gently into these topics with the hope that the reader will become keenly interested about the value of these topics to understand the nature and ecology of inland waters.

Learning Outcomes:

1. Describe the physical and chemical characteristics of inland waters and how various species have adapted to these habitats.
2. Understand how water movement, fluid dynamics, light, temperature, and chemicals structure biotic communities.

## 7. Aquatic Typologies

By relying on the patterns and processes in Aquatic Systems, this chapter outlines the nomenclature to classify and describe inland waters.

Learning Outcomes:

1. Understand how patterns and processes are used to classify Aquatic Systems.

### 8. Patterns and Processes

The combination of community dynamics and trophic interactions define Aquatic Ecosystems, but individuals negotiate these processes within genetically controlled physiological tolerances, morphological characters, and behavioral responses.

Learning Outcomes:

1. Summarize how individual tolerances, reproductive strategies, and trophic dynamics influence Aquatic Ecosystems;
2. Describe the contrasts between top-down and bottom-up controls in Aquatic Systems.

## Part III: Water in the Anthropocene

Part III focuses on waters in the anthropocene with a strong focus on management issues. Starting with supplies and quality of waters, the chapters integrate how inland waters are managed within the physical and chemical constraints of the matrix and how these impact patterns and processes in waters. Finally, I attempted to highlight a wide range of case studies that demonstrate both areas of concerns and potential restoration activities to improve environmental outcomes.

### 9. Development and Appropriation

In the context of a highly variable spatial and temporal patterns of rainfall, water is unevenly available. Thus human beings build infrastructure to store and move these waters as part of our development. Therefore projects to "develop" water have direct links to economic activity and political power. As a result, changes to the water cycle have profound implications to the sustainable use of water that considers a broad range of uses and values.

Learning Outcomes:

1. Describe how water management goals can be consistent with the Flood Pulse Concept;
2. Describe how ground-surface water interactions influence water supplies;
3. Summarize how Aquatic Ecosystems are modified by regulated rivers;
4. Evaluate how environmental flows might be used to improve ecological outcomes.
5. Apply Watershed Management Concepts to improve water supply management goals.

### 10. Water Quality and Catchments

Water is a resource, but can also be contaminated and a potential hazard. Whether from geologic sources, biological diseases, or anthropocentric pollution, various attempts are made to treat and reduce our risk and exposure to poor water quality.

Learning Outcomes:

1. Summarize the ecological processes that promote water-borne diseases;
2. Describe how Geochemistry and nutrients influence water quality;
3. Describe the fate, transport, and toxicity of pollutants in Aquatic Systems and their impacts;
4. Summarize efforts to assess and improve the water quality.

## 11. Biogeochemistry and Global Change

This chapter explores the physical, chemical, biological, and geological processes and reactions that govern the composition of and changes to Aquatic Systems. Depending on the redox, pH and chemical concentrations, chemicals that might that ultimately influence the patterns and processes in Aquatic Systems may be transformed.

Learning Outcomes:

1. Describe the patterns and processes that influence chemicals in Aquatic Systems, and
2. Summarize how biogeochemical cycles influence the use of Aquatic Systems.

## 12. Conservation and Restoration

Restoration ecology has become a well-developed field that provides numerous tools to improve the ecological structure and function of inland waters. With threats to species diversity, habitat quality, and ecosystem services, it is critical to appreciate the lessons learned from past restoration activities, and refine tools for future efforts. This chapter showcases a number of restoration efforts and how their success might be evaluated.

Learning Outcomes:

1. Describe the ecological consequences of invasive species to inland waters;
2. Summarize strategies to manage and control invasive species;
3. Describe the challenges and successes of efforts to restore wetlands, streams, and lakes in California.
4. Compare and contrast various definitions of restoration success;
5. Evaluate various monitoring methods to gauge the success of restoration.

## 13. Political & Watershed Boundaries

This chapter summarizes the regulatory and policy context for the allocation, use, treatment, oversight, and protection of inland waters. In a world where the water supply "pie" is shrinking, the institutional capacity must increase to meet increasing water demand and a changing climate. On the other hand, various partnerships have emerged to make significant progress to improve inland waters for the benefit of humans and nonhumans. Across the jurisdictions, several water-rights regimes, along with environmental legislation imposed by various state and federal agencies, create a dizzyingly complex political context; this primer will help readers to make sense of public policy and water.

Learning Outcomes:

1. Describe political conflicts and legal context for water supply;
2. Explain how endangered species protection influence the ecology of inland waters;
3. Describe the policy tools used to resolve conflicts and protect water supplies, water quality, and Aquatic Ecosystems.

## 14. Concluding Remarks

This chapter suggests that a robust approach can be built on three components.

Learning Outcome:

1. Describe three types of activities to protect and restore inland waters.

# Acknowledgments

I have been fortunate to be part of a generous community, which has been willing to contribute ideas, thoughts, and suggestions. And even before this book was an idea, the teachers who have led me here are the sources of my interests: Ms. Candland, Betty MacDonald, Stephen Gliessman, Micheal Josselyn, VT Parker, Jean Langenheim, Andy Fisher, and Daniel Press.

As the project became more defined, I was lucky to have a wide range of partners that provided ideas, contributed literature reviews, and edited prose to create a resource for those interested in California's inland waters. Although I take the responsibility (and blame) for the text, this has been a collaborative effort with outstanding students from CSU Monterey Bay and Pomona College.

The following students have made substantial contributions: Regelio Arenas, Ruth "Rose" Ashbach, Ryan Bassett, Brittani Bohlke, Cherie Crawford, Patricia Cubanski, Becky Chung, Miles Daniels, Holly Dillon, Janet Ilse, Pam Krone-Davis, Michele Lanctot, Sami Murphy, Pauline Perkins, Kirk Post, Adam "AJ" Purdy, Albert "Arnette" Young, and Erin Stanfield.

I give special thanks to several colleagues who have read early versions of the text and made substantial revision suggestions. These include Professor Andrew Fisher (UC Santa Cruz), Professor Emeritus Rick Hazlett (Pomona College), Professor Dwight Whittaker (Pomona College), Kathryn Hargan, Molly Detrich, and Adam Collins. In addition, I appreciate the hard work of Luyi Huang: redrawing figures and reviewing the copyright requirements for hundreds of figures.

To improve the text, I retained the services of Sarah Rabkin to provide editorial corrections and suggestions. Sarah's insight and capacity to wade through my text to see a clear path is exceptional. Without a doubt the text that flows well is the result of her efforts; I wish I had her assistance to revise more than just a few chapters. The prose will always need refinement, which I do regularly, but they will never reach the standard I seek. I am happy to include contributions, suggestions, additions, and corrections.

Finally, I thank my family and friends for their encouragement and willingness to watch me disappear into the text for hours as I tried to learn and write about complex topics that I felt ill-prepared to write. Thank you Frances, Dylan, Nina, Kari, Kaia, Albert, Casey, Nicole, Char, Deidre, Bowman, Aileen, Dave, Heidi, Rick, Heather, Tamara, Bryce, Zayn, Gerry, Jack, and Kay.

# Introduction

*The mind is not a vessel to be filled but a fire to be kindled.*
**Plutarch**

*All the water that will ever be is, right now.*
**National Geographic, October 1993**

The ecology and management of inland waters requires an interdisciplinary approach. As a reflection of how society is structured, failures to properly manage waters (reducing impacts of drought, floods, habitat degradation, and species diversity), requires us to think across disciplines to use tools to improve how we increase the resilience of inland waters and protect ecosystem services and habitat for a wide range of organisms.

## DELUGES AND DROUGHTS: WATER CYCLE AND HUMANS

### WATER AND THE BUILT ENVIRONMENT

As the saying goes, water is life. A truism since all cells are filled with water and water is required for every biological function. Another truism describes infrastructure as a function of human society, where much of that infrastructure is designed to maintain a supply of reliable clean water and protect ourselves from excesses and shortages.

In fact, our relationship to the Earth, depends on how we use infrastructure to mediate water resources. Do we withdraw water from streams and reduce fishery productivity? Do lakes receive waste products that the built environment is not designed to control. If we are to develop sustainable uses of water, the environmental outcomes of our built environment should be front and center in every decision we make about how water is 'developed' and used. And when we experience water-related disasters, we should question the effectiveness of how our society is structured and how it functions.

## HURRICANE HARVEY, MISSISSIPPI FLOODS, TREE MORTALITY IN CALIFORNIA

On a regular basis, some Huston residents experience flooding during heavy precipitation events. Thus the catastrophe of 2017s landfall of Hurricane Harvey was largely anticipated by those familiar with flood risks in the region. Nevertheless, with poor planning and zoning policies, residents had no capacity to be prepared for the rains that overwhelmed the city's drainage infrastructure (Fig. IV). Not only did the flooding become an immediate hazard, but it also led to water quality hazards even a year later.

During the spring and summer of 2019, the Illinois, the Missouri, the Arkansas, and the Mississippi Rivers reached flood stage (Fig. V). In a long-term effort to control annual flood waters, the United States government built levees and dams to reduce the risk of flooding. And when these efforts fail, Congress allocates more funding to build more flood-control structures. Meanwhile, all of these

**FIGURE IV**

Before and after the floods from tropical storm Harvey dumped over 2 feet of rain on Houston, Texas in three days. (Source: Shutterstock/Reuters/Business Insider).

**FIGURE V**

Source: NY Times.

rivers dams block fish migration, Riparian Wetlands are lost, and ecosystems services (e.g.flood storage, nutrient retention, riparian habitat, water purification) have become dramatically reduced, even as flooding events continue to occur.

In the west, over 150 million trees in the southern Sierra Nevada Mountains are dead. After the most extreme drought in over 1000 years, tree death was inevitable, but facilitated by bark beetles that attack water-stressed tress. Forest fires, loss of carbon sequestration, decreased public safety associated with falling dead trees will haunt the region in a state unprepared for the scale and scope of response needed. Moreover, wildfires as a result of fire suppression, over-stocked trees, and climate change will leave soils bare, increase erosion and sedimentation rates, and may increase flooding. Over the long-term, precipitation in the form of rain instead of snow may reduce the state's capacity to store water for the annual summer drought.

These examples demonstrate how poorly we appreciate the vagaries of water whether in excess or in scarcity. As coinhabitants of the planet with other organisms, it is critical that we learn to appreciate the complexities of water supply to provide drinking and irrigation water and habitat for fish and birds, while increasing the resilience to climatic perturbations. As we will see throughout this book, these are not natural disasters, but disasters of human making. Decisions having implications for the inhabitants of the landscape and aquascape well-being are hence required.

## RAINFALL DISTRIBUTION

California receives an average ~586 mm of rain per year. With a size of 423,970 $km^2$, the state enjoys 2.48 billion $m^3$ rainfall, the source of the state's inland waters each year. This precipitation does not fall evenly across the state or from year-to-year, creating dynamic gradients across the state, e.g., wetter in the north and dryer in the south, more rain in the winter and less in the summer. As weather storms arrive from the Pacific Ocean, western slopes tend to be wetter than eastern slopes, higher elevations usually receive more precipitation than lower elevations. These weather patterns also drive river hydrology, soil erosion, sediment transport, channel morphology, stream, and lake habitats.

With longer time horizons, we can appreciate how tectonics define the landforms that water flows across and through. Mountain ridges and waterfalls create migration barriers and unique species assemblages. Not only do aquatic communities also differ in headwater streams and floodplain rivers, but human settlement patterns also vary according riverine types. And with time, humans' modification of these waters to expand will also vary accordingly, and settlement decisions have defined how we negotiate co-inhabiting the planet—humans and nonhumans alike.

As our communities grow in size, the volume of water we appropriate often increases, but there is evidence that these withdrawals can level off, even as population increases (Fig. VI). These data suggest that there may be opportunities to further increase water-use efficiencies that might be used to improve or restore degraded inland waters, themes that we will explore in the latter half of the book.

## WHERE IS THE EARTH'S WATER?

When rain hits the soil surface, the water can evaporate, infiltrate, or runoff in surface waters. Indeed, by interacting with the land's surface, inland waters are directed by and modify landforms. Defined channels are cut into sloped hillsides or mountain slopes, which then combine to create streams, and some of which will fill basins to create lakes. But most of these waters account for a very small portion of the total amount of water on the planet (Table I).

Earth has ~1384 million $km^3$ of water. Since more than 97% of this water is stored in the oceans and another 2% is frozen in polar ice caps and glaciers, the amount of surface water available to humans is quite small: 0.6% of Earth's water is stored as liquid on land. Combined with groundwater, less than 1% of the water on Earth is available for human use.

Thus as a precious resource, critical for all organisms on the planet, our survival and the survival of many nonhuman species depend on how we manage these waters, even remote waters that we barely understand.

## A VANISHING CRYOSPHERE

Two percent of the planet's water is found in ice caps and glaciers. Climate change is gradually claiming glaciers throughout the world—the Arctic Ice Cap, the Antarctic Ice Sheets, and the glaciers of the Qinhai-Tibetan Plateau, Alaska, and Swiss Alps are also steadily disappearing (Fig. VII). In some cases, there are catastrophic slides with tragic results (Fig. VII). But overall, the Earth's Cryosphere provides important ecosystem functions, e.g., type of air conditioning for the Earth's climate, massive

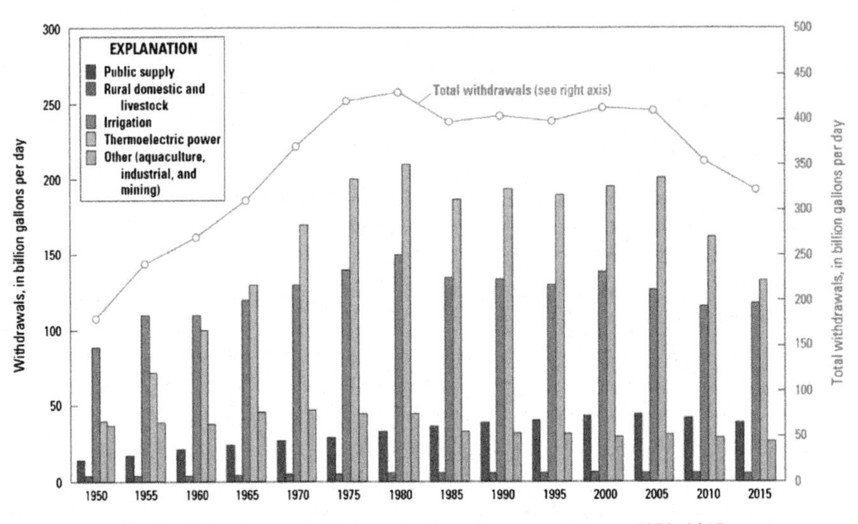


**FIGURE VI**

Water use in the United States (Source: U.S. Geological Survey et al., 2018).

**Table I Global pools of water. Source: Data from Baumgartner and Reichel (1975, p. 14). *Freshwater (and lakes and streams) include endorheic waters, which are often saline, thus these categories are a bit ambiguous.**

| Pool | Volume (km$^3$) | % Total Water | % Freshwater |
|---|---|---|---|
| Ocean | 1,348,000,000 | 97.4 | – |
| Freshwater | 36,120,000 | 2.6 | – |
| Icecaps and glaciers | 27,820,000 | 2.0 | 77.0 |
| Groundwater | 8,001,000 | 0.58 | 22.2 |
| Lakes and rivers* | 225,000 | 0.02 | 0.62 |
| Soil water | 61,000 | 0.004 | 0.17 |
| Atmosphere | 13,000 | 0.001 | 0.04 |

stores of water supplies, and even reducing flooding. The Qinhai-Tibetan Plateau's glaciers are the source of 9 of Asia's largest rivers and hydrology changes associated with accelerated glacier melting will affect over 1.5 billion people (Fig. VIII).

The Arctic Tundra is hummocky landscape of frozen peat soils and thousands of small lakes and ponds. But as the tundra warms, the peat soils are oxidized, emitting additional greenhouse gases ($CO_2$ and $CH_4$) thousands of small lakes and ponds have vanished from the landscape. In both cases, the long-term impacts of changes are poorly understood and will require teams of scientists to evaluate.

And closer to home, millions of lakes and rivers are subject to ice cover. In some cases, we have important records, e.g., the ice-on and ice-off records date back to 1443 for Lake Suwa, Japan. But we

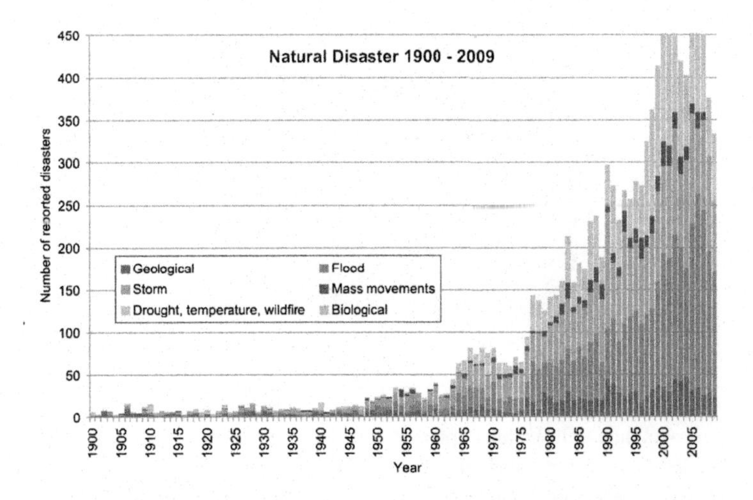


**FIGURE VII**

Number of natural disasters since 1900 (Source: Centre for Research on the Epidemiology of Disasters – CRED).

**FIGURE VIII**

Image of a glacier in the Swiss Alps with measured retreats on the order of 10s of meters per year (Source: https://www.earth.com/news/alps-lose-glacier-volume-2100/).

know very little about the ecology of these waters during ice cover, or what climate change will mean for these waters. We expect Aquatic Systems to differ when covered with ice, based on a wide range of factors including abiotic, geochemistry, and biogeochemistry, and biotic communities, but we know very little about the ecological processes under the ice. As the climate warms, shorter durations of ice cover may give rise to changes in anoxia risks, algae blooms, stream channel morphology, sediment transport, and community assemblages.

**FIGURE IX**

Death Valley pupfish (*Cyprinodon salinus salinus*). Once distributed throughout a mesic (or wet) Mojave Desert region thousands of years ago, the Death Valley Pupfish is now restricted to small permanent waters in what might considered an otherwise desolate and forbidding environment. As described in Chapter 1, we will be able to identify evidence for a mesic Mojave in recent geologic history that supported more widespread pupfish distributions.

# DEATH VALLEY PUPFISH AND WHITE MOUNTAIN TROUT

## PUPFISH IN AN ENDORHEIC BASIN

In the desert of eastern California, the Death Valley pupfish (*Cyprinodon salinus salinus*) is a curiosity. Living in a creek near the lowest point on the continent, this fish is adapted to shallow, hot, salty water where summer air temperatures are regularly above 49°C (120°F). This isolated population is found in a relatively inhospitable refugia, a small fragmented water that was once connected to a large terminal lake. Terminal lakes or endorheic basins have no outlet to the ocean exists. Since water can only leave by evaporation, salts accumulate over time, potentially reducing the number of species that can survive in increasingly saline waters. Although many terminal lakes, such as those in the Mojave Desert have been in decline since the close of the last ice age, their demise has been dramatically accelerated with agricultural water withdrawals. How we balance the human demand and water needs of biota in endorheic basins will be a daunting challenge for our descendants (Fig. IX).

## PAIUTE CUTTHROAT TROUT INTRODUCED TO FISHLESS STREAMS

In contrast restricted habitats of the desert pupfish, the Paiute cutthroat trout *(Oncorhynchus clarkii seleniris)* has been introduced to historically fishless creeks that drain White Mountain, CA, one of the highest mountains in the lower 48 states. Paiute Cutthroat are only native to Silver King Creek, a headwater tributary of the Carson River, which flows from the east side of the Sierra Nevada Mountains to Carson Sink, another endorheic basin. The fish was planted for sport fishing in what had been fishless streams. Although the introduction was done well before anyone could document its impact, there is no doubt the introduction had profound food-web changes to these streams that we are only beginning to understand. In fact, as we will learn in Chapter 12, species introductions disrupt inland waters throughout the world, displacing native species, altering food webs, and changing ecosystem processes, have some of the most destructive impacts to inland waters on local and global scales. To reduce the number of introductions and manage invasive species require Environmental Scientists to be rooted in science and policy.

## THE DISCIPLINES OF AQUATIC ECOLOGY

To understand the ecology of the *Death Valley pupfish* and *Paiute cutthroat trout*, we need to appreciate geologic and the climate records within their distributions; their natural history and ecology of close relatives; the ecological patterns and processes in which they are born, survive, and reproduce; and the impact of humans and their conservation status. In other words, the Environmental Scientist of inland waters will integrate principles in the Natural Sciences (e.g., Physics, Chemistry, and Mathematics); delve deep into Biology's subdisciplines (e.g., Physiology; Population Biology; Genetics; Evolution, and Community and Ecosystem Ecology), and Integrative Sciences (e.g., Geology, Biogeochemistry, and Climatology) to manage inland waters within a broad context, and root normative goals in Social Sciences and Humanities (Environmental Economics, Policy, and Ethics). This is a required, but not sufficient step to develop sustainable inland waters.

## PAST TRENDS AND FUTURE SCIENTISTS

Even as inland waters are changing, so are the sciences that are studying them. For example, the topics of focus within the field of Ecology are far from static. For example, in the 1980s, topics evaluating the role of life-history, plant reproduction, survivorship, disturbance, food webs and competition were prominent in ecological journal articles. But in the 2010s, these topics yielded to research focused on climate change, organism traits, genetics, anthropogenic effects, spatial and temporal scales, and community structure. Although disturbance certainly included the role of humans in the 1980s, the human dimension and role in the patterns and processes has become far more visible in research. In addition, research on gas fluxes and genetics were quite limited in their use in Ecology in the 1980s, but have become prominent with methodological advancements and funding.

Thus in spite of what we read in this or other texts, our understanding of inland waters is continually changing based on the available tool, research topic trends, and values that drive questions. Each generation will define how we will understand inland waters. Therefore it is key that students of inland waters avoid allowing the field to become anything, but a dynamic and creative endeavor.

Of course, how we as humans engage with our environment, either as separate from it or as part of it, has profound implications to what kind of world we will leave for future generations. As we reflect on recent disasters, e.g., Hurricane Harvey, Mississippi River Valley floods, or the severe drought in California, we are confronted with a simple truth: the built environment is quite sensitive to perturbations in the Water Cycle. What are lessons that we should take from these examples? What information do we need to reduce risks of disasters or improve the long-term survival of nonhuman organisms?

## THE INTERDISCIPLINARY PUZZLE MANIFESTO

It is my firm belief, we must be ready to learn and use as many appropriate tools as possible to address environmental issues. As with many disciplines, the interests we enter will certainly evolve and expand as we learn more about the topic. We may begin with a narrow topic of interest, but this can quickly require a robust understanding of several disciplinary topics. For example, an interest in duck behavior or steelhead migration, might turn into a surprising interest in Geology and Paleoclimatology; the fluid dynamics and light attenuation and photosynthesis; biogeochemistry at the surface water-sediment interface and comparative physiology; population modeling and isotope behavior in ecosystems and probability theory; and flooding and surface water-groundwater exchange. The challenge for Environmental Scientists is to use these disciplines judiciously and capitalize on areas that might provide insight. Without a doubt the role of the disciplinary sciences is a critical foundation, but for the Applied Scientists these are not enough—to improve management of inland waters for human services, restore the ecological processes, and maintain biological diversity is ultimately a question about how we manage human beings and how we construct the built environment.

Therefore Environmental Scientists, such as Geologists and Ecologists rely on teams with divergent and convergent interests. The impact of which is only limited by the combination of fields that can be brought to bear effectively. For each of us, the challenge then is to sort out which set of disciplinary puzzle pieces we can best use to accomplish the goal to define and promote ecological sustainabilities for inland waters.

## NEXT STEPS

### STUDY QUESTIONS

1. Why do extreme events have such an impact on landuse and planning?
2. How do Geology and Climate influence the status of *Cyprinodon salinus salinus*?
3. What disciplines are useful in understanding inland waters?

### PROBLEM SET

1. Learning about the waters in a region provides a context to think about "waters in a place." Generate a list of 40–50 streams in a region of interest and complete the following activities:
   a. List the streams and map them;
   b. Determine if they have monitoring associated with them, e.g., gauging, water quality monitoring, restoration, etc.
   c. Summarize the ecological issues in 3–5 of the streams (approximately one page (single-spaced) for each case study).
2. Find 10 Limnologists and evaluate their CVs to determine what type of disciplines have informed their research skills.

PART

# COEVOLVING AQUASCAPES

1

CHAPTER

# COEVOLUTION OF BIOTA, GEOLOGY, AND CLIMATE

# 1

*I write as though you could understand*
*And I could say it*
*One must always pretend something*
*Among the dying*
*When you have left the seas nodding on their stalks*
*Empty of you*
*Tell him that we were made*
*On another day*

**WS Merwin, "For a Coming Extinction"**

## CONTENTS



Ecology and Management of Inland Waters. https://doi.org/10.1016/B978-0-12-814266-0.00013-1

For over three billion years or approximately 60% of the Earth's 4.7 billion year existence, the planet's physical history has been inextricably linked with its evolving life forms. Earth's early atmosphere had no free oxygen, but included enough greenhouse gases to prevent the Earth's water to freeze, providing a conducive context for our life to evolve.

Early forms of life harnessed coupled reactions to generate energy and to reproduce. And as a rather dramatic evolutionary shift, some bacteria began producing oxygen as a waste product. Over the course of several hundred million years, the concentrations of oxygen increased and became toxic for many groups of bacteria.

In addition, living organisms altered the carbon cycle to influence methane and carbon dioxide concentrations, which has a direct role in controlling the Earth's temperature and climate. Thus, the composition of the Earth's atmosphere is a biological product, which can also control the Earth's climate that will in turn influence biological organisms (Fig. 1.1).

Beside the composition of the atmosphere, the Earth's climate depends on solar radiation, geographic positioning of the continents, geologic events, and biogeochemical cycles (for example carbon cycle). The result of these processes provides a feedback on evolutionary processes and the Earth's biodiversity, all of which have implications for inland water ecology and management.

This chapter explores how various planetary processes have selected for a wide range of species to invade inland waters while placing this evolution expansion within the context of major extinctions. Crossing disciplines to understand how ecosystems work and how they evolved is something the environmental scientists do on regular basis. In this case, we will focus on the geologic history of the earth to frame the development of inland waters with a long history. To appreciate the implications of the Earth's history for inland waters, we will aim at the following outcomes:

1. Describe the evolution of the Earth's atmosphere and the implications for life on Earth;
2. Recognize the limitations of the fossil record and summarize what they tell us about evolution of inland waters;
3. Summarize the causes of major extinction events, and describe why the current extinction process is unique in the Earth's history.

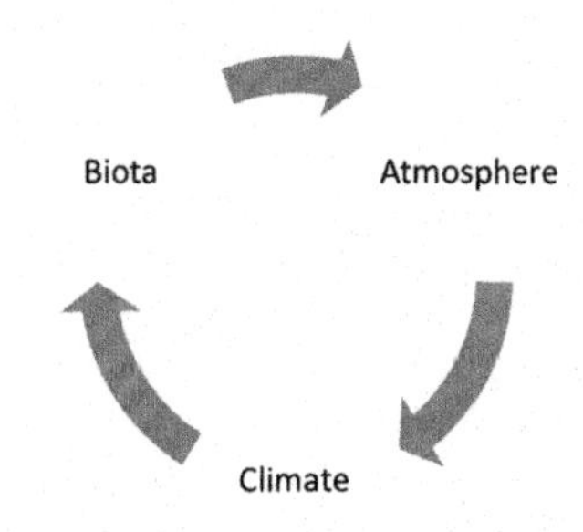


**FIGURE 1.1**

Simple model of the links between the Earth's biota, atmosphere, and climate.

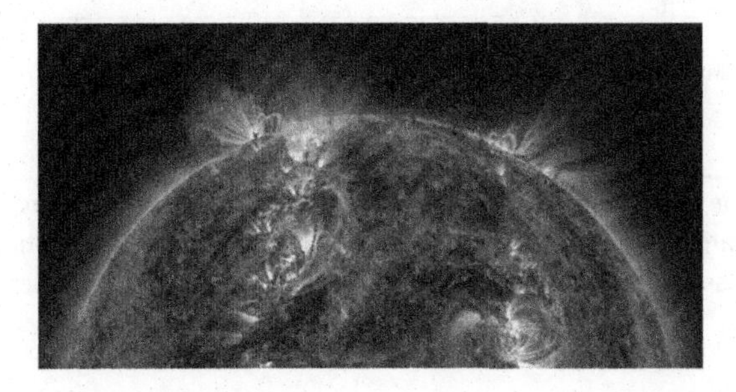

**FIGURE 1.2**

Image of the Sun's turmoil (Source: NASA). Solar flare on February 24, 2014 with a magnitude of X4.9. Flares are classified by their energy output: A-class (near background levels), followed by B, C, M, and X. Each letter represents a 10-fold increase in energy. An X has ten times more energy than M and 100 times more than C. Within each letter class, there is a finer scale from 1 to 9. The Solar and Heliospheric Observatory spacecraft recorded a solar flare as it erupted from the sun early on Tuesday, October 28, 2003, but it overloaded the sensors measuring it (X28!). The sun has been relatively quiescent in the last few decades, but we expect more sunspots and flares in the next decade.

## BIOTIC LINKAGES BETWEEN CLIMATE AND ATMOSPHERE

### STORY OF A STAR: A BRIEF HISTORY OF OUR SUN

Similar to evolution of life on Earth, our Sun has been developing over time: from the beginning of the universe to our present state, the changes of the Sun provide the Earth with a relatively unique opportunity for life to develop (Fig. 1.2).

Throughout Earth's history, our planet's surface temperature has remained within a surprisingly narrow range, despite receiving increasing energy from the Sun over the last few billion years.

The Sun was formed about 4.57 billion years ago and is about halfway through its 10-billion-year sequence of developmental stages. Theoretical models of the Sun's development suggest that 3.8 to 2.5 billion years ago, during the Archean period, it shone ~75% as brightly as it does now. In fact, the Sun did not provide enough energy to maintain water in a liquid form on Earth; and life, as we know it, would not have arisen. And yet, the geological record suggests that the Earth has remained within a

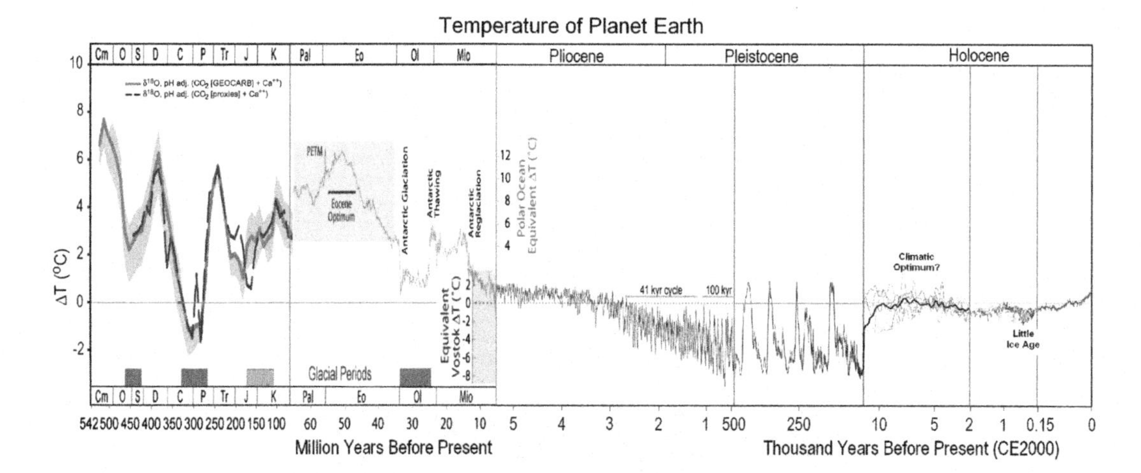


**FIGURE 1.3**

Estimated temperatures for the Earth since the Cambrian period (Cm) (Source: Greg Fergus (2014), https://commons.wikimedia.org/wiki/File:All_palaeotemps.svg#Data, and sources therein). Notice the variability prior to the last 20k years, which has a different timescale. In addition, temperatures have generally been declining since the Late Miocene (Mio).

10°C range throughout its history and that the Archean Period was warmer than current temperatures (Fig. 1.3).

The explanation may lie in the composition of Earth's ancient atmosphere. Compared to the current concentrations of greenhouse gases, carbon dioxide, methane and/or ammonia concentrations were quite high—enough to compensate for the smaller amount of solar energy and maintain a warm planet.

As the Sun continues its developmental sequence over the next billion years, the Earth's surface will likely become too hot for liquid water to exist, potentially ending terrestrial life (Fig. 1.4).

Even in this brief summary, we can appreciate how the Earth's temperature is sensitive to both the Sun's radiation and the composition of the atmosphere—and how these factors have allowed for the evolution of life as we know it.

## THE OXYGENIC CATASTROPHE

The atmosphere of the early Earth is poorly understood, but it is thought to have been dominated by reducing gases, that is, methane, carbon dioxide, and water, with virtually no free oxygen (Fig. 1.5). Although photosynthesis arose fairly early in the Earth's history, it took over a billion years to generate free oxygen in the atmosphere. The oxygen produced at first was almost instantaneously removed from the atmosphere by reducing minerals—most notably iron. Evidence of increasing oxygen concentrations is found in banded iron formations in the geologic record. These are striated layers of rust-colored rocks, that is, iron oxides that were formed under water. By approximately 2,250 mya (million years ago), the available iron was saturated (oxygenated), banded iron bed formation ended, and free oxygen concentrations began to accumulate in the atmosphere.

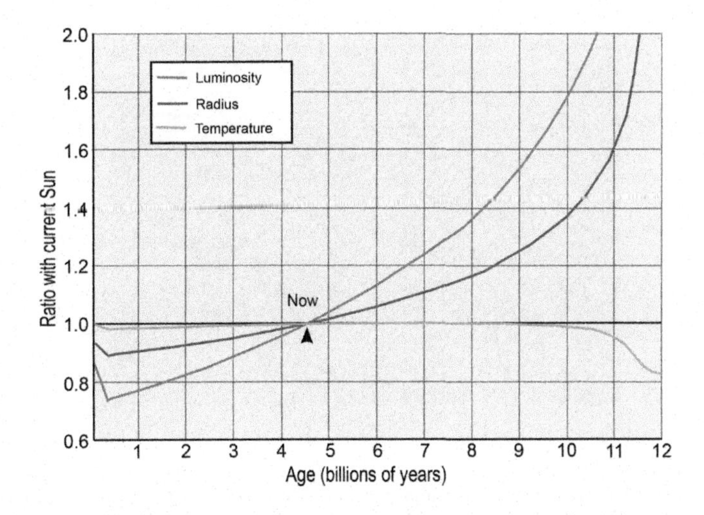


**FIGURE 1.4**

Sequence of Sun's stages. Through the main sequence, the Sun is gradually becoming more luminous (about 10% every 1 billion years), and its surface temperature is slowly rising (Source: unknown).

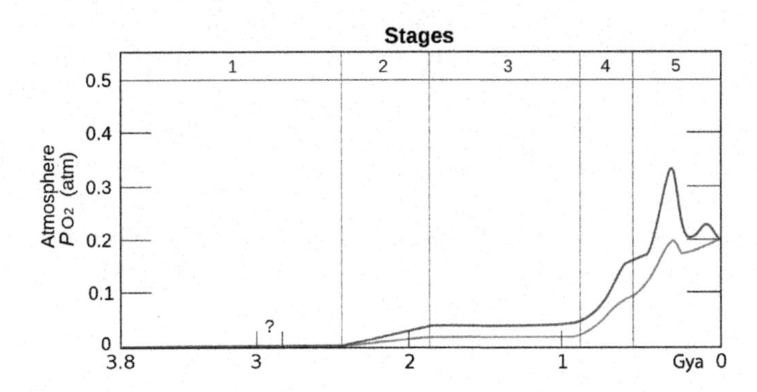


**FIGURE 1.5**

Reconstructed atmospheric oxygen concentrations (Source: Holland, 2006). Red (dark gray in print version) represents concentration in shallow oceans, and green (light gray in print version) represents oxygen in deep oceans.

Bacteria, probably a type of cyanobacteria, were the first organisms to carry out oxygenic photosynthesis. Stromatolites of fossilized oxygen-producing cyanobacteria have been found that date from 2.8 billion years ago, possibly as long as 3.5 billion years ago (Fig. 1.6). Cyanobacteria's ability to perform oxygenic photosynthesis converted the atmosphere from a reducing to an oxidizing environment. As a result, the Earth experienced a dramatic shift in the composition of life forms.

We can infer from reconstructions of the atmosphere and climate that the early Proterozoic Earth (3000 mya) was far from a hospitable place for modern life forms, where no atmospheric oxygen existed. However, for most early taxa, the oxygen was toxic. Because of its lethal impact on a biota

**FIGURE 1.6**

Living stromatolites at Lake Thetis, Western Australia. Stromatolite fossils have also been observed in several formations in Southern California (McMenamin, 1982).

largely adapted to anoxic conditions, the accumulation of oxygen in Earth's prehistoric atmosphere is called the "Great Oxygen Event" or "Great Oxygenic Catastrophe."

While the atmosphere became increasingly oxygenated, mechanisms to detoxify $O_2$ began to emerge. In general, biochemical mechanisms to prevent the toxicity rely on peroxides within peroxizomes, a subcellular organelle constituting a part of the evolution of eukaroyotes. But for more than a billion years, the cyanobacteria waste produced was poisonous to nearly every living organism. After life developed mechanisms to "manage" the toxicity of oxygen and its increasing atmospheric concentrations, the fossil record points to numerous periods of diversification. Whereas some bacteria were restricted and continue to be restricted to anoxic environments, others adapted to the new environment, including groups of taxa that created the oxygenated atmosphere.

Descendants of those prehistoric anaerobes (Archea) currently thrive in wetland soils, around geysers, in salt deposits, and within other low-oxygen settings, where they remain as important players in ecosystem processes. Their biogeochemical processes provide scientists hints at how ecosystems functioned early in the Earth's history, but also continue to influence the patterns and processes in contemporary inland waters.

UV radiation was probably a major constraint preventing marine organisms from invading land or inland waters. However, the production of atmospheric oxygen also produced stratospheric ozone that protects biota from harmful UV radiation. Rocks record an increase of biological diversity that coincides with the breakup of Rodinia (~750 mya), when oxygen concentrations reached the necessary threshold to create an ozone layer.

Thus we might consider the Earth's biota to have a central position in the composition of the Earth's atmosphere and climate (Fig. 1.7). Even as cyanobacteria "pollute" the atmosphere with oxygen, other taxa have capitalized on this reactive compound providing the mechanism to diversify organisms that continue to live in anaerobic (without oxygen) environments, and organisms that evolved to live in aerobic environments (with oxygen).

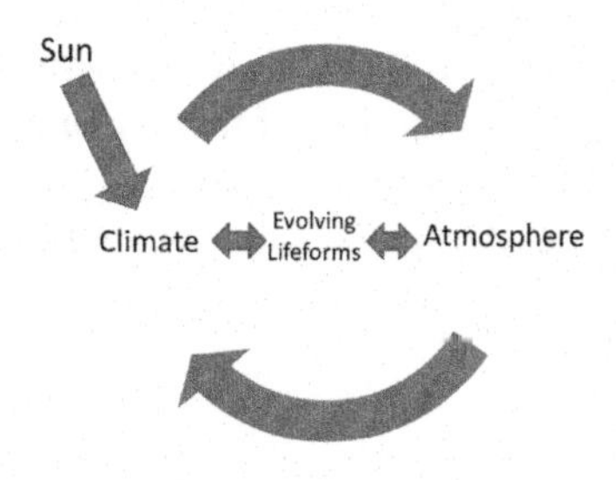


**FIGURE 1.7**

Modified conceptual model of how the Earth's biosphere works.

**Table 1.1 Recognized geochronology timespans. To learn more about these time-spans, their characteristics, and formations, see the Next Steps section.**

| Time Span | Number of Units | Common Examples |
|---|---|---|
| Eon | 4 Eons | Archean, Proterozoic, Phanerozoic |
| Era | 10 Eras | Paleozoic, Mesozoic, Cenozoic |
| Period | 22 Periods | Cambrian, Carboniferous, Permian, Triassic, Jurassic, Cretaceous |
| Epoch | 34 Epochs | Eocene, Oligocene, Miocene, Pliocene, Holocene |
| Age | 99 Ages | Tortonian, Piacenzian, Aquitanian |

# THE RECORD OF MAJOR EXTINCTIONS

## GEOLOGICAL FORMATIONS AS A RECORD

The history of the Earth is recorded in layers of rock that were formed in depositional environments (for example, oceans as in the case of the banded red-bed formations). Originally deposited in an horizontal fashion, the oldest records were at the bottom and youngest at the top. Though having experienced high pressure, elevated temperatures, and folding, these rock layers can tell us much about the early environment and preserve some organisms. Entire books have been written about the fossils within specific rock formations. For example, "Wonderful Life: the Burgess Shale and the Nature of History," was written by one of the most famous paleontologist, Stephen J. Gould.

Geologists developed time increments based on geologic markers, such as fossils in rocks. The system relies on a system of nested hierarchical units (Table 1.1). For example, the Aquitanian Age is that covers 23.03–20.44 Ma (Macratriinae) and is in the Miocene Epoch (23.03–5.333 Ma), the Neogene Period (23.03–2.58 Ma) and in the Phanerozoic eon (151–present Ma).

## PROTEROZOIC RECORDS

The Proterozoic Eon extended from 2500 Ma to 541 Ma, where fossils are scarce. To be more precise, few unaltered sedimentary rocks still exist from this time, and fewer still have fossils. Nevertheless, some early records exist and contain stratifor stromatilites, oncoids, and columnar stromatolites on the edge of the super continent Rodinia (~1100–~750 mya). These fossils were formed by cyanobacteria. Younger fossils document the many branches of multicellular organisms before the end of the Proterozoic Eon.

Besides the "Oxygenic Catastrophe," the geological record after the Preterozoic documents numerous catastrophic extinctions events.

## FIVE MAJOR EXTINCTIONS

One way to read the fossil record is as a series of catastrophic extinction events followed by species radiations. Paleontologists believe these extinction episodes were caused by a combination of phenomena and even use them to divide geologic intervals. Below, I summarize the five most dramatic extinction events, and will discuss the current 6th major event at the end of the chapter.

1. 450–440 mya: Ordovician-Silurian transition. Two extinction episodes separated by about 1 million years led to the loss of 60–70% of the species on Earth. Most scientists believe the extinctions were caused by a series of glaciations that lowered sea level and destroyed shallow marine habitat. The glaciations may have been caused by repositioning of the continents, specifically with Gondwana, a continent assembly composed of South America, Arabia, Africa, India, Australia, Madagascar, and Antarctica, migrating into colder climates over the South Pole (Fig. 1.8).
2. 375–360 mya: Near the Devonian–Carboniferous transition. In the later part(s) of the Devonian Period, a prolonged series of extinctions eliminated ~70% of all species. In addition, this extinction event may have lasted as long as 20 My (million years) and probably entailed a series of extinction pulses or phases with a variety of causes, including magnetism, global cooling due to carbon sequestration, possible asteroid impact, and an unfavorable arrangement of continents.
3. Approximately 252 mya: Permian–Triassic transition. This is the largest extinction episode on record and occurred over a short ~60,000 year period with 90–96% of species disappearing. This episode may actually include one to three distinct extinction pulses and causes. At some point, a spike in atmospheric $CO_2$ might have led to an abrupt injection of $CO_2$ into the ocean waters, reducing the pH and the preferential loss of calcified marine biota. More recently, however, researchers suggest that the $CO_2$ lead to global warming and temperature-driven hypoxia. Perhaps, it was a combination of effects.
4. 201 mya: Jurassic–Triassic transition. Between 70–75% of the species went extinct and recent evidence suggests the gases released from extensive volcanic activity and lava flow may have led to dramatic climate perturbations and major extinctions.
5. About 65.5 mya—At the end of the Cretaceous period, an asteroid hit the earth and released a great deal of sulfur, causing acid rain and global cooling, perhaps, a decade of planet-wide freezing temperatures. In any case, the episode resulted in an estimated loss of some 75% of the species living on the planet and demise of the dinosaurs.

Given the information presented so far, we can already appreciate how major extinctions are linked to climate, which is controlled by a combination of geologic events and extra-terrestrial processes. These events tell a chilling story: the episodic destruction of life forms on Earth in response to changing physical conditions. Thus we find a direct link between tectonics, climate, and biological diversity (Fig. 1.10).

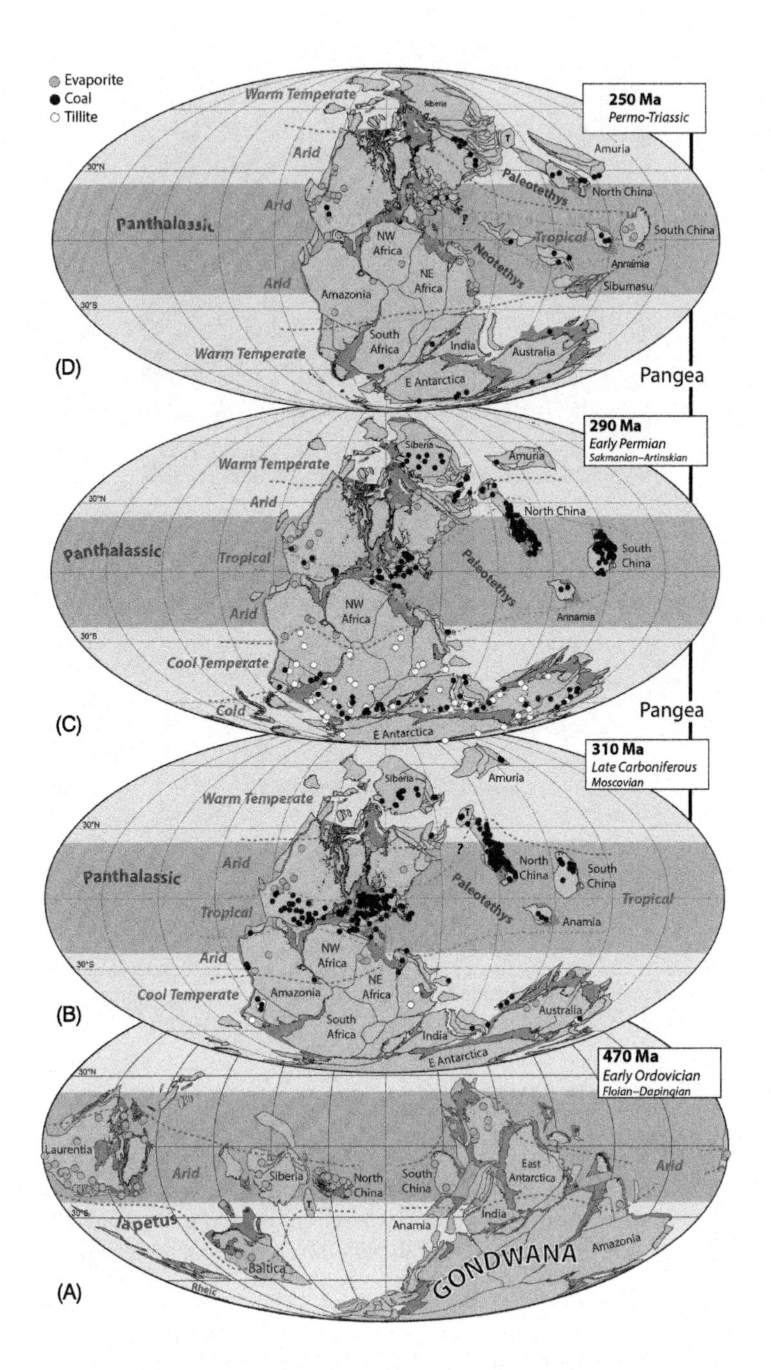


**FIGURE 1.8**

Ordovician–Triassic transition (Source: Torsvik and Cocks, 2016).

**FIGURE 1.9**

A Wyoming rock with an intermediate claystone layer that contains 1000 times more iridium than the upper (Neogene) and lower (Cretaceous) layers. Picture from the San Diego Natural History Museum.

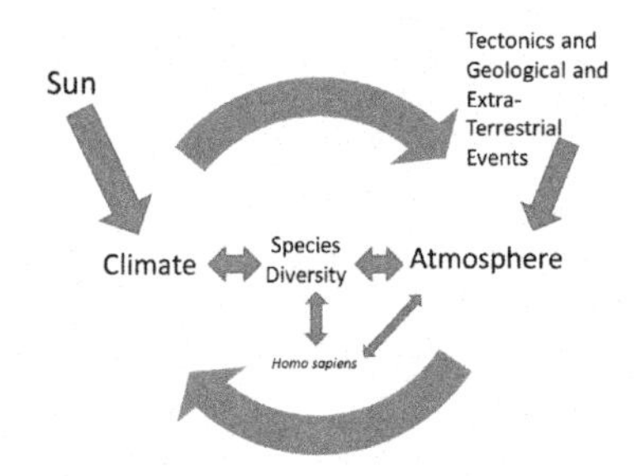


**FIGURE 1.10**

Conceptual model of major extinctions that include tectonics, geological events (e.g., volcanic eruptions), and extra-terrestrial asteroids.

# PALEOZOIC (541–252.17 MYA)

## THE EDICARIAN EPOCH

As researchers are learning, the Edicarian Epoch may include more taxa than we have realized. Although the Cambrian Period, to be covered in the next section, is known for a sudden diversification of the fossil record, it may be that the diversity was already developed, but remained sparse, hard to find, and even unpreserved.

## THE CAMBRIAN EXPLOSION AND INVASION OF FRESHWATERS

The Cambrian Period (541–485.4 mya) is marked by rapid diversification, capitalizing on a wide range of new extant taxa, e.g., shelled animals (542 mya), land plants (540 mya), jawless fish (530 mya).

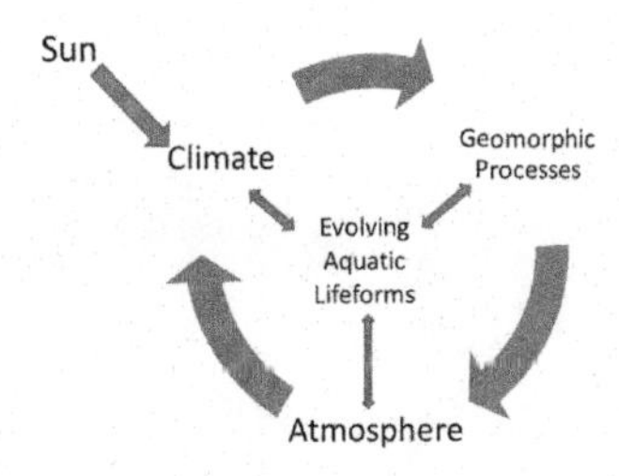


**FIGURE 1.11**

Simple diagram of the interrelatedness of climate, geomorphological processes, and biota. This simple model ignores the extra-terrestrial and major geological events, e.g., asteroids, volcanic eruptions.

Despite the evidence that moderately complex animals existed before (possibly long before) the start of the Cambrian, the pace of evolution was exceptionally fast in the early Cambrian. Possible explanations can be categorized into three broad categories of change: environmental (limited glaciation extent and/or events), developmental (evolution of *Hox* genes that allowed for segmented animal body plans), and ecological (increasing complexity of carbon cycle).

To explain the timing and magnitude of the explosion is beyond the scope of this text, but the fact that it co-occurred with dramatic changes in atmospheric composition, arrangement of continents, and temperature reflects the central theme of the chapter—the coupled nature of our planet—the climate system, geological processes, and evolving form of life.

## LAND PLANTS AND NEW RULES IN GEOMORPHOLOGY

Prior to the evolution of land plants, the Earth's surface was subject to high erosion rates and stream sediment transport loads. Stream channels were unstable—with high sediment volumes that altered drainage morphology in dramatic and frequent ways. Present-day analogs might include rivers down stream of recent glacial retreats, where trees and shrubs are not yet established.

The evolution of land plants was a game-changer for inland waters in several ways. First, by providing ground cover, plants reduced soil erosion rates dramatically. Second, by growing along stream margins, plants stabilized channel banks and defined a new relationship between stable streams and periodically inundated floodplains. Finally, as these terrestrial sources of organic matter entered aquatic systems, the trophic complexity of streams increased. With increasing complexity of the trophic structure, the number of freshwater taxa increased, as well as the number of fossils.

We might create a conceptual model that includes three processes: climate, geomorphology, and atmosphere, each influencing each other and evolving life forms connected to each (Fig. 1.11). This model demonstrates the complexity of finding a simple cause-and-effect relationship. Although extra-terrestrial events and processes can be identified, few physical or biological drivers can be considered independent and direct factors that define how the Earth's biosphere operates.

## MARINE TAXA INVADE FRESHWATERS

All of Earth's inhabitants have aquatic origins. These origins create a kind of "evolutionary inertia" bearing on life on Earth depending on liquid water. From the physical and chemical properties of

**FIGURE 1.12**

These U-shaped Arenicolites are found in fine-grained sandstone within an ancient storm dominated shoreface (Wall Creek Member, Frontier Formation, Powder River Basin, Wyoming). Note the diagnostic U-shape of the sand-infilled burrows. Caution: Arenicolites can be easily mistaken with Skolithos when only the vertical tube is visible (Source: Chuck D. Howell, Jr.).

water and its role as a solvent—water defines the existence of life on Earth. Thus every organism is fundamentally aquatic, whereas most of the evolutionary history is marine.

The oceans have had fairly consistent chemistry, meaning the evolving marine organisms were adapted to saline conditions compared to freshwaters. Furthermore, before plant species evolved, terrestrial organic food sources might have been limited relative to current conditions.

One of the oldest freshwater fossil animals was discovered in a formation approximately 520 mya, dating approximately 80 mya before the previous published record for bilaterians, organisms with bilateral symmetry. These have been identified as *Arenicolites* and *Skolithos* (Fig. 1.12). These worm-like taxa are relatively small compared to their marine relatives, possibly because of the physiological cost of maintaining a low osmotic pressure in the freshwater environments. Maintaining an appropriate osmotic pressure is a major hurdle for brackish and freshwater taxa and can be an important physiological expenditure. We will cover this physiological process in Chapter 6.

The invasion into freshwaters may have depended on the stream bed stability and reduce terrestrial erosion rates. Without these changes bilaterians might have been dislodged by scouring caused by stream flow or buried by sediments.

## THE ROCK CYCLE AND FOSSIL RECORDS

Unfortunately, the fossil record does not provide an unbiased sampling of prehistoric ecosystems. Fossils tend to be found in specific depositional environments, e.g., ocean and lake bottoms, and thus constitute a biased sample because of variable modes and prospects for preservation. For example, many fluvial/alluvial formations exists, but these tend to be fossil poor, perhaps because organisms and their remains tend to be washed away in an active channel before they can be fossilized, or their soft body parts fossilize poorly. Finally, almost no fossil records exists for seeps, springs, or subterranean rivers.

Finally, the fossil record contains numerous gaps, both spatial and temporal. There are relatively few lacustrine (lake) formations in the world to investigate—perhaps because lake beds are susceptible to drying or uplifting, hence vulnerable to erosion, and their records are lost. Older records are more likely to be destroyed in the rock cycle.

## FRESHWATER FOSSILS AND THEIR FORMATIONS

Should fossils be found in a formation, then two other questions arise. First, how does one determine if a fossil is from a freshwater, brackish, or marine taxon? Freshwater organisms might be carried downstream, deposited in the coastal zone, and misidentified as marine taxa. Alternatively, migratory species, whose guts contain euryhaline taxa may be found well upstream of a marine environment. Thus it helps to have an independent marker of a fossilized organism's habitat type. Many paleontologists work to develop these markers, which come from a diverse set of physical and chemical signatures, e.g., differences in strontium isotope ratios, and patterns of grain-size deposition (i.e., how they are sorted and layered in facies).

Second, the presence of a fossil may not be representative of the common taxa in an ancient aquatic community. For example, larger organisms are easier to see, but often rare. Hard body parts, such as skeletons, fossilize better than soft body parts, even though soft-bodied animals might dominate aquatic systems. Finally, we might identify "trace fossils"—evidence of an organism's activity in the absence of the animal itself—such as footprints, burrows, root cavities, and feeding marks. These activities suggest the presence of particular organisms, but tell us little about their importance in a community.

Even when freshwater formations are found, they often appear in transition zones that vary with time—that might be fully exposed when conditions dry and/or uplift or even submerge when sea levels increase.

## FOSSILS: CLUES ABOUT EVOLUTION

The Harding Sandstone Formation in Colorado contains a great number of fossils (i.e., fossiliferious). Diverse "agnathans," early vertebrate fish-like taxa, e.g., *Astraspis* (Fig. 1.13) and *Eriptychius* have been identified from this formation. Fish invasions into freshwater indicate that developing foodwebs were increasingly complex to support a wider range of taxa, but what type of habitat did vertebrates first evolve?

Although most researchers believe that the earliest vertebrates had marine origins, the Harding Sandstone Formation and others in Australia contradict this notion and suggest that the first vertebrates were freshwater taxa. However, the extent to which this is due to gaps in the marine fossil record remains an open question.

## CONTINENTAL MARGINS AS GATEWAYS TO FRESHWATER ECOSYSTEMS

Lakes of the Devonian along continental margins or in tectonic basins with occasional connection to marine waters created a gateway for animals to invade fresh waters and terrestrial habitats. For example, early scorpions (Branchioscorpionina) from the Early Devonian have been found in the Beartooth Butte Formation (Wyoming). These scorpions had gills, suggesting that scorpion ancestors were aquatic.

**FIGURE 1.13**

*Astraspis* (at the bottom) and the *Arandaspida* (at the top) (*Arandaspis*, *Sacabambaspis*, and a couple of others). However, these fossils have been found from different continents, so probably did not co-occur. *Astraspis* is from North America, and the *arandaspids* are all from South America or Australia.

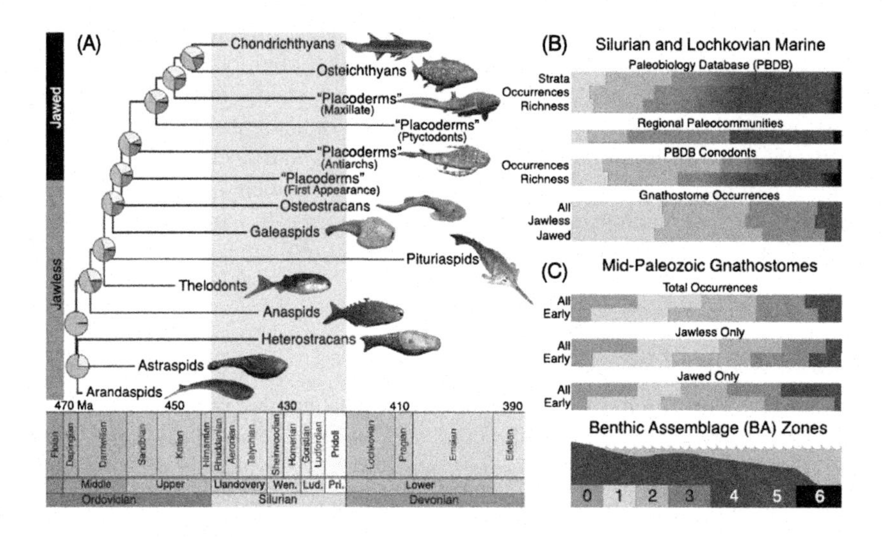


**FIGURE 1.14**

Coastal diversification in the Devonian. Note the hypothesized importance of zones close to the inland water sources (Source: Sallan et al., 2018).

The Late Devonian has increasing diversity of freshwater taxa (fish, arthropods, annelids, and mollusks). This increase may be linked to an increase in trophic complexity that resulted from the continuing diversification of land plants and subsequent addition of organic detritus in nonmarine and marginal-marine areas. In fact, some evidence suggests that many early jawed and jawless fish clades originated in enclosed coastal basins (Fig. 1.14).

**FIGURE 1.15**

Extinct reptile *Captorhinus aguti* Early Permian or Cisuralian (299.0 ± 0.8–270.6 ± 0.7 Mya), found in the Dolese Brothers quarry near Fort Sill, Comanche County, Oklahoma, United States. (Source: Didier Descouens (CC BY-SA 4.0)).

## TETRAPODS AND TERRESTRIAL ANIMALS

The ancestors of all tetrapods, four-limbed animals with backbones or spinal columns, began adapting to crawl and walk on land ~400 mya. Their strong pectoral and pelvic fins gradually evolved into legs, key evolutionary adaption for all four legged animals, including human beings.

## CARBONIFEROUS SWAMPY FORESTS

During the Carboniferous Period, much of the North American continent was covered by swampy forests that would eventually become fossilized coal beds—characteristic of Carboniferous deposits. Some of the most common fossils to be found in formations are the mollusks that began invading freshwaters during the Carboniferous Period.

The three most abundant families in terms of number of species are the Lymnaeidae (pond snails), the Planorbidae (ramshorn snails) and the Physidae (pouch or bubble snails). Descendants of these taxa are found in ponds, creeks, ditches, and shallow lakes almost worldwide.

One evolutionary innovation of the Carboniferous Period is the amniote egg. With its protective shell and fluid-filled membranes, the amniote egg allowed for the exploitation of the land by certain tetrapods. As a result, amphibians were the dominant terrestrial vertebrates during the Carboniferous Period (358.2–299 mya).

One branch of amphibians would eventually evolve into reptiles, the first fully terrestrial vertebrates. For example, an early reptile order cotylosauria fossils have been found in late Carboniferous formations, such as the Cutler Formation of New Mexico and Sangre de Cristo Formation of Colorado. These were striking, predaceous taxa indicative of freshwaters with enough biomass to support large predators (Figs. 1.15, 1.16).

**FIGURE 1.16**

Mesosaurs ("middle lizards") were a group of small aquatic reptiles that lived during the early Permian period, roughly 299 to 270 Mya (Source: Nobu Tamura's Paleoart Portfolio). Mesosaurs were the first aquatic reptiles, having apparently returned to an aquatic lifestyle from more terrestrial ancestors. However, just how terrestrial mesosaur ancestors had evolved remains uncertain.

## REINVASION OF MARINE HABITATS

By 270 mya the continents assembled into a new supercontinent, Pangea (Fig. 1.8). Extensive rainforests of the Carboniferous Period disappeared, leaving behind vast regions of arid deserts within the continental interior. Reptiles dominated the world of amniotes. The amniote, *Mesosaurus tenuidens*, with the capacity for live birth re-invaded the marine environment.

# MESOZOIC (252.17–66 MYA)

## REINVASION OF FRESHWATER

The Triassic Period began in the wake of the Permian–Triassic extinction event, which left the Earth's biosphere impoverished; it would take well into the middle of the Triassic Period for the Earth's biota to recover its former diversity. The vast supercontinent of Pangaea existed until the mid-Triassic, after which it began to gradually rift into two separate landmasses, Laurasia to the north and Gondwana to the south.

Through the Jurassic Period, the supercontinent Pangaea continued to its rifting process into two landmasses. This created more coastlines and shifted the continental climate from dry to humid, and many of the arid deserts of the Triassic were replaced by lush rainforests. Meanwhile, reptiles re-invaded aquatic habitats. The oceans were inhabited by marine reptiles, such as ichthyosaurs and plesiosaurs, while pterosaurs were the dominant flying vertebrates. In freshwaters, crocodylians made the transition from a terrestrial to an aquatic mode of life, then demonstrating that the origins of inland water taxa are quite diverse—from marine and terrestrial and marine, and then to freshwaters.

### CYANOBACTERIA: ANOTHER TOXIC EVENT

In 2017, Raymond Rogers presented evidence that a dense fossil formation (Maevarano Formation, Madagascar) may have been caused by blooms of harmful algal. Sauropods, crocodiles, and birds were victims of an event that resembles neurotoxin poisoning associated with cyanobacteria. Although the results need to be confirmed, this highlights the ecological impacts throughout history of cyanobacteria, in this case in a localized fashion.

## CENOZOIC ERA (66–0 MYA)

### PALEOGENE: PALEOCENE, EOCENE, AND OLIGOCENE

Within the Cenozoic Era are three Periods: the Paleogene (66–22.3 mya), the Neogene (23–2.58 mya), and the Quaternary (2.58–0 mya). These Periods are marked by dramatic changes to the Earth's climate, and continental arrangement are beginning to approach current patterns, but there are still dramatic changes afoot. For example, the transition between the Paleocene (66–56 mya)–Eocene (56–33.9) Epochs, high average global temperatures, about 55 mya. This period is referred to as the Paleocene–Eocene Thermal Maximum or PETM. While the Earth's temperatures were high, large portions of the ocean were probably inhospitable to marine life because of oxygen concentrations declined and sulfur reduction produced toxic $H_2S$. Of course, these hypotheses are based on the fossil record and still under scrutiny. In any case, the Earth's atmospheric composition and temperatures have certainly experienced changes over geologic time and can generate an environment that looks quite different to what we have experienced during the evolutionary history of the human species.

### PALEOCLIMATOLOGY AND FOSSILS

By analyzing the changes in fossilized diatoms and associate fauna, scientists document how the species composition and diversity changed during lake history as the climate became more arid. For example, changes in lake chemistry and subsequent changes in fauna in Lake Barstow, CA are indicative of lake shallowing (Fig. 1.17), which are associated with regional and global climate change and coincident with regional uplift.

Ecologically, grastropods are very diverse. Different species of freshwater mollusks favor various habitats. With modern analogous taxa, mollusk community associations can be used to infer paleo-habitat types, e.g., water depth, shoreline characteristics, and association with vegetation (Fig. 1.18). In Eastern Nevada, two mollusks associations were used to distinguish between different laucustrine (lake) habitat-types.

The Sheep Pass Formation is a geologic formation in Nevada preserving fossils dating back to the Paleogene period. In that formation, an association between several groups of mollusks, *Valvata–Hydrobia–Sphaeriidae* suggested low-energy, shallow, sublittoral, lacustrine environment that lacked emergent vegetation, whereas the Lymnaeidae–*Biomphalaria* association inhabited isolated ponds, in which emergent macrophytes were present. These analyses testify how abiotic factors influenced how various communities developed within an inland water.

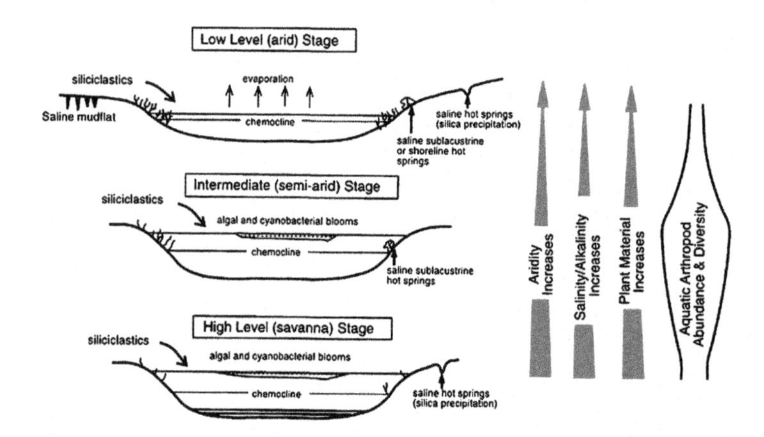


**FIGURE 1.17**

Model of the evolution of Lake Barstow over the time interval represented by the three concretion-bearing levels in the formation, showing the trends of increasing aridity, salinity–alkalinity, terrestrialization, and the decrease in faunal abundance. The lake goes through a high-level (savanna) stage and progresses into a final, low-level (arid) stage (Park and Downing, 2001).

## CLOSING OF THE CENTRAL AMERICAN SEAWAY

After the breakup of Gondwana ~180 mya, South America drifted north and made contact with the Caribbean plate. As South America scrapped along the Caribbean Plate, the Panama arc emerged from the sea and created a connection between North and South America. This connection had several effects: 1) The ocean connection between the Pacific and Atlantic near the equator was severed creating novel ocean circulation patterns; 2) global climate system created new patterns of precipitation and temperature, e.g., the development of the Mediterranean climate types, and 3) the formation of a land bridge between North and South America that allowed organisms to migrate from one continent to the other.

## MEDITERRANEAN-TYPE CLIMATE AND AQUATIC EVOLUTION

As we approach recent periods, we can easily appreciate how large-scale planetary process affect local regions. For example, the climate change associated with plate tectonics has direct implications for the species along the west coast of the United States. By focusing on the Mediterranean climate, we can appreciate how current climate regimes are dynamic and will help us frame human development and climate change with a regional context.

Mediterranean climates usually occur on the western side of continents between the latitudes of 30°and 45°. Summers are moderately warm and dry due to the domination of the subtropical high pressure systems, except in areas immediately adjacent to the coast, where summers are milder due to offshore cold ocean currents that can generate fog.

Along the west coast of the United States, lake sediments and the plant fossils from the middle Eocene Epoch indicate a subtropical to tropical rainforest climate. Rainfall was about 125–190 cm per

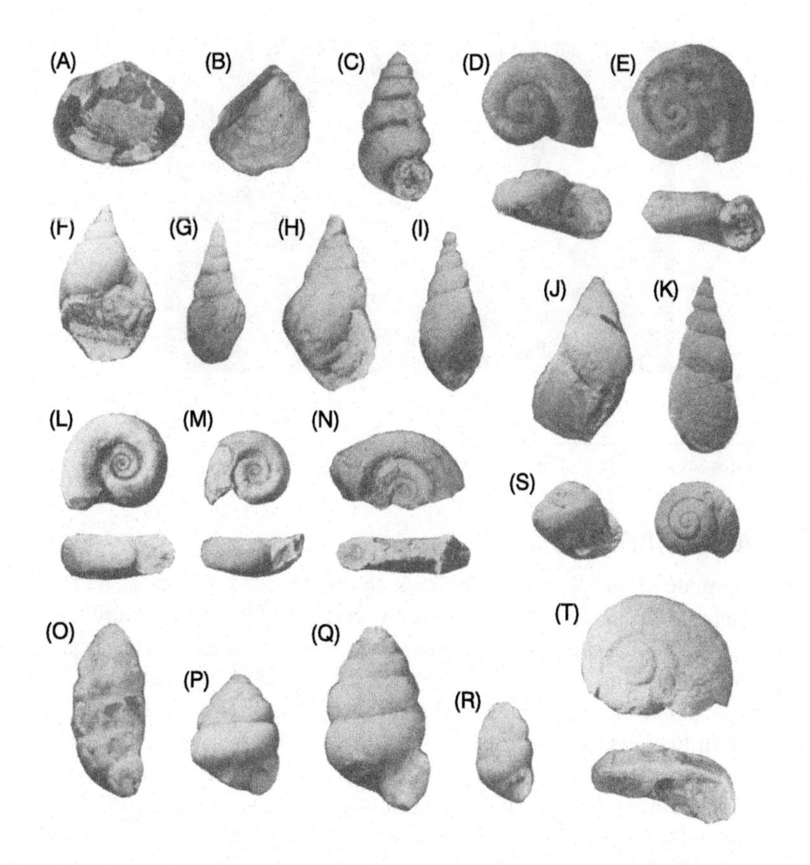


**FIGURE 1.18**

Molluscs from the Sheep Pass Formation. (A) *Sphaerium* sp. indet. (indeterminate). (B) *Pisidium* sp. indet. (C) *Hydrobia* sp. A. (D) *Valvata subumbilicata.* (E) *Valvata bicincta.* (F) *Lymnaea similis.* (G) *Lymnaeidae* gen. and sp. indet.: Form A. (H) *Lymnaea* sp. B. (K) *Physa longiuscula*(?). (L) *Biomphalaria aequalis.* (M) *Biomphalaria storchi.* (N) *Biomphalaria pseudoammonius.* (0) *Pupilla* sp. A. (P) *Chaenaxis* sp. A. (Q) *Vertigo*(?) sp A. (R) *Vertigo arenula*(?). (S) *Helix*(?) sp. A. (T) *Glypterpes rotundata.*

year and average annual temperature was about 20–25°C, in marked contrast to the present annual rainfall of 25 cm and average annual temperature of 16°C.

Profound summer drought began 15 to 14.5 mya (Miocene) in California. Seasonal drying gradually spread northward, while winter cold temperature regimes came from the north, selecting for cold- and drought-tolerant vegetation. Many of the warm/mesic-temperature plants of the early Miocene Epoch were extirpated. However, because much of California was geographically isolated, endemic species radiated creating a unique flora and fauna. The implications of this change have been documented in California's terrestrial vegetation, but few aquatic scientists have turned their attention to the effects on freshwaters during this climate shift and this appears to be an open area of research.

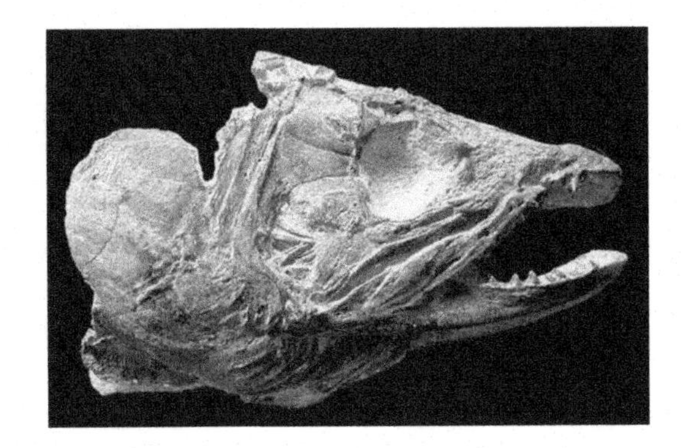

**FIGURE 1.19**

Miocene salmon fossil (Source: Eiting and Smith, 2007).

## CHANGING OCEAN CONDITIONS

Just as California's climate has undergone significant changes, ocean life has also experienced important shifts. For example, Miocene salmon had a more specialized feeding structure than extant salmon (Fig. 1.19). Fossilized late-Miocene Pacific salmon had unusually long, closely-spaced, and numerous gill rakers, plankton-straining structures in the pharynx. Middle Miocene *Oncorhynchus rastrosus* also has more numerous and more finely spaced gill rakers than living species. These fossils appear to be related to modern Sockeye salmon and Chum salmon, differing most in the numbers and morphology of gill rakers. Instead of the modern morphology that is suited for predators, these ancestors had the structural adaptations to be effective plankton feeders. These observations are consistent with oceanographic evidence for remarkably high plankton productivity in the North Pacific Ocean during the Middle Miocene, with closing of the ocean connection between North and South America. Thus we must keep in mind that habitat preferences, behavior, and resource use of ancestors may not align with our ecological understanding of current taxa.

## GREAT AMERICAN BIOTIC EXCHANGE

Perhaps, as early as 13–15 mya, the Central American seaway had closed, allowing the exchange of taxa between continents, which is called the Great American Biotic Interchange (GABI). For example, the exchange of freshwater fish between the continents is well documented. These changes serve as a prelude to the role of humans moving species around and one component of the ongoing 6th extinction event.

## GLACIATION: 40,000 AND 100,000 CYCLES

Over the last 1.6 mya, the Earth has experienced regular glacial and interglacial periods. The records suggest a 40,000-cycle early in the record, and then it switched to a 100,000-year cycle (Fig. 1.20). Although the cycles are usually explained by changes in the Earth's orbit around the sun or the Milankovitch cycles, what might explain the change in period? Since the Earth's orbit has not changed, researchers have been trying to explain these changes using a wide range of data sources. Recent ev-

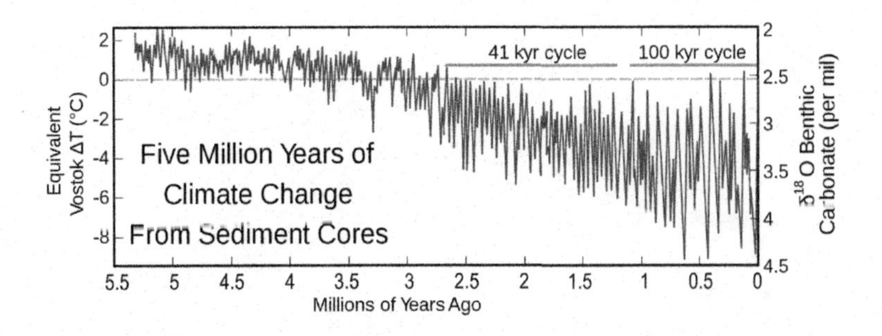


**FIGURE 1.20**

Oxygen isotopes can be used as proxies for temperature to document changes in the Earth's temperature (Voosen, 2018).

idence suggest that the speed of glacier movement might have changed due to their contact with the bedrock, collapse of the Atlantic Meridional Overturning Circulation, and perhaps a decline in atmospheric $CO_2$ concentrations. Again, as with the PETM, we see that the earth system is quite sensitive to the several factors that have profound implications to the evolution of the Earth's inhabitants.

## WELCOME TO THE ANTHROPOCENE

### A NEW EPOCH

Between 400,000 years ago and the second interglacial period in the Middle Pleistocene, around 250,000 years ago, the first Homo sapiens were present on the earth. The history of humans relative to the history of the world is quite small. In spite of ourselves, our history is insignificant compared to the planet's history, but at the same time our history may be the most consequential to the inhabitants that we share the Earth with (Fig. 1.21).

Although the term, Anthropocene, has been controversial and many geologists object to the term, it has become a part of the lexicon to describe the planetary process that are now dominated by humans.

Humans dominate global nutrient cycles, including nitrogen and phosphorous. Human's have reduced the capacity of ozone to protect the Earth's inhabitants from dangerous UV radiation and increase the concentration of $CO_2$ that threatens to heat the planet even more. Nuclear detonations and failed nuclear reactors have release radiative materials into the biosphere, and novel organic compounds have contaminated soils, streams, lakes, estuaries, and oceans. Concentrated areas of humans and cows and poultry and pigs generate high concentrations of wastes that is treated in some areas, and not in others. In direct ways, human settlements have been at the expense of many sensitive species, while the importation, both on purpose or accidental, and has created novel communities that often result in dramatic changes in species composition and even extinctions.

### SIXTH EXTINCTION

Relative to our chapter, the Anthropocene is likely to cause an extinction event not seen since the Cretaceous Period. For inland waters, this scenario has already been well on its way for several decades.

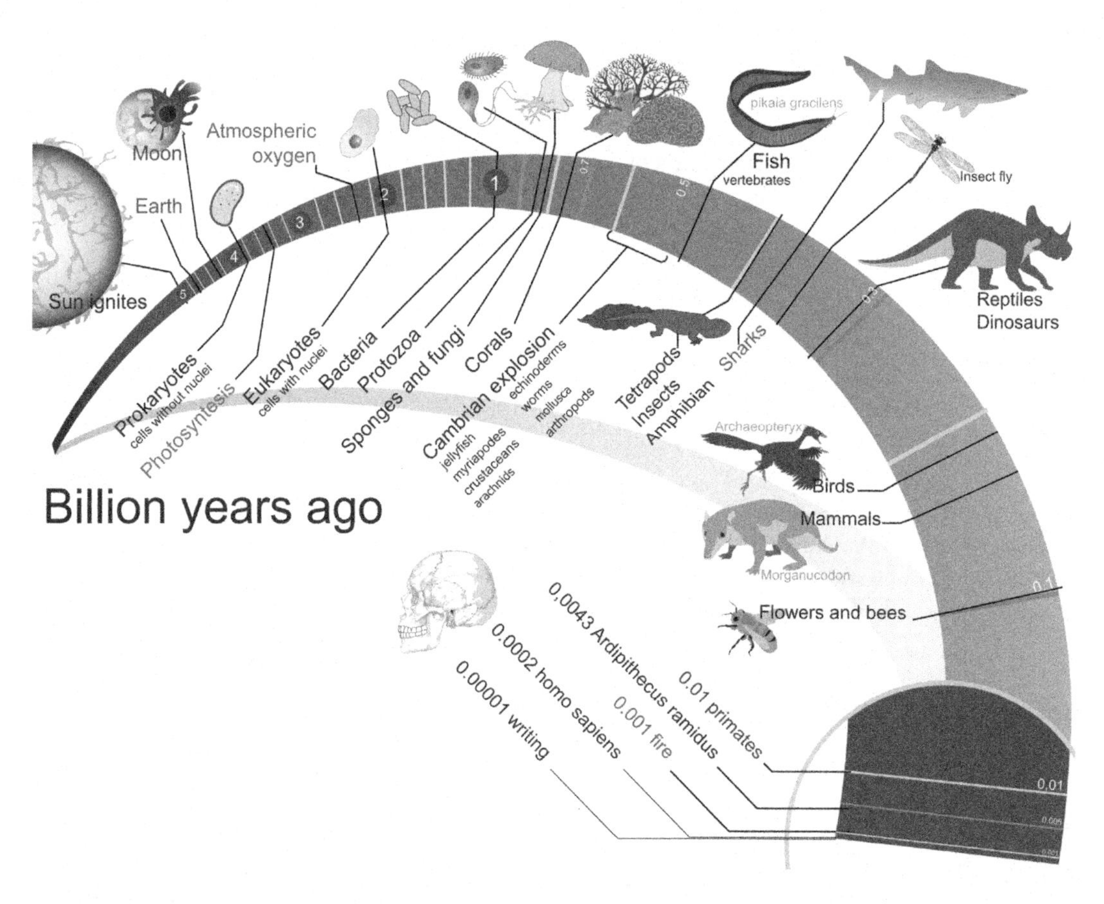


**FIGURE 1.21**

An evolutionary timeline documents the recent diversification of the primates in the context of Eukaryotic diversification staring 2 mya (Source: Wikicommons).

Freshwater fish and mussels have already been extirpated from tributaries of the Mississippi and rivers draining the Appalachian mountains due to dams and introduced species. Amphibians populations have been plummeting world-wide for a host of reasons, but commonly by the introduction of fungal diseases.

In the west, the competition for water is played out in legal battles, where fish, frogs, and other species do not have the same standing in court as humans or corporations.

## WHERE DO HUMANS FIT?

The controversy about the term Anthropocene, parallels the problem of how humans fit in the environment. Are we part of or separate from nature? Although this might sound like an academic distinction

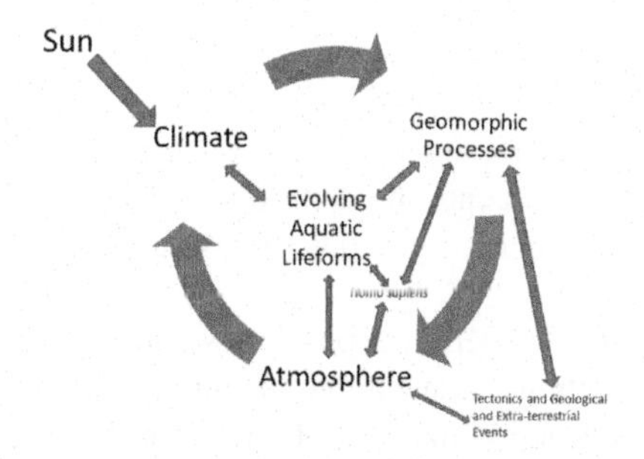


**FIGURE 1.22**

Homo sapiens are both part of and separate from our ecosystems.

without a practical application, the question has value in how we approach the management of our environment, and for our purposes, inland waters.

Without a doubt, we are part of the system, but we have a unique role and have profound effects, potentially on par with role of cyanobacteria to disrupt the entire Earth's biosphere. Thus for the purposes of this text, human beings will be both part of and separate from the aquatic ecosystems we will be learning about (Fig. 1.22). By privileging humans, I will argue that it is paramount to do our best to protect our aquatic systems, because they are both the basis of our evolutionary roots and are a fundamental resource on which we rely.

## NEXT STEPS

### STUDY QUESTIONS

1. What kinds of questions about aquatic ecosystems can the fossil records help us answer?
2. Describe 3 major events that allow for diversification of inland waters.

### LITERATURE RESEARCH PROJECTS

Though this chapter provides basic guidance for navigating the complexities of the Earth's geologic history, it does gloss over important details. To improve your understanding, I suggest you use the following set of exercises to probe deeper into the relationship between atmosphere, climate, and the Earth's evolving taxa.

1. Select a region on the planet that has a relatively good fossil record. Describe the fossil record with five examples related to freshwater systems of the region. Be sure to describe the limitations of our knowledge and what formations help us understand our aquatic systems.
2. Provide a description about how geology, climate, and evolving taxa have interacted to create the aquatic systems we now have in California.
3. Understanding the geologic and evolutionary history of the Earth requires knowledge of Earth's geologic timeline. Create flash cards for the following: Eons (Hadean, Archean, Proterozoic, and Phanerozoic); Eras (Eoarchean, Paleoarchan, Mesoarchean, Neoarchean, Paleoproterozoic, Meso-

proterozoic, Neoproterozic, Paleozoic, Mesozoic, and Cenozoic); Periods (Tonian, Cryogenian, Edicarian, Cambrian, Ordovician, Silurian, Devonian, Carboniferous, Permian, Triassic, Jurassic, Cretaceous, Paleogene, Neogene, and Quaternary); Epochs (Paleocene, Eocene, Oligocene, Miocene, Pliocene, Pleistocene, Halocene, and Anthropogene), and their dates and some description of the planet's characteristics during each geologic timespan. You may find mnemonic approaches to help you memorize these.

4. Coal was the result of Carboniferous Period deposits in swampy forests. How were these extensive forests created? Was there something unique about the taxa or the environment to limit decomposition of this material? Is coal production like that of the Carboniferous still possible? Are there coal deposits in California? If so, where and how were these formed? Note: This is an active area of research, so you find some ambiguity in the conclusions.
5. Using various research resources, investigate a freshwater formation that occurs in a region of the world. Develop a list of important freshwater taxa, and construct a simple food web.
6. Strontium isotope ratios ($^{86}Sr/^{87}Sr$).

CHAPTER

# ARCHEOLOGY OF INLAND WATERS 2

*Of all our planet's activities—geological movements, the reproduction and decay of biota, and even the disruptive propensities of certain species (elephants and humans come to mind)—no force is greater than the hydrologic cycle.*

**Richard Bangs & Christian Kallen, River Gods**

*All the rivers run into the sea; yet the sea is not full; unto the place from whence the rivers come, thither they return again.*

**Ecclesiastes 1:7**

*I've known rivers:*
*I've known rivers ancient as the world and older than the flow of human blood in human veins.*
*My soul has grown deep like the rivers.*
*I bathed in the Euphrates when dawns were young.*
*I built my hut near the Congo and it lulled me to sleep.*
*I looked upon the Nile and raised the pyramids above it.*
*I heard the singing of the Mississippi when Abe Lincoln went down to New Orleans, and I've seen its muddy bosom turn all golden in the sunset.*
*I've known rivers:*
*Ancient, dusky rivers.*
*My soul has grown deep like the rivers.*

**Langston Hughes (1902–1967), The Negro Speaks of Rivers**

## CONTENTS




Ecology and Management of Inland Waters. https://doi.org/10.1016/B978-0-12-814266-0.00014-3

In 1913 the Los Angeles Aqueduct was completed and began delivering water to the City of Los Angeles from the Owens River. With a modest allotment, ranchers and irrigators still enjoyed an adequate supply of water, but after learning that additional water supplies were needed, the City began taking all of the water in the river. Reaching a boiling point for the Owen Valley residents, a group of ranchers blew up segments of the canal, as a protest and perhaps getting water back. As a vivid example of water as contested commodity, we need to appreciate the socioeconomic, cultural, and political dimensions of water are confronted with the acknowledgment that water is political.

In the distant past, nomadic people needed water for drinking, feared floods, and used water environments for fishing and hunting. After becoming sedentary agriculturists dug wells, irrigated land, and built levees for flood protection—to reduce resource insecurities. Infrastructure developments require well-organized societies, which become symbols of progress and identity, like the aqueducts of the Roman Empire, canals in China, and dams in the United States.

Monuments documenting water technology provide a historical perspective in our development of water, where water becomes "as a resource" provides safe drinking supplies, irrigation allotments, transportation routes, managed fisheries, concentrated power generation, elitist aesthetics, and monetized recreation. When water fails to deliver these products, we are confronted by the contradictions of modernity.

The human condition depends on our planet's inland waters. An effective use of water can impart success, whereas an ineffective use can lead to disaster. If nothing else, like all life forms on Earth, humans require water— water that is both reliably available and free of contaminants and pathogens. Meeting these goals is, perhaps, one of the greatest challenges of human kind. In this chapter, we introduce and analyze case studies to expose the evolving nature of our relationship with water.

After reading this chapter, you should be able to

1. Describe how water use has been central to the development of human societies;
2. Explain how the geography of a place provides specific opportunities and constraints in human-water relations;
3. Describe unintended consequences of water development projects;
4. Summarize examples aimed to improve the sustainable use of water.

## DRINKING WATER AND WASTE

If there is a universal use of water for humans it starts with drinking water. Yet, securing safe and reliable drinking water supplies continues to be out of reach for over 30% of the world's population. In fact, it begs the question if there were ever periods of time that humans had access to safe and reliable drinking water supplies.

Part of the question is answered by the extent that people have access to proper waste treatments, which is ~30%! In a rapidly developing planet, increasing concentrations of human settlements means increasing human waste and limited ecosystem capacity to process the waste. Thus as human density increases, the society must improve the infrastructure with plumbing that separates drinking waters from waste waters.

## THE WATER CYCLE

Water falls to the Earth with relatively low concentrations of salts—although slightly higher when rain clouds form over oceans, due to aerosols of salt produced by the ocean. The precipitation is starting point for freshwaters and falls to the surface as snow or rain in land or on water. On land, water flows from higher elevation to lower elevation basins. The catchment area, the area that facilitates the

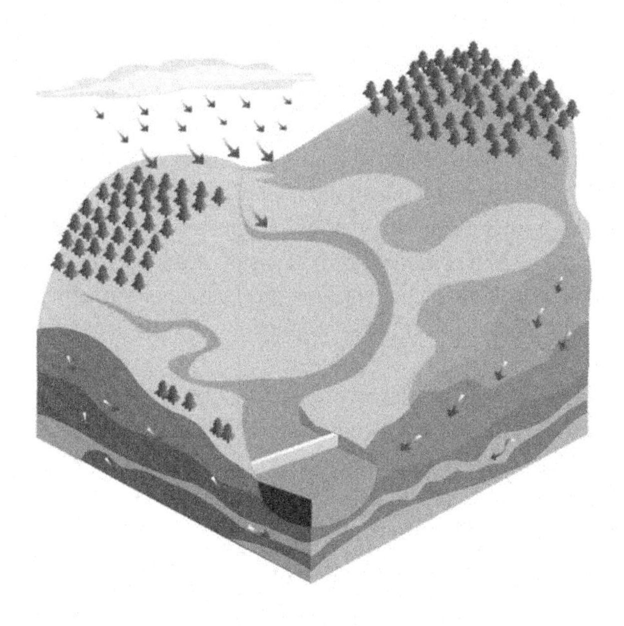

**FIGURE 2.1**

Sketch of a watershed or catchment (Courtesy of Luyi Huang).

movement of surface water to a basin is called a watershed (Fig. 2.1), although many English speaking countries might refer to these as catchment areas. As the water interacts with the landscape, e.g., soils, geologic formations, vegetation, and human activities, the water picks up additional salts and chemicals and carries them to the basin's outlet.

These processes define the waters available for drinking water. We cannot drink seawater, thus rely on the water cycle to provide relatively dilute concentrations of salts.

Water returns to the atmosphere via evaporation or evapotranspiration (through the leaves of plants). As part of this evaporation process, water fractionates the relative abundance of stable isotopes.

Both Hydrogen and Oxygen have a few stable isotopes—atoms with different numbers of neutrons, thus different weights. The isotopes are stable because they are not in the process of decaying or giving off radiation (radiogenic). Some isotopes are very rare, so we will look at the ones that are commonly measured.

Hydrogen usually has no neutrons, but it can also have one and remain stable. Hydrogen atom with one neutron is referred to as Deuterium and can be symbolized as $^{2}H$ or D with different relative abundances: 99.98% H and 0.0156% D. Oxygen has three stable isotopes, which has 8, 9, or 10 neutrons, symbolized as $^{16}O$, $^{17}O$, and $^{18}O$, respectively. They also have differing relative abundances on Earth: 99.76% = $^{16}O$, 0.04% = $^{17}O$, and 0.20% = $^{18}O$. Of course, composed of $H_2O$ (two Hydrogen atoms and one Oxygen atom), water can have several combinations of isotopes with differing molecu-

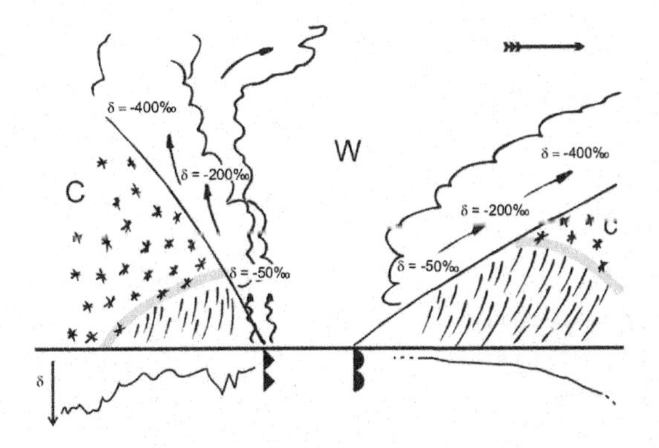


**FIGURE 2.2**

Water fractionation modified from Sodemann (2006).

lar weights (MW) and **average** natural abundances.[1] Because some of these combinations are so rare, we can effectively ignore them.

Returning to fractionation, it is relevant to note that it is similarly related to mass differences between the nuclei of isotopes, but is associated with incomplete and unidirectional processes, such as evaporation and diffusion. In general, the lighter isotope will react faster and will become concentrated in products.

For water, the higher the mass number, the lower the vapor pressure. Thus $^{16}O$ and H preferentially enter the vapor phase, whereas $^{18}O$ and D preferentially concentrate in the liquid phase. Consequently, in evaporation, water vapor is enriched in $^{16}O$ and H, whereas the remaining liquid water is enriched in $^{18}O$ and $^{2}H$. More specifically, $H_2^{18}O$ is enriched in liquid water by 1% relative to its concentration in water vapor at the same temperature.

The fate of water can be "followed" using isotopic abundances as it completes the water cycle (Fig. 2.2). Various pools have signatures that reflect the fractionation processes (Table 2.1): evap-

[1] Note the differences between what might be more easily detected and relative fractionation potential. Note that these averages are so sensitive to fractionation that the relative abundances are not found in any given system. Sorted by relative abundance.

| Isotope | MW | Relative Abundance (%) |
|---|---|---|
| $H_2^{16}O$ | 18 | 99.72 |
| $H_2^{18}O$ | 20 | 0.20 |
| $H_2^{17}O$ | 19 | 0.040 |
| $HD^{16}O$ | 19 | 0.016 |
| $HD^{18}O$ | 21 | $3.1 \cdot 10^{-5}$ |
| $HD^{17}O$ | 20 | $6.3 \cdot 10^{-6}$ |
| $D_2^{16}O$ | 20 | $2.4 \cdot 10^{-6}$ |
| $D_2^{18}O$ | 22 | $4.9 \cdot 10^{-9}$ |
| $D_2^{17}O$ | 21 | $9.7 \cdot 10^{-10}$ |

**Table 2.1 Stable isotopes of water.**

| Natural Reservoir | $\delta^{18}O$‰ | $\delta D$‰ |
|---|---|---|
| Ocean Water | -6 – +3 | -28 - +10 |
| Arctic Sea Ice | -3 – +3 | 0 – +25 |
| Marine moisture | -15 – -11 | -100 – +75 |
| Lake Chat | +8 – +16 | +15 – +50 |
| Alpine Glaciers | -19 – -3 | -130 – -90 |
| (Sub)Tropical precipitation | -8 – -2 | -50 – -100 |
| Mid-latitude precipitation | -10 – -3 | -80 – -20 |
| Mid-latitude snow | -20 – -10 | -160 – -80 |

oration, precipitation, freezing, melting, sublimation, etc. These fractionation processes allow us to evaluate the residence time for water in each pool—oceans, atmosphere, glacial ice, alpine lakes, etc.

Before completing the water cycle, water interacts with the lithosphere or biosphere in a myriad number of ways. Water in glaciers or groundwater might be stored for hundreds of thousands or millions of years. Water might flow through creeks and streams and into major rivers and estuaries before ending up in the ocean, where the average time a water molecule remains is about 3000 years. In contrast, the average time in the atmosphere is about 9 days.

## RELIABLE WATER SUPPLIES

Singapore receives approximately 2.3 meters of rain per year. This is over twice as much rainfall as Seattle and Dallas, Texas, both of which receive about 852 mm of rain per year. Whereas we might think of Seattle as a wet city, the rainfall in Dallas is surprising when we think of the region as arid. More surprising yet is that Singapore's government acknowledges that their water supply is tenuous, because storing the water is extremely difficult in small country (Fig. 2.3). Stored water for human use relies on lakes and reservoirs and groundwater. If the geology of a region does not have water bearing formations, i.e., aquifers, to develop, then topography might be used to store water behind dams. If these characteristics are not inadequate or the funding is absent to build big dams, communities might rely on surface waters, which can be notoriously finicky. For example, Cape Town, South Africa nearly reached a complete shut down of its municipal water supply in 2018, demonstrating its vulnerability even with reservoirs.

Of course, the water supply in many regions relies on groundwater. One of the most obvious, but poorly captured by policy makers, is the fundamental relationship between surface and ground waters. In part, due to pumping and climate change, groundwater elevations are changing. The GRACE satellite mission has done well to document these changes by analyzing changes in the Earth's gravity. Throughout the book, we will reference the relevance of the relationships between surface and groundwater as an ecologically important and economic value, especially as aquifers are overpumped (Fig. 2.4).

**FIGURE 2.3**

Upper Peirce Reservoir in Singapore.

## SOCIAL COSTS OF LEAD PLUMBING

Plumbing to improve water supplies has been a marker of development and advances to democratize indoor plumbing has been a key aspect of economic development. Lead (Pb) was convenient to use for plumbing—for many centuries, lead was the favored material for water pipes, because its malleability made it practical to work it into the desired shape. In fact, its use was so common that the word "plumbing" is derived from the Latin word for lead— plumbum (Fig. 2.5).

Use of lead for water pipes may have been the most damaging feature of the advanced water technology of ancient times. Even before the health hazards of ingesting lead was understood, the lead-related health problems, such as stillbirths and high rates of infant mortality, was linked to lead piping. Nevertheless, lead water pipes were still widely used in the early 20th century, and remain in many households.

Despite the Romans' common use of lead pipes, their aqueducts rarely poisoned people. Unlike other parts of the world, where lead pipes cause poisoning, high calcium concentrations created a layer of plaque that prevented the water contacting the lead itself in Roman plumbing.

And yet the use of Pb in water pipes is a reoccurring public health threat. Even in highly developed countries, the historic use of Pb pipes remains problematic. Old lead pipes used in the water system of Flint, Michigan has been linked to a public health crisis. The root cause of the Flint lead crisis was corrosion. For 50 years, Flint had purchased its water from Detroit, 90 miles away. However, in 2014, it switched water sources to save money without appreciating how the slightly more acidic water of the Flint River would impact the city's pipes.

Furthermore, officials did not use common corrosion control methods that Detroit and many other cities use in their water systems, e.g., adding phosphates to the water to help keep lead from dissolving into the water flowing through the pipes. The corrosive water pumping underneath Flint quickly ate away at the protective layer inside the city's old lead pipes, exposing bare lead to the water flowing through them. When the city switched water supplies, a protective layer of rust began to be stripped away from the interior of the pipes, strongly discoloring the water and leaching the lead from that rust into the water (Fig. 2.6).

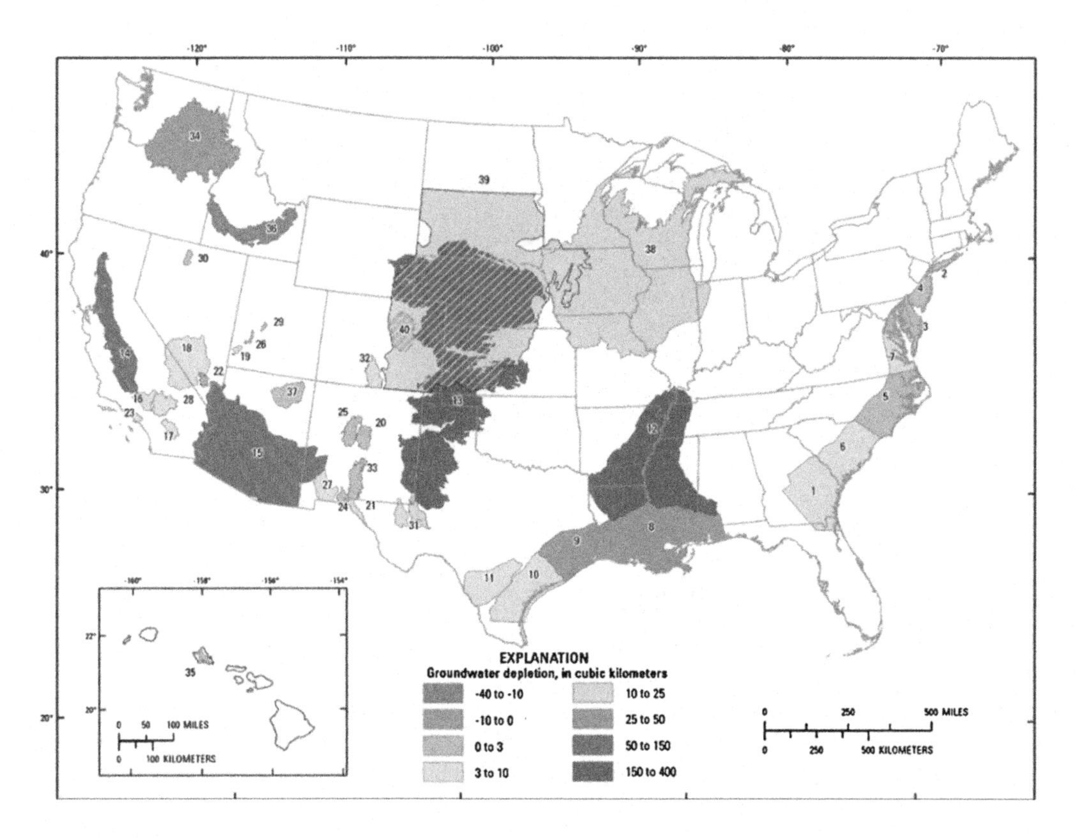


**FIGURE 2.4**

Map of the United States (excluding Alaska) showing cumulative groundwater depletion, 1900 through 2008, in 40 assess aquifer systems or subareas (Source: USGS).

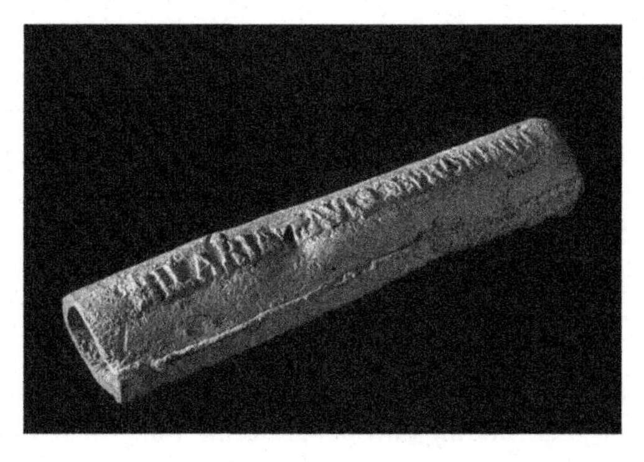

**FIGURE 2.5**

Lead water pipe, Roman, 1–300 CE. The inscription cast into the side of the pipe indicates that the work was undertaken by a team under an "Imperial Freeman Procurator Aquarum" — an official in charge of maintaining the water supply. In some cases, piped water supplies in the Roman empire were quite complex and sophisticated.

**FIGURE 2.6**

Domestic water supply pipes from Flint, Michigan.

The most vulnerable population, children, suffer both short- and long-term affects, including the loss of memory, eye-hand coordination, and appropriate weight gain. Some of the people responsible were criminally charged for an assortment of crimes. As environmental scientists, we must be vigilant to protect vulnerable populations by working hard to understand the history and science of the lessons to manage infrastructure appropriately.

## WASTES AND WATERBORNE DISEASES

Of course, for many it is not merely access to water that is important, but water free from water-borne diseases. From a public health perspective, the threats posed by water-borne diseases are always a concern—but these risks are unevenly distributed across the globe.

It is not likely that we would have much concern about the fate of human waste if it was not associated with disease, but understanding this relationship has been unevenly appreciated across various cultures, and our capacity to reduce the risks associated with human waste has also been a process in human development.

Waterborne diseases are caused by pathogenic microorganisms that most commonly are transmitted in contaminated freshwater. Infection commonly results during bathing, washing, drinking, in the preparation of food, or the consumption of food that is infected, caused by a range of organisms (Table 2.2). Various forms of waterborne diarrheal disease probably are the most prominent examples, and affect mainly children in developing countries; according to the World Health Organization, such diseases account for an estimated 3.6% of the total DALY global burden of disease, and cause about 1.5 million human deaths annually. The World Health Organization estimates that 58% of that burden, or 842,000 deaths per year, is attributable to unsafe water supply, sanitation, hygiene, and cleanliness.

## HYDROLOGY: APPLIED SCIENCE FOR PROGRESS

Hydrology is the study of water flow. Using Newtonian physics, hydrologists related water velocity (m/s), water flow ($m^3/s$), velocity head ($v^2/2g$), and hydraulic head ($\psi + z$), where m is meters, s is second, $v$ is velocity (m/s), $\psi$ is difference in elevations, and $z$ is the bottom of water table.

**Table 2.2 Major water-borne illnesses.**

| Disease | Agent | Source | Symptoms |
|---|---|---|---|
| Amoebiasis | Entamoeba histolytica | Nontreated drinking water, flies in water supply | Abdominal discomfort, fatigue, weight loss, diarrhea, bloating, fever |
| Cryptosporidiosis | *Cryptosporidium parvum* | Water filters and membranes that cannot be disinfected, animal manure, seasonal runoff of water. | Flu-like symptoms, watery diarrhea, loss of appetite, substantial loss of weight, bloating, increased gas, nausea |
| Cyclosporiasis | *Cyclospora cayetanensis* | Nontreated drinking water | Cramps, nausea, vomiting, muscle aches, fever, and fatigue |
| Giardiasis | Giardia lamblia | Untreated water, poor disinfection, groundwater contamination, campgrounds where humans and wildlife use same source of water. Beavers and muskrats create ponds that act as reservoirs for Giardia. | Diarrhea, abdominal discomfort, bloating, and flatulence |
| SARS (Severe Acute Respiratory Syndrome) | *Coronavirus* | Manifests itself in improperly treated water | Symptoms include fever, myalgia, lethargy, gastrointestinal symptoms, cough, and sore throat |
| Hepatitis A | *Hepatitis A virus* | Can manifest in water and food | Symptoms are only acute and include fatigue, fever, abdominal pain, nausea, diarrhea, weight loss, itching, jaundice and depression. |
| Poliomyelitis (Polio) | *Poliovirus* | Enters water through the feces of infected individuals | Symptoms vary, from delirium, headache, fever, and occasional seizures; to spastic paralysis, non-paralytic aseptic meningitis; or even paralysis or death |

Henry Darcy, a French hydraulic engineer interested in purifying water supplies using sand filters, conducted experiments to determine the flow rate of water through the filters. As a member of the Corps, Henry built an impressive pressurized water distribution system in Dijon following the failure of attempts to supply adequate freshwater by drilling wells. The system carried water from Rosoir Spring 12.7 km away through a covered aqueduct (watercourse) to reservoirs near the city, which then fed into a network of 28,000 meters of pressurized pipes delivering water too much of the city. The system was fully closed and driven by gravity, and thus required no pumps with just sand acting as a filter. Later, he conducted column experiments that established what has become known as Darcy's law; initially developed to describe flow through sands, it has since been generalized to a variety of situations, and is in widespread use today.

Darcy's law that discharge rate ($Q$) is a simple proportional relationship between the instantaneous discharge rate through a porous medium, the viscosity of the fluid, and the pressure drop over a given

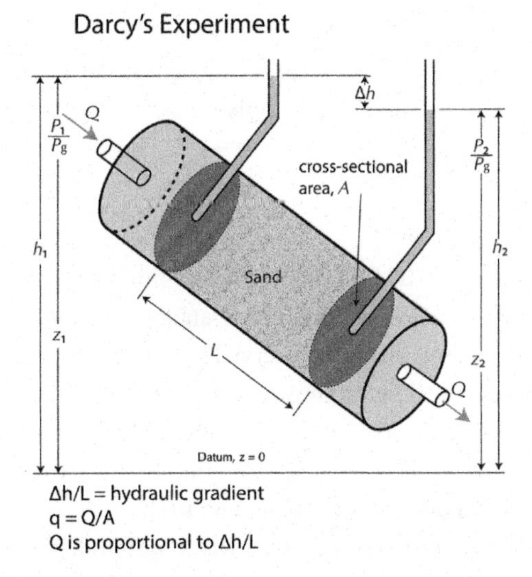


**FIGURE 2.7**

Darcy used sand filters to develop equations for the pressure changes that have been used to design water treatment facilities and groundwater models (drawn by Luyi Huang).

distance:

$$Q = -\frac{\kappa A\,(p_{\mathrm{b}} - p_{\mathrm{a}})}{\mu L}. \tag{2.1}$$

The total discharge, $Q$ (units of volume per time, e.g., $m^3/s$) is equal to the product of the intrinsic permeability of the medium, $\kappa$ ($m^2$), the cross-sectional area to flow, $A$ (units of area, e.g., $m^2$), and the total pressure drop $\rho_b - \rho_a$ (pascals), all divided by the viscosity, $\mu$ (Pa·s) and the length, over which the pressure drop is taking place $L$ (m). The negative sign is needed because fluid flows from high pressure to low pressure. Note that the elevation head must be taken into account if the inlet and outlet are at different elevations. If the change in pressure is negative (where $\rho_b > \rho_a$), then the flow will be in the positive $x$-direction. There have been several proposals for a constitutive equation for absolute permeability, and the most famous one is probably the Kozeny equation (also called Kozeny–Carman equation) (Fig. 2.7).

Dividing both sides of the equation by the area and using more general notation leads to

$$q = -\frac{\kappa}{\mu}\nabla p, \tag{2.2}$$

where $q$ is the flux (discharge per unit area, with units of length per time, m/s) and $\nabla p$ is the pressure gradient vector (Pa/m). This value of flux, often referred to as the Darcy flux or Darcy velocity, is not the velocity which the fluid traveling through the pores is experiencing.

It might be hard to see how these equations might revolutionize the science of water—but these are the foundation of water treatment facilities that have been used to reduce the risk of water-borne diseases and allow urban centers to become increasingly dense.

## FISHERIES AND FOOD

Obtaining fish as food dates back over 100,000 years for hominids. Prehistoric artifacts—fishing hooks, fishing nets, and discarded fish bones—are ubiquitous ancient human populations. In East Timor, a 42,000 year old fish hook documents pelagic fishery. There is no doubt that humans have been exploiting inland fishes and mollusks well before this documented example.

Currently, 94% of the developing world rely on fish diets. And 6% of the world's protein comes from fish, whereas in some cases, some countries are composed of over 20% fish diets. The most common species include salmon, talapia, carp, catfish, and trout.

In one example, almost all fisheries in the Zambezi River system have experienced severe declines in catch rates, loss of larger, most valuable fish species, and increased use of environmentally damaging active fishing gears. The fisheries of the Barotse, Caprivi, and Kafue floodplains, and lakes Kariba (Zambian sector), Malawi and Malombe are all fished down. The concept of balanced harvesting with moderate effort has no relevance to these African inland fisheries, where rapid human population growth and lack of alternative livelihoods for small-scale fishers means they have no choice but to continue fishing despite dwindling returns. In some areas, e.g., Liuwa Plain National Park in Zambia and conservancies in Namibia, comanagement with local communities has potential for success, but other fisheries, e.g., Lake Malombe in Malawi, are so severely fished down that there is no prospect of recovery without radical restructuring of exploitation patterns coupled with habitat restoration.

## CROP DOMESTICATION AND IRRIGATION

Neolithic periods saw the importance of three cereals—emmer, einkorn and barley, however, other crops were soon domesticated as well: flax, peas, chickpeas, bitter vetch, and lentils. In the corridor around the Fertile Crescent and later in Europe, domestication may have occurred over a relatively short period of between 20 and 200 years.

Early irrigation schemes, which promoted crop surpluses was the likely start of hydromodification (Fig. 2.8).

## ARID CLIMATES: PRE-COLOMBIAN COMMUNITIES

Like many cultures in arid climates, Native Americans developed sophisticated irrigation systems to improve the reliability of water supplies and maintain domesticated crops. In the south-west, Hohokam people canals to bring river water to fields dated back to 2000 BCE. Rainwater-capture systems, which were well developed by the Pueblo peoples, is another example of social organization and the development of water infrastructure (Fig. 2.9).

**FIGURE 2.8**

Mesopotamia irrigation system (Source: Persian Qanat Iran Heritage).

**FIGURE 2.9**

Rainwater was captured by early Puebloans, which was directed to small plots to grow various crops (Source: National Park Service).

## IRRIGATION AND LEGAL SYSTEMS

The first reservoirs were built for irrigation. But to be effective, in whatever environment, these reservoirs required agreements between users to maintain the reservoirs and ensure an adequate distribution. Ancient irrigation codes impress modern irrigation specialists because of their longevity. The Spanish agricultural cooperatives were probably built on a foundation of social networks built around common lands and irrigation communities. With high levels of human capital and the existence of a wide layer of middle size farms, cooperative irrigation systems developed.

**FIGURE 2.10**

Mission of San Javier: appearing lush and green in the midst of the season's exceptionally dry weather (Source: www.livingrootsbaja.org).

## MIGRATIONS, COLONIZATION, AND TRANSFORMATIONS

The expansion and migration of human populations has been a constant process, each group bringing with them certain cultural practices. For example, hunters who have reached new regions, e.g., North America, may have reduced the size of megafauna and introduced new species. Not only have species introductions been one of many ecological transformations, the landscape was altered as people brought certain cultural ideas and ideologies with them. For example, the native peoples and the landscape of Baja California were transformed by colonization by Spanish colonialists. As a mix of practical and religious meaning, the Jesuit priest missionaries appreciated surface waters, called oases, which were used to support introduced crops and development of predictable water supplies. As a result, these oases became sacred places and now support heritage crops, biodiversity, and traditional foods (Fig. 2.10).

In California Alta (nominally the state of California), missions were built with the same colonizing goals that displaced native peoples, while also developing irrigation systems to support the mission crop production. For example, the Los Angeles Mission was one of the first projects to alter the Los Angeles River that would culminate with the replacing the channel with concrete levies.

## IRRIGATION SCHEMES: DISTRIBUTION AND RESERVES

The Dujiangyan irrigation scheme has served China for 18 centuries. Like the Roman aqueducts, this system provides cultural symbols of organizational capacity and planning that have become emblematic of cultural organization. Originally constructed around 256 BC by the State of Qin as an irrigation and flood control project, the system's infrastructure is on the Min River (Minjiang), the longest tributary of the Yangtze. Originally, the Min rushed down from the Min Mountains, but slowed abruptly after reaching the Chengdu Plain, filling the watercourse with silt, which made the nearby areas extremely prone to floods. Li Bing, then governor of Shu for the state of Qin, and his son headed the construction of the Dujiangyan, which harnessed the river using a new method of channeling and dividing the water rather than simply following the old way of dam building (Fig. 2.11). The Dujiangyan irrigates over 5300 square km of land in the region.

**FIGURE 2.11**

The Dujiangyan irrigation system (Sichuan, China) was originally constructed around 256 BC by the State of Qin as an irrigation and flood control project, and is still used today (Source: duechina.com).

## EXORHEIC AND ENDORHEIC BASINS

Historically, there have been highly variable Pleistocene Lakes in Great Basin and many of them have gone dry with the melting of the ice-sheets. In the anthropocentric, the rate of drying has been increasing dramatically. For example, the Aral Sea has been declining rapidly due to agricultural diversions, and the North American's endorheic lakes face the same plight, e.g., Walker Lake and Pyramid Lake (Fig. 2.22).

The outlet of many watersheds drain into the ocean; these watersheds are called exorheic. However, many waters cannot drain to the ocean because they are surrounded by mountains. These basins are called endorheic. In fact, much of the continental United States contain basins that do not drain into the ocean. Mono Lake (California), the Great Salt Lake (Utah), Pyramid Lake (Nevada), Devils Lake (North Dakota), and even Death Valley (California) are examples endorheic basins in the United States. Well known examples of endorheic basins in other countries include the Aral Sea (Kazakhstan / Uzbekistan) and Lake Chad (Chad).

Because endorheic basins have no outlet, water exits the basin by evaporation. Thus the salts or other chemicals dissolved in the water remain in the water, where their concentrations will increase for thousands of years or more. The balance between inflow and outflow in these basins depends on the regional climate, and as you might guess, are very vulnerable to climate changes or water diversions by humans.

In both surface and groundwater sources, the balance between inputs and outputs is a key indicator of the sustainability of their use. If lakes are declining or groundwater withdrawals exceed recharge, then the use of the basins are out of balance, indicating unsustainable water use. Water withdrawals from the Aral Sea watershed is one of the most dramatic examples of unsustainable water use, where salt concentrations exceeded the physiological limits of the fish, leading to the collapse of the fishery (Fig. 2.12).

# WATER AND TRANSPORTATION

## THE GRAND CANAL

The Grand Canal or the Beijing–Hangzhou Grand Canal is the longest canal in the world (1776 km). Starting at Beijing, it passes through Tianjin and the provinces of Hebei, Shandong, Jiangsu, and Zhejiang to the city of Hangzhou, linking the Yellow River and Yangtze River. The oldest parts of the canal

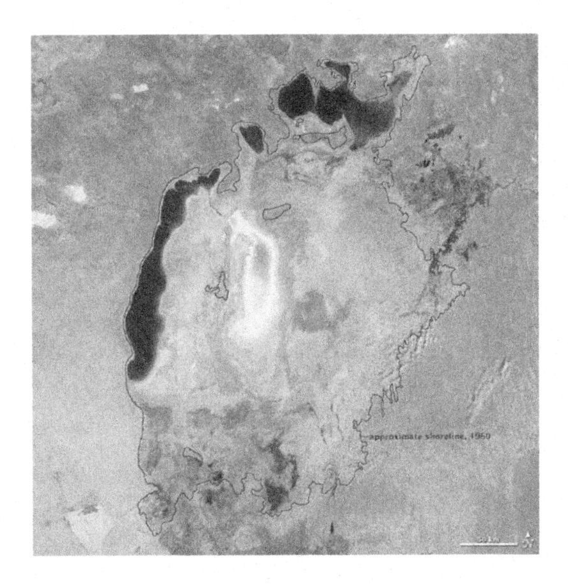


**FIGURE 2.12**

Aral Sea has declined due to water diversions aimed to improve the agricultural productivity in the region (Source: NASA). As if often the case, balancing the ecological value and human efforts to develop economically is a constant tension.

date back to the 5th century BC, although various dynasties expanded, rebuilt, and altered the canal over time. The canal was designed to transport goods between the provinces, but could also be used to flood attacking armies in low lying areas.

## COLONIZATION: SUEZ, PANAMA, AND NICARAGUA CANALS

The idea of constructing a canal connecting the Mediterranean and the Red Sea was first proposed by French engineers during Napoleon Bonaparte's occupation of Egypt, but construction did not commence until 1859 and was completed 10 years later. For the next 87 years, it remained largely under British and French control, and Europe depended on it as an inexpensive shipping route for oil from the Middle East. Since its construction, the canal reenforced colonial power, a symbol for the aspiration for sovereignty and proximate causes of armed conflicts for regional and geopolitical aims—the Suez Canal is more than just a ditch filled with water in the desert.

In addition, the canal has had ecological impacts. For example, the Bitter Lakes, which were hypersaline natural lakes, were linked by the canal. At first, because of the high salinity, they prevented Red Sea species from invading the Mediterranean. However, the lakes gradually equalized with canal water from Red Sea, and plants and animals from the Red Sea colonized the eastern Mediterranean. Invasive species that originated from the Red Sea and introduced into the Mediterranean by the canal have become a major component of the Mediterranean ecosystem and have serious impacts on the ecology, endangering many local and endemic species. About 300 species from the Red Sea have been identified in the Mediterranean, and there are probably others yet unidentified.

Not to be outdone by the Europeans, the United States supported Colombian rebels and an independent Panama to construct the Panama Canal. The treaty imposed on the newly independent Panama ensured that the profits went to US businessmen and investors with little return to Panama. The embittered Panamanians finally won control of the Panama Canal in 1977, and the United States formally withdrew from the Canal Zone, which had been a US territory.

The project planners then had to find a way to fill this area above the sea with water. They found such a good opportunity in the Chagres River, which they used to build a dam. The dam flooded a massive area close to the river, and thus helped create the Gatun Lake, which now forms the main part of the canal (Fig. 2.13). Since the canal is above the sea level, it needs an effective transit mechanism that would lift and lower the vessels as they pass along the waterway. This mechanism comes in a system of locks which work at several stages of the canal's length, and do just that—lift and lower the ships. To operate properly, the canal has to use vast amounts of water, which is collected during the rainy season. The Gatun Lake is also very important as a water source for the canal's operation during the dry season. The isolation, completed by 1914, of Barro Colorado Island from mainland Panama by the waters of Gatun Lake, accelerated extinction on this island, because populations threatened by local extinction were now much less likely to be reinforced or reestablished by colonists from elsewhere. In addition, introduced species have reshaped the food web of the lake (see Chapter 3).

In June 2013, Nicaragua's National Assembly approved a bill to grant a 50-year concession to finance and manage the project to the private company headed by a Chinese billionaire. The concession can be extended for another 50 years once the waterway is operational. Media reports suggest the project would have been delayed or even possibly canceled, but such projects seem to have the capacity to be resurrected depending on the political and economic context.

Scientists are concerned about the project's environmental impact, as Lake Nicaragua is Central America's key freshwater ecosystem, where endemic freshwater sharks live (Fig. 2.14). Construction of a canal using the San Juan River as an access route to Lake Nicaragua was first proposed in the early colonial era. The United States abandoned plans to construct a waterway in Nicaragua in the early 20th century after it purchased the French interests in the Panama Canal. By May 2017, no concrete action had been reportedly taken to construct the canal and doubts were expressed about its financing.

## WATER AND POWER

### INDUSTRIALIZATION AND WATER

Industrialization requires power. Although coal could be used to create steam and drive mechanical force, the use of wind and water dates back even further. A waterwheel is a machine for converting the energy of flowing or falling water into useful forms of power, often in a watermill (Fig. 2.15). A waterwheel consists of a wheel (usually constructed from wood or metal), with a number of blades or buckets arranged on the outside rim, forming the driving surface. Most commonly, the wheel is mounted vertically on a horizontal axle, but can also be mounted horizontally on a vertical shaft, for example, the tub or Norse. Vertical wheels can transmit power either through the axle or via a ring gear, and typically drive belts or gears; horizontal wheels usually directly drive their load.

Waterwheels were still in commercial use well into the 20th century, but they are no longer in common use. Uses included milling flour in gristmills, grinding wood into pulp for papermaking,

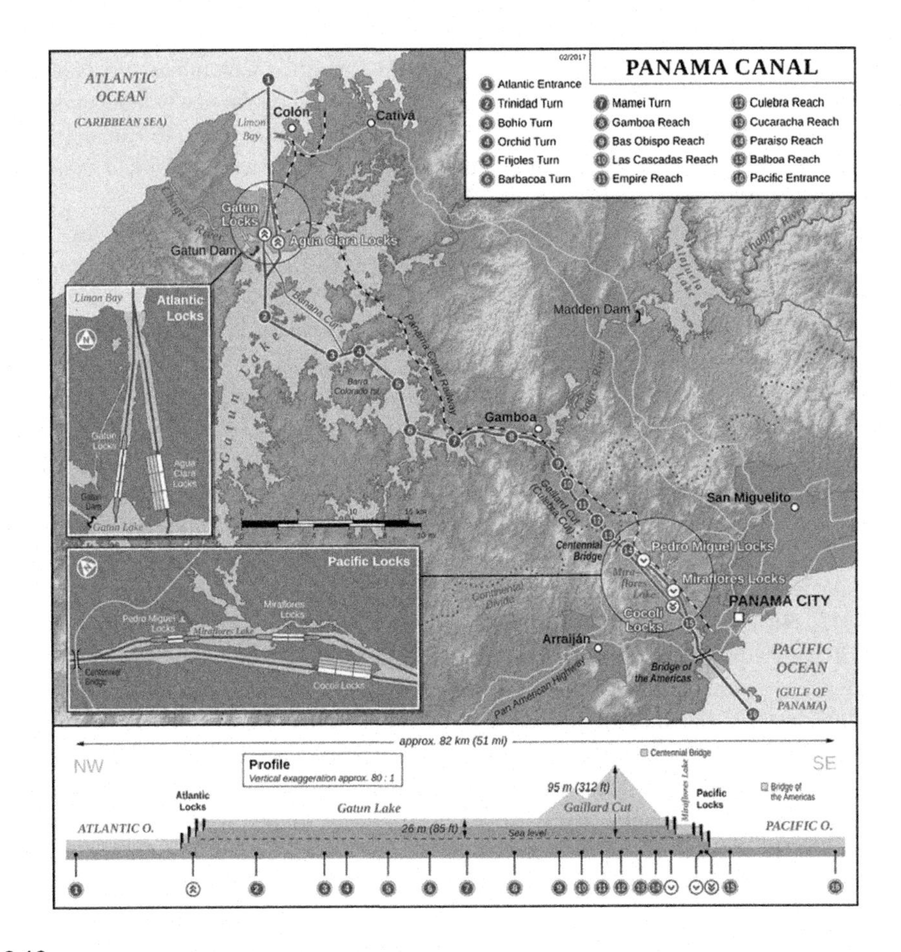


FIGURE 2.13

Barro Colorado Island is located in the man-made Gatun Lake in the middle of the Panama Canal. The island was formed when the waters of the Chagres River were dammed to form the lake in 1913. When the waters rose, they covered a significant part of the existing tropical forest, but certain hilltops remained as islands in the middle of the lake.

hammering wrought iron, machining, ore crushing, and pounding fiber for use in the manufacture of cloth.

Some waterwheels are fed by water from a mill pond, which is formed when a flowing stream is dammed. A channel for the water flowing to or from a waterwheel is called a millrace. The race bringing water from the mill pond to the waterwheel is a headrace; the one carrying water after it has left the wheel is commonly referred to as a tailrace.

In the mid to late 18th century John Smeaton's scientific investigation of the waterwheel led to significant increases in efficiency, supplying much needed power for the Industrial Revolution.

**FIGURE 2.14**

Image of a bull shark. Lake Nicaragua, despite being a freshwater lake, has sawfish, tarpon, and sharks. Initially, scientists thought the sharks in the lake belonged to an endemic species, but has been identified as the bull shark, *Carcharhinus Leucas*, a shark that regularly enters freshwaters (Source: Ben Team).

**FIGURE 2.15**

*De re metallica* was a book that described the state of the art of mining, refining, and smelting metals by Georg Bauer, whose pen name was the latinized Georgius Agricola.

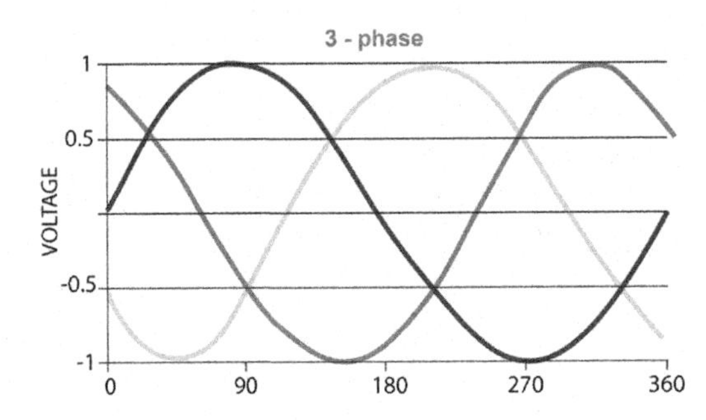


**FIGURE 2.16**

Single and two-phase AC synchronous motors were inefficient and cumbersome. This application of single phase AC meant that motors had to be synchronized with the generator during startup. You can imagine how difficult this was with miles between the two machines, let alone the use multiple motors in various different locations (drawn by Luyi Huang).

## GENERATING MECHANICAL POWER

Hydropower or waterpower is power derived from the energy of falling water or fast running water, which may be harnessed for useful purposes. Since ancient times, hydropower from many kinds of watermills has been used as a renewable energy source for irrigation and the operation of various mechanical devices, such as gristmills, sawmills, and textile mills.

## GENERATING ELECTRICAL POWER

In the late 19th century, hydropower became a source for generating electricity. Cragside in Northern England was the first house powered by hydroelectricity in 1878; one of the first commercial hydro-electric power plant was built at Niagara Falls in 1879, and powered street lamps installed in the city of Niagara Falls by 1881. The development of the electrical generators and motors allowed for the spatial separation between electrical generation and the use of electricity to do work. In 1890 Cyrus Baldwin (the first president of Pomona College, Claremont, CA) initiated the development of power generator using San Antonio Creek. However, direct current (DC) could not be transmitted over the 14 mile distance to Pomona. Nonetheless, Tesela's experiments with alternating current and transformers in Germany was used to design a transmit power a record 28 miles to San Bernardino, CA.

The designer, Almarian Decker, however, wanted to design the power station to run three-phase alternating current (Fig. 2.16). However, Westinghouse refused to build a three-phase AC generator calling it "inpractible" for the San Antonio power station, Redlands' Mill Creek. Mill Creek No 1, the first commercially viable three-phase power plant in the world, began its power generation September 7th, 1893, powering irrigation pumps for citrus groves in the California Desert.

**FIGURE 2.17**

Three Gorges Dam is a hydroelectric dam that spans the Yangtze River, Hubei province, China (Source: Wikipedia).

Since the early 20th century, the term hydropower has been used almost exclusively in conjunction with the modern development of hydroelectric power. International institutions, such as the World Bank, view hydropower as a means for economic development.

The Three Gorges Dam is a hydroelectric gravity dam that spans the Yangtze River in the Hubei Province, China. The Three Gorges Dam is the world's largest power station in terms of installed capacity (22,500 MW). In 2014 the dam generated 98.8 terawatt-hours (TWh) and had the world record, but was surpassed by Itaipú Dam (Paraná River, Uruguay) that set the new world record in 2016 producing 103.1 TWh (Fig. 2.17).

As well as producing electricity, the dam is intended to increase the Yangtze River's shipping capacity and reduce the potential for floods downstream by providing flood storage space. However, due to flooding of archaeological and cultural sites, leading to the displacement of some 1.3 million people, and causing significant ecological changes, including an increased risk of landslides, the dam has been controversial both domestically and abroad.

## THE HYDROGRAPH: DISCHARGE RECORDS TO REDUCE RISKS AND PROMOTE DEVELOPMENT

The sources of stream flow include rain, snow, glacier melt waters, and springs or seeps along hillslope gradients or geological discontinuities. The flow might be seasonal, perennial, or periodic. Steam reaches flow through steep canyons or across broad flood plains. Over the course of their lifetime, a stream might pass through a range of habitat types: meadows, forests, chaparral, savannas, grasslands, and tidal marshes. At their terminus, they might drain into the ocean (exorheic) or into landlocked basins (endorheic).

Because stream flows depend on a number of sources, their flow rates can vary dramatically, but often depend on precipitation as an important driver of variation. Of course, even the idea of a normal rainfall year is a widely accepted misnomer—no yearly patterns reflects "the average." In spite of some dominant seasonal trends, rain events patterns vary dramatically from year to year. The variation in stream behavior then influences the variations in the ecology of streams. Understanding the flood-return intervals, for example, allow planners to reduce the risk of catastrophic flooding—of course, the record here is uneven and we will follow this up more in Chapter 9.

The comparison between the Eel, Sacramento, and Mojave rivers is a good example of how stream flows vary. The Eel River drains temperate rain forests and discharge peaks during the winter rainy season; the Sacramento drains from mountain headwaters, whose discharge is dominated by snow melt and then modulated by the construction and operation of Shasta dam starting in the 1940s, and the Mojave River's intermittent flow depends on desert rainfall events in streambed that is often dry (Fig. 2.18).

As a keystone to understand river dynamics, hydrographs have been used to design hydroelectric dams, estimate reservoir storage and release criteria, and irrigation water availability. In effect, the hydrograph is scientific equivalent of religious icons: representing a set of values and specialized knowledge with veiled view of the future (assuming that river flow will fallow patterns in the past).

## GREEN REVOLUTION: POWER AND IRRIGATION

An important link in the Green Revolution is the linkage between energy and food production. The TVA (Tennessee Valley Authority) is the ultimate symbol to link these two human production projects. The Tennessee Valley Authority was enacted by congress to create a network of dams across state jurisdictions and used to generate electricity. During WWII, the energy was harnessed to produce munitions (nitrate for explosives). Once the war was over, TVA continued to make nitrate, but now used as fertilizers to increase agricultural production. Some will argue that WWII actually created the supply for fertilizers that was then looking for a market, and increasing acres of farmland began receiving this product. Unfortunately, the United States now suffers from a nearly universal problem of over fertilized farmlands, and the loss of excess nitrate promotes algal blooms that have negative impacts on surface- and groundwaters.

The Green Revolution project was a development model designed to be exported as a package to developing countries that included high-yielding varieties, fertilizer inputs, chemical pest controls, and irrigation. When the Ashwan Dam was built in the Nile River, it was the symbol of modernity for Egypt and participation in the Green Revolution project. But just as the environmental movement was developing, the ecology and public health costs cast a long shadow on the project. Built to irrigate and generate electricity, the dam construction forced the removal of important ancient archeological sites; impounded water increased the rate of schistosomiasis, irrigated waters increased the rate of soil salinization in crop plantings, and reduced water flow into the eastern Mediterranean Sea reduced fishery production.

## ENERGY FOR MONOPOLISTIC TRANSNATIONALS

Malaysia, having diversified sources of energy—crude oil, coal, natural gas, and hydroelectric—has argued that the construction of 12 hydropower dams will decrease Malaysia's dependence on crude oil. Recently, the "Four-Fuel Diversification" strategy has given way to the "Five-Fuel Diversification" strategy to include renewable energy sources. Certainly, these are positive steps, but the construction of hydroelectric capacity was designed on the assumption that energy demand and GDP were closely linked. Unfortunately, the optimistic expectation that GDP would continue to grow at 8%, and then

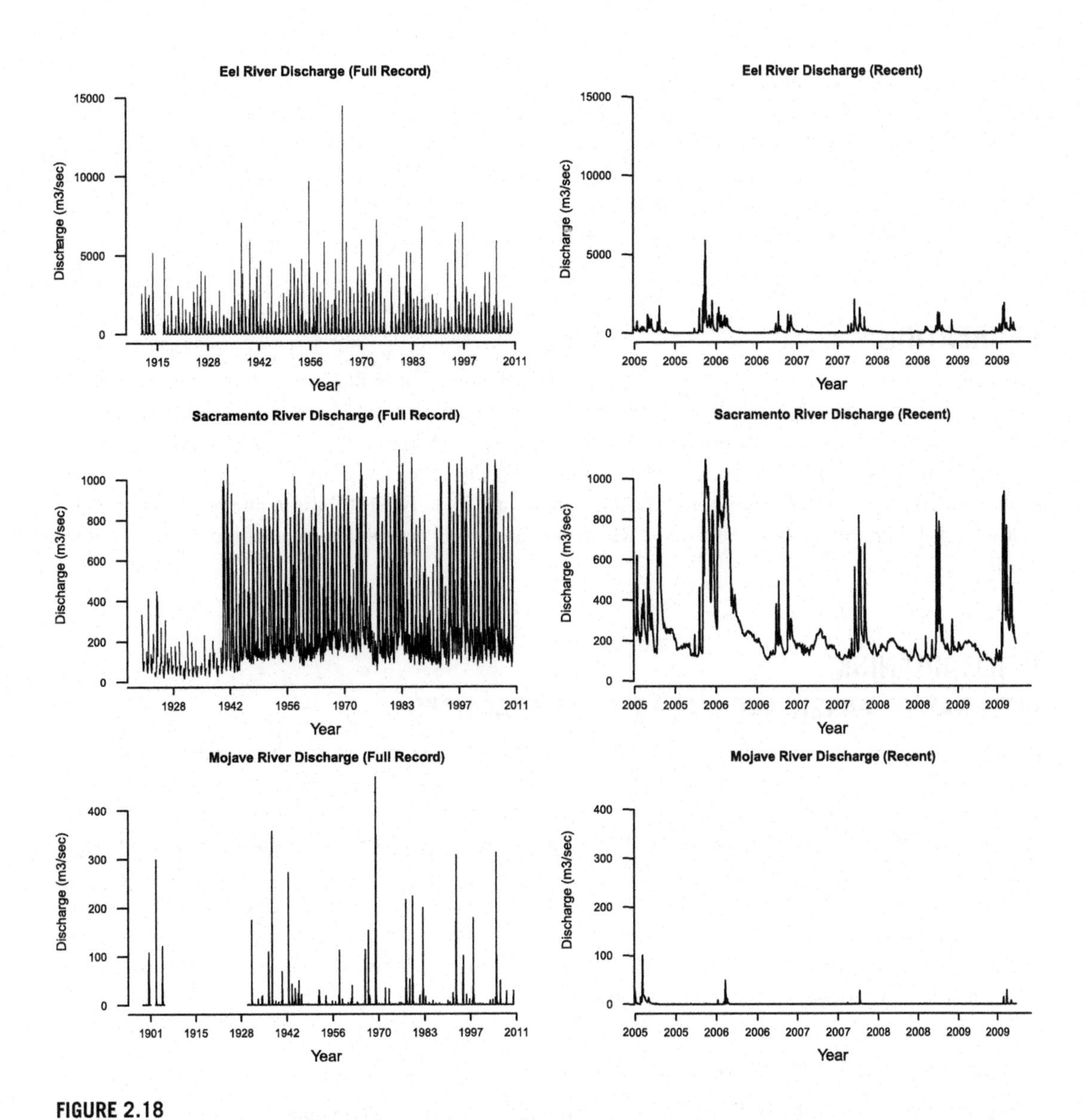


**FIGURE 2.18**

Discharge is a common way to evaluate waterflow and these "hydrographs" are standard tools for hydrologists (Source: USGS).

require 166 Mega tons of oil equivalents (Mtoe) by 2020 is far from the reality. Instead, Malaysia's energy demand is only 51.6 Mtoe in 2013. To boost the demand, the government has worked tirelessly to attract (with cut-rate energy prices) various energy-intensive, heavy industries, most notably bauxite processing for aluminum.

Energy scarcity and energy poverty describe how populations have inadequate access to energy, usually in terms of electricity. Currently, most estimate that about 5–10% of the population is located in inaccessible locations for the power grid—thus, the Bakun and Baram are inappropriate projects for these communities. In other words, the projects like the Bakun and Baram were not designed for those who might be described as energy poor; if addressing energy poverty was the goal, the government would develop very different projects, i.e., small scale and widely distributed energy, e.g., minihydro.

## RECREATION

Recreation adjacent to and on waters has a long cultural history, but in the 20th century, recreation was monetized. For inland waters, these activities include fishing, motorized boats, jet skis, and whitewater rapid adventures. According to the United States Bureau of Reclamation, recreation and tourism is the largest industry within the western states and second largest U.S. employer. Outdoor recreation in the United States is a $350 billion industry, with approximately $140 billion attributable to public lands and $40 billion to public waterways. Recreation and travel combined is one of the world's largest businesses.

## UTILITARIANISM

### ALTERNATIVE MARSHLAND MANAGEMENT PRACTICES

For modern Europeans, a marshland has rarely been considered to have much economic value, but more of a place to be drained and then grazed or ploughed and farmed. But for the peoples of ancient South America, the rivers of the Euphraties, and much of Southeast Asia, marshland was clearly "home", rice productions, or desirable because of the wide range of wildlife attracted to these watery environments.

For example, in the Pananal and Beni region of Bolivia, whereas Europeans created canals to drain landscapes, ancient South Americans created canals to flood the landscape and lived on artificial islands and lagoons (Fig. 2.19). These canals were highly engineered, where some followed contours of the land, and others followed the natural topography.

### EXTRACTION AND MINING OF WATER

Viewing the planet as a resource has become the bedrock of resource economics, where the use of resources can be viewed from an anthropogenic lens. In other words, the planet's resources have value inasmuch as they have value to human beings. On the surface this sounds reasonable way to think about sustainability if we include maintaining the resource for future generations, and consider that humans also value noneconomic uses of the planet.

In the case of ground water use, the application of these concepts can become a bit hazy. For example, how should we evaluate groundwater mining, which is the removal or withdrawal of water in the natural ground over a period of time that exceeds the recharge rate of the supply aquifer. It is also called "overdraft" or "mining the aquifer."

**FIGURE 2.19**

Pre-Columbian fish weirs and ponds in the savannas of Baures, Bolivia (Source: Dan Brinkmeier, Field Museum of Natural History).

Groundwater is contained in specific rock units called aquifers. Water, ultimately from rain or snow, percolates downward directly from rain, or from a riverbed or lakebed, through soil, sediment, and rock, following the route of least pressure, to reach a level where it is saturated. It is then ground- water, occupying the microscopic spaces between the rock particles in the aquifer. In natural circumstances an aquifer is close to equilibrium in its water content, with recharge balancing outflow.

The water level in a natural aquifer is called the water table. Although it may rise and fall from season to season and year to year, the water table usually varies round some average depth. If the watertable reaches the ground surface, water will tend to ooze out, as a natural seep or spring. In the end, every drop of groundwater eventually leaves the aquifer by outflow as a natural spring, or as seepage into a lake, river, or the sea, or pumped out of a well, but by that time it has been replaced by other water. Waterflow above or below ground follows physical laws that are well understood. In general, groundwater flows very slowly compared with the unconfined flows that are familiar in rivers and streams: rates are more of the order of feet per day rather than feet per second.

Hydrogeology is the study of aquifers and the water contained in them. It is crucial in assessing the impact of human activities on groundwater and in planning for the wise use of water in the future, thus falls within a clearly utilitarian approach.

## WATER AND ECOSYSTEM SERVICES

As environmental sciences have developed, so has our understanding of the social, ecological, and economic value of inland waters. These might fall into a category of "ecosystem service", where we integrate the human interests with ecosystems processes. The service is to human beings, thus reenforces an anthropocentric value-system. We might contrast this to an ecocentric value-system, where the nonhuman world has inherent value, irrespective of what we think we might value. There is much debate about these concepts, their importance, and epistomological roots. Nevertheless, it is important to appreciate the limitations of the anthropocentric view, where ecosystem services might not be a robust way to value our inland waters.

Evaluating water is a multidimensional endeavor.

**Structural metrics**

- Biological diversity
- Native riparian vegetation width
- Floodplain presence/width
- Canopy cover
- Oxygen level
- N and P concentrations
- Pollutant concentration
- Organic matter
- Temperature
- Mean annual flow and depth
- Turbidity
- Channel morphology
- Stream bead substrate

**Functional metrics**

- Productivity/reproduction, migration, trophic status
- pollutant removal rates
- hydraulic retention
- photosynthetic active radiation
- Biochemical Oxygen Demand and whole stream metabolism
- Nutrient cycling/flux rates
- Pollutant removal or sequestration
- decomposition rates
- thermal regime (magnitude, duration, and timing)
- flow regime (magnitude, duration and timing)
- Sediment flux
- Channel migration, erosion rate
- Stream bed mobility

# AESTHETICS & EXISTENCE VALUE

## WILD AND SCENIC

In 1972, California passed the Wild and Scenic River System Act, which was strengthened by the National Wild and Scenic River Act in 1982 as amended from the 1968 act. The act attempted to strike a balance between dam and designating permanent protection important for free-flowing rivers. To accomplish this, the act prohibits federal support for construction of dams or other instream activities that would disrupt the river's free-flowing condition, water quality, or resource values. However, designation does not affect existing water rights, or the existing jurisdiction of states and the federal government over waters as determined by established principles of law.

On signing the Wild & Scenic Rivers Act, President Lyndon Johnson said:

*In the past 50 years, we have learned—all too slowly, I think—to prize and protect God's precious gifts. Because we have, our own children and grandchildren will come to know and come to love the great forests and the wild rivers that we have protected and left to them ... An unspoiled river is a very rare thing in this Nation today. Their flow and vitality have been harnessed by dams and too often they have been turned into open sewers by communities and by industries. It makes us all very fearful that all rivers will go this way unless somebody acts now to try to balance our river development.*

The act has three designations,

**FIGURE 2.20**

Carmel River during a storm event (Source: https://www.cbsnews.com/news/evacuations-flash-flood-warnings-downpours-in-already-drenched-northern-central-california/). There has been a tremendous amount of resources allocated to restore the river. Many of these efforts have been successful.

***Wild River Areas*** rivers or sections of rivers that are free of impoundments and generally inaccessible, except by trail, with watersheds or shorelines essentially primitive and waters unpolluted. These represent vestiges of pre-columbian America.

***Scenic River Areas*** rivers or sections of rivers that are free of impoundments, with shorelines or watersheds still largely primitive and shorelines largely undeveloped, but accessible in places by roads.

***Recreational River Areas*** rivers or sections of rivers that are readily accessible by road or railroad, that may have some development along their shorelines, and that may have undergone some impoundment or diversion in the past.

As of August 2018, the National System protects 13,416 miles of 226 rivers in 41 states and the Commonwealth of Puerto Rico; this is less than 0.25% of the nation's rivers. By comparison, more than 75,000 large dams across the country have modified at least 600,000 miles, or ~17%, of US river system.

## LAND USE AND WATER: CARMEL RIVER — A RUINED RIVER?

Along the coast of California is the Carmel River. This relatively small river used to support a healthy population of steelhead fish and a diverse riparian vegetation. However, efforts to reduce flooding and promote development in the floodplain initiated the construction of levees. Agriculture, housing developments, and golf courses began competing for water supplies. The riparian vegetation has become significantly degraded due to the lowered watertable and streamflow fails to support a robust steelhead run. The cumulative impacts of water diversions, dam building, and developments in the floodplain are reaping dramatic ecological changes (Fig. 2.20). But just as in the case of the Aral Sea, to address the human impacts on these ecological systems, we will need thoughtful approaches to create sustainable ecosystems.

## NAMING SPECIES BEFORE THEY ARE LOST: LINNAEUS AND EO WILSON

Wilson, whose expertise is ants, has been a strong advocate to protect biodiversity. Along with Robert MacArthur, he described how islands could maintain a greater number of diverse species. That idea was the foundation for nature reserve design, which he describes as islands in a sea of land, what is termed as island biogeography.

Approximately 2 billion species have been given scientific names, a process started by Carl Linnaeus, a Swedish biologist in 1735. Based on what we know today, we estimate 10 billion species exists. Unfortunately, based on the tradition of protecting species, unless the species is named, it does not exist—at least for conservation purposes.

With an estimated rate of anthropogenically caused extinction, between 100 and 1000 times above background, 1/2 of the species on Earth may become extinct by the end of the 2100 century. As Wilson and others argue, the only effective way to protect these species is to evaluate the role of expanding reserves to protect biodiversity. For Wilson, this is a call to improve the science, and a call to develop the political will to create socially just political structures designed to avoid a catastrophic loss of biodiversity.

## ECOLOGY OF SCARCITY: INFRASTRUCTURE AND HUMAN DEVELOPMENT

### ROMAN AND MAYAN INFRASTRUCTURE

Urban hydraulic systems started to develop in the Bronze Age, and particularly in the mid-third millennium BCE, in an area extending from India to Egypt. But on the island of Crete, where the Minoan civilization was flourishing, a new level of hydrologic infrastructure included the construction and use of aqueducts, cisterns, wells, fountains, bathrooms and other sanitary facilities, which might be recognized as a contemporary advance lifestyle. Capitalizing on these technologies, the Romans developed high engineering skills and extended these technologies on large-scale projects throughout their large empire.

After the fall of the Roman Empire, the concepts of science and technology related to water resources retrogressed. Water supply systems and water sanitation and public health declined in Europe. Whereas Islamic cultures, on the periphery of Europe, had religiously mandated high levels of personal hygiene, along with highly developed water supply, sewerage, and adequate sanitation systems, Europe acquired again high standards of water supply and sanitation only in the 19th century.

On the North American continent, the Lowland Maya civilization flourished from 1000 BCE to 1500 CE in and around the Yucatan Peninsula. Known for its sophistication in writing, art, architecture, astronomy, and mathematics, this civilization is still obscured by inaccessible forest, and many questions remain about its makeup. Although some scholars suggest that the Maya Lowlands contained small city-state centers ruled by warring elites, recent data suggest a regional network of densely populated cities with complex integrative mechanisms. Thus in contrast to the idea that settlements were supported by a relatively sparse rural population practicing swidden farming and limited role from intensive agriculture, recent research suggests that these populations were supported by a regional agricultural economy of great complexity.

The landscape included intensive agriculture that included a complex network of channels designed either to draw water away from natural streams toward infrequently flooded areas, or to drain those same areas during major floods. Large and small channels intersected at regular intervals, forming nested grids within "channelized" fields.

Overall, it is important to note that a wide range of civilizations have modified waterflow for millennia, and in some cases with intensive practices. Thus our current activities are far from unique and in many cases, extension of the practices done by our ancestors.

## WATER: PROBLEMS

All solutions of water problems may be sorted into nonstructural, structural, and mixed measures. Nonstructural measures include regulation and insurance. Structural measures consist of combinations of the four categories of structures:

- those that transfer water in space;
- those that change the water regime in time;
- those that change water power potential; and
- those that change water quality.

Modern water resources planning uses the principles of advanced economics in matching water demand and water supply by selecting and sizing a set of structures as the water resources system.

## MAKING WATER PREDICTABLE: INFRASTRUCTURE

Of the top 10 urban centers, three California regions are found on a list that take more water from watersheds quite distant: Los Angeles, San Francisco, and San Diego. On a per capita basis, Los Angeles, San Francisco, and San Diego residents consume among the greatest amount of water from cross boundary sources (Table 2.3).

What are the ecological affects of these transfers? How do these centers of economic and political power affect the waters in the source waters and people who rely on them?

Without a doubt, one of the most famous example of cross-basin controversies comes from the conflict of the Los Angeles Aqueduct that imported water to Los Angeles from the Owens Valley as described in the opening of Chapter 2, but there are plenty of other examples—each with its own problematics (Fig. 2.21).

## DROUGHT, MIGRATION, AND CONFLICT

Food security depends on reasonably predictable weather climates. Unfortunately, many believe the variation in weather is likely to increase as a result of climate change (longer droughts, higher frequency of extreme events, such as snow, floods, hurricanes). Although definitive conclusions that extreme weather events are increasing in frequency is partially understood, some indicators are starting to reveal themselves.

An extended drought in Syria caused migration to the city centers of the country. Becoming overwhelmed by the social services need, limited employment opportunities, and a relatively unresponsive government, devolved into a civil war, these problems and processes and the subsequent civil war in Syria is relatively undisputed. However, recent analysis suggest that the drought was part of a larger pattern of climate change in the region. If these results are robust, the war and associated refugees

**Table 2.3 Cross-boundary water transfers and populations of large urban regions. Transfers refer to cross boundary transfers and are in units of liters/day/person (modified from McDonald et al., 2014).**

| Urban Region | Country | Population (2010) | Transfer |
|---|---|---|---|
| Los Angeles | USA | 13,223,000 | 673 |
| Boston | USA | 4,772,000 | 693 |
| Marumbai | India | 19,422,000 | 165 |
| Karachi | Pakistan | 13,500,000 | 187 |
| Hong Kong | China | 7,053,000 | 347 |
| Alexandria | Egypt | 4,440,000 | 523 |
| Tianjin | China | 8,535,000 | 255 |
| Tokyo | Japan | 36,933,000 | 59 |
| San Francisco | USA | 3,681,000 | 547 |
| San Diego | USA | 3,120,000 | 462 |
| Ahmandabad | India | 6,210,000 | 219 |
| New York | USA | 20,104,000 | 67 |
| Tel Aviv | Israel | 3,319,000 | 369 |
| Pretoria | South Africa | 1,468,000 | 829 |
| Chennai | India | 8,523,000 | 133 |
| Algers | Algeria | 2,851,000 | 375 |
| Aleppo | Syria | 3,068,000 | 346 |
| Athens | Greece | 3,382,000 | 306 |
| Cape Town | South Africa | 3,492,000 | 285 |

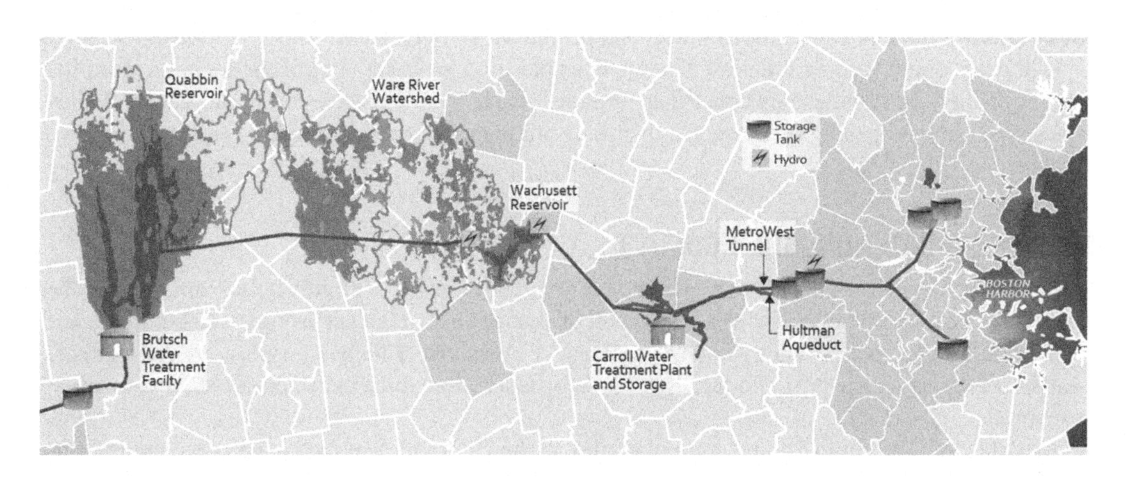


**FIGURE 2.21**

Boston water system.

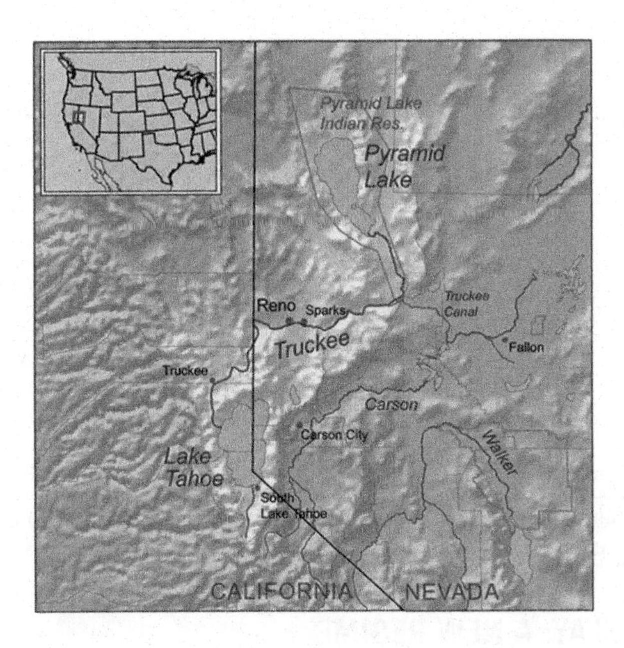


**FIGURE 2.22**

Pyramid Lake Watershed: An Endorheic Basin (Source: City of Reno).

represent an important framework: that climate change may mediate conflict and cause climate-change refugees.

## SHRINKING PIE?

With an increasing demand on water resources, it is tempting to think of water as a Malthusian-like resource, where the 'development' of water supplies might be linear and population growth (or demand) for water is increasing exponentially. If framed this way, water is a fixed resource. Thus we view water allocation as a zero-sum game—water used for one use preclude other uses. Some argue that the nonhuman, ecological values are destroyed by water development projects, whereas others complain that their livelihoods are threatened by environmental regulations designed to protect natural systems. We face a pragmatic choice, but potentially false dichotomies are not always helpful (Fig. 2.23).

Instead of thinking about water as a limited resource and allow the politics of scarcity to dominate the discourse, I argue that we can base our relationship with water as participants in an ecological play, where political and economic processes coexists with inland water ecosystems, and our goal is to learn how they align. Without being naive, we are not going to ignore the real and fundamental conflicts in society, but try to take a broad view by understanding the science of water. With this framework, we would predict that conflicts of interest in water resources activities will increase the required administrative, arbitration, and market decisions to resolve conflicts using a discursive approach, which we will expand upon on Chapter 14.

**FIGURE 2.23**

"No water, No jobs." Common road signs along Interstate 5 running through the San Joaquin Valley. The are symptoms of a complex power struggle for the control and potential use of water in California (Source: Arnett Young).

## EVOLUTIONARY PLAY: A NEW REGIME

### SOCIAL AND ENVIRONMENTAL JUSTICE

Who would have known that the oil price shocks in the 1970s could potentially displace nearly 20,000 people from the Kayan, Kenyah, and Penan communities in Sarawak rainforests some 50 years later?

To further increase the portfolio, Malaysia also began evaluating sites for hydroelectric generation and identified the rivers in Sarawak as early as the 1960s, but concrete planning only occurred after 1979 when the country identified its vulnerability to oil supplies, and realized that the correlation between energy and development suggested that for the country to grow, more energy generation capacity was key (Fig. 2.24).

Finally, as a justification for most dams, the flood risk for downstream residents is usually reduced. However, it might be more accurate to say that risks are displaced by new ones. With "flood protection", the floodplain is colonized. However, the capacity of the dam to prevent flooding is often overstated because of extreme with unknown frequencies. When this happens, the residents of the floodplain exeperience unanticipated flooding. Of course, there are other examples where the dam fails completely and then the losses tend to be catastrophic. In might be better to say that the flood risk is altered. This is particularly useful when you consider the impact of the Bakun dam on the Dayak, whose settlements were flooded by the dam. After public opposition originally halted the dam projects in the early 2000s, Malaysia turned its attention to the Bakun dam, which was slated one of the 11 dams after the Baram dam was completed. The Bakun hydroelectric dam, completed in 2011, serves an excellent example of the failed promises of big dam projects: displaced residents remain antagonistic to their resettlement and dramatic changes to the river's ecology are undocumented.

In the case of the Bakun Dam, the indigenous people displaced by the project struggle to eke out a living over a decade after they were resettled (Fig. 2.25). About 10,000 people were resettled to the town of Asap in 1998, surrounded by areas licensed for oil-palm plantations. Dayak people expected to be compensated with land and housing, but the discrepancy between the actual compensation is

**FIGURE 2.24**

Bakun power house.

**FIGURE 2.25**

Bakun dam.

dramatic. Instead of getting two houses, each family got one. Instead getting 22 acres of land, they received 3 acres. Many went into debt to pay for additional housing. Many received only three acres of rocky, sloped or sandy soils that are too small, of too poor quality to generate a living, and too far from Asap to manage effectively. On their traditional lands, the Dayak could fish in the river, hunt,

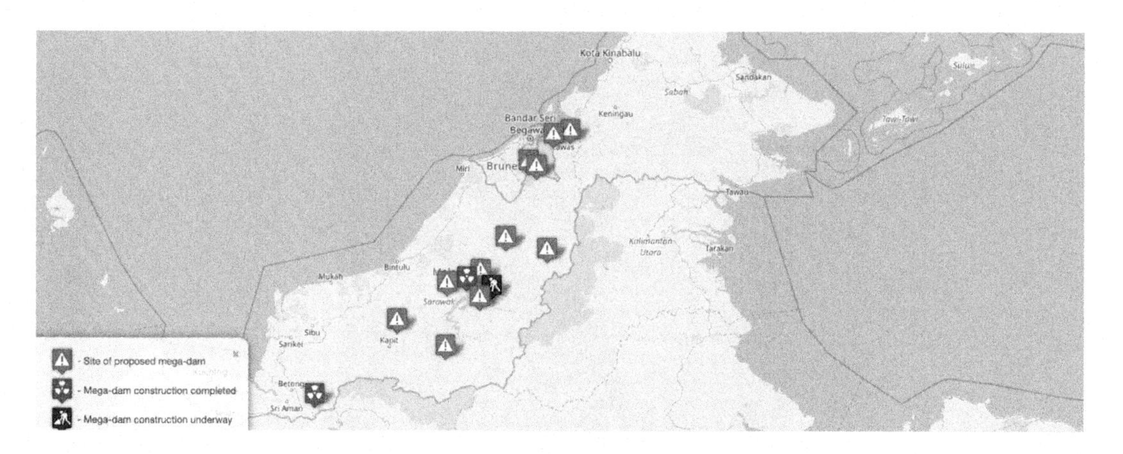


**FIGURE 2.26**

Malaysia's plan for hydroelectric generation known as SCORE (Sarawak Corridor of Renewable Energy) (map by Luyi Huang).

and gather forest products. In the resettlement areas, they have no access to forests. They went from a food-secure to food-insecure status.

For the Dayak people of the Baram, the lessons learned by the resettlement of the Bakun River watershed provides a dire warning—of failed promises for compensation, loss of social cohesion, and the irreversible forfeiture of cultural heritage.

Moreover, in the Bakun and Baram river watersheds, the people of Sarawak have poignantly demonstrated the socioecological disruption. For the time being, the construction of the Baram Dam has been halted. But dam plans seem to have several bouts of reincarnation, so the long-term outcome will probably always remain uncertain.

The alignment with environmental scientists to provide better information to stakeholders to make informed decision about the environmental benefits and costs of development is a good example, demonstrating that compromises are possible even by governments with limited democratic control.

The conflict between development goals to build dams for hydroelectricity and indigenous peoples in Sarawak was set in motion in the 1970s. The justification for hydroelectric development has been a dominant narrative in developed and developing countries for decades (Fig. 2.26). However, counter narratives have arisen as ecologists have documented how dams change river geomorphology, water quality, habitat value, and access.

The evaluation of projects for their potential social and environmental impacts is evaluated through SEIA reports in Malaysia. Yet Sarawak Energy Berhad, the state energy monopoly, claimed that the assessment could not be completed because of protests, releasing the following statement:

> *Sarawak Energy's ability to commence and complete the feasibility studies and SEIA reports to ensure community issues and points of view are taken into account have been disrupted by the ongoing protests. While we respect the right of individuals and organizations to express their point of view, it should be done in a manner that is lawful and does not place their safety or the safety of others at risk. The behavior of the NGOs protesting at Baram in the past has breached both these basic principles.*

**FIGURE 2.27**

Dayak, native peoples of Borneo, listen to a discussion of the Baram Dam, which threatens to flood their homeland (Source: Tom White).

In the case of the Dayak communities, the alignment between social and environmental justice and ecological protection converges on the 'Stop the Baram' campaign (Fig. 2.27). Evidence is on their side: socio-ecological impacts of the dam would be devastating for their way of life and dramatic changes to the river ecology. With international support from Rivers International and thoughtful engagement with various groups of interest, the Dayak effectively used the media and a blockade to force the government to (temporarily) shelve the Baram Dam.

When announcing the government's decision, Chief Minister, Datuk Patinggi Tan Sri Adenan Satem, articulated a view that the Baram protesters would regret their victory:

> *There have been many protests and blockades by the people who voiced their disagreement to the building of the Baram dam. If you don't want the dam, fine. We will respect your decision. I hope you understand the impact for refusing it, as you will be missing out on related projects which are beneficial, such as roads and other necessities.*
>
> *One day, you will find that not building the dam has given some disadvantage and as a result of this, you suffer. That is in your own hand. It is your decision.*

To keep these development projects going, the Malaysian government now has pinned its hopes on construction on one of the other 12 dam projects— Baleh Dam located in the upper Rejang basin in central Sarawak. The rainforest in this part of Southeast Asia has some of the highest rates of plant and animal endemism, i.e., species which are found only in that region, and the ecological damage from the proposed dam will be considerable, even though the severity of its impact cannot be known at this time. As the Dayak people's successful resistance to the Baram River dam project reveals, those who live within the Baleh watershed will need to be vigilant and persistent in their opposition if they hope to protect the human and ecological communities that inhabit that region.

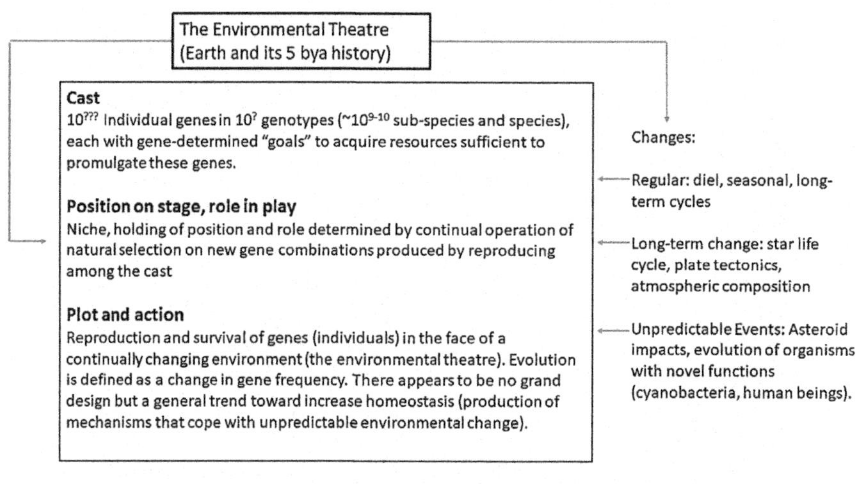


**FIGURE 2.28**

Diagram of Hutchinson's "Environmental Theatre" that includes the hierarchy of time scales (Source: Moss, 2010, Figure 1.1, based on Hutchinson, 1965).

## NEW RULES AND INSTITUTIONS: A REGIME FOR MANAGING INLAND WATERS

A civilization may be conceived as a collection of various infrastructures. The 25–30 main purposes in water resources activities compose a large part of these infrastructures, e.g., drinking and irrigation water, hydroelectric and transportation, fisheries and recreation, etc. Controversies between water resources development and protection of the environment will increase until new methods for their resolution are designed. Aging of hydraulic structures and water resources systems already poses many difficult problems for their revitalization. Pressures mount to extract maximum benefits from existing systems before building new ones. Society will continue to press for the decrease of some risks from water-related structures. Cleaning polluted water environments, especially aquifers, will be on the main agenda of water activities in the first half of the 21st century.

Water and water rights will be considered as market commodities. Priorities in using sources of water may switch due to the impact of various types of pollution. The climatic change phenomenon will have a large influence on water resources planning and development in future civilizations.

It is this context that we begin our journey into inland waters. The stage for the species in these systems present an evolutionary stage and ecological play. Following the conceptual framework of for aquatic systems from Fig. 2.28, we will insert social systems into the framework and consider how we as humans are part of these systems, as we develop our knowledge of water science and ecology of inland waters.

# NEXT STEPS

## CHAPTER STUDY QUESTIONS

1. As human population densities increase, water supplies become more complex. Describe some of the drivers that make complexity a requirement.
2. Agriculture and aquaculture rely on freshwater sources of water. How does water supply and quality affect these two food sources. What are the commonalities and differences in water use and management between agriculture and aquaculture?
3. Water can be thought of as an ecosystem service for transportation. What are the pros and cons of thinking of water in this way?
4. What are the limits of hydropower?

PART

# THE ECOLOGICAL THEATER AND THE EVOLUTIONARY PLAY

2

CHAPTER

# THE PLAYERS: EVOLVING AQUATIC SPECIES

# 3

*From the first growth of the tree, many a limb and branch has decayed and dropped off; and these fallen branches of various sizes may represent those whole orders, families, and genera which have now no living representatives, and which are known to us only in a fossil state. As we here and there see a thin, straggling branch springing from a fork low down in a tree, and which by some chance has been favored and is still alive on its summit, so we occasionally see an animal like the* Ornithorhynchus *(Platypus) or* Lepidosiren *(South American lungfish), which in some small degree connects by its affinities two large branches of life, and which has apparently been saved from fatal competition by having inhabited a protected station. As buds give rise by growth to fresh buds, and these, if vigorous, branch out and overtop on all sides many a feebler branch, so by generation I believe it has been with the great Tree of Life, which fills with its dead and broken branches the crust of the earth, and covers the surface with its ever-branching and beautiful ramifications.*

**Charles Darwin, On the Origin of Species**

## CONTENTS



**Ecology and Management of Inland Waters. https://doi.org/10.1016/B978-0-12-814266-0.00016-7**

In the early 1970s, dramatic changes were occurring in a tropical lake in Panama. The bufferfly peacock bass (Fig. 3.1), a South American fish, known as *Cichla ocellaris,* was introduced to Gatun Lake and a whole group of predators dramatically declined—a single species introduction had irreversibly altered the entire food web. To understand the effects on the lake, we need to know who is who—which players evolved in the lake, and which were introduced.

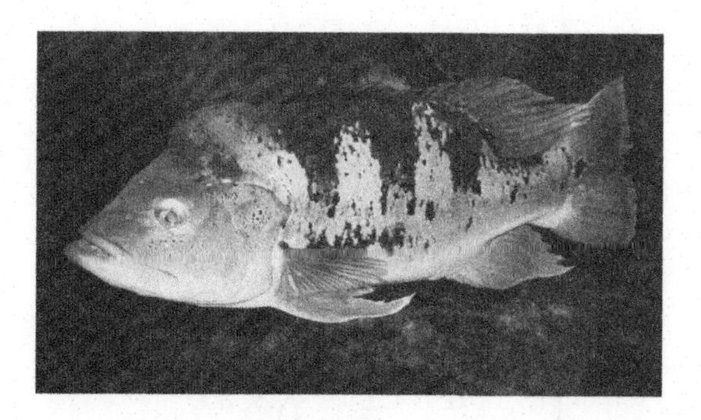

**FIGURE 3.1**

Cicla ocellaris (Source: https://en.wikipedia.org/wiki/Cichla_ocellaris). *Cicla ocellaris* is a member of the family Cichlidae, which is one of the largest vertebrate families. Absent from Asia, most of the species are found in South America, Africa, India and Madagascar, suggesting their ancestors diverged when these landmasses were still associated with Gondwana supercontinent. However, recently data questions this hypothesis that this family dispersed after the break up of the continent (Friedman et al., 2013).

The history of life on Earth is a history of natural selection, where extant species reflect reproductive success: success is contingent on various deterministic and random processes. As populations of species diverge as a result of reproductive isolation and environmental change, a range of diversity has developed that we are only beginning to understand.

For millions of years, species have been invading new habitats, e.g., inland waters, but now these unprecedented invasions are facilitated by humans that create new species combinations. To appreciate these changes, we will apply revolutionary concepts to understand radiation of biological diversity and how these help us define native and exotic species. These actors participate in an ecological drama and ultimately play a significant role in how we coexist on the planet.

This chapter summarizes major divergences in taxa—with a specific interest in characteristics that may confer adaptive value for life in inland waters. For environmental scientists, the tree of life allows us to organize the evolutionary relationships among taxa and appreciate how species lineages radiate through geologic time, thus generating biodiversity. Furthermore, with a better understanding of the biodiversity, we can identify the ecological foundation in the patterns and processes of aquatic systems.

In addition, we summarize some characteristics of species introductions and how these processes are unique to inland waters on time scales that we can best appreciate.

After reading this chapter, we will be able to

1. Explore the evidence that demonstrates that species diversity lies at the microscopic level;
2. Summarize major branches of evolutionary divergence that has created the Earth's biodiversity;
3. Describe how hypothesized ancestral relationships help us appreciate the complex divergences in the taxa that have invaded inland waters;
4. Synthesize the tension between native and introduced species, while appreciating the history of ecological patterns of invasions, range expansions, relative to the current pace of introductions.

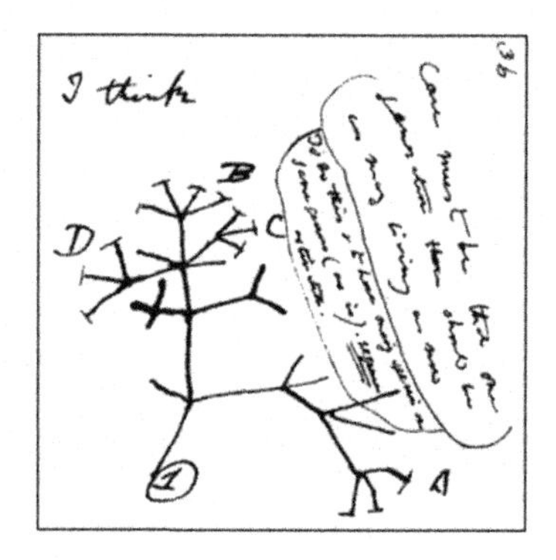


**FIGURE 3.2**

One of the first sketches of a phylogenetic tree was made by Charles Darwin as he developed his theory of evolution (Source: WikiMedia).

How or why *Cicla ocellaris* was introduced to Lake Guhan is a relatively trivial question to answer. However, in terms of policy and management, the issue is anything but trivia. But to understand the ecological and evolutionary implications for species introductions requires extensive field work, population modeling, and an understanding of selection processes change. Before we launch into these concepts that will take several chapters to fully appreciate, let us begin with a basic understanding of the family tree of inland waters.

# ENDEMIC AND EXOTIC TENSIONS

## PHYLOGENY AND THE FOSSIL RECORD

To appreciate the concept of endemism, we will first explore what evolutionary biologists are learning about the evolutionary processes and how these processes have promoted biodiversity across space and through the Earth's history. Let us begin with a discussion of ancestry.

A phylogeny describes the genealogical relationships between species—based on shared ancestry—which can be represented as a tree-like model. Depicting evolutionary relationships using the tree metaphor dates back to Darwin and his early sketches (Fig. 3.2). Inspired by this concept, biologists have been using this tree metaphor to graphically illustrate the relationship between taxa. The tree of life continues to be revised with new discoveries, such as the work of Carl Woese that subsequently separated Archea from bacteria. Combined with morphological characteristics, the fossil record, molecular biology analyses, and bioinformatics, our understanding of these relationships have made important strides.

For environmental scientists, the tree of life allows us to organize the evolutionary relations of taxa, while appreciating historical "inertia" with respect to a taxon's ecological relations. Furthermore, the fossil record provides the evolutionary context, albeit a obscure one, that links the origin of the Earth to the developing ecological system.

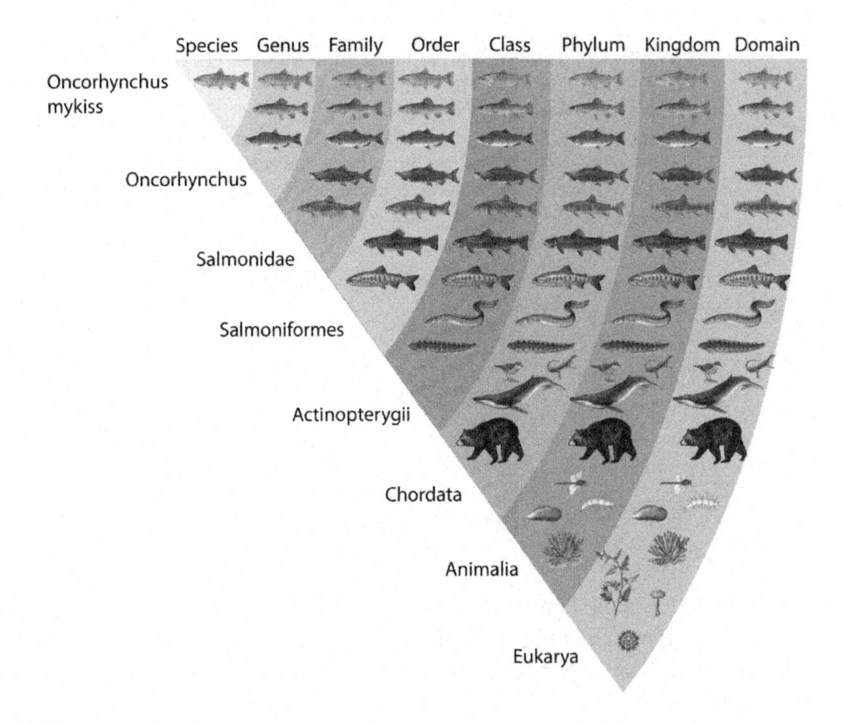


**FIGURE 3.3**

Hierarchical classification of steelhead, *Oncorhynchus mykiss*.

## SYSTEMATICS AND CLASSIFICATION SCHEMES

Depending on the group of taxa, biologists—or more precisely systematists—classify species within larger groupings based on shared characteristics (character states) and evolutionary relationships. In general, the goal for systematists is to develop a classification scheme, in which taxa will be grouped according to shared evolutionary ancestry, referred to as monophyletic. However, as more analytical tools are applied, we are confronted with historically defined groupings that have some resemblance to each other, yet have different lineages. These groups are called polyphyletic. With more research, scientists hope to reconcile the evolutionary relationships of these organisms and accurately classify them. But it is a work in progress, and some groups are rather unstable. Thus as we march through the diversity of inland waters, take time to appreciate that these groupings are hypothesized relationships and subject to continual scrutiny with conflicting conclusions—a healthy indicator of science.

These groupings are hierarchical in the sense that lower groupings are nested into higher-level groupings (Fig. 3.3). For example, the phylum is a high-level grouping for animals, which cascades down to class, order, family, and genus. For example, the zebra mussel is a member of the Mollusca phylum, Bivalvia class, the Veneroida order, Dreissenidae family, and *Dreissena* genus. The species name, composed of the genus and specific epithet (descriptor/modifier), is *Dreissena polymorpha*. Because the genus and specific epithet are Latin, they are italicized, but only the genus is capitalized unless the specific epithet is a proper noun. Over the years, biologists have developed identifier suffixes

**Table 3.1 Table of taxonomic rank suffixes. The naming conventions are overseen by groups of taxonomists who specialize in different groups. Because of the difference in the naming practices and historical conventions, significant inconsistencies exists and numerous exceptions, but when followed, these suffixes provide some indications of the taxonomic rank.**

| Category | Suffix |
|---|---|
| Phylum (animal) | |
| Division (plant) | |
| Order (plant) | iopsida |
| Order (plant) | ales |
| Family (animal) | idae |
| Family (plant) | aceae |
| Sub-family | eae |

for each category (Table 3.1). For the family-level grouping the suffix is "idae" for animals and "aceae" for plants. Finally, we should note that phylum and genus are singular, whereas their plural forms are phyla and genera. These rules can be hard to remember at first, but if you pay attention to how people use them, you will get used to it.

Because of the relative lack of stability at the higher taxonomic ranks, we face a complex task of developing a cogent structure. At this point, we rely on several compelling, but fluid classification systems. Importantly, the traditional hierarchies have become destabilized, where categories, such as phylum and kingdom have become terribly confusing because some include unrelated taxa and are polyphyletic. These classification categories will probably be replaced with new organizational structures just as the five Kingdom scheme has become outdated. For our purpose, we will describe monophyletic groupings or "clades" without tying them to classification ranks.

## ZEBRA AND CUMBERLAND MONKEYFACE PEARLY MUSSELS

The Zebra and Cumberland monkeyface pearly mussels are examples of freshwater mussels that evolved within distinct mussel families, the Dreissenidae and Unionidae. In this case, these two families have wildly different diversity patterns. The Dreissenidae only has two genera world-wide, whereas the Unionidae have over 50 genera in North America alone. Although these two mussels share a common ancestor, they diverged in the Mesozoic (252–66 mya), and their ranges became geographically distinct. It is critical to appreciate that one of these families did not evolve from the other—instead they share a common ancestor, just as humans did not evolve from extant chimpanzees, yet we share a common ancestry in the Hominidae family.

The zebra mussel (*Dreissena polymorpha*) is a small freshwater mussel in the Dreissenidae family (Fig. 3.4). Although their shells are shaped similar to those of true mussels, they are not closely related. Based on fossil records, morphological observations, and genetic analysis, zebra mussels are more closely related to venus clams and probably evolved in lakes of southern Russia and Ukraine and in the Ural, Volga, and Dnieper rivers. The zebra mussel was accidentally transported around the world

**FIGURE 3.4**

Image of zebra mussel, *Dreissena polymorpha* (Source: USGS).

in ship ballast water. Since the ballast water is low in salinity, the mussels could remain viable in ships traveling to new continents, where they continue to expand their range. Because the species evolved in the Eastern Europe, its presence in the Great Lakes is categorized as an exotic species. Furthermore, because its capacity to rapidly spread and displace natives, the zebra mussel is considered as an invasive species (Fig. 3.5).

In contrast, the Cumberland monkeyface pearly mussel (*Theliderma intermedia*) is native to Tennessee and Virginia. Like other true mussels, it has larvae called glochidia that attach to the gills of fish to develop into juvenile mussels and hitch rides to locations within the tributary system (Fig. 3.57). The fish hosts for this species include the streamlined chub (*Erymystax dissimilis*) and blotched chub (*Erymystax insignis*).

The midwest and southern part of the United States have one of the most diverse freshwater mussel fauna in the world. Due to their reproductive isolation in mountain streams in the Appalachian and tributaries in the Mississippi basin, these mussels have diverged relatively recently (a few million years ago) and survived recent glaciation in refugia streams.

Nevertheless, *Theliderma intermedia* is a federally listed endangered species, which means that the entire species might be lost without legal protection (Fig. 3.6). Whereas populations in the Powell River and Duck River appear to be viable, the Cumberland pearly mussel has been extirpated from the Elk River (Tennessee and Alabama). The range loss makes the species less resilient to extinction, demonstrating the importance of maintaining quality habitat that mirrors the natural range.

With these two mussel examples, we appreciate both common ancestry and divergence, reflecting different selection processes. Although both now inhabit waters in the United States, their populations dynamics are radically different. The Cumberland pearly mussels is rare and endangered, whereas the zebra mussel has rapidly expanded and become a major nuisance species. Now we will further dig the history of species radiation and divergence.

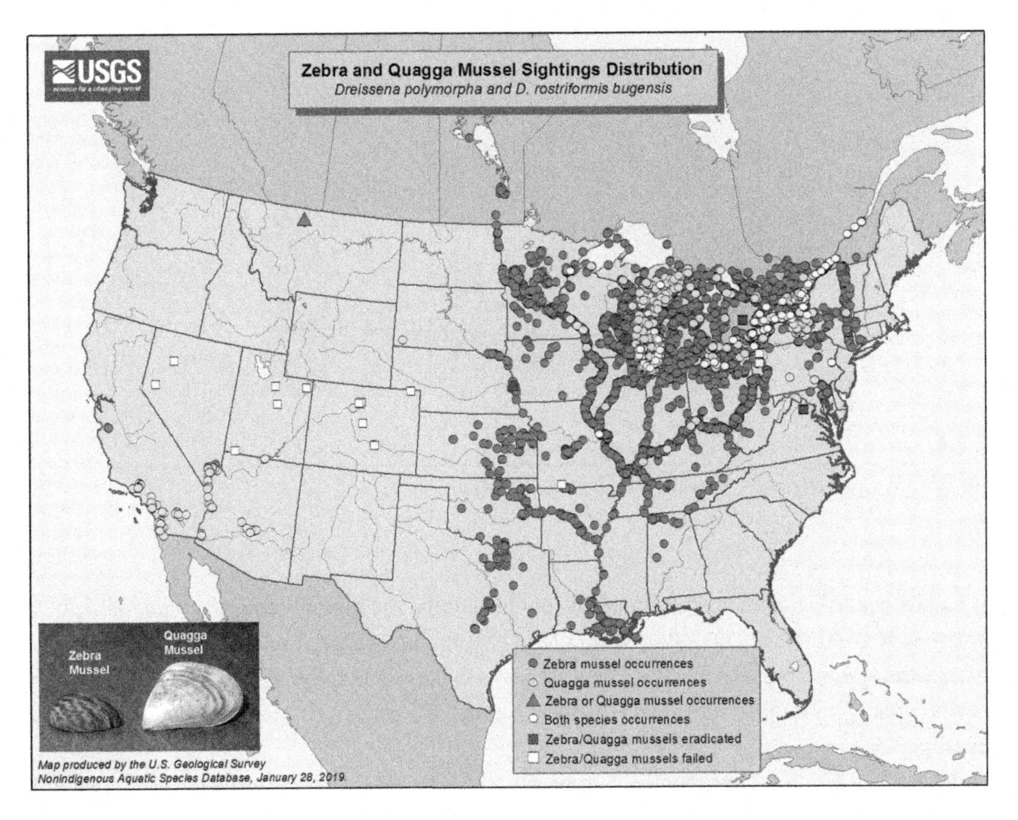


**FIGURE 3.5**

Recent zebra mussels distribution (Source: United States Geological Survey Noindigenous Aquatic Species Program, https://nas.er.usgs.gov/queries/FactSheet.aspx?speciesID=5). Since the 1980s, zebra mussels became established in the Great Lakes and the Hudson River and are commonly found on the bottom of ships. As filter-feeders, zebra mussels remove algae from the water column that had been an important energy for fishes.

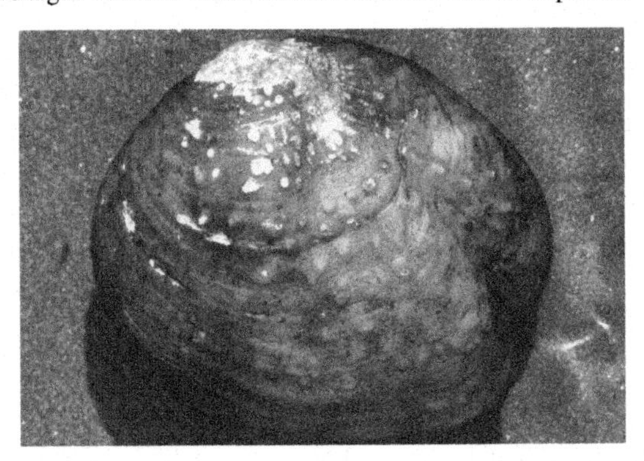

**FIGURE 3.6**

Cumberland monkeyface pearly mussel, what a great name! (Source: Wikipedia).

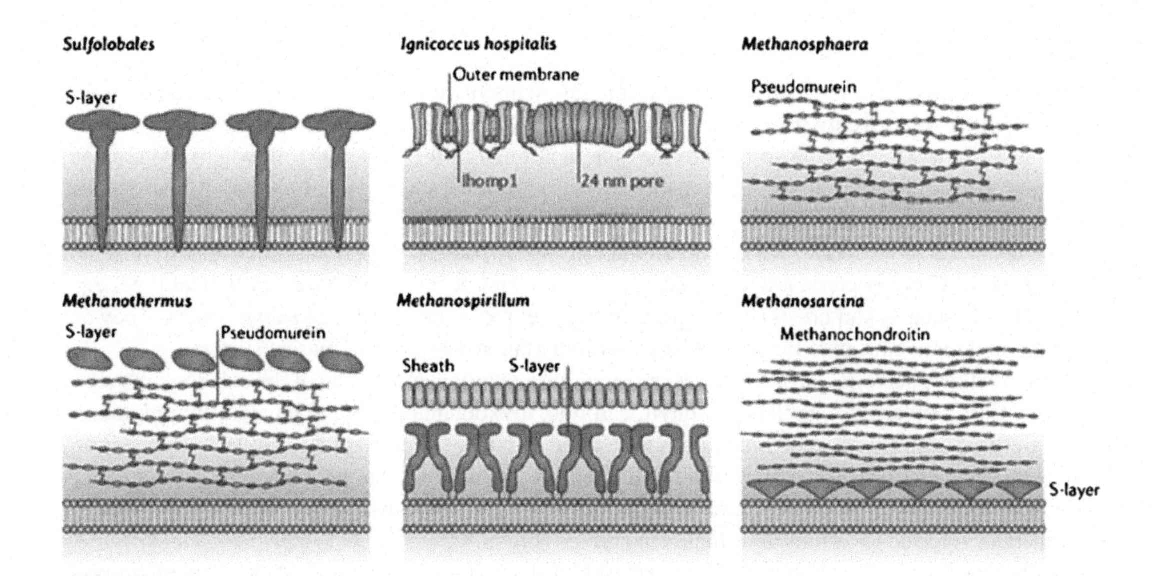


**FIGURE 3.7**

Archea cell wall. Structurally archaea also have a cell wall like bacteria to assist the cells in maintaining its chemical equilibrium, however they are composed of glycosylated proteins instead of peptidoglycan. This change provides osmoprotection and environmental protection in a vast variety of demanding habitats (Source: https://sites.google.com/site/melanieswitzerbvb101/home/general-archaeal-adaptations).

## PROKARYOTES AND HORIZONTAL GENE TRANSFER

### BACTERIA AND ARCHAEA

Both bacteria and archaea are single-celled prokaryotes, which means they have no nucleus or other membrane-bound organelles within their cells. Based on years of research, we understand the differences in the composition of their cell walls: Archaea have cell walls composed of Pseudopeptidoglycan, whereas bacteria cell walls are Peptidoglycan / Lipopolysaccharide; this difference among many demonstrates the early divergence between these two domains.

Archaea and bacteria reproduce asexually by the process of binary fission, budding and fragmentation, whereas only bacteria also have the ability to form spores to remain dormant for years.

Often referred to as extremophiles, Archaea were thought to be specialized for extreme and harsh environments, such as hot springs, salt lakes, marshlands, hydrothermal vents, gut of ruminants and humans, thus having a wide range of adaptations to live in these environments (Fig. 3.7). But they are ubiquitous and found in soil, oceans, and inland waters.

Both of these groups are extremely important in ecosystems and that includes aquatic systems. For example, some of these organisms alter the redox environment and thrive in anaerobic environments, which we will explore in Chapter 11.

## BACTERIA AND CYANOBACTERIA

Bacteria are single-celled organisms with a simple internal structure that lacks a nucleus, and contains DNA that either floats freely in a twisted, thread-like mass called the nucleoid, or in separate, circular pieces called plasmids. Ribosomes are the spherical units in the bacterial cell, where proteins are assembled from individual amino acids using the information encoded in ribosomal RNA. Aquatic bacteria are diverse and important in water column and sediments in every type of inland water.

Cyanobacteria is a diverse, widely distributed group of bacteria that obtain their energy through photosynthesis. Cyanobacteria are distributed worldwide and occur in almost every habitat type, e.g., surface of soil (such as soil crusts), hot springs, snow, ice, rocks, and fresh and saline waters. They are especially abundant eutrophic freshwaters and as phytoplankton blooms in marine systems

Although these bacteria are prokaryotes, an outer pigmentation area inside of the cell contains chromoplasm. Chromoplasms contain chlorophyll, carotine, myxothanhophyll and phycocyanin pigments, which are used to capture solar radiation and convert it into potential energy. Under a light microscope, this organizational structure resemble organelles, thus these taxa were originally classified as blue-green algae, but cyanobacteria are bacteria and not algae. Moreover, algae itself is polyphyletic, and generally referred to as photosynthetic eukaryotes (Fig. 3.8).

As described in Chapter 1, cyanobacteria altered Earth's atmosphere and may have created dinosaur graveyards. Thus cyanobacteria structure ecosystems in numerous ways by influencing aquatic biogeochemical cycling, producing toxic chemicals, and altering foodweb structure.

## PLASMIDS, VIRUSES, PRIONS, ETC.

Rogue DNA, RNA, proteins roam the biosphere. In some cases these macromolecules are part of plasmids or viruses that stabilize the DNA and RNA genes and facilitate their entry into cells and to reproduce. Although not technically living organisms, these particles can have profound ecological impacts.

For example, plasmids are composed of coding DNA strands that can replicate independently inside a cell. They are most commonly found in bacteria and archaea. Plasmids often carry genes that confer antibiotic resistance and the ability to breakdown complex and even synthetic toxic chemicals.

Viruses are small infectious agents that replicate inside the living cells of other organisms. Viruses can infect all types of life forms, from animals and plants to microorganisms. Viruses are composed of proteins that encapsulate a DNA or RNA genome, and may have relatively simple or elaborate structures (Fig. 3.9). Cyanophages, viruses that "attack" cyanobacteria, have been implicated in transferring genes that can increase photosynthetic efficiency in cyanobacteria, also implicated in controlling cyanobacteria blooms (see Section Bacteria and Bacteriophage Infections and Coevolution, page 157).

In addition, prions are infectious agents composed protein material. A protein as a stand-alone infectious agent stands in contrast to all other known infectious agents, such as viruses, bacteria, fungi, and parasites, all of which contain nucleic acids (DNA, RNA, or both). Prions have been linked to zoonotic diseases that resembles Cteutzfedt–Jakob disease in human populations.

## HORIZONTAL GENE TRANSFER

Horizontal gene transfer is the movement of genetic material between unicellular and/or multicellular organisms in contrast to "vertical" transmission of DNA from parent to offspring. The importance

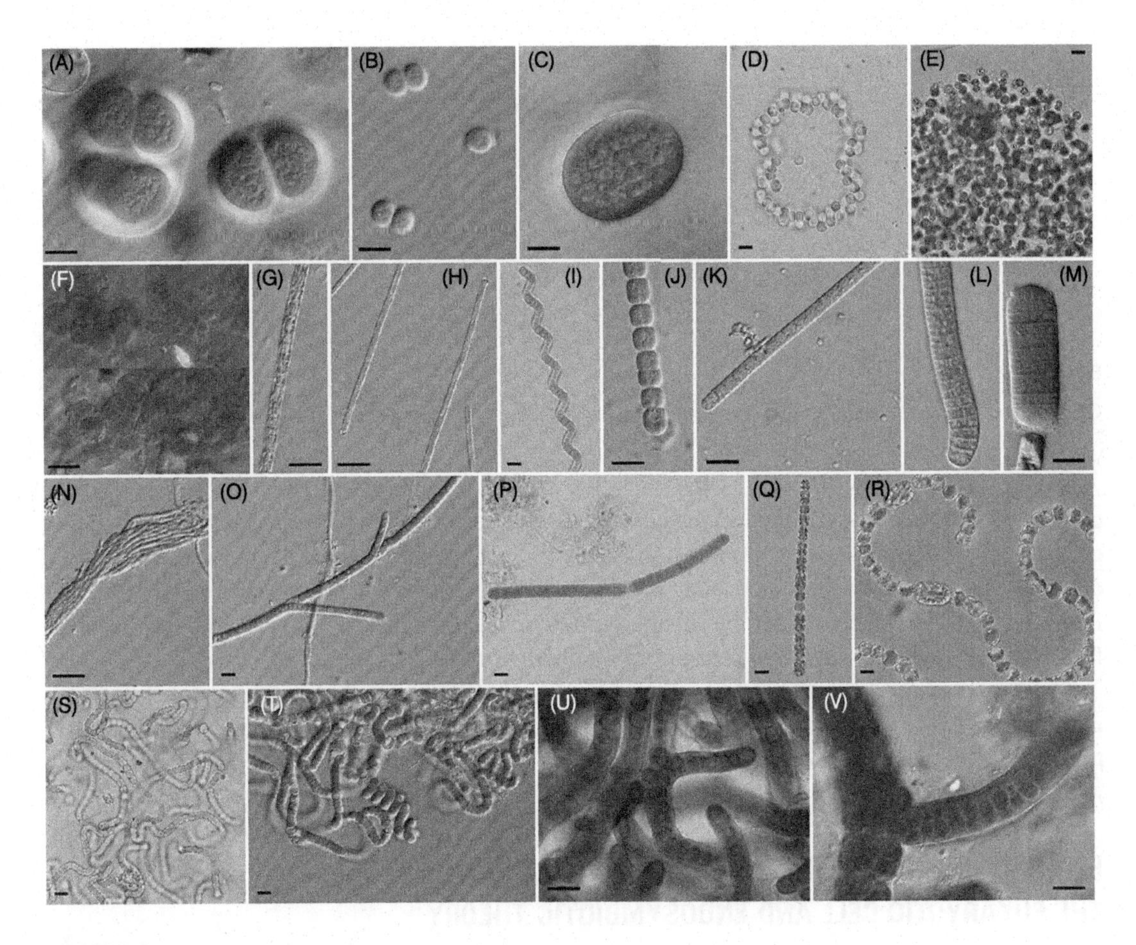


**FIGURE 3.8**

Morphologic variability in Cyanobacteria. The morphology can be described as simple unicellular, filament, branched filament, filament aggregates, sheets or mats. Some species might have different morphologies depending on the environmental conditions. Illustration of morphological diversity in cyanobacteria. Groups (orders). I. Chroococcales: (A) *Chroococcus subnudus*, (B) *Ch. limneticus*, (C) *Cyanothece aeruginosa*, (D) *Snowella litoralis*, (E) *Microcystis aeruginosa*. II. Pleurocapsales: (F) *Pleurocapsa minor*. III. Oscillatoriales: (G) *Planktothrix agardhii*, (H) *Limnothrix redekei*, (I) *Arthrospira jenneri*, (J) *Johanseninema constricum*, (K) *Phormidium* sp., (L, M) *Oscillatoria* sp., (N) *Schizothrix* sp., (O) *Tolypothrix* sp., (P) *Katagnymene accurata.*, IV. Nostocales: (Q) *Dolichospermum planctonicum*, (R) *Dolichospermum* sp., (S) *Nostoc* sp., (T) *Nodularia moravica*. V. Stigonematales: (U, V) *Stigonema* sp. Scale bar A–U = 10 lm, V = 20 lm. (Color figure online) (Source: Dvořák et al., 2015.)

horizontal gene transfer is just beginning to be appreciated. In fact, among single-celled organisms horizontal gene transfer is perhaps the dominant form of genetic transfer. For example, genes in the diatom *Phaeodactylum tricornuntam* probably came from a bacterium in the last 90 million years. Capitalizing

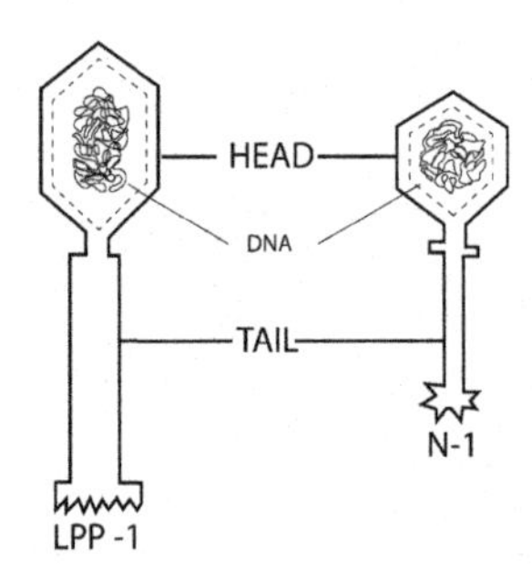


**FIGURE 3.9**

Cyanophage, a virus that attacks Cyanobacteria, has similar structure as other viruses (drawn by Luyi Huang).

on plasmid and virus particles and conjugation, horizontal gene transfer plays an important role in the evolution of bacteria and eukaryotes that can degrade novel compounds, such as human-created pesticides and the compounds associated with the evolution, maintenance, and transmission of virulence. In fact, it might be useful to consider the aquatic systems as a soup of genetic material with the potential for sharing and the most important mechanism for prokaryotic and eukaryotic evolution.

We should acknowledge the problematics of biological phylogeny and evolutionary ancestry when gene transfer is so promiscuous. The tree of life's branches might be connected to each other via gene transfer across a wide range of unrelated taxa.

## EUKAROYOTES SYSTEMATICS AND EVOLUTION

### THE EUKARYOTIC CELL AND ENDOSYMBIOTIC THEORY

Eukaryotes have more complex cell structure than bacteria and archaea. Specifically, they have internal membrane-bound organelles and a nucleus enclosed by the nuclear envelope. In many cases they will have other organelles that preform specialized functions, such as mitochondria, where respiration occurs and/or chloroplasts, where photosynthesis is carried out (Fig. 3.10). These organelles may also contain their own DNA, which begs the question, why is there additional DNA within these organelles? And how did it get there?

According to endosymbiotic theory, plastid organelles (e.g., chloroplasts) in plants and various eukaryotic algae have evolved from cyanobacteria via endosymbiosis, where an archeae cell engulfed a functional cyanobacterium. Meanwhile the origin of mitochondria is probably an engulfed bacterium. Thus this theory explains the presence of DNA inside these organelles and evolution of prokaryotes.

Over time, various Archaea, prokaryotes and even eukaryotes have been engulfed (via independent events) in various combinations and range of outcomes, promoting a great deal of diversity at the microscopic level and the capacity to colonize new habitats. For example, as proto-photosyntheic organisms, they were no longer dependent on energy sources, e.g., hydrothermal vents and could begin expanding into shallow, surface waters.

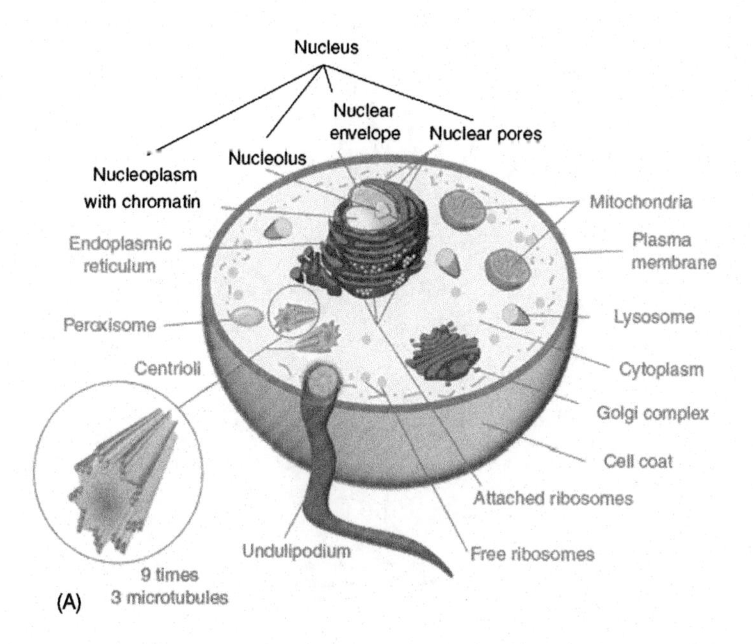


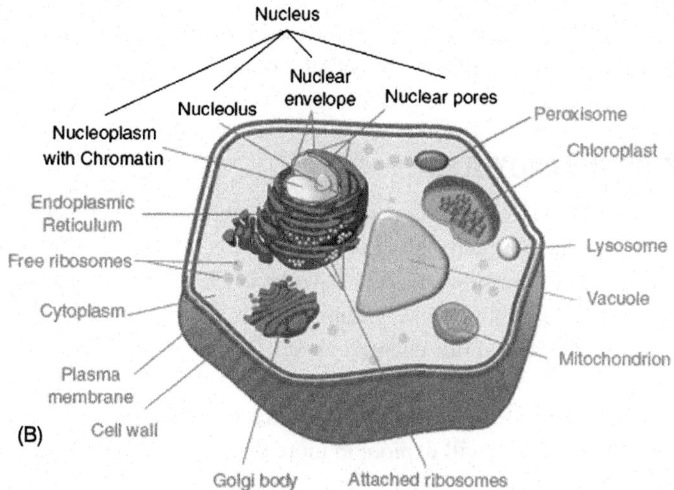


**FIGURE 3.10**

Animal and plant eukaroytoic cells have complex organelles and structures (Source: http://scitechconnect.elsevier.com/symbiosis/). In some cases, we have a pretty good idea of how these organelles evolved because of shared gene sequences with prokaroyotes.

**FIGURE 3.11**

A radically new phylogeny of the eukarotes. For environmental scientists, we should appreciate how profoundly diverse the microscopic taxa compared to plants and animals that comes to mind when we think about biodiversity.

## PHYLOGENETICS OF EUKAROYOTES

Our understanding of evolutionary of eukaroyotes has made great strides in recent years. In fact, except for a few pesky groups, we have a pretty good idea of what is related to what, and about how long ago they diverged. Recent classification schemes have tacitly established new monophyletic groups (Fig. 3.11). Although members within these groupings still shift, with new analysis tools, we can certainly appreciate a central theme in this chapter: most of the taxonomic diversity is at the microbial scale.

To organize the groups in this text, we will rely on a recent grouping that simplifies life into five megagroups and nine groups that we will explore in more detail as warranted (Table 3.2).

## PROTIST MEGAGROUPS

### ARCHAEPLASTIDA

Archaeplasti ("ancient plastids") are organisms acquired as the result of the fusion between a bacterium and a symbiotic cyanobacterium. Archaeplasti includes Glaucophyta, a relatively obscure freshwater group; Rhodophyta, a diverse group, but only has a few freshwater families, and the highly studied

**Table 3.2 Taxonomic groupings based on an alternative grouping (Source: He et al., 2014). He et al. (2014) did not include the Rhizaria because of a lack of representatives in the databases to analyze, but I included it with the SAR megagroup.**

| Megagroup | Group | Key Characteristic | Examples |
|---|---|---|---|
| Archaeplastida | Viridiplantae | Cellulose in their cell walls, primary chloroplasts derived from endosymbiosis with cyanobacteria; chlorophylls a and b; lack phycobilins | Chlorella |
| SAR | Stramenopila | Two equal flagella | Diatoms, Thalassiori |
| SAR | Alveolata | Cortical alveiola-flattened sac-like structures | Plasmodium, Dinoflagellates, Ciliates |
| SAR | Rhizaria | Mostly unicellular eukaryotes and non-photosyntethic. Described from rDNA sequences, they vary considerably in form, having no clear morphological distinctive characters, but for the most part they are amoeboids with filose, reticulose, or microtubule-supported pseudopods. Many produce shells or skeletons, which may be quite complex in structure, and these make up the vast majority of protozoan fossils | |
| Discoba | Jakobida | | Jakoba |
| Discoba | Discicristata | | Eugena |
| Metamonada | Rtortamonads, diplomonads, and possibly the parabasalids and oxymonads | | |
| Amorphea | Amoebozoa | | Polyshodylium |
| Amorphea | Obazoa: Fungi | One flagella | Phycomyces |
| Amorphea | Obazoa: Holozoa | | Branchiostoma |

Chloroplastida. The Chloroplastida, also known as Viridiplantae, includes both the green algae and land plants (Fig. 3.12).

Viridiplantae (literally "green plants") are eukaryotic organisms made up of the green algae, which are primarily aquatic, and the land plants, which emerged from within the green algae. More than 350,000 species of Viridiplantae exist. Both green algae and plants play a key role in aquatic systems from stabilizing stream channels (see Section Land Plants and New Rules in Geomorphology) to providing sources of organic matter (see Section Autocthnous and Allocthonous Sources of Organic Matter). Viridiplantae have cell walls composed of cellulose and lignin (wood) that are structurally difficult to breakdown, but provide important habitat for many aquatic species.

Plants play an important role in aquatic systems—although with a few important exceptions, mostly in fresh- and brackish waters. Plants often define the aquatic system, e.g., marshes, swamps, mangroves, riparian forests, and sphagnum bogs. Based on recent phylogenies, the text refers the following groups: nonvascular plants (bryophytes), lycophyes, ferns (spores), gymnosperms (cone bearing plants), basal angiosperms (flowering plants), moncots, and eudicots (e.g. Fig. 3.13).

We will be returning to plants throughout the text, so we will turn now to other groups. Nevertheless, the diversity of plants and their role in aquatic systems is impossible to understate. They affect

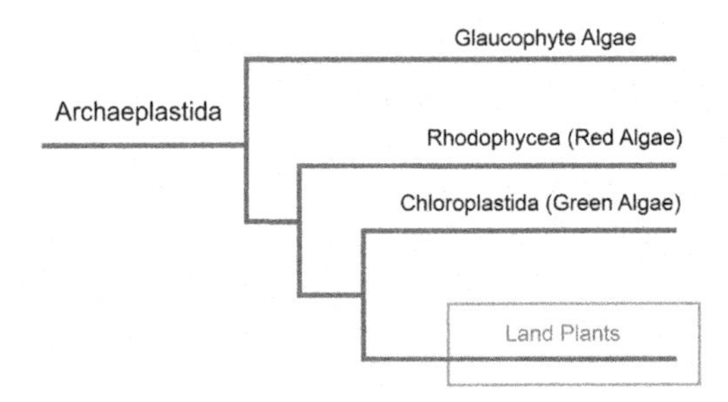


**FIGURE 3.12**

Archaeplasti phylogeny (Source: Layla Khalili). Green algae traditionally excludes the land plants, rendering them a paraphyletic group. Green algae and plants descended from a common evolutionary ancestor. However the term is not very meet the criteria for monophyletic since it does not include all descendant groups, i.e., land plants. As we learn more about the evolution of various taxa biologists are often confronted with these ambiguities. In general, I recommend using these ambiguities to explore scientists learn how to describe the natural world with a vocabulary that is developing.

**FIGURE 3.13**

Image of *Amborella* trichopoda (Source: Scott Zona). *Amborella* is a monotypic genus endemic to Grande Terre, of New Caledonia.

water supply and quality; species diversity; food web dynamics; productivity; and biogeochemistry, etc. (Fig. 3.14).

## SAR: STRAMENOPILES, ALVEOLATA, AND RHIZARIA

The megagroup SAR includes three lineages that are each hugely diverse and speciose in their own right: Stramenopiles, Alveolata, and Rhizaria (SAR is an acronym for these three groups).

***Stramenopiles*** are distinguished by their characteristic rigid tubular flagellar hairs (the group name means "straw hairs"), although these have been lost in many species and several whole subgroups (Fig. 3.15). Stramenopiles include a wide range of photosynthetic and nonphotosynthetic taxa.

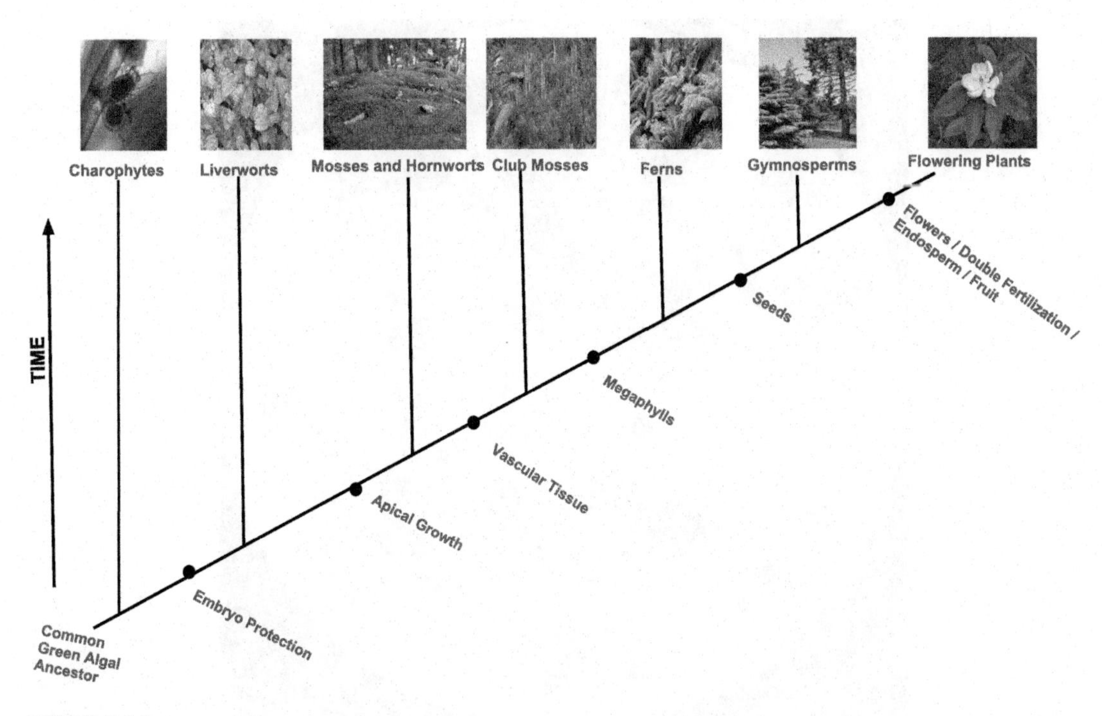


**FIGURE 3.14**

Diversification and family tree of the vascular plants (Source: Lauren Prue).

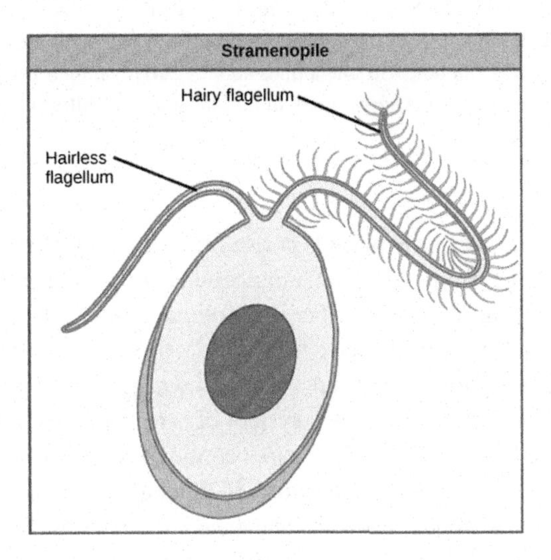


**FIGURE 3.15**

Image of Stramenopile flagella, each with different characteristics.

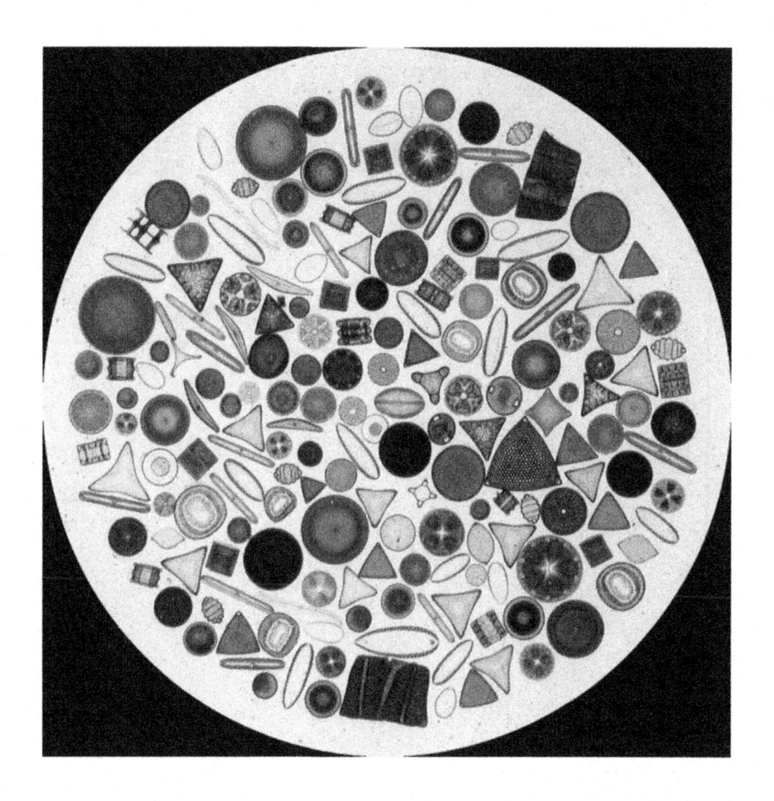

**FIGURE 3.16**

Diatoms come in diverse shapes and sizes (Source: GNU Free Documentation License, Version 1.2). Diatoms have plastids that came from red algae, a member of the Archaeplastida. Their cell walls are composed of hydrated silica dioxide. Although the diverged from other Stramenopiles about ~250 mya, their early fossil record has been relatively scant; more recent records have been extremely important source of information about climate and water quality in lake sediments.

Photosynthetic stramenopiles have plastids derived from a Rhodophyta donor and form a monophyletic group. Ranging from the giant multicellular kelp to the unicellular diatoms, stramenopiles play a key role as plankton. For example, diatoms (Bacillariophyta) have unicellular/colonial forms with bipartite siliceous "cell walls" (frustrules) (Fig. 3.16).

***SAR: Alveolata*** encompasses three of the most well-known groups of protists, Apicomplexa, Dinoflagellata, and Ciliophora. They share a system of sacs underneath the cell membrane called alveoli. These flattened alveoli are packed into a continuous layer just under the membrane and typically form a flexible thin skin-like structure. In dinoflagellates they often form armor plates. Apicomplexa is quintessentially parasitic and cause diseases in humans and animals (e.g., Malaria (*Plasmodium*), Toxoplasmosis (*Toxoplasma gondii*), and Cryptosporidiosis (*Cryptosporidium parvum*)). *Cryptosporidium parvum* caused the largest waterborne-disease outbreak documented in the United States, making 403,000 people ill in Milwaukee, Wiscon-

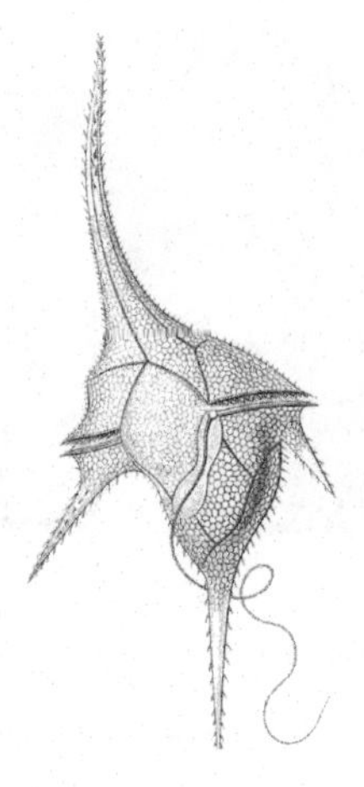

**FIGURE 3.17**

*Ceratium hirundinell* species of dinoflagellates. Most species of Ceratium are unicellular organisms and have armored plates, two flagella, and horns. Ceratium has a global distribution and can cause harmful algal blooms. (Source: Wikicommons).

sin. Because it is resistant to all practical levels of chlorination, it is one of the most important waterborne pathogens in developed countries.

Dinoflagellata includes conspicuous and important component of plankton and function as autotrophs and/or grazers (many are mixotrophic). Dinoflagellates are involved in several phenomena of great ecological importance, such as harmful algal blooms, symbioses with reef-forming corals, and important parasitic associations with animals, or with other protists (Fig. 3.17).

As a third main group of Alveolates, the Ciliophora are a diverse group of mainly freshwater heterotrophic ciliates. Most have large numbers of cilia (i.e., arrays of coordinated eukaryotic flagella), which in many species cover almost the entire cell. New environmental sequencing studies indicate that the full diversity of ciliates is far from uncovered.

***SAR: Rhizaria*** is the most morphologically diverse of the higher-order lineages. Named for the rootlike appearance of their "pseudopods", the group is comprised of heterotrophic amoebae, flagellates, amoeboflagellates, and some spore-forming parasites and unusual algae. No set of morphological features unite Rhizaria to the exclusion of other eukaryotes; nevertheless, molecular phylogenetic analyses suggest they are a distinct taxon.

As one type of Rhizaria, Radiolarians are large, often abundant, marine amoebae with microtubule supported pseudopodia (Fig. 3.18). Abundant in marine waters, their role in inland waters is unknown.

## DISCOBA

Discoba contain groups known as the Discicristata and Jakobida; as a portmanteau of Discicristata and Jakoba, the Discoba is often referred to as escavates, which has traditionally relied on a classification scheme based on their flagellar structures.

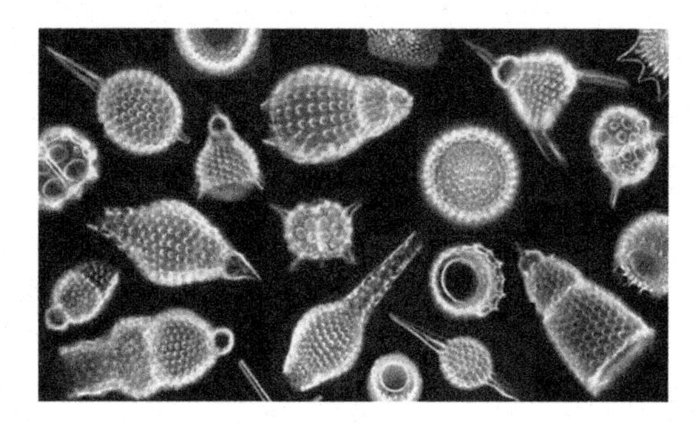

**FIGURE 3.18**

The Radiolaria are protozoa of diameter 100–200 μm in size that produce skeletons, typically with a central capsule dividing the cell into the inner and outer portions of endoplasm and ectoplasm. The skeleton is usually made of silica (Source: Wikipedia).

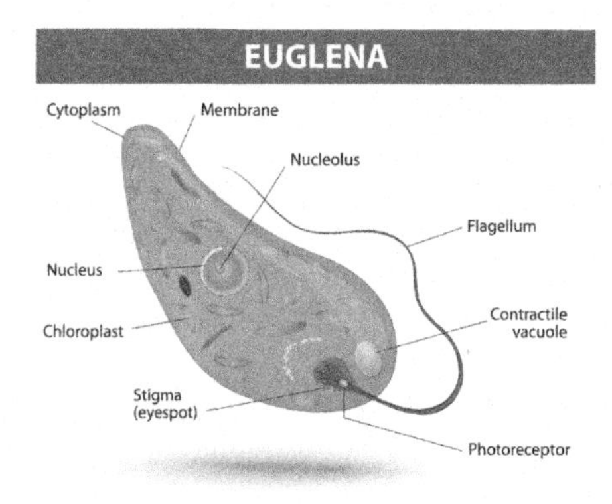


**FIGURE 3.19**

*Euglena gracilis* is a species of single-celled Eukaryote algae in the genus *Euglena*. They have a secondary chloroplast and can feed by photosynthesis, heterotrophy or phagocytosis. They have a highly flexible cell surface, allowing them to change shape from thin cells up to 100 μm long to spheres of approximately 20 μm. They have two flagella, only one of which emerges from the flagellar pocket (reservoir) in the anterior of the cell, and can move by swimming, or by so-called "euglenoid" movement across surfaces. *Euglena gracilis* has been used extensively in the laboratory as a model organism, particularly for studying cell biology and biochemistry.

The Euglenozoa are a large group within the flagellate excavates and Descicristata, and include a variety of common free-living species, as well as, a few important parasites, some of which infect humans. *Euglena gracilis* is probably the most recognizable, but hardly representative of the diversity of the Discoba (Fig. 3.19).

Although jakobids were not recognized as a taxon until the 1990s, and their diversity remains under-studied, they have recently attracted considerable attention because of their plesiomorphic, bacterial-like mitochondrial genomes. They include a collection of amoebae, flagellates, or "amoeboflagellates,"

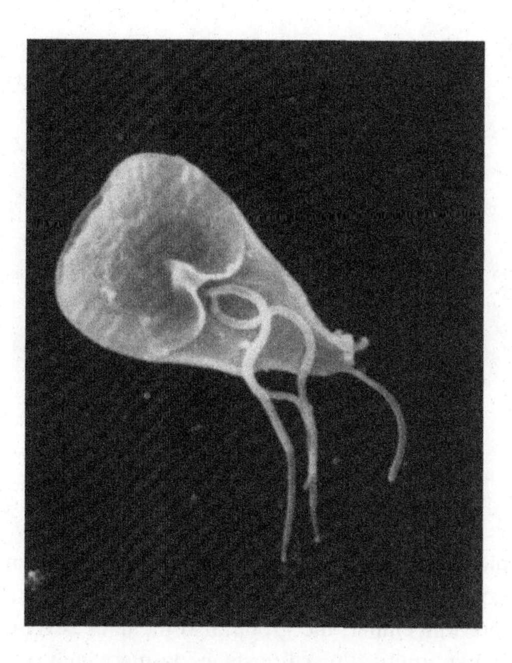

**FIGURE 3.20**

*Giardia lamblia*. As a parasite, it attaches to the epithelium cells in the gut and causes gastrointestinal pain and diarrhea (Source: CDC / Janice Haney Carr http://phil.cdc.gov/PHIL_Images/8698/8698_lores, Public Domain, https://commons.wikimedia.org/w/index.php?curid=825607).

i.e., lifecycles that include both amoeba and flagellate cell-type. Many are predatory species that glide over surfaces, whereas some are free-living or parasitic flagellates.

## METAMONADA

The metamonads is a group of anaerobic protozoa, almost all of which are flagellates. *Giardia lamblia* is a Metamonada implicated with public health. As a member of the Diplomonadida, which are mostly "doubled" cells with two nuclei and flagellar apparatuses, *Giardia lamblia*, also known as *Giardia intestinalis*, is a flagellated parasite that colonizes and reproduces in the small intestine, causing giardiasis. Chief pathways of human infection include ingestion of contaminated water (Fig. 3.20).

## AMORPHEA: AMOEBOZOA

The Amorphea assemblage unites two huge clades: 1) "Obazoa" composed of animals, fungi, and their immediate protist relatives and 2) "Amoebozoa", a grouping of heterotrophic protists.

***Amoebozoa*** mostly (though not entirely) consists of organisms that are amoebae for much or all of their lifecycle. Molecular genetic analysis supports Amoebozoa as a monophyletic clade and includes many of the best-known amoeboid organisms, such as *Chaos* (Fig. 3.21), *Entamoeba, Pelomyxa,* and the genus *Amoeba* itself. Free-living species are common in both salt and freshwa-

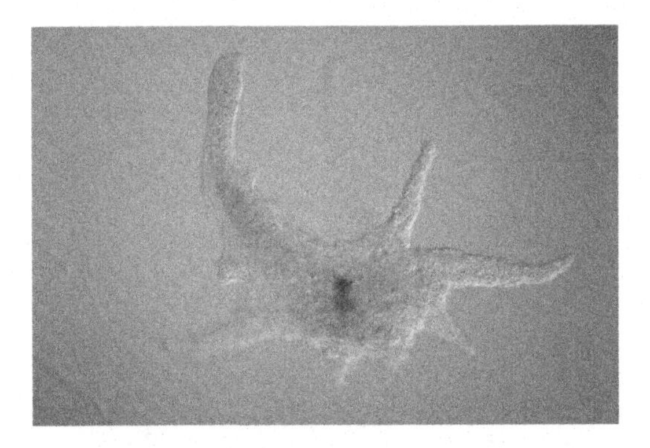

**FIGURE 3.21**

*Chaos carolinense* (Source: Wikipedia: Tsukii Yuuji).

ter, as well as in soil, moss, and leaf litter. Some live as parasites or symbiotes of other organisms, and some are known to cause disease in humans and other organisms. Whereas the majority of amoebozoan species are unicellular, the group also includes several varieties of slime molds, which have a macroscopic, multicellular life stage, during which individual amoeboid cells aggregate to produce spores.

***Amorphea: Obazoa*** As a sister group to the Amboebozoa, the Obazoa include Animals (Metazoa) and Fungi, which are closely related to one another. Molecular phylogenetics support a monophyletic origin of the fungi with four monophyletic groups (Fig. 3.22).

Aquatic fungi, traditionally called hyphomycetes, fall within the ascomycete lineage. These saprophytic organizms eat dead and decaying organic matter. For example, after leaves fall from trees and enter streams, these fungi penetrate through the leaf surface and spread through tissue. With extracellular enzymes that break down leaf tissue, e.g., cellulose, the leaves are also made palatable to invertebrates. Leaves colonized by fungi have a lower C:N ratio and more nutritious for macroinvertebrates than leaves without fungi (Fig. 3.23).

## METAZOA: PROTOSTOMES AND DEUTEROSTOME

Animals are generally considered to have emerged within flagellated eukaryota, which is reenforced by our understanding that the Amoebozoa are a sister group and provide the foundation animal systematics (Fig. 3.24). Metazoans are animals and share the following characteristics: similarly oriented cells connected via specialized junctions (epithelia); extracellular matrix composed of fibrous proteins; diploid cells with gametic meiosis that yield haploid eggs and sperms, followed by embryonic development. They are divided up into two main groups: protostomes and deuterostomes. These categories are based on patterns of embryonic development.

The lack of a cell wall in Metazoans may have a profound impact on the development of and ecosystem. Without a cell wall, animal cells can ingest extracellular particulates and even cellular material through phagocytosis. This evolutionary novelty becomes the foundation for animals to exploit

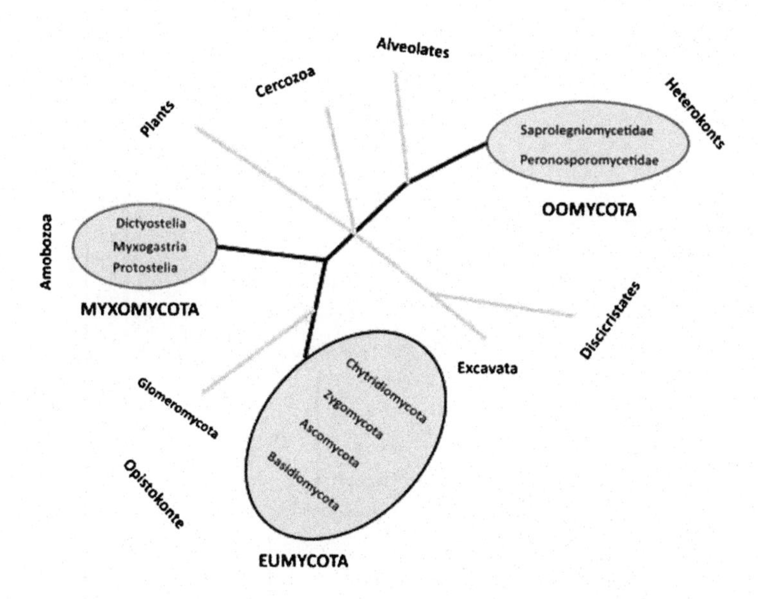


**FIGURE 3.22**

Classification scheme based of four types of fungi (Source: Sime-Ngando): Chytridiomycota (chytrids), Zygomycota (bread molds), Ascomycota (yeasts and sac fungi), and the Basidiomycota (club fungi). Placement into a division is based on the way in which the fungus reproduces sexually. The shape and internal structure of the sporangia, which produce the spores, are the most useful character for identifying these various major groups.

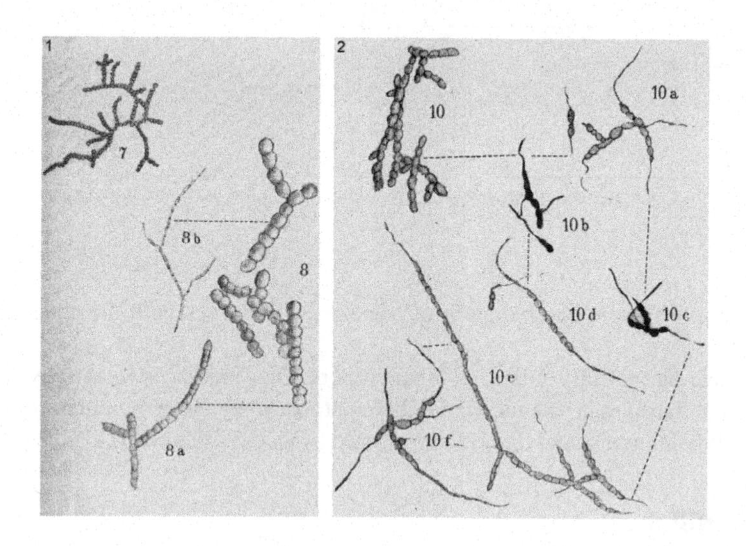


**FIGURE 3.23**

Morphology of Hyphomycetes (Source: Maximilian J. Telford).

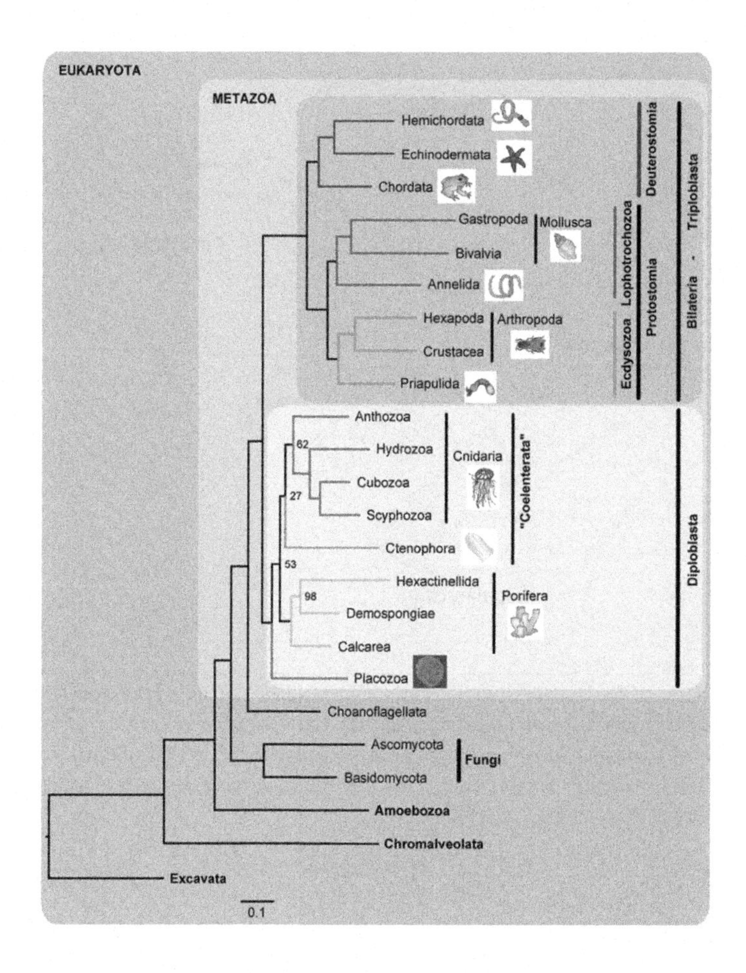


## FIGURE 3.24

Maximum likelihood phylogenetic tree of Metazoan relationships using the concatenated data matrix (Source: Schierwater et al., 2009).

prey. By adding trophic levels, the radiation of Metazoans coincides with the development of more complex food webs.

Few animals have successfully colonized inland waters. They include several Ecdysozoans (Arthropoda, Onychophora, Tardigrada, Nematoda, and Nematomorpha), three Spiralian groups (Annelida, Mollusca, and Platyhelminthes) and one Deuterostome (Craniata) (Table 3.3).

## PORIFERA: SPONGES

Sponges come in an incredible variety of colors and an amazing array of shapes. Whereas most sponge diversity is marine, the Spongillidae is an extant family of freshwater, whose fossil record begins in the Cretaceous.

**Table 3.3 Organizational structure of the Metazoan covered in the text.**

| Group | Cross-referenced names |
|---|---|
| Protostomes: Porifera | Sponges |
| Protostomes: Cnidaria | Jellyfish |
| Protostomes: Myxozoa | Myxozoa |
| Protostomes: Ecdysozoans | Crustaceans, Arachnids, Insects |
| Protostomes: Spiralia | Mollusks (snails and clams), Annelids, and Platyhelminthes |
| Deuterostomes: Craniata | Fish, Birds, Amphibians, Reptiles, Mammals |

**FIGURE 3.25**

Freshwater sponge, *Spongilla lacustris*.

*Spongilla*, one of 23 genera in the Spongillidae, attach themselves to rocks and logs and filter water for various small aquatic organisms, such as protozoa, bacteria, and other free-floating pond life (Fig. 3.25). Unlike marine sponges, freshwater sponges are exposed to far more adverse and variable environmental conditions; thus they have developed a dormancy mechanism. When exposed to excessively cold or otherwise harsh situations, these sponges form gemmules (Fig. 3.26), highly resistant "buds" that remain viable after the rest of the sponge has died. When conditions improve, the gemmules "germinate" and a new sponge is born.

## CNIDARIA: JELLYFISH

Although the majority of jellyfish are marine taxa, freshwater jellyfish can play an important role in the food web dynamics of lakes. They have two life-history stages, poly and medussa forms (Fig. 3.27). Both polyp and hydromedusa forms use nematocysts (stingers) to capture prey.

Like other cnidarians, *Craspedacusta sowerbyi* is an opportunistic predator, feeding on small organisms that come within its reach. Conspicuous swarms of hydromedusae appear sporadically, but are only one part of the animal's lifecycle. During these times, *C. sowerbii* predates zooplankton, including

**FIGURE 3.26**

*Spongilla lacustris,* gemmules (Source: Lamiot).

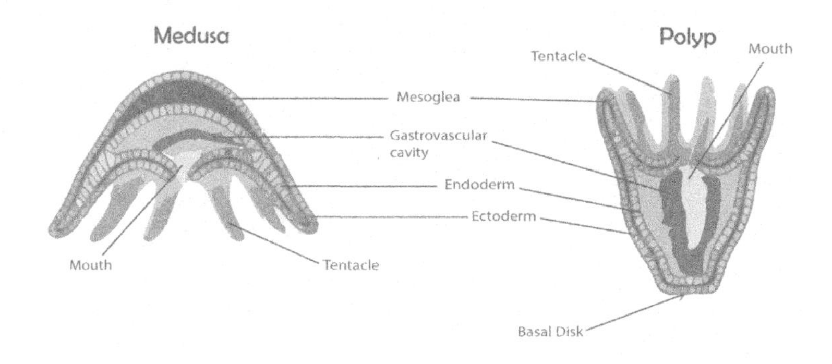


**FIGURE 3.27**

Poly and medusa of sponges (Source: Slobodkin and Bossert, 2010).

daphnia and copepods. Prey is caught with their stinging tentacles. Drifting with its tentacles extended, the jelly waits for suitable prey to touch a tentacle. Once contact has been made, nematocysts on the tentacle fire into the prey, injecting poison, which paralyzes the animal, and the tentacle itself coils around the prey. The tentacles then bring the prey into the mouth, where it is digested.

## MYXOZOA

The Myxozoa are aquatic parasitic animals and were probably derived from jellyfish. With 2,402 species described, many have a two-host lifecycle, involving a fish and an annelid worm or bryozoan. For example, *Ceratomyxa shasta* is a virulent myxosporean parasite of salmon and trout in the Pacific Northwest (Fig. 3.28).

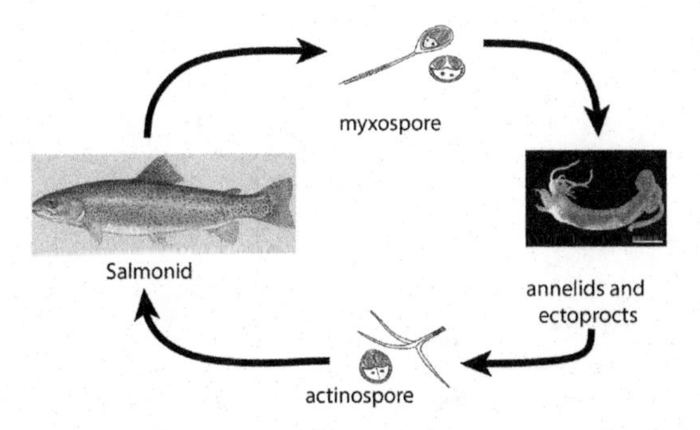


**FIGURE 3.28**

Myxospore lifecycle that parasites two very different taxa (Source: Atkinson and Bartholomew, 2010). The myxospore stage is consume by a polycheate. Polycheates are members of Spiralia along with mollusks and platyhemeites). *Ceratomyxa shasta* becomes (somehow) a actinspore, which is then consumed by a craniata salmon.

# ECDYSOZOA: EXOSKELETON SHEDDERS

## INLAND WATER GROUPS: FOUR OUT OF EIGHT ECDYSOZOA

This monophyletic group of metazoans molt their cuticle (exoskeleton) during their lifecycle and include eight majors groupings (Fig. 3.29). Whereas most of the diversity is marine, four groups have successfully invaded inland waters: nematodes (Nematoda), horse hair worms (Nematomorpha), tardigrades (Tardigrada), and arthropods (Arthropoda).

## NEMATODA

The 28,000 described species of nematodes barely covers the estimated total of 1,000,000 species in this diverse group. Although we have a poor understanding nematode diversity, they can be very important in aquatic systems (Fig. 3.30).

Natural nematode communities influence bacterial activity; microbivorous or fungivorous nematode activity influences microbial biomass and nutrient cycling in the benthos. Some may have very specific roles in ecosystems and broad ecosystem affects. For example, nematode grazing has been associated with nutrient mineralization that can stimulate bacteria activity and increase sediment respirate rates. Although stream habitats may have over 150 species and over 200 individuals per ml, their basic life history parameters (e.g., generation time, eggs per female) are largely unknown, thus nematode ecology deserves more research for inland waters and beyond.

## NEMATOMORAPHA

Approximately 350 species of nematomorphs or horsehair worms have been described, the majority (21 genera) are freshwater-type. They are long and slender worms that aggregate and copulate in the water. Microscopic larvae hatch from eggs and infect a variety of freshwater animals, in which they encyst. To complete their lifecycle, gordiids must make a transition from aquatic to terrestrial environments, thus need to be consumed by terrestrial animals to reproduce (Fig. 3.31).

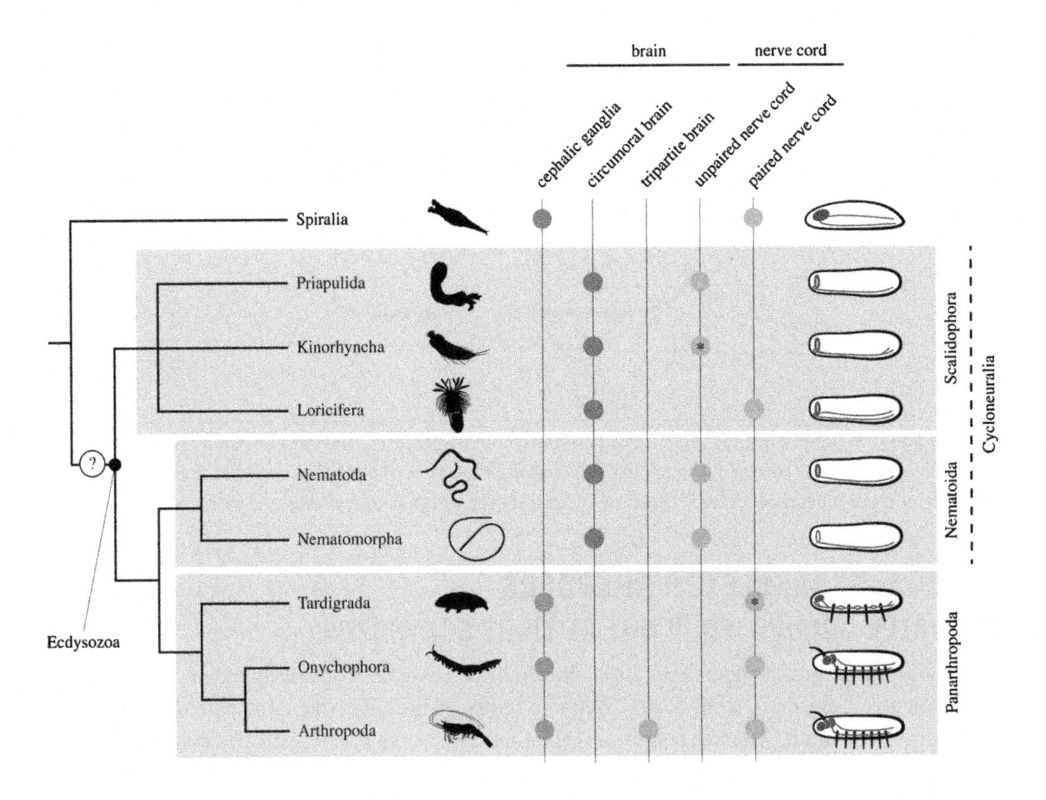


**FIGURE 3.29**

Ecdysozoa phylogentic tree based on their nervous system (Source: Martín-Durán et al., 2016). It is common to find phylogentic trees that vary based on the data used. Remember, no dataset is reliable alone; it is best to combine multiple sources of evidence to make a good judgment.

## TARDIGRADES

Tardigrada is a phylum closely allied with the arthropods. They are usually less than 500 μm in length, have four pairs of lobe-like legs and are either carnivorous or herbivores. Although most of the 900 plus described tardigrade species live in the thin film of water on the surface of moss, lichens, algae, and other plants, 13 genera are truly aquatic. Their role in aquatic ecosystems is largely unknown (Fig. 3.32).

## ECDYSOZOA: ARTHROPODS

### CRUSTACEANS AND HEXPODS

As members of the ecdysozoa, the arthropods represent one of the most visible and successful inland water clades. Just as other ecdysozoa, arthropods molt as part of their lifecycle. Arthopods include insects, spiders, and allied taxa with ~1.2 million species. The most diverse arthropod group is found

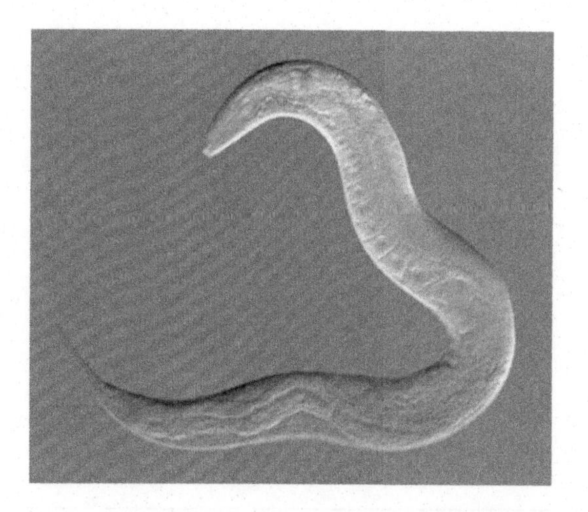

**FIGURE 3.30**

Aquatic nematode with a filamentous cyanobacterium (Source: Wikicommons).

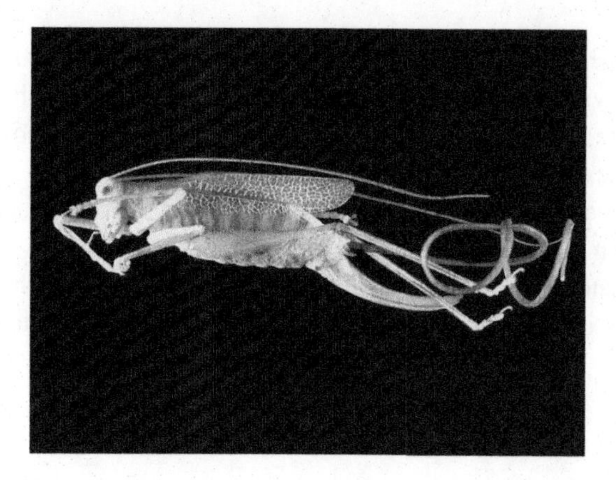

**FIGURE 3.31**

Spinochordodes in Meconema (Source: Wikipedia).

in the insect order, Coleoptera (> 380,000 species), which represents about 38% of all species in 39 insect orders, see page 112.

The phylogeny of arthropods has always been messy. But recent analyses place insects ("Hexapoda") nested within crustaceans (Fig. 3.33). In other words, the traditional concept of crustacean is paraphyletic, because it has not included insects. Based on this revision, the next few sections have been organized using reasonably stable categories (Table 3.4).

**FIGURE 3.32**

Image of Tradgrade or water bear (Source: Science Photo Library).

## CHELICERATA

Chelicerata include horseshoe crabs, scorpions, spiders, and mites. They originated as marine animals, possibly in the Late Ordovician period. Some extant remain marine, such as horseshoe crabs and possibly sea spiders, but there are 77,000 air-breathing chelicerates described and approximately 500,000 remain unidentified species.

Although some air-breathing spiders have adapted to aquatic systems, water mites are more common. Like spiders, water mites have eight legs and soft bodies (Fig. 3.34).

The distribution of water mites depend on abiotic and biotic factors, such as temperature fluctuations and habitat stability and heterogeneity. They are parasitic or predacious or both, depending on their life history stage.

## OLIGOSTRACA

The oligostraca include ostracods and burchihcura species. Ostracoda is sometimes known as the seed shrimp because of their appearance. Their oval bodies are flattened from side to side and protected by a bivalve-like, chitinous or calcareous valve or "shell".

Ostracods are small crustaceans, typically around 1 mm in size, but varying from 0.2–30 mm and have bilateral symmetry. Their distribution range and ecology of various species are often based on water parameters (i.e., temperature, dissolved oxygen, pH, conductivity) and reflect geographical isolation.

One family in the Burchihcura, crustaceans ectoparasites,commonly known as Carp lice or fish lice, is parasitic on fish, with a few on invertebrates or amphibians. With spiny proboscis armed with suckers, they attach to aquatic taxa as juveniles to feed on mucus and sloughed-off scales, or pierce the skin and feed on the internal fluids (Fig. 3.35).

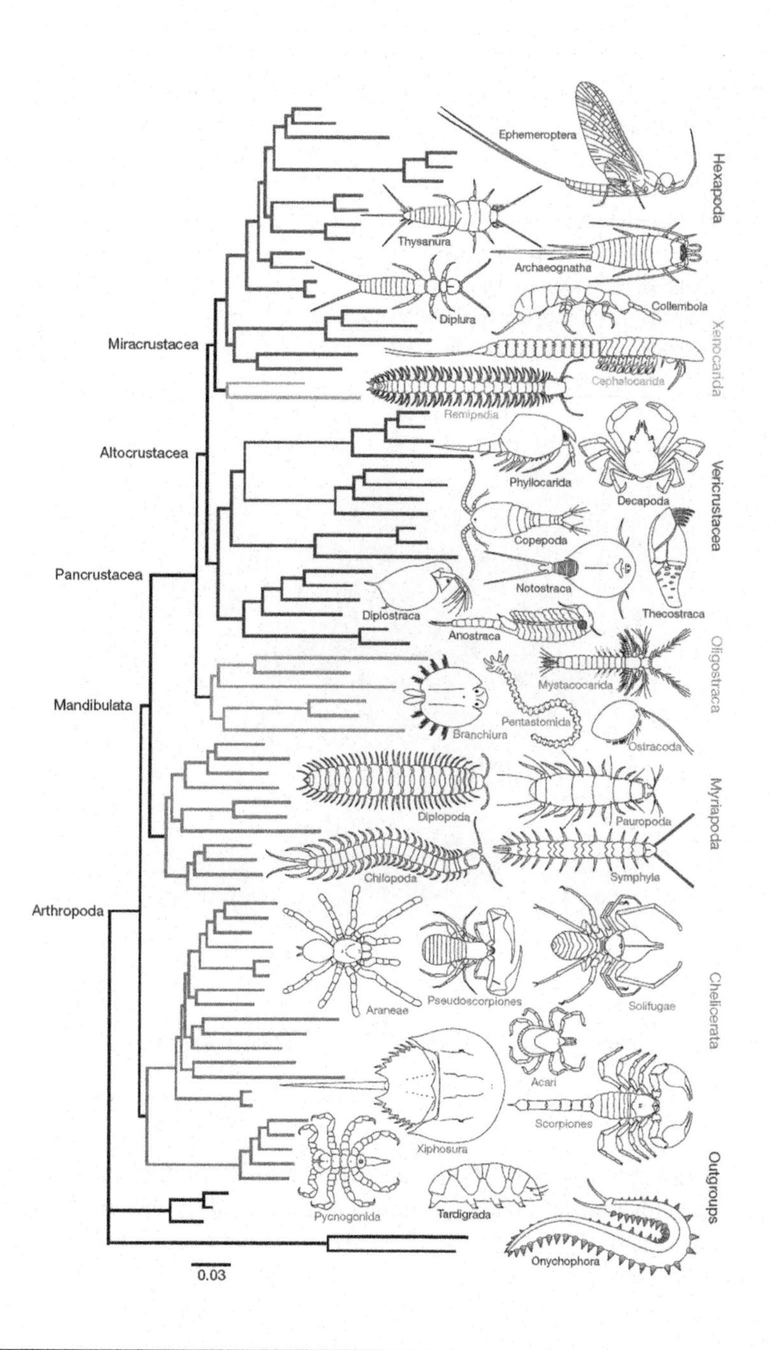


**FIGURE 3.33**

Crustacean phylogeny that includes the hexapods (Source: Regier et al., 2010).

**Table 3.4 Table of arthropod groups used in the following sections.**

| Clade | Common inland water group(s) | Examples |
|---|---|---|
| Chelicerata | | Spiders, water mites |
| Oligostraca | | Ostrocods, Burchihcura |
| Vericrustacea | | |
| Vericrustacea | Anastoca/Cladecera, etc. | |
| Xenocardia | Remipedia, Cephalocarida | |
| Hexapoda | Odonata (Dragonflies) | |
| Hexapoda | Tricoptera (Caddisflies) | |
| Hexapoda | Plecoptera (Stoneflies) | |
| Hexapoda | Diptera (True Flies) | |

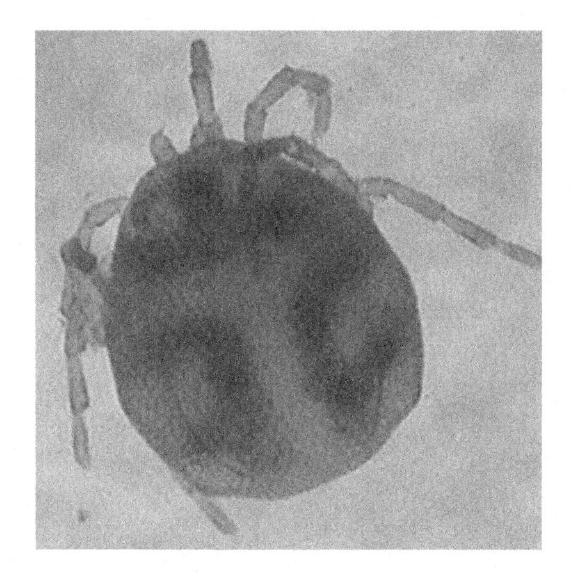

**FIGURE 3.34**

*Arrenurus* sp., a genus of water mites within the family Arrenuridae (Source: Discover Life).

## VERICRUSTACEA

Vericrustaceans includes several groups: Anostraca (fairy shrimp), Phyllopoda, Copepoda, and Malacostraca. They are often filter feeders, ingesting mainly unicellular algae and organic detritus, including protists and bacteria. As part of the larger zooplankton community, they can become quite abundant and even dominate the water column as consumers of algae and bacteria, and as prey for fish, birds, and other aquatic predators.

***Anostraca*** have 20 body segments, bearing 11 pairs of leaf-like phyllopodia (swimming legs), and the body lacks a carapace (Fig. 3.36). They swim "upside-down" and feed by filtering organic particles from the water or by scraping algae from surfaces.

**FIGURE 3.35**

*Argulus* sp., an ectoparasite—parasites that attach to the exterior of their hosts, shown here attached to stickleback, Gasterosteidae (Source: Michal Grabowski).

They live in vernal pools and hypersaline lakes across the world, including pools in deserts, in ice-covered mountain lakes and in Antarctica. Fairy shrimp can serve as indicator of species richness.

Brine shrimp (*Artemia salina*) live in saline lakes, often in endorheic basins, where they can be geographically isolated, producing highly endemic genotypes.

***Phyllopoda*** include Cladocera (water fleas), Spinicaudata (clam shrimp), Laevicaudata (clam shrimp), and Notostraca (tadpole shrimp).

Cladocera includes the genus *Daphnia*. *Daphnia* are a planktonic crustaceans (Fig. 3.37). As primary consumers, Cladocera played a central role the Mesozoic Lacustrine Revolution, when a major change in the freshwater ecosystem changed with the appearance of predator-prey relationships that cladocerans effectively exploited. The Cladocera exhibit a striking morphological stasis over millions of years, yet they have overcome significant ecological barriers and they continue to show remarkable plasticity.

Nostraoca or tadpole shrimp are omnivores living on the bottom of temporary pools and shallow lakes. Notostracans are omnivorous, eating small animals, such as fishes and fairy shrimp. They occur in freshwater, brackish water or saline pools, as well as in shallow lakes, peat bogs and wetlands. In Californian rice paddies, the species *Triops longicaudatus* is considered a pest, because it stirs up sediment, preventing light from reaching the rice seedlings (Fig. 3.38).

***Copepoda*** are typically 1–2 mm long, with a teardrop-shaped body and large antennae. Some polar species can reach 1 cm in length! Like other crustaceans they have an armored exoskeleton, but copepods are so small that in most species this thin armor, and the entire body, is almost totally transparent. Copepods have a compound, median single eye, usually bright red and in the center of the transparent head. Common genera include Cyclops, Mesocyclops, Acartia, Diaptomus, Calanus, Megacyclops, Neocalanus, Tropocyclops, and Acanthocyclops.

When conditions are good, copepods populations can increase quickly due to flexible life history characteristics. Food quality can influence life history traits and reproduction in *Boeckella*

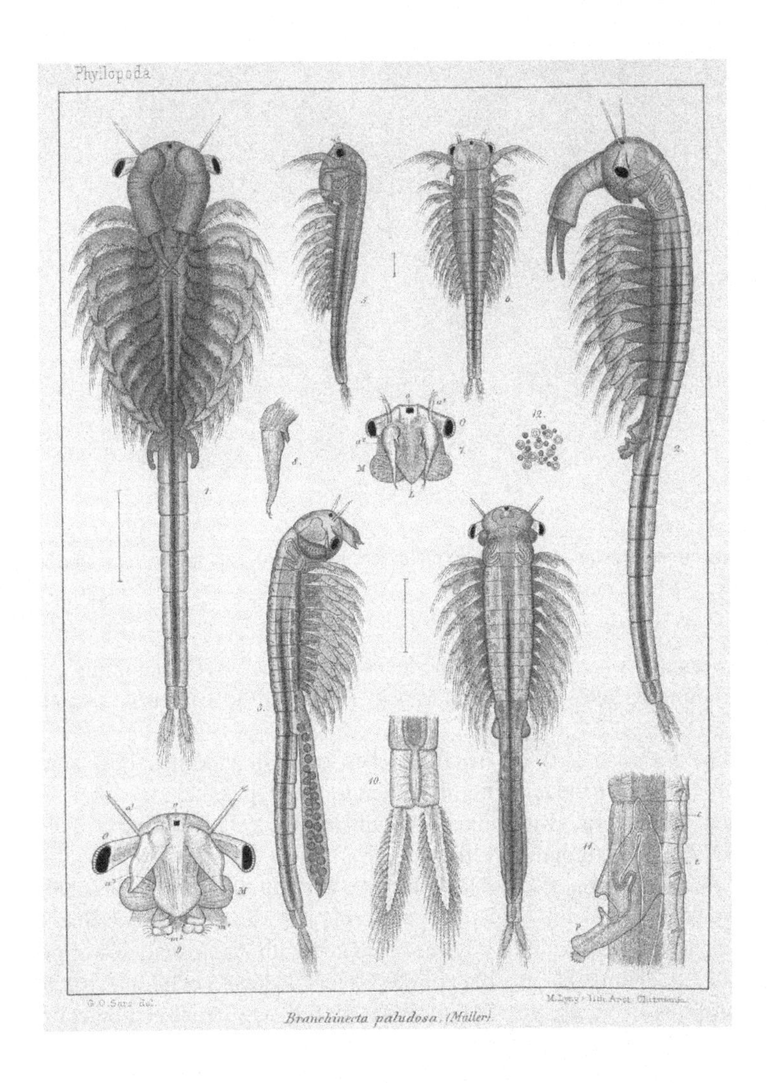


**FIGURE 3.36**

Sketches of fairy shrimp; showing female with eggs (Source: Wikicommons).

*triarticulata* (Fig. 3.39), making it a challenge to understand population dynamics. In another example of copepod reproductive flexibility, cyclopoids (e.g., *Cyclops bicuspidatus*, *Cyclops leuckarti*, and *Cyclops bicolor* can develop into resting stages or enter a dormant condition as a response to environmental conditions.

***Malacostraca*** include Amphipods, Thermosbaenacea, Mysids, and Isopods. Malacostraca is the largest of the six classes of crustaceans, containing about 40,000 living species, divided among 16 orders. Its members display a great diversity of body forms and include crabs, lobsters, crayfish, shrimp, krill, woodlice, amphipods, mantis shrimp and many other, less familiar animals.

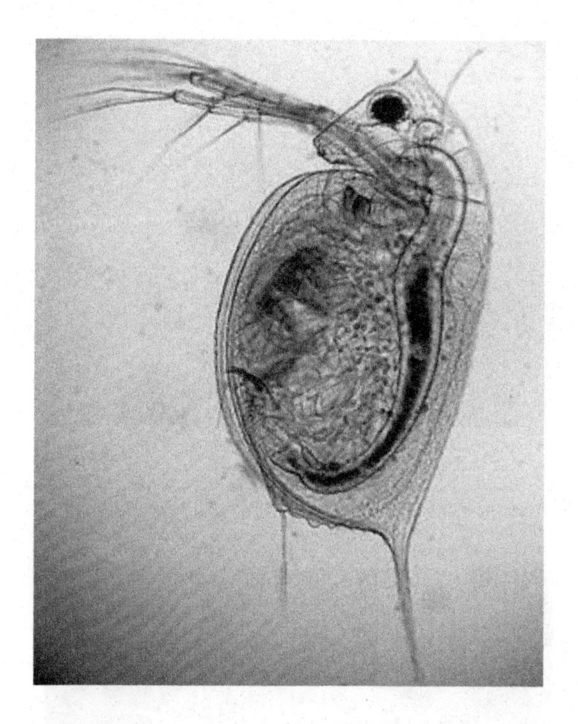

**FIGURE 3.37**

*Daphnia* sp. (Source: Wikipedia).

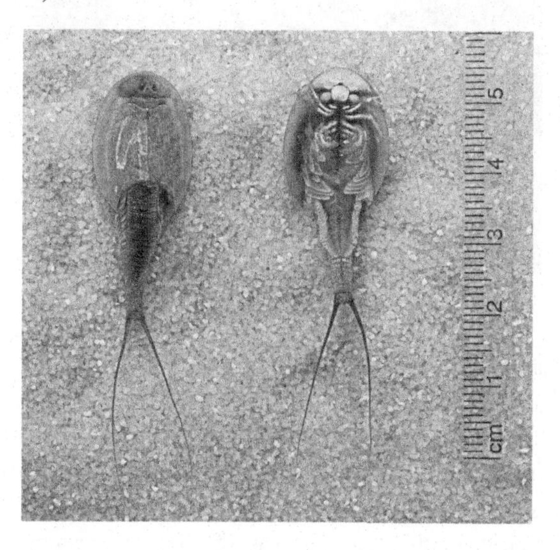


**FIGURE 3.38**

*Triops longicaudatus*, tadpole shrimp, is a freshwater crustacean of the order Notostraca, resembling a miniature horseshoe crab (Source: Micha L. Riese).

**FIGURE 3.39**

The Copepoda, *Boeckella triarticulata* (Source: Centre for Biodiversity Genomics).

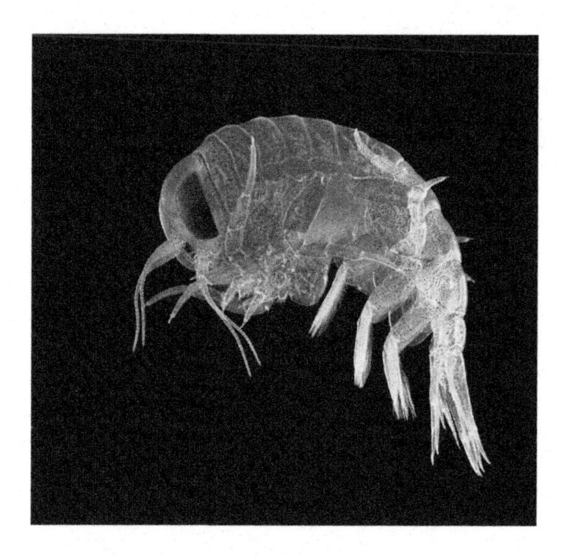

**FIGURE 3.40**

Hyperia sp. of the order Amphipoda (Source: Uwe Kils).

***Amphipoda*** is an order of malacostracan crustaceans with no carapace and generally with laterally compressed bodies (Fig. 3.40). Of the 7,000 species, 5,500 are classified into one suborder, Gammaridea, often referred to as a freshwater shrimp.

***Thermosbaenacea*** is a group of crustaceans that live in thermal springs in freshwater, brackish water, and anchialine habitats. Due to their troglobitic lifestyle, thermosbaenaceans lack visual pigments and are therefore blind (Fig. 3.41).

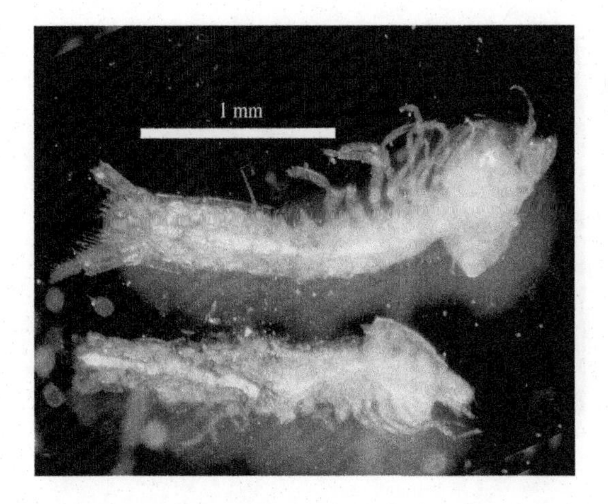


**FIGURE 3.41**

Tethysbaena ophelicola (Source: N. Ben Eliahu).

**FIGURE 3.42**

*Hemimysis anomala* (Source: S. Pothoven). The mysid's head bears a pair of stalked eyes and two pairs of antennae. The thorax consists of eight segments each bearing branching limbs, the whole concealed beneath a protective carapace and the abdomen has six segments and usually further small limbs.

***Mysida*** is a small, shrimp-like crustacean (Fig. 3.42). Their common name opossum shrimp stems from the presence of a brood pouch ("marsupium") in females. The fact that the larvae are reared in this pouch and are not free-swimming characterizes the group.

***Isopoda*** includes woodlice and their relatives. Isopods live in the sea, in freshwater, or on land. Females brood their young in a pouch under their thorax. Isopods have various feeding syndromes: some eat dead or decaying plant and animal matter, others are grazers, or filter feeders, a few are predators, and some are internal or external parasites, mostly of fishes. Aquatic species mostly live on the benthos, but some can swim for a short distance (Fig. 3.43).

**FIGURE 3.43**

*Eurydice pulchra*, not a few taxon! (Source: Hans Hillewaert). All have rigid, segmented exoskeletons, two pairs of antennae, seven pairs of jointed limbs on the thorax, and five pairs of branching appendages on the abdomen that are used in respiration.

***Decopoda*** or decapods (literally "ten-footed") include many familiar groups, such as crayfish, crabs, lobsters, prawns, and shrimp. Most decapods are scavengers.

In spite of a perplexing distribution (Fig. 3.44), most of the diversity is found in Canada and the USA, where some 338 native crayfish taxa have been described (Fig. 3.45).

***Thecostraca*** Barnacles—no freshwater members.

## XENOCARIDIA

Xenocaridia comprises two groups that were recently discovered: Remipedia and Cephalocarida. The Remipedian *Speleonectes atlantida* is an eyeless crustacean. It was discovered in August 2009 in the Tunnel de la Atlantida, the world's longest submarine lava tube on Lanzarote in the Canary Islands off the west coast of North Africa. Like other remipedes, the species is equipped with venomous fangs. Although the Xenocaridia has no freshwater species, it is a sister group to Hexapoda and a good reminder that even high-level groups remain undescribed.

# ECDYSOZOA: HEXAPODA

By considering that the Hexapoda are crustaceans, it may be tempting to think of them as aquatic. However, based on the morphology and fossil record, the first hexapods were probably terrestrial—invading the land from the marine environment (Fig. 3.46). This evolutionary transition from aquatic to terrestrial life required adaptations in locomotion, breathing, reproduction, and mechanisms for food capture. The complexity of these changes is an active area of research.

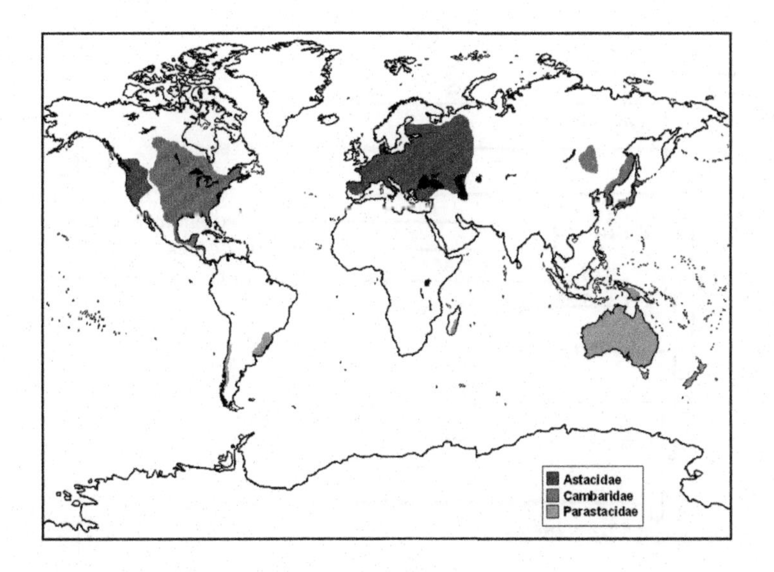


**FIGURE 3.44**

Map of crayfish distribution (Source: Courtesy of Carnegie Museum of Natural History). The Southern Hemisphere (Gondwana-distributed) family Parastacidae lives in South America, Madagascar, and Australasia, and is distinguished by the lack of the first pair of pleopods. Of the other two families, members of the Astacidae live in western Eurasia and western North America and members of the family Cambaridae live in eastern Asia and eastern North America.

**FIGURE 3.45**

Fillicambarus (Source: Keith A. Crandall).

Although there is a general form of the Hexapoda, each order, family, and species has morphological variations. With six legs, the hexapod include insects, as well as, other groups such as the Collembola.

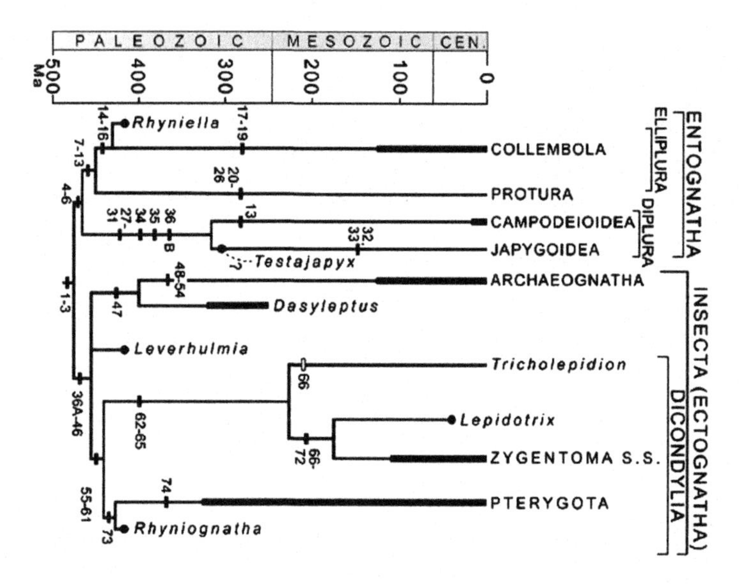


**FIGURE 3.46**

Morphologically based phylogeny of recent lineages of Hexapoda and of early, Paleozoic fossils (Grimaldi, 2010). Within the Hexopoda, two related groups, Entognatha and Ectognatha, probably split around ~480 mya (Fig. 3.46). Entognatha have no wings and contain the Collembola (springtails), Diplura, and Protura, which evolved independently from insects and each order. Within the Ecotognatha, the major extant lineages include Archeognatha (jumping bristletails), Zygentoma (e.g., silverfish), and Pterygota (winged insects).

In spite of the terrestrial origin, many insect groups have obligate aquatic larvae life-history stages and emerge as adults that can fly, mate, and then lay eggs in or near surface waters, so their larvae can develop on the bottom of a lake or stream.

## COLLEMBOLA: SEMIAQUATIC HEXAPOD

Collembola are not insects, but have a similar body form (Fig. 3.47). A subset of the 700 Collembola species are semiaquatic, in other words, they do not have specialized adaptations as a fully aquatic taxon. Nevertheless, they are commonly found in the neuston, i.e., the air-water interface. For example, *Podura aquatica* is exclusively aquatic species, but lives on the surface of the water as a scavenger.

The squat, bluish, grey *Podura aquatica* has a large, flat furcula, which allows it to effectively jump on water without breaking the water surface tension. Being sensitive to water surface disruptions, *Podura aquatica* can locate and feed on detritus on the surface of still waters.

## ORDER: EPHEMEROPTERA AND GILL MORPHOLOGY

In contrast to the Collembola, mayflies or Ephemeroptera, are adapted to water in the larval stage. Ephemeroptera are often the most abundant insect in streams. The larvae (instars) molt in the water 10–20 times, which might take few months up to a year or more to complete, whereas the adult stage

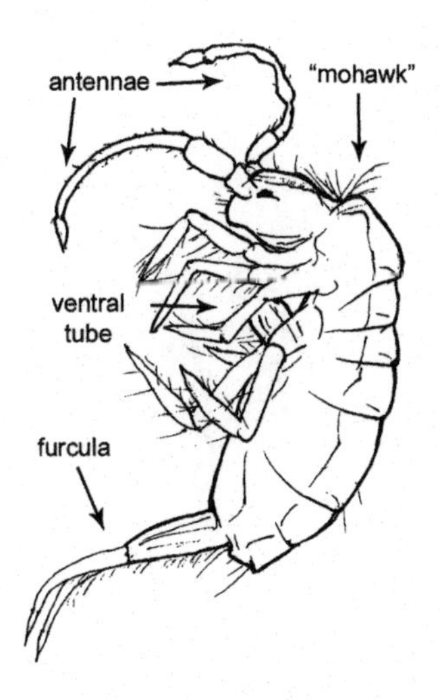


**FIGURE 3.47**

Sketch of collembola morphology (Source: Courtesy of Derek Tan). Some of the taxa have a bifurcate jumping organ (furcula), which is held in place by a tenaculum.

can last less than a few hours to a few days. Larvae can be rapid swimmers, but generally crawl around on the substrate. In the last instar, dark wingpads are visible. When they emerge they may have a short subimago (immature) stage before becoming imago (reproductively mature adults). Without functional mouthparts, these adults are designed to disperse and mate.

Most mayfly nymph have seven pairs of conspicuous and distinctive tracheal gills on the abdomen that have been implicated in gas exchange (i.e., oxygen for respiration) and salt balance. Gills may be located on the ventral, lateral, or dorsal side of the abdomen and on abdominal segments—1 through 7—although they may be missing from one or more segments (Fig. 3.48).

## ORDER: ODONATA AND INSECT DISPERSAL

A Paleozoic formation in Oklahoma from the Permian boast of "giant dragonflies." *Meganeuropsis permiana* had a water-dependent life history stage and a wing span up to 71 cm! Although not directly related to modern dragonflies, it holds the record as the largest insect.

Nevertheless, dragonflies and damselflies are large insects and members of the Odonota. Of all the aquatic insects, these may be the most charismatic from a human perspective. Cultural references to dragonflies through art and literature are extensive. In addition, conservation efforts for dragonflies are becoming more widespread and well-known, and we will highlight these in Chapter 12.

Adults are most often seen near bodies of water. Many adults range far from water. Males tend to emerge and establish a territory near water to attract a female. Females generally occupy areas outside the margins of waters until they are ready to mate (Fig. 3.49). After mating, eggs are laid in water or on vegetation near water or wet places that develop into instars, with approximately 9–14 molts that are (in most species) voracious predators on other aquatic organisms, including small fishes. The larvae grow and molt and ultimately transform into reproductive adults.

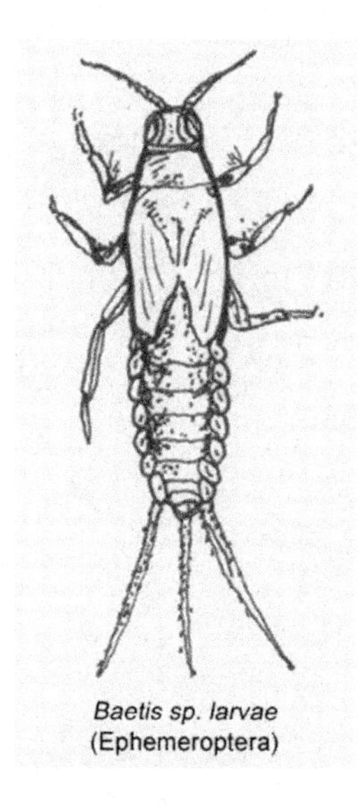


**FIGURE 3.48**

Baetis sp. in the ephemeroptera (Source: Halvard Hatlen). Most possess three tail-like structures at the end of their bodies, composed of two long cerci and one caudal filament. Some species, for example genus *glsEpeorus* sp., have only two cerci.

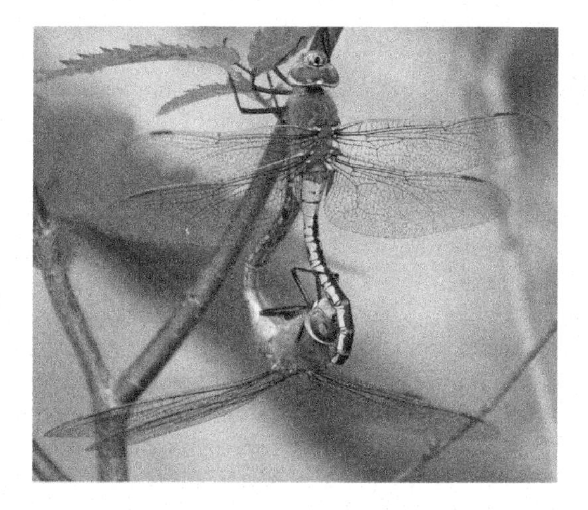

**FIGURE 3.49**

*Anax junius*, common green darner, in the "wheel" mating position (Source: Henry Hartley). The common green darner can migrate, but can also be a year-around resident.

**FIGURE 3.50**

*Limnephilus flavicornis* made use of snail shells to build its case (Source: Wikipedysta:Głodny).

Odonata and the other aquatic insects depend on migration to find suitable habitats. In ephemeral waters or to avoid competition, invertebrates are carried downstream in what is called behavioral drift. As adults these insect can fly upstream to new streams. But dragonflies can also migrate seasonally. Dragonfly migrations have been noted since the 18th Century. They can travel at speeds up to 48 km/h, which rivals many birds. They usually remain relatively close to inland waters and avoid open water. These migrations probably have an important role, but we do not know much about where or why they migrate, factors that deserve more research.

## ORDER: TRICOPTERA–HOUSE BUILDERS AND PREY AVOIDANCE

Tricoptera or caddisflies larvae can be found in lotic and lentic habitats. Many caddisfly larvae build protective cases around their bodies that consist of fine particles, leaves, pebbles, and other materials (Fig. 3.50). Some species of caddisflies can be identified by the patterns and the materials used for their cases. Most species living in portable cases are herbivores and scrape algae and bacteria from rocks, whereas free-living Tricopterans are often predacious on other arthopods. Case-building caddisflies can have a substantial grazing effect on the benthic and other algae, and after avoiding predation can export that production to the surrounding terrestrial habitat, linking the aquatic and terrertial ecosystems.

## ORDER: PLECOPTERA AND COLD WATER HABITATS

Plecoptera (stoneflies) have distinct north and south hemisphere lineages, in spite of some mixing across the equator. Stoneflies are believed to be one of the most primitive groups of Neoptera, with close relatives identified from the Carboniferous and Lower Permian geological periods. The modern diversity, however, apparently is of Mesozoic origin.

In contrast to most Ephemeroptera that have three tail-like appendages, the Plecoptera have two circi (Fig. 3.51). Plecoptera are herbivores and are generally found in cold fast-moving streams or well-aerated lakes.

Along with the Ephemeroptera, Tricoptera, the Plecoptera are often good indicators of cool, well oxygenated waters. In fact, these taxa are used as indicators of high water quality, and their abundance is quantified as the EPT index (Ephermoptera, Plecoptera, Tricoptera).

**FIGURE 3.51**

Golden stonefly larvae, Plecoptera, Perlidae (Source: Böhringer Friedrich).

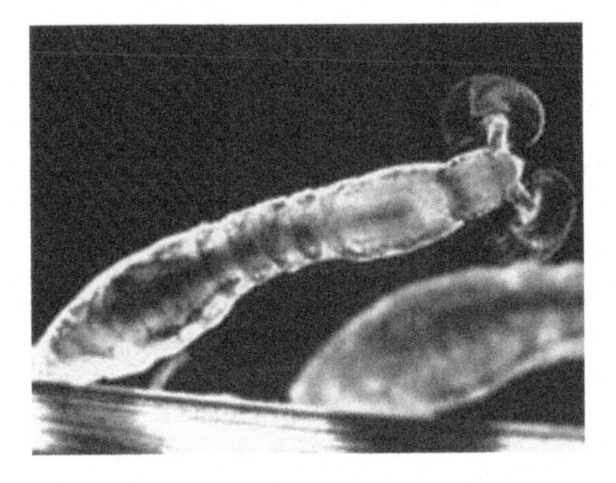

**FIGURE 3.52**

*Simulium* sp. (Source: Warren Photography) transmits Onchocerciasis, also known as river blindness, is a disease caused by infection with the parasitic nematode *Onchocerca volvulus*. Symptoms include severe itching, bumps under the skin, and blindness. It is the second-most common cause of blindness due to infection, after trachoma.

## ORDER: DIPTERA—DETRITIVORES AND DISEASE VECTORS

Diptera are known by entomologists as "true flies" and possess a pair of wings on the mesothorax and a pair of halteres (modified, tiny wings), derived from the hind wings. The following groups: midges, black flies (Fig. 3.52), and mosquitoes have important roles in aquatic ecosystems. As larvae Diptera play an important role in the processing of organic matter, whether capturing FPOM in the water column or processing organic matter in the benthos. As adults, however, a few taxa are vectors of malaria, west nile virus, some of the most pressing public health concerns.

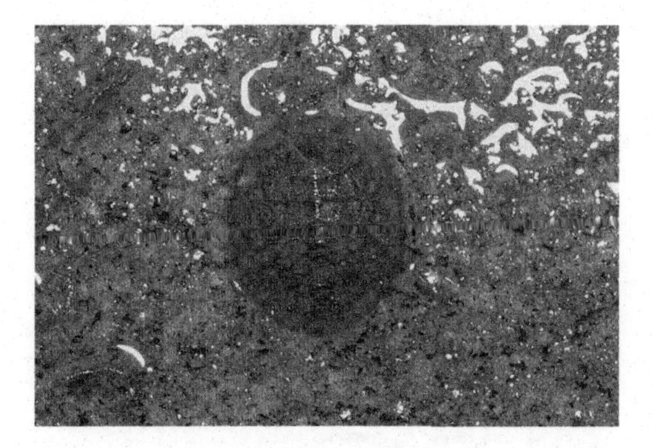

**FIGURE 3.53**

Image of water-penny bettle (Source: TheAlphaWolf). Water-penny beetles are a family (the Psephenidae) of 272 species (in 35 genera) of aquatic beetles found on all continents, in both tropical and temperate areas. The young, which live in water, resemble tiny pennies. The larvae feed—usually nocturnally—on algae on rock surfaces. The presence of water penny larvae in a stream can be used as a test for the quality of the water, as they are pollution-sensitive. They cannot live in habitats where rocks acquire a thick layer of algae, fungi, or inorganic sediment. Therefore their presence along with other diverse phyla signifies good-quality water. They are around 6 to 10 millimeters in length.

## MEGALOPTERA: BIG PREDATORS

Megaloptera (alderflies, dobsonflies, and fishflies) are relatively unknown insects across much of their range, due to the adults' short lives. The larvae are carnivorous and can feed on small invertebrates, such as crustaceans, clams, worms, and other insects. The larvae possess strong jaws that they use to capture their prey, but grow slowly, taking anywhere from 1 to 5 years to reach the last larval stage. When they reach maturity, they crawl out onto land to pupate in damp soil or under logs. For many species the adults never feed and live only a few days or hours. Dobsonfly larvae, commonly called hellgrammites, are often used for angling bait in North America.

## COLOEPTERA: BEETLES

A water beetle is a generalized name for any beetle that is adapted to living in water at any point of its lifecycle. Most water beetles can only live in freshwater, although there are a few marine species that live in the intertidal zone or littoral zone. There are approximately 2000 species of water beetles (Fig. 3.53).

Many adult water beetles carry an air bubble, called the elytra cavity, underneath their abdomens, which provides an air supply, and prevents water from getting into the spiracles. Others have the surface of their exoskeleton modified to form a plastron, or "physical gill," which permits direct gas exchange with the water. Some families of water beetles have fringed hind legs adapted for swimming, but most do not. Most families of water beetles have larvae that are also aquatic; some have aquatic larvae and terrestrial adults.

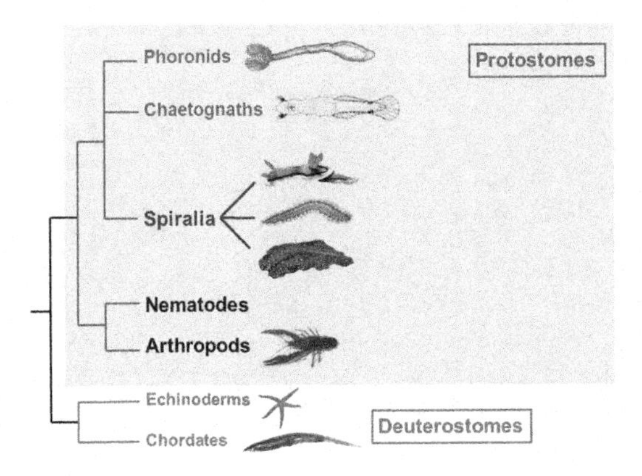


**FIGURE 3.54**

Diagram of the spiralia group of taxa and relatives (Source: Rosamond Marshall).

## OTHER AQUATIC INSECTS

With 23 insect orders, this summary of 8 is far from adequate. Other orders also have important members in the aquatic system that include the Hymenoptera (bees and wasps), Lepidoptera (butterflies and moths), and Neuroptera (lacewings). Furthermore, because of their importance in bioassessment (Chapter 10) and trophic interactions (Chapter 6), aquatic entomology is a compelling area to work.

## SPIRALIA

## PROTOSTOMES: SPIRALIA

The last group of protostomes to cover and have inland water members are the Sirialia. In general, they share patterns in early embryonic development and include molluscs, annelids, platyhelminths, and diverge from other Protostomes (Fig. 3.54). Although their greatest diversity lies in the marine environment, many groups are found in freshwaters and terrestrial ecosystems.

The mollusc composes the largest group of invertebrate animals with around 85,000 recognized and extant, mostly marine species. They are highly diverse, not just in size and in anatomical structure, but also in behavior and in habitat. The gastropods (snails and slugs) are the most numerous in terms of classified species and account for 80% of the molluscs.

## GASTROPODS

Gastropod molluscs are one of the most diverse groups of animals. Estimates of total extant species range from 40,000 to over 100,000, but may be as high as 150,000. Ecologically, gastropods play an important role as grazers—often grazing periphyton and epiphyton. In some cases, they can dominate the benthos, e.g., the New Zealand mudsnail, *Potamopyrgus antipodarum,* has dramatically altered species composition where it has invaded.

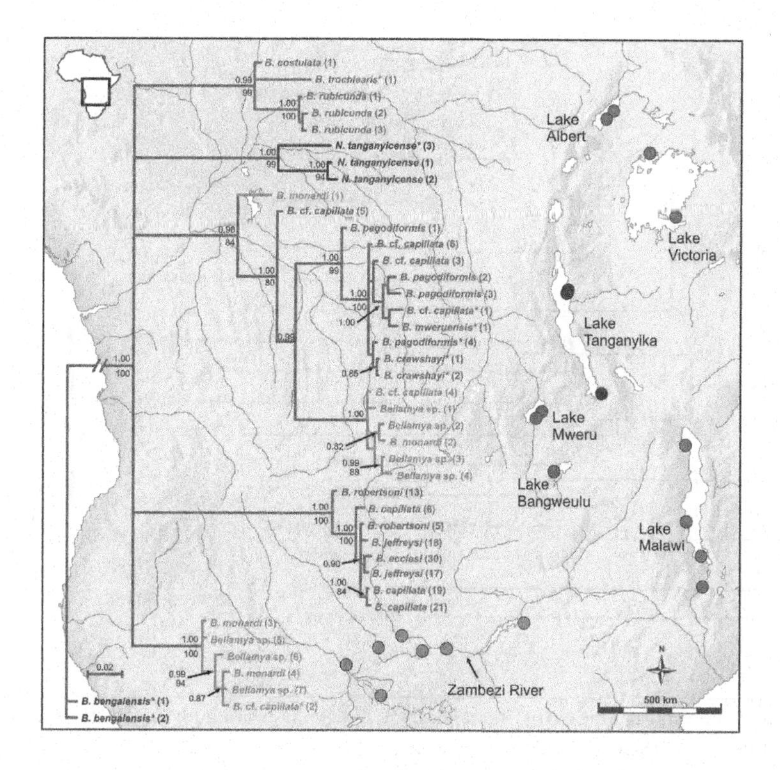


**FIGURE 3.55**

Bellamya speciation in Lake Malawi (Source: Schultheß et al., 2010).

The gastropod shell is an evolutionary success story. It protects soft body from predators, but may also have forced the development of the craniata—skulls that allow muscles to leverage a powerful jaw to crush the gastropod shells. Thus the success of the shell, drove skull development, a characteristic passed to human ancestors.

Gastropods have figured prominently in palaeobiological studies (see Chapter 1). For example, *Bellamya* is an important lentic gastropod genus found in Kenya, Tanzania, and Uganda. Changes in climate changes in the African rift-lake, Malawi, led to the extinction of all but one evolutionary lineage of the Pliocene Bellamya fauna in the lake (Fig. 3.55). After severe Pleistocene low stand and reflooding of the lake may have allowed a spectacular speciation processes in the modern endemic Bellamya radiation in Lake Malawi.

## BIVALVES

Like Grastropods, most Bivalves or mussels and clams are marine taxa, but a number inhabit freshwaters or inland waters. The systematics of mussels and clams suggests that 6 main groups exist (Fig. 3.56).

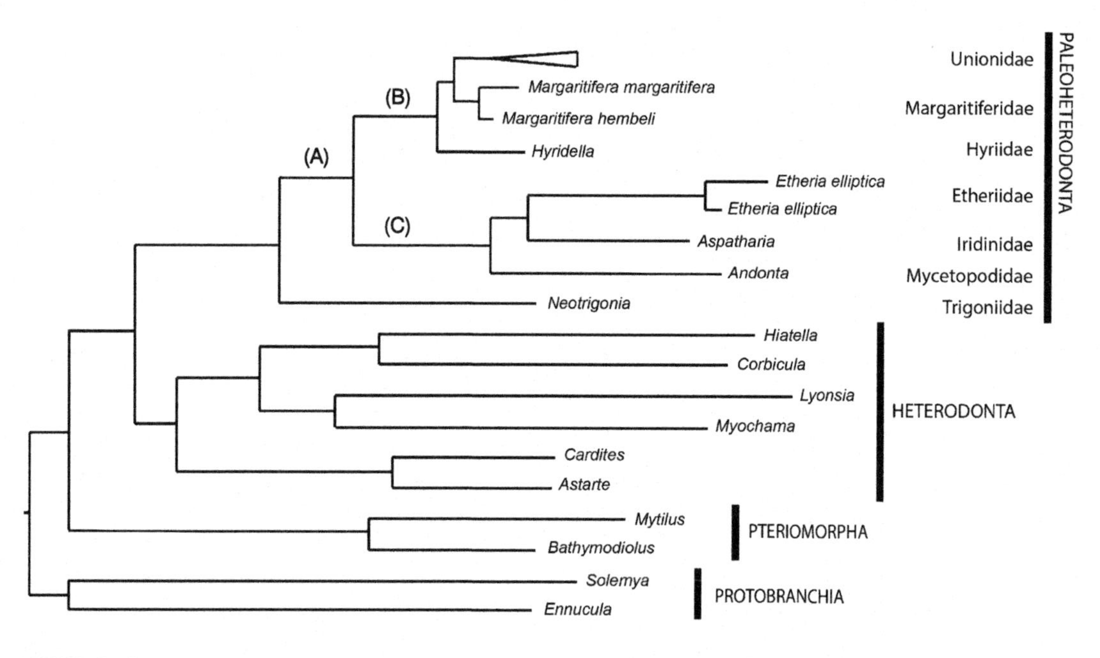


**FIGURE 3.56**

Bivalve phylogeny. (A) Freshwater mussels (Unionoida); (B) Glochidia-bearing mussels (Unionoidea); (C) Lasidia-bearing mussels (Etheroidea) (Source: Pfeiffer et al., 2019).

Freshwater mussels and clams have two shells connected by a hinge-like ligament. Mussels and clams live in a variety of freshwater habitats, but are most prevalent in stream and rivers. Adult sizes range from 5 mm to 22 cm. The wide variety of shapes and colors are reflected in species, such as purple wartyback, pink heelsplitter, and threeridge. And of course, we have already been introduced to the Cumberland monkeyface pearly mussel.

On the stream bottom, they are sometimes only noticeable by two small siphons, which are used to draw and expel water. When dislodged, a large muscular foot is used to move through the stream bottom. To reduce the changes of being dislodged, mussel habitat value can be measured as a function of shear stress (see Section Navigating Shear Stress: Freshwater Mussels).

In addition, exotic species are regularly encountered in estuarine habitats. As described at the beginning of the chapter, the zebra mussel, *Dreissena polymorpha*, is a famous invader in North America.

The lifecycle of the freshwater mussel is rather complex and depends on river currents carrying sperm and fish to disperse an early life stage, glochidia, that attaches to fish gills (Fig. 3.57). Thus mussels are able to fertilize eggs and disperse upstream in lotic systems in spite of being a lousy swimmer. To optimize growth, developing mussels select habitat with access to food via water currents, but without being swept away by shear stress and lose the ground gained as glochidia.

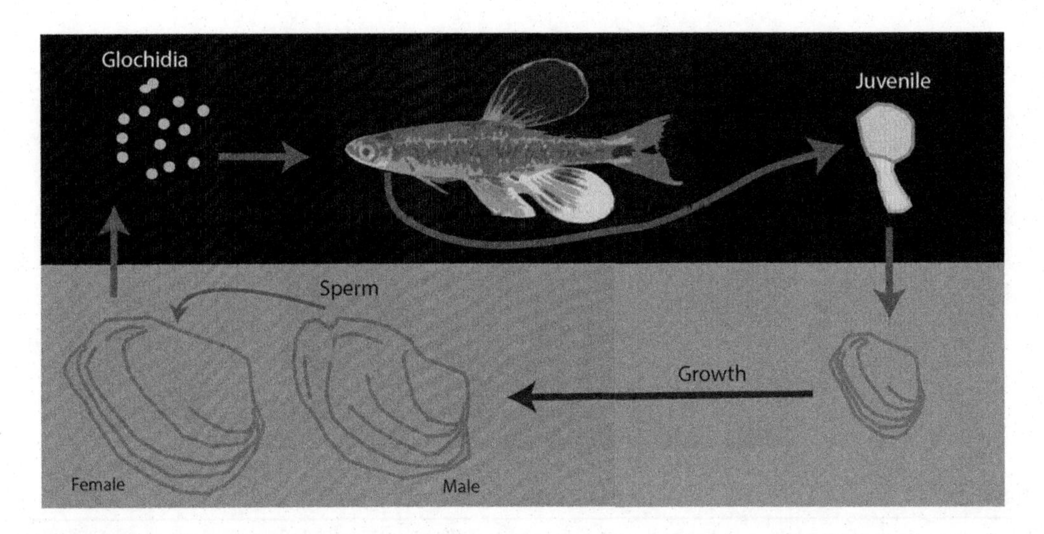


**FIGURE 3.57**

Mussel lifecycle (Source: Rostovskaya Marina). Unlike other animals that move and actively search for a mate, the sedentary mussel depends on the river currents and fish to reproduce. The process begins with the male releasing sperm, and the female located downstream drawing sperm in through her intake siphon so internally held eggs can be fertilized. Numbering in the 100s to hundreds of thousands, the fertilized eggs develop into glochidia within her gills. Once mature, they are released into the water column to begin the second part of their lives—attaching to the gills, fins, or scales of freshwater fishes.
Over several weeks, glochidia begins to develop gills, a foot, and other internal structures to become juveniles. Now fully transformed, but still microscopic, each juvenile will drop off the fish and begin its life on the stream bottom. Unbeknownst to the fish, it has just served as a taxi transporting the young mussel into new habitat away from its parent. If the mussel is lucky enough to grow into an adult, it may live 20–100 years or more, depending on the species.

## ANNELIDS, PLATYHELMINTHES, AND OTHER SPIRALIA

Ringed worms (Annelida), flat worms (Platyhelminthes), and a few other Spiralia can easily be overlooked in aquatic systems, because they are hard to sample from interstitial muds. But leeches and Oligochaetaes can be important animals in inland waters. For example, some leeches feed on vertebrate blood and invertebrate hemolymph, whereas most are predatory and eat other invertebrates. Oligochaetes occur in interstitial muds in lake and stream bottoms and probably play an important role in nutrient cycling, but few studies have qualified their importance.

*Schistosoma* is a genus of trematodes, commonly known as blood-flukes and a Platyhelminthe. To complete its lifecycle, Schistosoma must pass through a mammal and a snail host. This parasitic flatworm is responsible for schistosomiasis. After malaria, schistosomiasis is one of the most devastating parasitic disease infecting hundreds of millions worldwide (Fig. 3.58).

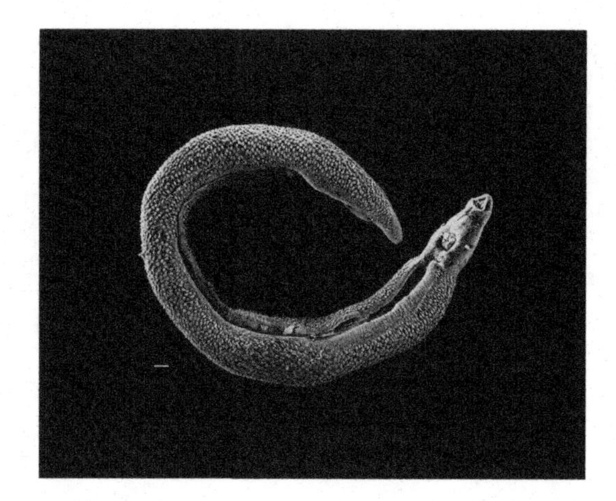

**FIGURE 3.58**

Electron micrograph of an adult male *Schistosoma* parasite worm. The bar (bottom left) represents a length of 500 μm (Source: David Williams).

# DEUTEROSTOME

## DEUTEROSTOMES AND FRESHWATER ANIMALS

Deuterostomes ("second mouth") are distinguished from protostomes by their embryonic development; for example, the first opening (the blastopore) becomes the anus in deuterostomes, whereas the first opening becomes the mouth in protostomes (Fig. 3.59). The divergence between these developmental patterns probably occurred over ~700 mya.

There are two major clades of Deuterostome, but only the Vertabrata have freshwater members. And within the Vertebrata, only the Chorates, which include Cyclostomes (Myxini (hagfish) and Petromyzontida (lampreys and allies)), and Gnathostomata (jawed vertebrates), have freshwater members. Ultimately, the jawed-fish would diverge into tetrapods and mammals (Fig. 3.60).

## JAWS AND INLAND WATER INVASIONS

Historically, vertebrates were thought to have evolved in the oceans, but this assumption has become something of a controversy. Some argue that the development of the jaw and backbone coincided with the invasion of freshwater.

Whereas the timing and mechanism of the jaw development has not been resolved, it has been hypothesized that the development of the jaw was driven to improve predatory success. However, our understanding of evolution of the jaw keeps changing, relying on newly discovered fossils in China, using more sophisticated understanding of the genetic controls, and analyzing detailed patterns embryo development from a wide range of extant taxa.

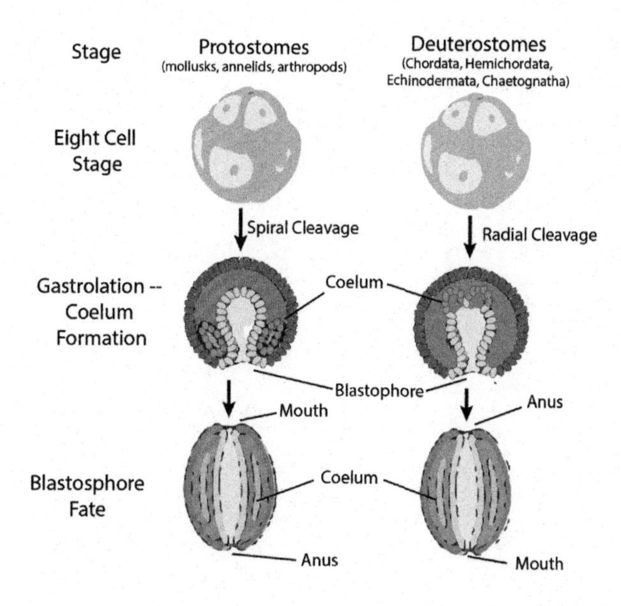


**FIGURE 3.59**

Duetersome and Protosome Comparison. Deuterostomes are also known as enterocoelomates because their coelom develops through enterocoely (Source: Jerry Coyne).

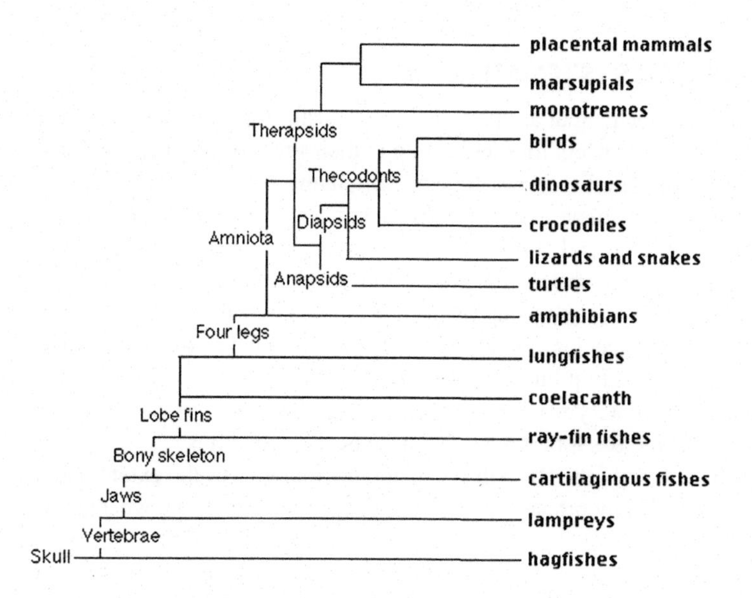


**FIGURE 3.60**

The vertebrate clade includes several taxa that occur in inland waters (Source: Courtesy of John W. Kimball). Of the extant jawless fishes, only lampreys have freshwater taxa.

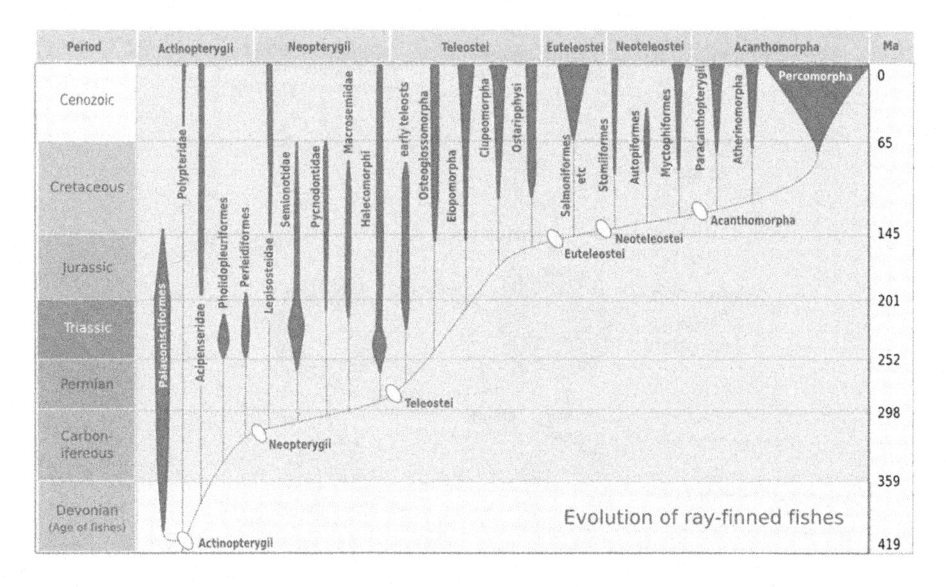


**FIGURE 3.61**

Evolution of ray-finned fishes, Actinopterygii, from the Devonian to the present as a spindle diagram. The width of the spindles are proportional to the number of families as a rough estimate of diversity. The diagram is based on Benton, M.J. (2005). Vertebrate Palaeontology, Blackwell, 3rd edition, Fig. 7.13 on page 185.

## EVOLUTION AND FISH SYSTEMATICS

Given the novel developments that are part of the successful repertoire in the Deuterostomes, we can begin to highlight specific barriers for these group in their invasion of inland waters. Since most of the planet's vertebrate species are fish, we will begin with them.

Fish are cold-blooded aquatic animals with backbones, gills, and fins. Many are torpedo-shaped (fusiform) and can travel efficiently through water, but fish also come in a variety of other shapes—from flattened and rounded, as in flounders, to vertical and angular, as in sea horses (Fig. 3.62).

Over 40% of the world's extant fishes occur in freshwaters, which accounts for ~208 fish families. Of the planet's 27,977 fish species, about 11,952 can be classified as freshwater dwellers.

Fish are generally divided up into cartilaginous (sharks and rays), ray-finned fish (Actinopterygii) and lobe-finned fish (Sarcopterygii). Ray-finned fish have fins with webs of skin supported by bony or horny spines ("rays"), whereas lobe-finned fish have fleshy, lobed fins.

Ray-finned fish evolved in the late Siluarian, and the majority of the extant fish are ray-finned fish. Even with their success, many branches have been pruned by extinction, whereas others radiated after extinction events (Fig. 3.61).

A member of the lobe-finned fish, the lungfish's greatest diversity was in the Triassic period; fewer than a dozen extant genera exist. The lungfish evolved the first proto-lungs and proto-limbs. The first tetrapodomorphs, which included the gigantic rhizodonts, had the same general anatomy as the lungfish, which were their closest kin, but they appear not to have left their water habitat until the late Devonian Epoch (385–359 Ma), with the appearance of tetrapods (four-legged vertebrates).

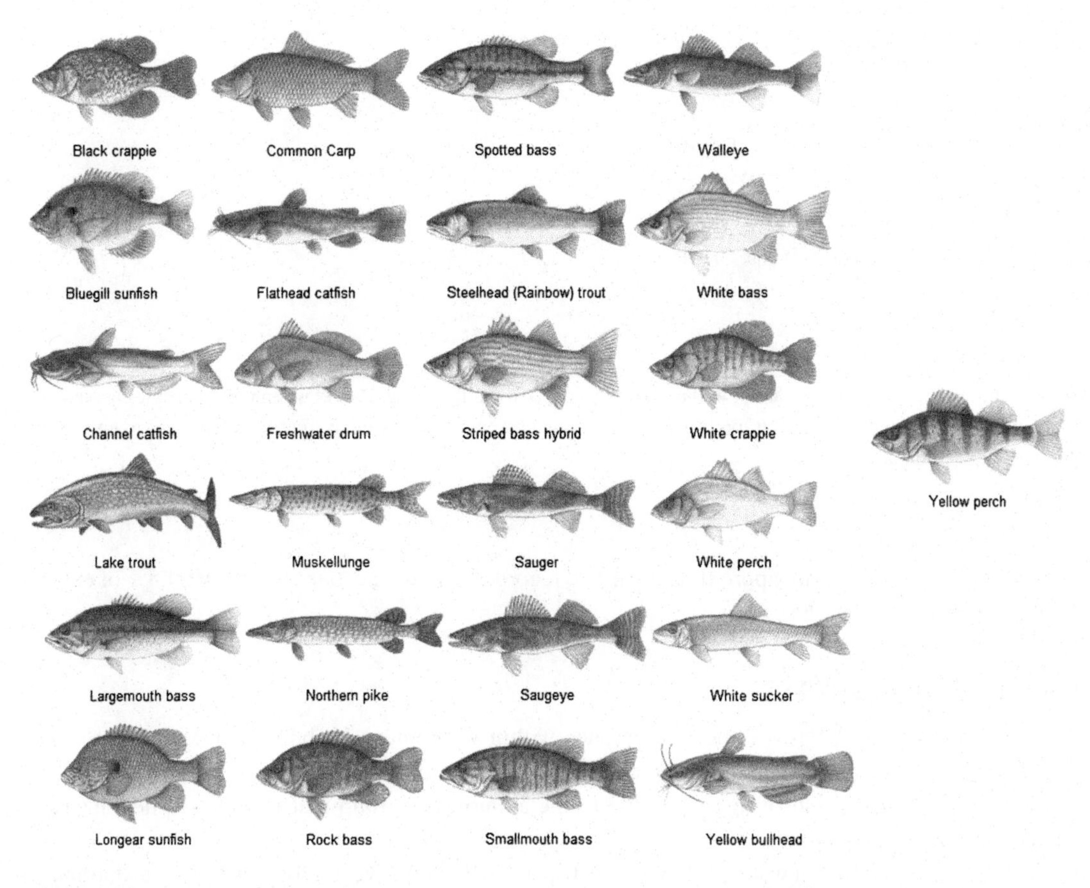


**FIGURE 3.62**

Common freshwater lake fish caught by anlgers (Source: Pinterist).

## TETRAPOD AND AMPHIBIANS

Animals with four appendages or tetrapod fossils appear by the late Devonian, 367.5 mya (Fig. 3.63). Although tetrapods and fish share a common ancestry, the specific aquatic ancestors and exact process by which they colonized the earth's land continues to be an active area of research and debate among palaeontologists.

Like modern amphibians, early tetrapods depended on aquatic systems for much of their life history. In particular, their eggs and early life history stages were water-dependent. Modern salamanders and frogs vary in their dependency on water. Some species are fully aquatic throughout their lives, some take to the water intermittently, and others are entirely terrestrial as adults.

Salamander diversity is most abundant in the Northern Hemisphere and most species are found in the Holarctic ecozone, with some species present in the Neotropical zone. Frogs are widely distributed, ranging from the tropics to subarctic regions, but the greatest concentration of species diversity is in

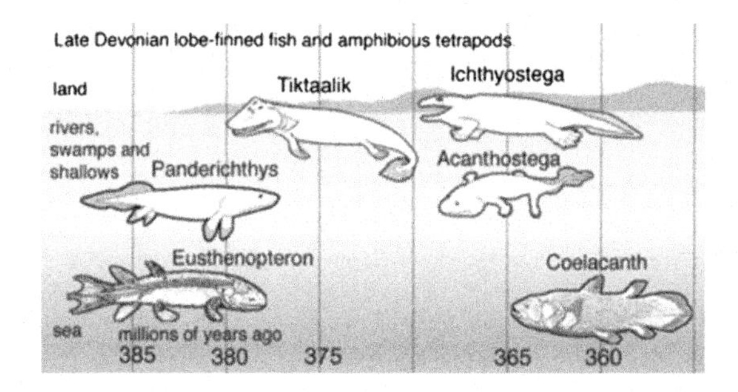


**FIGURE 3.63**

The evolution of tetrapods began about 400 million years ago in the Devonian Period when the earliest tetrapods evolved from lobe-finned fishes. Tetrapods include all living and extinct amphibians, reptiles, birds, and mammals (Source: Dave Souza).

tropical rainforests. There are approximately 4,800 recorded species, accounting for over 85% of extant amphibian species (Fig. 3.64).

## AMNIOTES AND REPTILES

The amniotic egg represents a critical divergence within vertebrates: enabling animals to reproduce on dry land. The egg has a shell to help prevent drying, and a series of membranes that surround the developing chick. This kind of egg is unique to the amniotes, a group that includes turtles, lizards, birds, dinosaurs, and mammals.

Free from reproducing in water, the amniotes dispersed globally, eventually to become the dominant land vertebrates in the Carboniferous Period. Very early in the evolutionary history of amniotes, basal amniotes diverged into two main lines, the synapsids and the sauropsids, both of which persist into the modern era.

One group of amniotes diverged into the reptiles, which includes lepidosaurs, dinosaurs, crocodilians, turtles, and extinct relatives, whereas another group of amniotes diverged into the mammals and other now extinct groups. The amniotes include the tetrapods that further evolved for flight—such as birds from among the dinosaurs, and bats from among the mammals (Fig. 3.65).

The reptiles associated with inland waters include the following orders: Squamata (snakes and lizards), Testudines (turtles), and Crocodilia (crocodiles, alligators, and caiman).

Snakes are often associated with aquatic habitats with a quite "flexible" body design. In fact, various lineages have radiated from lizards with legs to snakes without legs. The evolution of lizards to snakes has occurred independently some 26 times!

*Agkistrodon piscivorus* is a venomous snake, a species of pit viper, found in the southeastern United States. This is the world's only semiaquatic viper, usually found in or near water, particularly in slow-moving and shallow lakes, streams, and marshes. The snake is a strong swimmer and has even been seen swimming in the ocean, and successfully colonized islands off both the Atlantic and Gulf coasts.

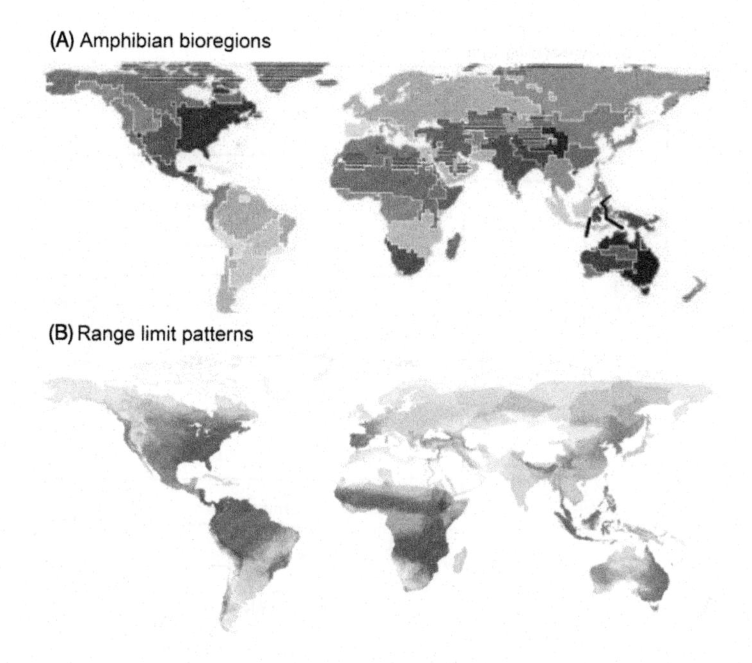


**FIGURE 3.64**

Results from the network analyses for the world's amphibians (Source: Vilhena and Antonelli, 2015). (A) Amphibian biogeographical regions of the world determined from geographical range data. Similar colors (gray shades in print version) indicate membership to a higher-level clustering, in this case equivalent to realms. The analysis used a resolution of two degree grid cells. (B) Species range limits colored by region. Geographically close and neighboring regions were given contrasting colors to highlight boundaries and boundary mixing.

Turtles are common in many types of aquatic ecosystem and some regions are known for their turtle diversity. For example, ~30 species of native freshwater turtles exists in Alabama. Their decline has been for collection for Asian foods, traditional medicines and for pet markets (see Fig. 3.66).

Crocodiles or true crocodiles are large aquatic reptiles that live throughout the tropics in Africa, Asia, the Americas and Australia. The modern American alligator, *Alligator mississippiensis,* is well represented in the fossil record of the Pleistocene and easily recognizable, and feared because of human attacks (Fig. 3.67).

The transition included the sister group of ancestors to birds, theropoda, and their close relatives the sauropodomorpha. (see Fig. 3.68).

## AVIFAUNA

Birds have a wide range of aquatic dependencies—ducks and herons are well-adapted to the aquatic life, some specialized in riparian forests and their association with water is more about the vegetation and cover as in the case of the Least Bell's Vireo. In addition, some groups depend on the aquatic

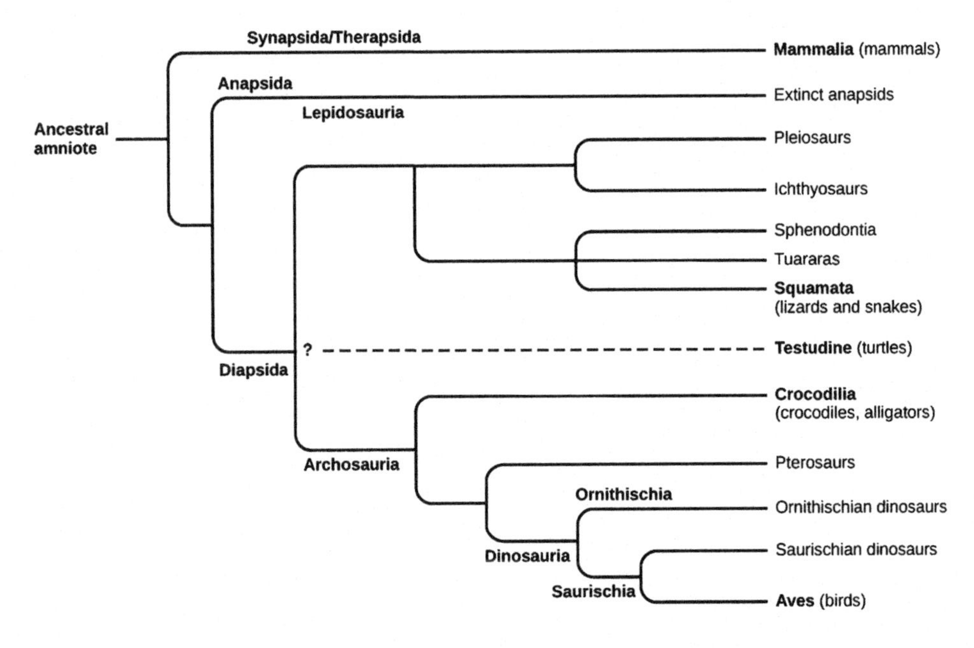


**FIGURE 3.65**

Hypothesized phylogeny of the amniotes (Source: lumenlearning). Amniotes are tetrapod vertebrates that include reptiles, birds, and mammals. In constrast to fish and amphibians, that lay their eggs in water, the amniotes lay their eggs on land or retain the fertilized egg within the mother.

**FIGURE 3.66**

Alabama red-bellied turtle, *Pseudemys alabamensis*, is one of the most endangered in the state and is the official state reptile (Source: Courtesy of Bill Summerour).

**FIGURE 3.67**

Image of American Alligator (Postdlf). Members of the alligator superfamily first arose in the Late Cretaceous. Leidyosuchus of Alberta is the earliest known genus. Fossil alligatoroids have been found throughout Eurasia, the animals having radiated there when land bridges across both the North Atlantic and the Bering Strait connected North America to Eurasia during the Cretaceous and Tertiary periods. The Chinese alligator probably descended from a lineage that crossed Beringia during the late Tertiary.

ecosystem in different ways: for example, kingfishers are usually thought to live near rivers and eat fish, whereas some insectivores feed on aquatic adult insects.

In this section, we will cover several themes that can help us appreciate how various birds live in or associated with water.

***Dabbling Ducks: Combating Thermal Challenges–Insulation and Physiology*** The family Anatidae includes the ducks and most duck-like waterfowl, which includes ducks, geese, and swans. Probably the most well known aquatic bird is the mallard or wild duck (*Anas platyrhynchos*), which breeds throughout the temperate and subtropical Americas, Europe, Asia, and North Africa, and has been introduced to many other parts of the world. As members of the Anatidae, a subfamily, we often refer to these as dabbling ducks, which feed mainly on vegetation in shallow bottoms by upending on the water surface, or grazing, and only rarely dive (Fig. 3.69).

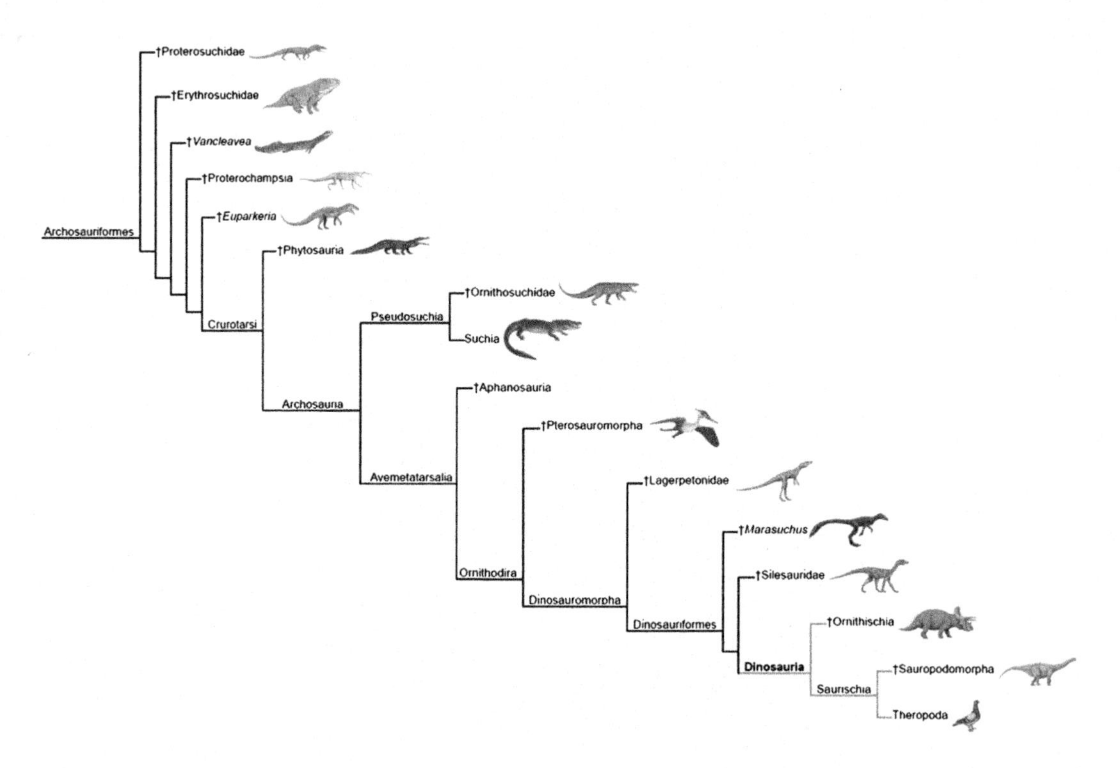


**FIGURE 3.68**

Archosauriformes (Source: S.J. Nesbitt).

**FIGURE 3.69**

Anatidae Mallard (female on left and male on right (Source: Richard Bartz).

As warm blooded animals, water creates a thermal challenge; they must maintain relatively high body temperatures, which is on average ~43°C for mallards. Yet, the heat lost to water can be substantial. How do birds maintain their body temperatures given the potential for heat loss?

**FIGURE 3.70**

Willow Flycatcher (Source: U.S. Fish and Wildlife Service). Adults have brown-olive upperparts, darker on the wings and tail, with whitish underparts; they have an indistinct white eye ring, white wing bars and a small bill. The breast is washed with olive-grey. The upper part of the bill is grey; the lower part is orangish.

Birds rely on feathers and oils to trap insulating air between the water and their skin. Nevertheless, the high metabolism requires birds to feed regularly.

***Willow Flycatcher, Yellowbilled Cuckoo and Least Bells Vireo—Riparian Forests*** We generally associate aquatic birds with slow-moving rivers and lakes, but there are numerous birds associated with riparian forests as well. Different trees and plants, such as willows, willow thickets, oaks, cottonwoods, alders, and dead sycamores, provide tree cover and nesting sites. Often, birds, such as the yellow-billed cuckoo (*Coccyzus americanus*), prefer more dense riparian zones, covered with bushes. The Willow Flycatcher (*Empidonax traillii*) lives in dense willos and is a small insect-eating bird (Fig. 3.70).

## MAMMALS

Mammals, in particular, placental mammals diversified after the Cretacious-Paleogene boundary (66 mya) (Fig. 3.71).

Numerous mammals are associated with inland waters from casual users for drinking or bathing; semi-dependent animals for foraging and hunting; and fully dependent and unable to live in terrestrial ecosystems. Between these different users, a vast number of mammals are users of aquatic systems, but for the time being, we can limit our introduction to two examples:

***Beavers*** The beaver, Rodentia, is one of the most geomorphologically important mammals in aquatic systems. Beavers are known for building dams, canals, and lodges (homes). The beaver (*Castor* spp.) is a large, primarily nocturnal, semiaquatic rodent. They are the second-largest rodent in the world (after the capybara). Their colonies create one or more dams to provide still, deep water to

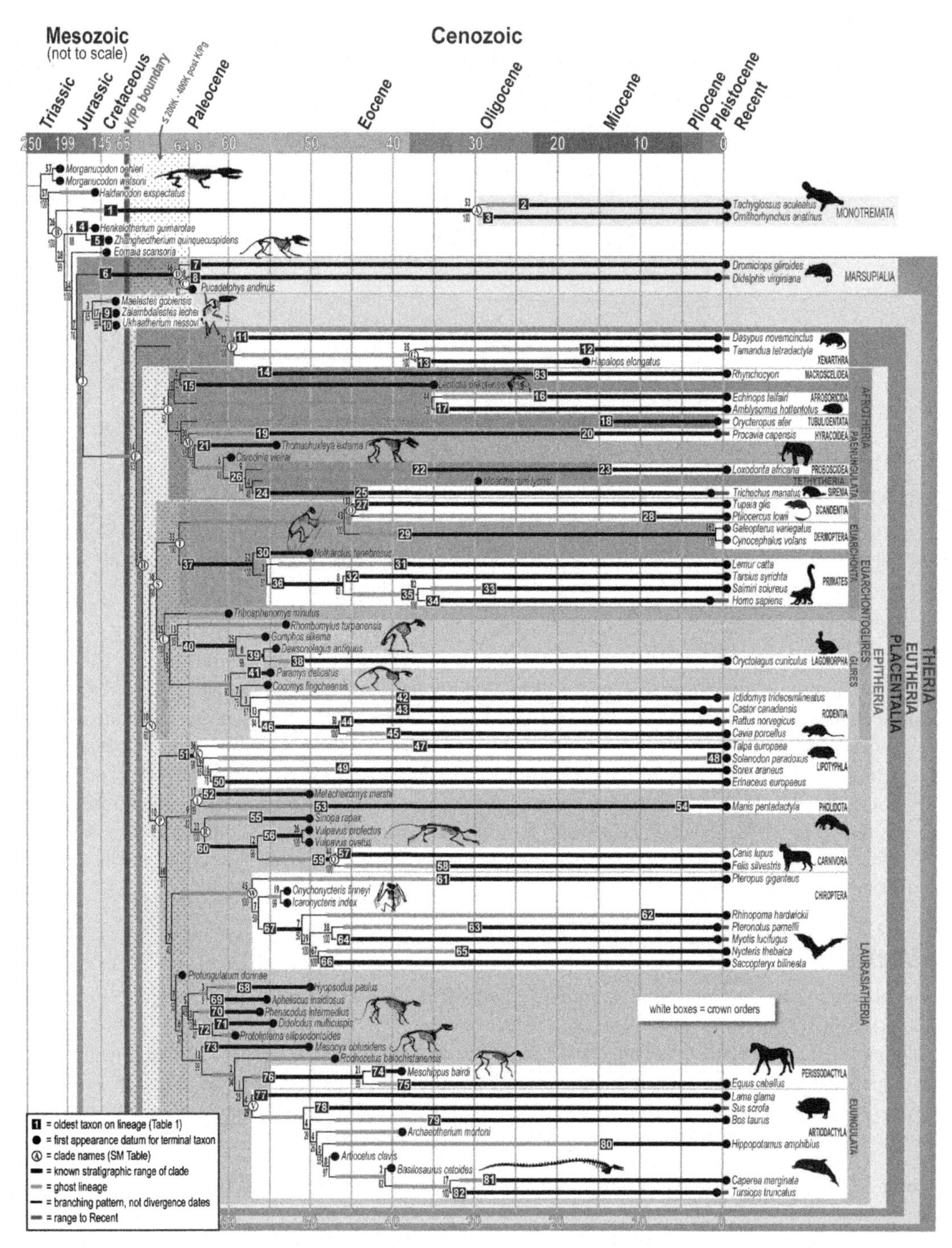


**FIGURE 3.71**

Phylogeny of mammals (Source: O'leary et al., 2013).

**FIGURE 3.72**

Beaver lodge (Source: WikiCommons). Beavers always work at night and are prolific builders, carrying mud and stones with their fore-paws and timber between their teeth. Because of this, destroying a beaver dam without removing the beavers is difficult, especially if the dam is downstream of an active lodge. Beavers can rebuild such primary dams overnight, though they may not defend secondary dams as vigorously. Beavers often create a series of dams along a river.

protect against predators, and to float food and building material (Fig. 3.72). The North American beaver population was once more than 60 million, but as of 1988 was 6–12 million. Effectively, ecosystem engineers, they alter the geomorphology of streams and Beaver management (and control) activities try to limit their work in streams, which affect flooding.

Beavers build dams and dens to create protection against predators and to provide easy access to food.

***Freshwater dolphins*** River dolphins feed primarily on fish. River dolphins, alongside other cetaceans, belong with even-toed ungulates, and their closest living relatives the hippopotamuses, having diverged about 40 million years ago. Just like their whale relatives, river dolphins produce a variety of vocalizations, usually in the form of clicks and whistles. River dolphins are fully aquatic mammals that reside exclusively in freshwater or brackish water and are restricted to certain rivers or deltas (Fig. 3.73). This makes them extremely vulnerable to habitat destruction, water quality degradation, and by catch.

## SPECIES RADIATION, POPULATIONS GROWTH, AND COMPETITION

The ancient Lake Biwa (Japan) has historically experienced substantial changes in the lake size. As lake-size has increased, organisms have access to great habitat diversity. In this case, genetically distinct gastropod lineages independently colonized Lake Biwa and rapidly radiated into 15 extant *Biwamelania* species. Thus the species diversification seems to be associated with habitat diversification. However, at the population level, what drives diversification?

**FIGURE 3.73**

In 2012 Bolivia's parliament declared that the pink river dolphin or "Bolivian bufeo" (*Inia boliviensis*) is considered part of the countries national heritage (Source: World Wildlife Fund).

Ecologists have theorized that when shared resources are limited, individuals compete with each other. When population size increases, the resources become limited and populations approach the carrying capacity of the ecosystem. By competing for the same resources, phenotypes will partition the resources to avoid direct competition. Overtime, the population might specialize on specific resources, become reproductively isolated, and diverge. In fact, some believe competition drives resource partitioning, i.e., species evolve to avoid competition. Although competition theory has been controversial, it remains a central dogma in ecology to explain species interactions; we turn to that topic in Chapter 4.

# NEXT STEPS

## CHAPTER STUDY QUESTIONS

1. Describe 10 major evolutionary adaptions that might infer fitness in inland waters.
2. What is the evidence that demonstrates that species diversity lies at the microscopic level?
3. Describe three major branches of evolutionary divergence that has created the Earth's biodiversity.
4. Describe how hypothesized ancestral relationships help us appreciate the complex divergences in the taxa that have invaded inland waters.
5. Describe the tension between native and introduced species, while appreciating the history of ecological patterns of invasions, range expansions, relative to the current pace of introductions.

## PROBLEM SETS

1. The classifications schemes used in this text, namely Table 3.2 and Fig. 3.11, embrace inconsistencies. There are several ways to negotiate these. Ignore and pretend they do not exist, or figure out why the discrepancies exist or reconcile them. Develop an analysis of why these dependencies exists based on the peer-reviewed literature and justify which strategy you prefer.
2. Using Fig. 3.62, draw a phylogenic tree for the taxa pictured. Describe some of the characteristics of the families represented in the chart. Although you should consider more recent treatments,

**Table 3.5 Classes of fish and families of fish that occur in California (sensu stricta, Nelson et al., 2006). Inland fish families are from Moyle (2002).**

| Class | Description | Number of CA Families |
|---|---|---|
| Myxini (Hagfish) | Jawless, marine taxa | None |
| Petromyzontida (Lamprey) | Jawless, anadromous | 1 |
| Placodermi | Extinct | None |
| Chondrichthyes (Cartilaginous fishes) | Sharks and Rays | None |
| Acanthodii | Extinct | None |
| Actinopterygii (Ray-finned fishes) | High diversity | 22 |
| Sarcopterygii | Lungfish, Ceocanths, Tetrapods | None |

**Table 3.6 Ray-finned Taxa.**

| Taxa (various ranks) | Distribution | California |
|---|---|---|
| Cladista (Birchirs) | Africa | NA |
| Chondrostie (Kaluga, Huso dauricus, H. huso) | ? | ? |
| Chondrostie (Paddlefish, *Polyodon spathula*) | Mississippi Drainage | ? |
| Chondrostie (*Psephurus gladius*) | China (piscovorious) | ? |
| Lepisosterformes (Gars) | North, Central America, Cuba | NA |
| Amiformes (Bowfins) | Eastern N. America | NA |
| Osteoglossomorpha (Mooneyes) | North America | NA |
| Osteoglossomorpha (Bonytongues, featherfin knifefish, elephant fish) | North America | NA |
| Elopmorpha (Freshwater and spaghetti eels) | ? | NA |
| Clupeomorpha (Herrings) | Worldwide | I |
| Anotophysi (Knerias, mudheads) | ? | ? |
| Cyprinoformes (Minnows, suckers, etc.) | ? | ? |
| Characiforms (Darters, catfish, etc) | worldwide | N, I |
| American knifefishes | | |
| Siluriformes, catfish | | |
| Freshwater Smelts (Osmeridae, Retropinniaae, Galaxiidae) | | |

consider using Kocher and Stepien (1997) as starting place, but keep Mooi and Gill (2010) as it cites a "crisis" in fish systematics.

## LITERATURE RESEARCH ACTIVITIES

1. Select a region in the world that interests you. Look up what groups of aquatic species have a particularly high amount of endemism. Evaluate the hypothesizes that explain the endemism, compare related taxa and how their biogeography contrasts with the group you have selected.
2. Select 3 of the taxa in Table 3.5, and determine the earliest fossils found in the state and how the range allowed these fish to be part of California's freshwater taxa.

## ADVANCED APPLICATIONS

There are numerous taxa that occur in freshwaters (Table 3.6). These taxa are widely distributed, but the table is missing key aspects. Use the literature to fill in the missing information from the table.

CHAPTER

# THE RULES: POPULATION GROWTH AND COMPETITION 4

*One general law, leading to the advancement of all organic beings, namely, multiply, vary, let the strongest live and the weakest die.*

**Charles Darwin, The Origin of Species**

## CONTENTS



Algal blooms are caused by the rapid increase in phytoplankton population size. Carried by the current, phytoplankton are composed of algae and/or cyanobacteria; and when conditions are right, their density can increase at an exponential rate, i.e., an algal bloom, which can lead to hypoxia or human health hazards.


Ecology and Management of Inland Waters. https://doi.org/10.1016/B978-0-12-814266-0.00017-9

Can we predict these algal blooms by understanding their reproductive rates? Intuitively, we know that the population size will not reach infinity, thus it must attain some limit. But why? Can we predict this population limit? How do other species affect this limit? Finally, to what extent does natural selection influence these population dynamics and capacity to compete in aquatic ecosystems?

Modeling phytoplankton and zooplankton populations has a long history in aquatic ecology. The earliest equations that describe population dynamics are based on model organisms in laboratory studies, which Alfred Lotka and Vito Volterra independently developed, the standard populations models.

As groups of interbreeding organisms, populations level is where natural selection is often observed and provides analytical tools to help us predict how aquatic systems are structured. As a way to quantify survivorship, growth, and reproduction, population models are central in how we understand ecological systems.

After reading this chapter, you should be able to

1. Use density dependent and independent equations to predict populations sizes;
2. Evaluate how intra- and interspecific competition can influence population growth;
3. Understand the role of competition within ecological and evolutionary time scales; and
4. Summarize the limitations of competition.

## POPULATION SIZE AS A FUNCTION

Phytoplankton populations change with time and can be summarized as a function of the initial population size ($N_0$) and time ($t$). It may be useful to reflect on this simple but powerful equation:

$$N = f(N_0, t). \tag{4.1}$$

Our next step is to define the function with some simple biological realities: individuals are born ($B$) and die ($D$), a natural phenomenon that effects the population size. Thus the population at a specific time ($N_t$) will increase with the number birth and decrease with the number of deaths. Thus we can summarize populations sizes after a discrete time period ($t + 1$) as

$$N_{t+1} = N_t + (B - D). \tag{4.2}$$

Assuming a starting populations of 10, one birth per individual and 20% mortality, we can estimate the population dynamics using Eq. (4.2) and produce Table 4.1 and Fig. 4.1.

By converting birth and deaths to a per individual rate—$b$ for birth rate and $d$ for death rate—we can develop a model based on observations, such as these:

$$N_{t+1} = N_t(b - d) = RN_t. \tag{4.3}$$

Eq. (4.3) describes simple exponential growth. Where b and d are independent of $N$, the equation is referred to as the Malthusian model. However, we usually combine $b - d$ and refer to these as $R$, the net growth rate.

Now, as noted above, this equation is based on discrete time intervals. Alternatively, we might evaluate population as a continuous function. For many readers, the words "discrete" and "continuous"

**Table 4.1 Population growth rate using Eq. (4.2).**

| Time | $N_t$ | $B$ | $D$ | $B-D$ | $R$ |
|---|---|---|---|---|---|
| 0 | 10 | 10 | 2 | 8 | 0.8 |
| 1 | 18 | 18 | 4 | 14 | 0.78 |
| 2 | 32 | 32 | 6 | 26 | 0.81 |
| 3 | 58 | 58 | 12 | 46 | 0.79 |
| 4 | 104 | 104 | 21 | 83 | 0.80 |
| 5 | 187 | 187 | 37 | 150 | 0.80 |
| 6 | 337 | 337 | 67 | 270 | 0.80 |
| 7 | 607 | 607 | 121 | 486 | 0.80 |
| 8 | 1093 | 1093 | 219 | 874 | 0.80 |
| 9 | 1967 | 1967 | 393 | 1574 | 0.80 |
| 10 | 3541 | 3541 | 708 | 2833 | 0.80 |

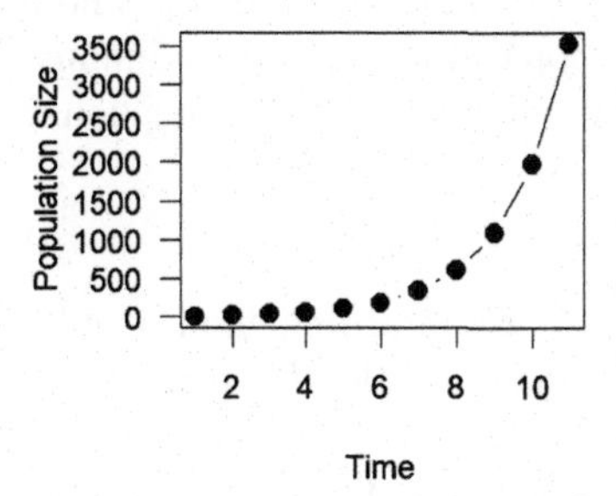


**FIGURE 4.1**

Exponential population growth, using Table 4.1 for data.

functions harken to memories of calculus, where a simple differential equation can be used to describe population growth:

$$dN/dt = rN, \tag{4.4}$$

where $dN/dt$ is the instantaneous growth rate of the population and $r$ is referred to as the intrinsic rate of increase. The rate of change is based on an infinitesimally small time or an approximately instantaneous rate. Our next step is to solve for $N(t)$, which requires some more calculus,

$$\int_{N(0)}^{N(t)} \frac{dN(t)}{N} = r\int_0^t dt, \tag{4.5}$$

whose solution is $N(t) - ln(0) = rt - r0 = rt$. After exponentiation of both sides and rearranging the equation, we are left with

$$N_t = N_0 e^{rt}. \tag{4.6}$$

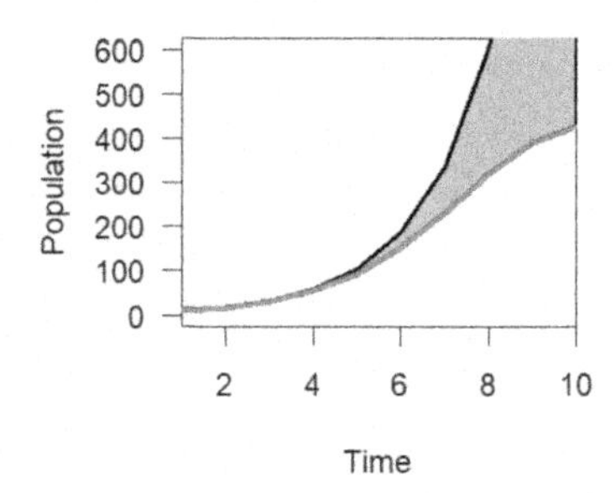


**FIGURE 4.2**

Exponential growth and growth with the sigmoidal constraint of carrying capacity. The shaded areas might be considered as where species overshoot their carrying capacity and would respond with reduced population growth.

Eq. (4.6) describes exponential population growth that has no limits—in other words, where $r$ is positive, there is nothing to stop the population size to go to infinity. Obviously, we have not observed this in nature, so we need to add some complexity to this equation to add realism.

## DENSITY DEPENDENCE

The Malthusian model does not specify resource limitations that biologists observe in nature. In fact, after reading Malthus' "An Essay on the Principle of Population," Pierre-Franşois Verhulst published an equation to describe population growth proportional to both the population size and the available resources. The logistic growth equation defines the change in population size with time as a function of the growth rate ($r$), and initial population ($N$), and carry capacity ($K$),

$$\frac{dN}{dt} = rN\left(1 - \frac{N}{K}\right), \tag{4.7}$$

where $\frac{dN}{dt}$ refers to instantaneous growth rate. The choice of the letter K came from the German word "Kapazitätsgrenze" (capacity limit), whereas $r$ came from rate. With specific parameters, this model can describe populations with more realism than exponential growth alone (Fig. 4.2).

In the early 20th Century, the equation was rediscovered several times and became known as the law of population growth.

In spite of this relatively elegant equation, estimating the parameters ($r$ and $K$) in real systems is not a trivial activity. To appreciate the steps to apply these equations to aquatic systems, refer to the "next steps" section.

## MAXIMUM SUSTAINED YIELD

With the reoccurring experience of fishery collapses, population biologists and economists have applied the logistic growth to the fishery management with marginal success. Based on the assumed goal to maximize the fishery (i.e., harvest), scientists developed the maximum sustainable yield (MSY) concept. Theoretically, MSY predicts the largest yield (or catch) that can be taken from a fish stock over an indefinite period. In other words, the population size will be in equilibrium:

$$N_1 = N_0 + B - D + I - E. \tag{4.8}$$

The simplest way to model harvesting is to modify the logistic equation so that a certain number of individuals is continuously removed:

$$\frac{dN}{dt} = rN\left(1 - \frac{N}{K}\right) - H, \tag{4.9}$$

where H represents the number of individuals being removed from the population, i.e., the harvesting rate. When H is constant, the population will be at equilibrium when the number of individuals being removed is equal to the population growth rate.

The equilibrium population size under a particular harvesting regime can be found when the population is not growing, that is, when $\frac{dN}{dt} = 0$. This occurs when the population growth rate is the same as the harvest rate:

$$rN\left(1 - \frac{N}{K}\right) = H. \tag{4.10}$$

Fundamental to the notion of sustainable harvest, the concept of MSY aims to maintain the population size at the point of maximum growth rate by harvesting the individuals that would normally be added to the population, allowing the population to continue to be productive indefinitely. As observed in Fig. 4.2, at low populations the growth rate is not dramatically constrained by resources, thus population growth is density independent and remains near its maximum. Even at intermediate population densities, represented by half the carrying capacity, reproduction rates at their maximum rate. At this point, called the maximum sustainable yield, surplus of individuals can be harvested because population growth is maximized and the large number of reproducing individuals maintained.

When the population exceeds this size, density dependence begins to limit as the population size approaches the carrying capacity. When the population reaches this size, no surplus can be harvested and yield drops to zero.

Fig. 4.3 shows how growth rate varies with population density. For low densities (far from carrying capacity), there is little addition (or "recruitment") to the population, simply because there are few organisms to give birth. At high densities, though, there is intense competition for resources, and growth rate is again low because the death rate is high. In between these two extremes, the population growth rate rises to a maximum value ($N_{MSY}$). This maximum point represents the maximum number of individuals that can be added to a population by natural processes. If more individuals than this are removed from the population, the population is at risk for decline to extinction. The maximum number that can be harvested in a sustainable manner, called the maximum sustainable yield, is given by this maximum point.

MSY is extensively used for fisheries management. Unlike the logistic model, MSY has been refined in most modern fisheries models and occurs at around 30% of the unexploited population size. This fraction differs among populations, depending on the life history of the species and the age-specific selectivity of the fishing method.

However, the approach has been widely criticized as ignoring several key factors involved in fisheries management and has led to the devastating collapse of many fisheries. As a simple calculation, it ignores the size and age of the animal being taken, its reproductive status, and it focuses solely on the

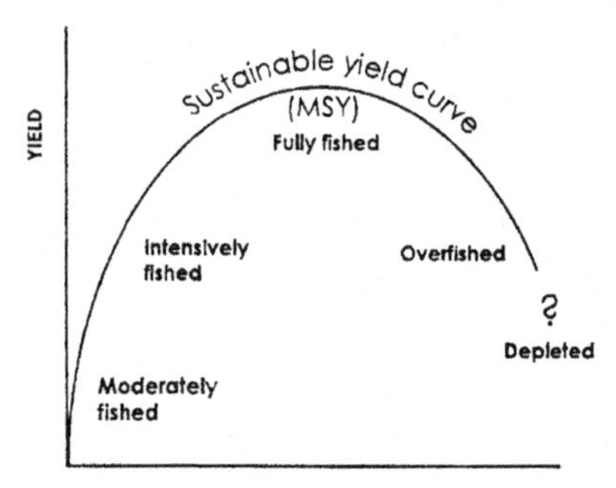


**FIGURE 4.3**

Maximum sustained yield. (Source: Food and Agriculture Organization of the United Nations, 1996, S.M. Garcia, The Precautionary Approach to Fisheries and its Implications for Fishery Research, Technology and Management: An Updated Review, http://www.fao.org/3/W1238E02.htm. Reproduced with permission.)

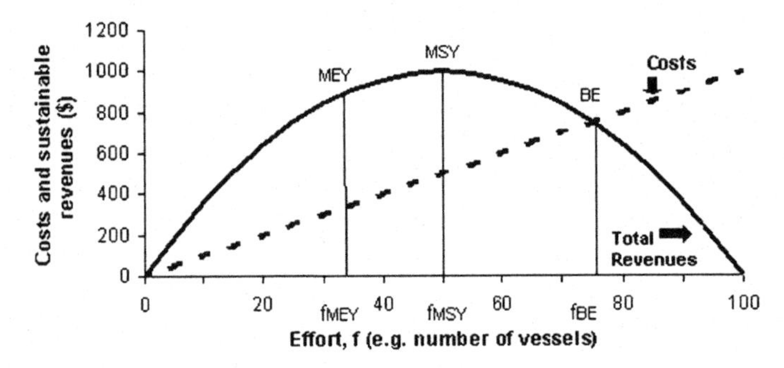


**FIGURE 4.4**

Maximum economic yield (MEY), maximum sustainable yield (MSY), and equilibrium biomass (BE) estimates. (Source: Food and Agriculture Organization of the United Nations, 2002, Kevern L. Cochrane, A Fishery Manager's Guidebook – Management Measures and Their Application, http://www.fao.org/3/y3427e/y3427e07.htm. Reproduced with permission.)

species in question, ignoring the damage to the ecosystem caused by the designated level of exploitation and the issue of bycatch. Among conservation biologists, it is widely regarded as dangerous and misused.

Finally, the most damning criticism of MSY is how it ignores the economics of the fishery. In particular, how fishing effort changes with yield. For example, if an open access fishery is fully fished, allowing new entries into the fishery, as the yields decline, fishers increase their effort to maintain profits in the face of declining catch. As effort increases, the fishery becomes increasingly overfished with a strong potential to collapse.

Thus economists have modified the equations to capture some of these problems, defining optimum sustainable yield and maximum economic yield. The maximum sustainable yield is usually higher than the optimum sustainable yield and maximum economic yield (Fig. 4.4).

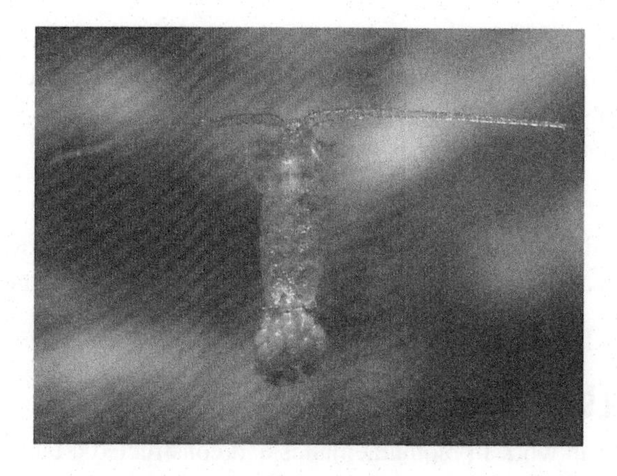

**FIGURE 4.5**

*Eudiaptomus gracilis* female with eggs (Source: http://www.flickriver.com/photos/petermaguire/4621945868/).

# INDIVIDUAL-BASED MODELS

## MODELING DAPHNIA FECUNDITY AND POPULATIONS DYNAMICS

An alternative way of thinking about population growth is considering that various stages have various states: age, size, number of eggs, and age of eggs. In 1954 a German limnologists, Elster, published a paper using the ratio of eggs per female and the development time to estimate the birth rate of *Eudiaptomus gracilis*. This small copepod and member of the crustacea has a wide distribution (marine, brackish, and freshwaters) and serves as a good organism for population modeling (Fig. 4.5).

By using the eggs produced per female and egg development time, we can estimate the birth rate ($B$):

$$B = E/D, \tag{4.11}$$

where $B$ is the birth rate (i.e., number of newborn (between $t$ and $t+1$) / population size at $t$), $E$ is the ratio of eggs per female, and $D$ is the egg development time. Although $D$ is rarely constant, which creates bias, we can use this model as an estimate population growth.

Using the birth rate to calculate the instantaneous growth rate ($r$) using the exponential growth rate model

$$N_t = N_0 e^{rt}, \tag{4.12}$$

where $N_0$ is the size of the initial population at time zero, $N_t$ is the population size at time $t$, $d$ is the instantaneous rate of mortality rate, and $b$ is the instantaneous birth rate (which can be calculated from $B$, i.e., $b = ln(1 + B)$). As you can probably guess, the model relies on instantaneous rates, which is an assumption that holds because of populations reproducing frequently enough.

When $r$ and $bD$ are relatively large, the $E/D$ can diverge considerably, thus they suggest

$$E/D = (e^{bD} - 1)/D. \tag{4.13}$$

Now, we can link this to environmental variables, i.e., food! For example, we might measure phytoplankton concentrations in the water column as a source of food as measured in mg $L^{-1}$ carbon. And to capture the development time, use temperature as a proxy:

$$b = ln[(C_t/N_t) + 1]/D. \quad (4.14)$$

The model is simulated in fixed time steps (usually 0.1 day) over a period of several days up to a few months. The time scales are selected with respect to the egg development time, which is about 4.4 days at 15C°.

## LESLI MATRIX MODELS

In 1939 Patrick Leslie began work in biomathematics and constructed a table to understand how key life history strategies played out in populations dynamics. Leslie used matrix algebra to model discrete age classes. For example, an insect population composed of individuals are eggs, juveniles (instars), mature (usually just females) can be represented with three classes. Using the rate of transition between each of life history stages, matrix models of populations calculate the growth of a population.

Let us assume that there is a population of fish that can live for up to three years and reproduce during the second and third year. We can define a time-discrete model of the growth of this population using a Leslie matrix. $f_x$ is the fecundity of age class $x$ (number of offspring per individual), and $s_x$ is the probability that an individual in age class $x$ will survive until age class $x+1$. The matrix representing our population (linear system) is

$$\mathbf{L} = \begin{pmatrix} o & f_2 & 0 \\ s_1 & 0 & 0 \\ 0 & s_2 & 0 \end{pmatrix}$$

and is called a *Leslie* matrix, named after P.H. Leslie who published his work in this regard in 1945. Take a few minutes to identify the location for each parameter. This matrix can be used to predict the stable age distribution and to calculate the population growth rate after the age distribution is established.

At any given time, a field survey of this population can give a measure of the abundances in each class $x$, which we will note $n_x$:

$$\mathbf{n} = \begin{pmatrix} n_1 \\ n_2 \\ n_3 \end{pmatrix}.$$

## WORKING AN EXAMPLE

Larval mayfly, which are ecdysozoans, are born from eggs as molt to increase in size. Then they go through a metamorphosis and emerge as subimagos, and then develop into adults. For example, Okanagan Mayfly (*Hexagenia limnbata*) in the Ephemeroptera might lay 100 eggs. We will assume that 50% (or a probability of 0.50) that the eggs will hatch and that each instar (nymph) stage will make it to the next and to the adult stage (Fig. 4.6). We specify 7 stages (eggs, instars, imago, and

**FIGURE 4.6**

*Hexagenia limnbata*, Okanagan mayfly.

**Table 4.2 Population parameters for life history classes.**

| | |
|---|---|
| egg -> 1st instar | 0.50 |
| 1st -> 2nd | 0.80 |
| 2nd -> 3rd | 0.50 |
| 3rd -> 4th | 0.50 |
| 4th -> imago | 0.50 |
| imago -> adult | 0.60 |
| adult (female) | 0.33 |

adult), where the females die after laying their eggs, so the survival is zero. The number of eggs laid can be quite large, we will estimate a 100 eggs/female. See Table 4.2 as a summary of these parameters.

So let us model the following life history stages: eggs, instars, imago, and adult. The Lesli Matrix to capture this model is

$$\mathbf{L} = \begin{pmatrix} 0 & 0 & 0 & 0 & 0 & 0 & 100 \\ 0.80 & 0 & 0 & 0 & 0 & 0 & 0 \\ 0 & 0.50 & 0 & 0 & 0 & 0 & 0 \\ 0 & 0 & 0.50 & 0 & 0 & 0 & 0 \\ 0 & 0 & 0 & 0.50 & 0 & 0 & 0 \\ 0 & 0 & 0 & 0 & 0.50 & 0 & 0 \\ 0 & 0 & 0 & 0 & 0 & 0.60 & 0 \\ 0 & 0 & 0 & 0 & 0 & 0 & 0 \end{pmatrix}$$

Again to simplify our example, we will assume an ephemeral stream with the presence of eggs to start the population. With an initial population of $\mathbf{n}^0 = \mathbf{n} = (100, 0, 0, 0, 0, 0, 0, 0)$, that is of 100 eggs, and no other life stages present.

With each stage, we have fewer individuals until the female successfully reproduces (Table 4.3).

**Table 4.3 Example of population parameters.**

| Stage | Eggs | 1st | 2nd | 3rd | 4th | Subimago | Adult | Female | Eggs |
|---|---|---|---|---|---|---|---|---|---|
| Probability | NA | 0.80 | 0.50 | 0.50 | 0.50 | 0.50 | 0.60 | 0.33 | NA |
| $N_x$ | 100 | 80 | 40 | 20 | 10 | 5 | 3 | 1 | 100 |

After one season, we predict 100 eggs laid by the 1 female, in what appears to be a stable population. The key attribute of this modeling approach is that one can determine if there is one age class, where the species is most vulnerable and preventing a viable population. The principal drawback of stage-structured modeling is the potentially large number of parameters that must be computed.

### THEORETICAL CONSIDERATIONS

Leslie Matrix models are one of the most best known way to describe population growth and projected age distribution. It is generally applied to populations with an annual breeding cycle. It is also used in population ecology to model the changes in a population of organisms over a period of time. In the model, the population is divided into groups based on age classes or life-history stages.

However, age is not an accurate predictor of birth or death rates in many organisms. For example, fish fecundity (egg production) is often correlated with size and or weight, where age might be a poor predictor. Thus fish size is a better predictor of birth rate than age, and one of the reasons "fisher managers are interested in protecting, BOFFs, "big old fat females", because their fecundity might be orders of magnitude than other small fish.

Demography depends on physiological stages and development into these stages is not consistent among individuals. Some individuals exhibit retarded or accelerated development or regression. There are subclasses of the population that have different demographic characteristics.

## POPULATION MODELS WITH TWO COMPETING SPECIES

### LOTKA–VOLTERRA EQUATIONS

Populations do not exist in isolation, but in a community of interacting and competing species. Thus, we can extend the Logistic Equation (Eq. (4.7)) for more than one species and track how growth rates might differ when the populations share resources.

When two species occupy the same area and presumably use the same resources, their growth rates might depend on the other's populations size. Thus instead of sigmoidal for population growth, where $(K - N)/N$, we specify how each population influences $K$, $(K - N_1 - N_2)/K$, where $N_1$ and $N_2$ influence the growth potential. Although this addition is pretty straightforward, it assumes that species 1 and species 2 have equal effects on $K$. To account for differential effects, coefficients are used to adjust for each of these effects. Thus we can specify an equation to track populations as

$$\frac{dN_1}{dt} = r_1 N_1 \frac{K - N_1 - \alpha_{12} N_2}{K} \tag{4.15}$$

and

$$\frac{dN_2}{dt} = r_2 N_2 \frac{K - N_2 - \alpha_{21} N_1}{K}, \tag{4.16}$$

where $\alpha_{12}$ is the effect of species 2 on species 1, and $\alpha_{21}$ is the effect of species 1 on species 2. As we further extend the model, biologists also relax the assumption that $K$ is the same for both species. Instead, we differentiate $K_1$ and $K_2$ and include a coefficient for how each species influences it own $K$, i.e., $\alpha_{11}$ and $\alpha_{22}$:

$$\frac{dN_1}{dt} = r_1 N_1 \frac{K_1 - \alpha_{11} N_1 - \alpha_{12} N_2}{K_1} \tag{4.17}$$

and

$$\frac{dN_2}{dt} = r_2 N_2 \frac{K_2 - \alpha_{22} N_2 - \alpha_{21} N_1}{K_2}. \tag{4.18}$$

These equations have been extremely powerful to describe populations dynamics between two species.

## ISOCLINES AND SPECIES INTERACTIONS

The implications for Eqs. (4.17) and (4.18) can be depicted graphically using state-space graphics. Before turning to these figures, let us define the conditions for zero growth for $N_1$ and $N_2$. For $N_1$'s growth to be zero, $K_1 - \alpha_{11} N_1 - \alpha_{12} N_2 = 0$. After rearranging the equation, we can determine that the x-intercept is $K_1$ and y-intercept is $K_1/\alpha_{12}$, and can be plotted as an "isocline."

Fig. 4.7 displays the potential growth of $N_1$ (x-axis) in the presence of $N_2$ (y-axis). $N_1$ population growth depends on its starting point relative the isocline, the diagonal line (where population growth is zero), $dN_1/dt = 0$. If the population size of $N_1$ is greater than the isocline, then $K_1$ will limit the population growth and the population will decline. But note that $K_1$ depends on $N_2$. Alternatively, if $N_1$ is less than the carrying capacity, then competition for resources does not limit growth, and $N_1$ increases.

A similar graphic can be developed for $N_2$'s population, but graphed on the y-axis (Fig. 4.8).

Figs. 4.7 and 4.8 illustrate population growth for one isocline at a time. We will now turn to figures where both N1 and N2 isoclines are included—these will help us appreciate the outcomes of interspecific competition relative to both isoclines.

Now, let us look at how the species interact together in these state-space graphics, but evaluate four scenarios by adjusting the x- and y-intercepts (Fig. 4.9). In the Scenario A, species 1 wins, whereas species 2 wins in Scenario B. By the joint movement of the two populations, one of the species is driven out of the community, whereas the winner reaches its carry capacity ($K_1$ for Scenario A and $K_2$ for Scenario B).

In Scenario C, the isoclines of the two species cross one another (Fig. 4.9). In this scenario, no stable equilibrium point exists, and the outcome depends on the initial abundances of the two species, but there is always one winner and one loser. Note: $K_1 > K_2/\alpha_{21}$ and $K_2 > K_1/\alpha_{12}$.

Finally, in the Scenario D the isoclines cross one another, but in this case $K_1 < K_2/\alpha_{21}$ and $K_2 < K_1/\alpha_{12}$ (Fig. 4.9). In this case, their joint trajectories always head toward the intersection of the isoclines. Rather than out-competing one another, the two species coexist at this stable equilib-

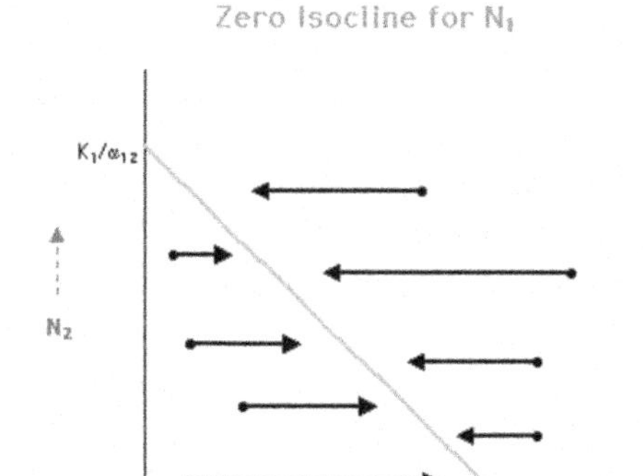


**FIGURE 4.7**

Each point in a state-space graph represents a combination of abundances of the two species. Any given point along, for example, N1's zero isocline, represents a combination of abundances of the two species, where the species 1 population does not increase or decrease.

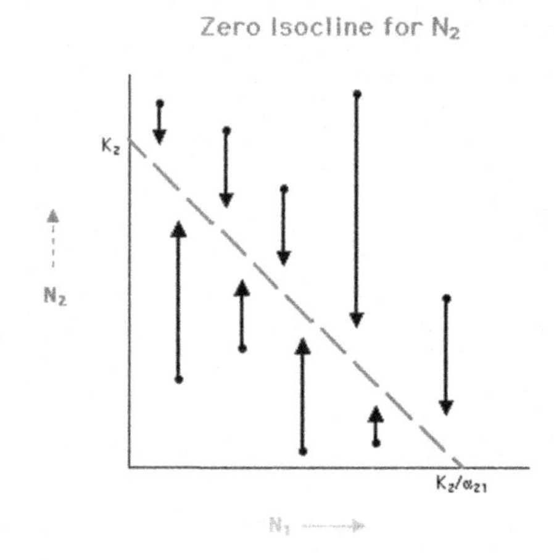


**FIGURE 4.8**

The isocline intersects the graph on the y-axis at $K_2/\alpha_{12}$ when the carrying capacity of N1 is filled by the equivalent number of individuals of N2, and no individuals of N1 are present. The intersections of the isocline for N1 are essentially the same, but on different axes.

rium point. Regardless of the initial abundances, these community reaches a stable point—and since communities have coexisting species; this scenario parallels our observations.

The Lotka–Volterra model of interspecific competition are exploratory models in biological systems. Their applicability to real-world ecosystems is quite limited. The assumptions of the model:

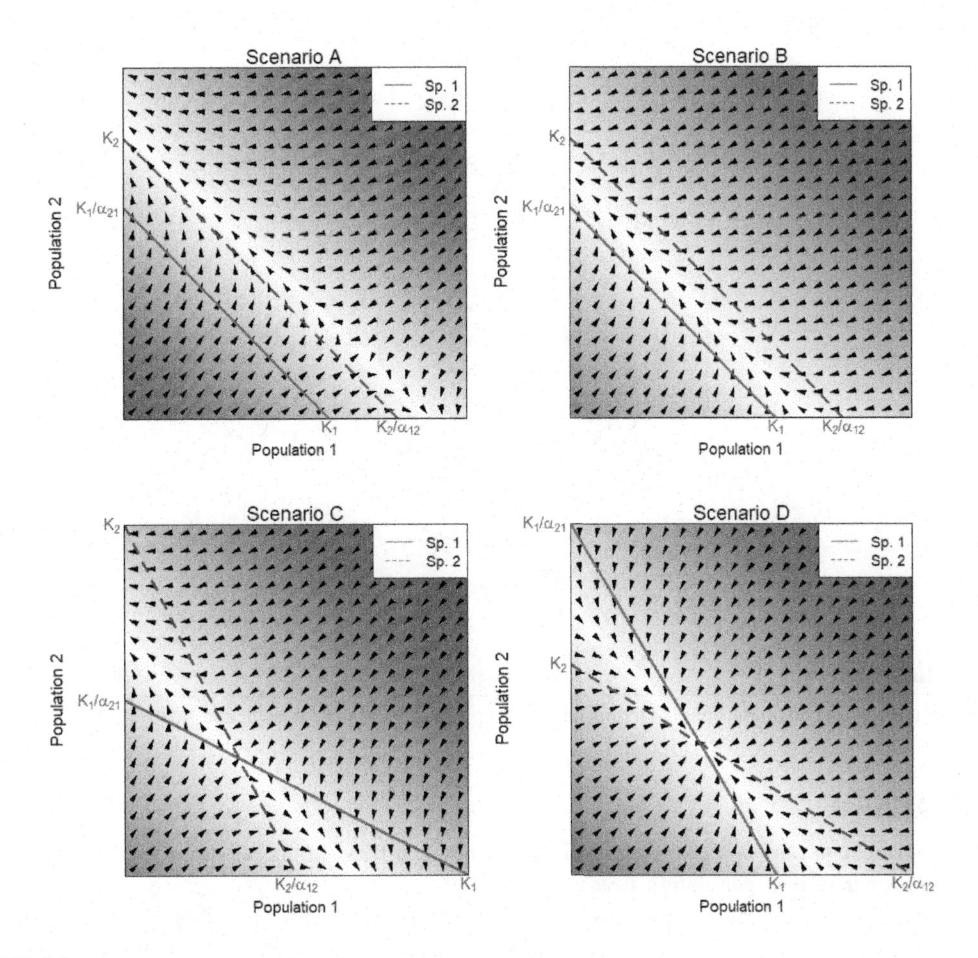


**FIGURE 4.9**

Depending on the value of the parameters, the outcome of two species can vary dramatically. Note: Scenario D was easily adopted by ecologists; it reenforced a view of nature that is in equilibrium.

- No migration;
- Carrying capacities are constant;
- Competition coefficients are constant.

It is hard to believe any ecosystem could meet these assumptions. Furthermore, estimating the competition coefficients is nearly impossible, especially when more than two species are interacting; then it is an impossibility. In spite of the our attachment to these models, they serve an important heuristic purpose: they describe how density dependence can be modeled for competing species; provide a framework for various outcomes based on competitive abilities $\alpha$, and suggest mechanisms for species divergence linking ecological and evolutionary time scales.

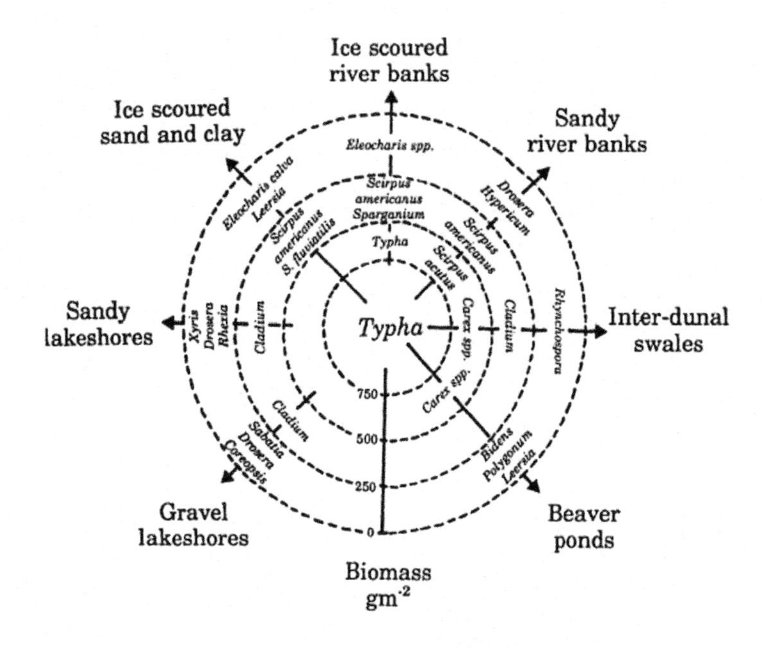


FIGURE 4.10

Typha dominance depends on the abiotic circumstances and processes (Source: Wisheu and Keddy, 1992).

## FLUCTUATING *K*

Although these equations describe the population dynamics, they do not capture the mechanisms of population growth or how population sizes decline.

In competition theory, the competing species must be competing for limited resources—implying that they do so at a set point. In reality, resources fluctuate and the limitations vary through time and space. Thus it is more likely that species respond to a range of signals and resources than to simple estimates of $K$. For example, *Typha* spp. (cattails) interaction (dominance or coexistance) depends on the abiotic factors, such as the type of disturbance (Fig. 4.10).

## COMPETITIVE EXCLUSION AND OVERLAPPING NICHE

### NICHE THEORY

When biologists tested competition theory, the results were often that one species excluded another, and these results led to the theory of competition as articulated by Gauss:

*Two species cannot coexist unless they are doing things differently.*

That was rephrased by Hutchinson as

*No two species can occupy the same ecological niche.*

Later, ecologists extended this into the competitive exclusion principle, where species that complete overlap could not coexist. As we shall describe below, many consider this to be a central driver in evolution.

## THE PARADOX OF PLANKTON

In 1961 G. Evelyn Hutchinson published a paper outlining a fundamental problem in ecology asking why many species coexist in spite of their overlapping resource use? According to competitive exclusion principle, the best competitor should dominate and exclude, i.e., drive poor competitors, out of the system.

## MECHANISMS THAT CONTROL POPULATION GROWTH

There are, of course, more complicated models than what we have introduced in this chapter, but let us turn to another way to consider populations dynamics in the context of trophic interactions.

In a simplistic sense, there are bottom up and top down forces that control populations sizes. Let us begin with three equations that can be used to define some bottom up forces, in this case, nutrients that might stimulate phytoplankton growth.

## DIFFERENTIAL RESPONSES TO NUTRIENTS

## ENZYME KINETICS

There are three main equations used to model the growth of phytoplankton: The `Michaelis-Menten relationship`, the `Monod Equation`, and the `Droop equation`. We will use these to model inorganic nutrient substrates to determine how species vary in their growth rates relative to nutrient concentrations. As it turns out, there is something inherent about taxa that determine their capacity to turn carbon, nitrogen, phosphorus, etc., into biomass. Using ratios of various nutrients can be used to predict the efficiency of trophic relations and will be further discussed in Chapter 8.

Using the Michaelis–Menten equation, a speed of a chemical reaction, with the form of Substrate ⟶ Products, can be estimated:

$$V = V_{max} * [S]/(K_s + [S]), \tag{4.19}$$

where the speed of production or velocity $V$ can be modeled based on $V_{max}$ (maximum rate), and where $K_s$ is a rate limiting factor and $S$ is concentration of substrate (Fig. 4.11).

The enzymatic relationship for uptake rates can be used to determine the speed that a limiting nutrient is transported into a cell, which then might be used to predict growth. To extend this concept, a variant on the Michaelis–Menten equation was developed to model growth based on the external concentrations of nutrients, using the Monad equation. Because of the simplicity of this equation, it provides a pragmatic link between nutrient concentrations and growth rates.

One way to identify what is limiting growth is to reduce concentrations of various nutrients systematically, one at a time. When this produces little or no effect, the nutrient that is being tested is probably in excess. If there is a definite effect, the nutrient is very likely the "growth-limiting nutrient," a topic further developed in Chapter 8.

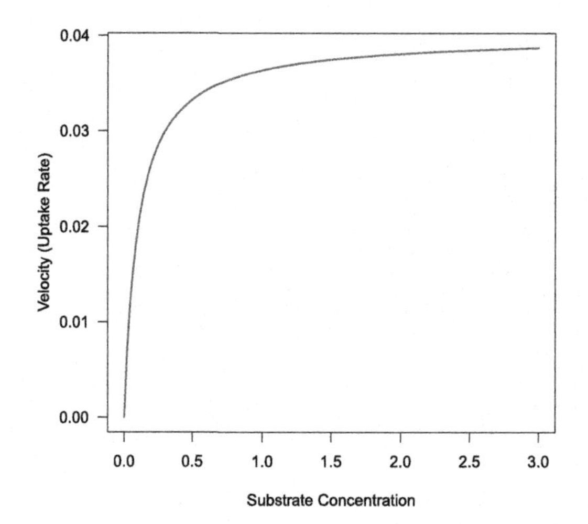


**FIGURE 4.11**

Modeling nutrient uptake using Michaelis–Menten equations.

The equation related to the above-mentioned method is named after Jacques Monod, who proposed using an equation of this form to relate microbial growth rates in an aqueous environment to the concentration of a limiting nutrient. The Monod equation has the same form as the Michaelis–Menten equation, but differs in that it is empirical, whereas the latter is based on theoretical considerations. Using the Michaelis–Menten form, the Monad equation, predicts the rate of growth for a specific functional group in the ecosystem:

$$\mu = \hat{\mu} S/(K_s + [S]), \tag{4.20}$$

where $S$ is the concentration of the limiting nutrient; $\mu$ is the specific growth rate coefficient; $\hat{\mu}$ is the maximum specific growth rate, and $K_s$ is called the half-saturation coefficient and produces a very similar result as the Michaelis–Menten model (Fig. 4.12).

Furthermore, the applicability of the Monod model is doubtful because luxury uptake of nutrients and storage for later growth may lead to a temporal uncoupling between reproductive rates and dissolved nutrient concentrations. When evaluated with more precision, Droop found that intracellular concentration or cell quota of a limiting nutrient predicted growth rates. To further refine Droop equation, $\mu$ is better modeled as a function of the intracellular "cell quota" $Q$ and the "subsistence quota", $q$:

$$\mu = \mu_{max}\ (1 - q/Q]). \tag{4.21}$$

However, measuring these parameters is not practical in the field. Moreover, the $\mu_{max}$ is the maximum theoretical growth rate does not occur in reality, hence the model is too simplistic.

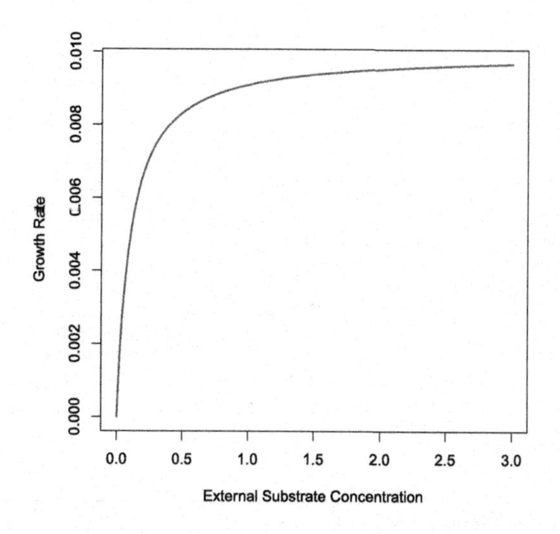


**FIGURE 4.12**

Relationship between external nutrient concentration and growth rate using the Monod equation.

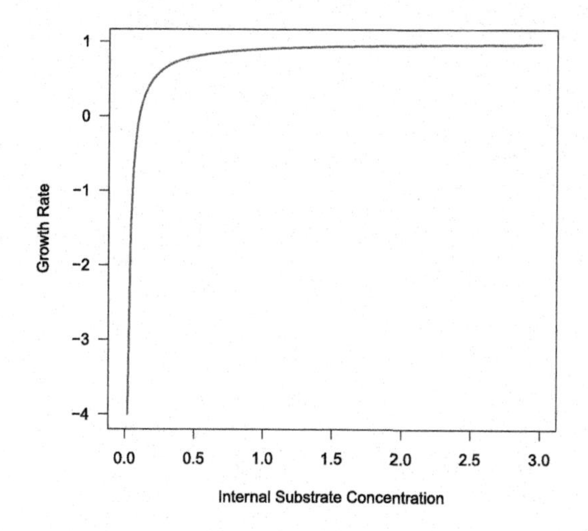


**FIGURE 4.13**

Relationship between internal nutrient concentration and growth rate using the Droop equation.

The applicability of the Monod model is only warranted under steady state conditions as realized in chemostat cultures. On the other hand, the applicability of the Droop equation is less restricted to steady state conditions, but the cell quota of individual species cannot be measured easily under natural conditions (Fig. 4.13).

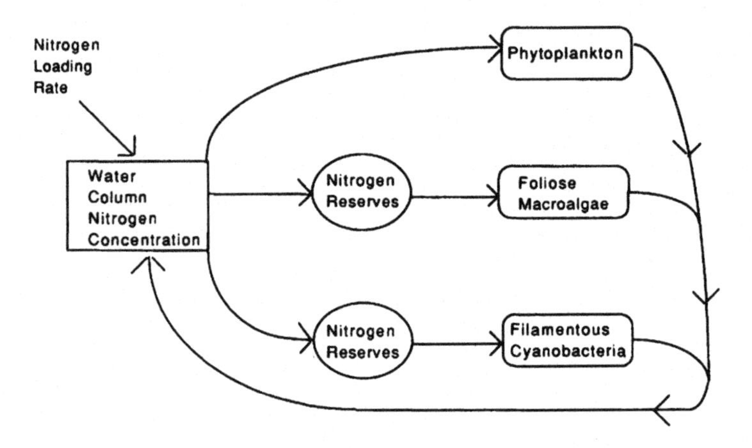


**FIGURE 4.14**

Nitrogen loading and algae growth (Source: Fong et al., 1994).

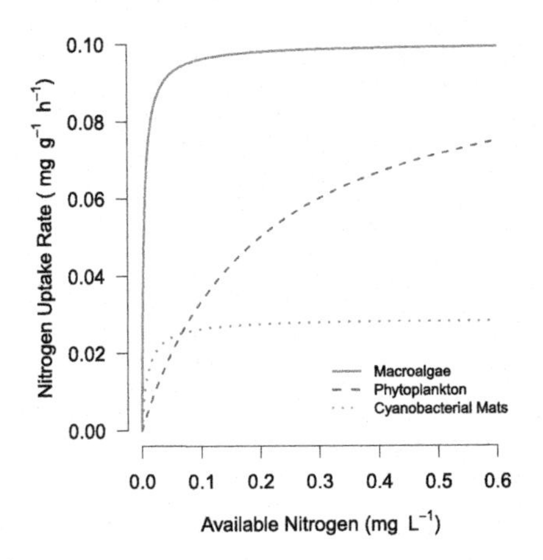


**FIGURE 4.15**

Uptake of nitrogen and growth as a result of N (Source: Fong et al., 1994).

## APPLYING NUTRIENT KINETICS TOWARD GROWTH PREDICTION

Laboratory measurements were used to define the species parameters for each of these equations: `Michaelis-Menten relationship`, `Monod Equation`, and `Droop equation`. For example, these equations were modified to model algae growth in Southern Californian coastal lagoons based on nitrogen.

Based on the conceptual model of nitrogen use in the coastal lagoons (Fig. 4.14), the differential uptake rate of nutrients could be modeled. Based on the Michaelis–Menten equation, the growth rates respond to nitrogen additions (Fig. 4.15) and explain how various species might respond with differing nutrient concentrations in terms of population growth.

Unfortunately, seston consists of a mixture of many phytoplankton species, protozoa, bacteria, and detritus. Each component may have a different cellular stoichiometry. The restricted applicability of the Monod equation and the technical difficulties in measuring the independent variable of the Droop equation have been considered a major obstacle in predicting the extent of nutrient limitation of natural phytoplankton populations.

## ECOLOGICAL VERSUS EVOLUTIONARY TIME SCALES

### ECOLOGICAL CRUNCH AND COMPETITION

Intra- and interspecific competition may drive evolution to partition resources, thus avoid competition. This mechanism may drive biological diversification. Regulatory competition experiments demonstrate how species can partition resources and become independently reproducing populations. However, on the geologic scale, the same process can be interpreted as range expansion and habitat diversification following ecological catastrophes, such as extinction events. Meshing the time scales and processes remains an area of research with more questions than answers.

### CICHLIDS AND AFRICAN RIFT ZONE LAKES

The African cichlid fish radiations are the most diverse extant animal radiations. Whereas some fish have gone through major radiations, there is evidence that numerous small radiations provide better detail on how these fish have diversified. For example, it appears that speciation rate declines through time as niches get filled up during adaptive radiation: young radiations and early stages of old radiations are characterized by high rates of speciation.

Furthermore, the number of species in cichlid radiations increases with lake size, supporting the prediction that species diversity increases with habitat heterogeneity, but also with opportunity for isolation by distance.

## NEXT STEPS

### CHAPTER STUDY QUESTIONS

1. Test density-dependent and density-independent equations to predict populations sizes using a wide range of parameters.
2. Describe how intra- and interspecific competition can influence population growth.
3. Describe how the implications of competition within ecological and evolutionary time scales differ?
4. What are the limits of competition.

### PROBLEM SETS

1. Consider basic sampling designs that you might need to estimate the population growth using Eq. (4.7) for the following taxa:

    - *Microcystis aeruginosa*
    - *Bursilla monhystera*

- Epeorus sp. (Ephemeroptera)
- *Salix* spp.
- *Alligator mississippiensis*

## LITERATURE RESEARCH ACTIVITIES

1. Using database tools, determine how the number and type of articles on competition have changed since 1990 for the following taxa: freshwater insects, amphibians, freshwater fish, and freshwater mammals. Are the trends parallel or divergent? Justify your answer using the literature.
2. Population modeling has become an important activity to predict and manage fisheries. Select three lakes where fishery modeling has been done and compare and contrast the data sources, models use, and management recommendations. Describe what you learn about modeling based on this research.

CHAPTER

# THE ROLES: BEYOND COMPETITION 5

*Every time we exterminate a predator, we are in a sense creating a new predator.*
**David Rains Wallace, The Untamed Garden and Other Personal Essays**

*Life is as dear to a mute creature as it is to man. Just as one wants happiness and fears pain, just as one want to live and not die, so do other creatures.*
**Dalai Lama 2012**

## CONTENTS



**Ecology and Management of Inland Waters. https://doi.org/10.1016/B978-0-12-814266-0.00018-0**

As a member of neotropical knifefish order (Actinopterygii: Gymnotiformes), the electric eel (*Electrophorus electricus*) inhabits calm, muddy bottom or stagnant waters in the Amazon and Orinoco River basins in South America. It has a long body and swims by undulating its elongated anal fin and can produce electric fields for navigation, communication, attack, and defense. (See Fig. 5.1.)

**FIGURE 5.1**

Image of the electric eel (*Electrophorus electricus*) (Source: Steven G. Johnson). The electric eel produces electric fields for a wide range of functions.

The electric eel has specialized organs composed of electrocytes that can generate a shock up to 860 volts and 1 ampere of current (860 watts) for two milliseconds. Although potentially lethal to many species, the discharges usually stun prey, e.g., invertebrates, fish, and small mammals. However, eel food preferences depend on their life-history stage. First-born hatchlings may eat eggs and embryos from later clutches (i.e., cannibalism), while juveniles eat invertebrates, such as shrimp and crab. This example demonstrates the ambiguities of competition. Does cannibalism reduce competition or is it a predator–prey trophic strategy or both? However we characterize these relationships, these dynamics help structure the foodweb in these South American rivers.

Whereas competition may influence population dynamics, exclude weaker competitors, and promote niche differentiation, we also recognize that numerous other types of species interactions occur in

**Table 5.1 Type of asymmetric interactions between two species.**

| Interaction | Effect of Sp. 1 on Sp. 2 | Effect of Sp. 2 on Sp. 1 |
|---|---|---|
| Exploitation | Positive | Negative |
| Competition | Negative | Negative |
| Mutualism | Positive | Positive |
| Commensalism | None | Positive |
| Amensalism | None | Negative |
| Nuetralism | None | None |

streams, lakes, wetlands, etc. Predation, cannibalism, disease, and parasitism are commonly observed factors that affect population dynamics.

After reading this chapter, you should be able to

1. Use predator–prey equations to evaluate the role of trophic interactions to predict population dynamics;
2. Summarize how sexual selection can influence population dynamics, e.g., predator/prey and infection vector and host;
3. Describe succession and how communities change with time;
4. Evaluate various lotic and lentic systems using energy transfer and nutrient flux in ecosystems.

# ENVIRONMENT: BIOTIC FACTORS

## TYPES OF SPECIES INTERACTIONS

Whereas the common use of the word "environment" refers to everything around us, ecologists identify two broad environmental factors: 1) abiotic factors: the physical and chemical factors, and 2) biotic factors: interacting organisms. When evaluating an area, we analyze a combination of abiotic and biotic factors to better understand the patterns and processes in ecological systems. The biotic factors include numerous types of interactions, including competition, discussed in Chapter 4. Although inter- and intraspecific competition may play an important role in population growth, other asymmetric interactions also structure communities in visible and dramatic ways (Table 5.1).

# EXTENDING POPULATION BIOLOGY

## PRIMARY PRODUCTION AND HERBIVORY

Periphyton is a combination of algae and bacteria that grow on stream bottom substrates. The growth of the algae is not only controlled by the competition for light and nutrients, but also by herbivory. Herbivory is a type of exploitation as it reduces algae (or plant) biomass, but have other effects too, e.g., increasing the diversity of the algae species, increasing productivity by reducing self-shading, and stimulating growth by releasing nutrients. But these changes can also depend on the herbivore species.

For example, in a laboratory study testing which herbivores impact algae assemblages, the mayfly *(Centroptilum elsa)* had little effect on periphyton biomass and chlorophyll *a* concentrations. Whereas

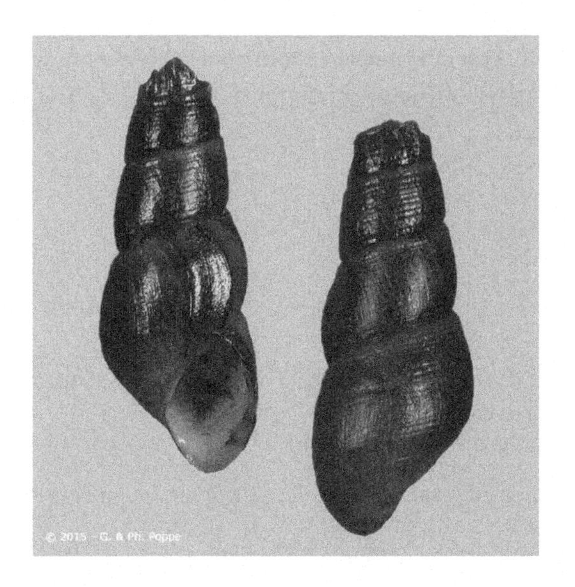

**FIGURE 5.2**

*Juga silicula shastaensis* (Source: G. & P. Poppe).

the snail (*Juga Silicula*) (Fig. 5.2) reduced periphyton biomass and chlorophyll *a* by nearly 50%, reduced the cyanobacteria population and some diatom species, and increased rate of primary production by ~25%. These results demonstrate that herbivores can not only have an ecosystem affect, but specific species can have community assemblage implications.

## PREDATOR–PREY DYNAMICS

Predator–prey models describe some of the most visible interactions in biological systems. In a broad sense, species compete, evolve, and disperse as they seek resources to sustain their existence. But resources, such as plant- or animal-based food, can be modeled as a predator–prey system and another type of exploitation.

Depending on the setting, these interactions can be described as resource-consumer, plant-herbivore, parasite-host, or susceptible-infectious interactions. These loss-win interactions dominate ecosystems and, in fact, some competitive interactions may be predator–prey interactions in disguise.

First, let us define a predator population equation in the absence of prey. Without food resources, we might expect the predator to decline exponentially, as described by the following equation:

$$dP/dt = -qP. \tag{5.1}$$

This equation uses the product of the number of predators ($P$) and the predator mortality rate ($-q$), to describe the rate of decrease of the predator population ($P$), with respect to time ($t$).

In the presence of prey, however, this decline is opposed by the predator birth rate, which is determined by the predator consumption rate of prey. The consumption of prey is the product of four

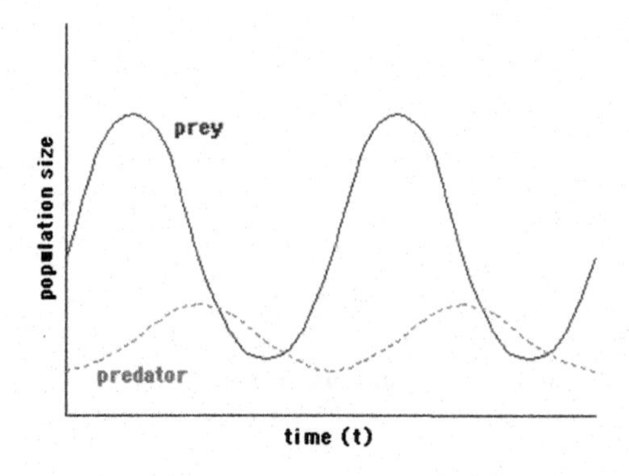


**FIGURE 5.3**

Predator–prey dynamics (Source: M. Beals, L. Gross, and S. Harrell).

parameters: number of predators ($P$); number of prey ($N$), the attack rate ($a'$), and the predator's efficiency to convert food into offspring ($c$):

$$dP/dt = ca'PN - qP. \tag{5.2}$$

The entire term, $ca'PN$, tells us that increases in the predator population is proportional to the product of predator and prey abundance. As predator and prey numbers ($P$ and $N$, respectively) increase, their encounters become more frequent, whereas the actual rate of consumption will depend on the attack rate ($a'$).

Turning to the prey population, we would expect that without predation, the numbers of prey would increase exponentially. The following equation describes the rate of increase of the prey population with respect to time, where $r$ is the growth rate of the prey population, and $N$ is the abundance of the prey population:

$$dN/dt = rN. \tag{5.3}$$

We saw the same equation in Chapter 4. However, with predation, the prey population is prevented from increasing exponentially. The term for consumption rate from above ($a'PN$) describes prey mortality, and the population dynamics of the prey can be described by the equation

$$dN/dt = rN - a'PN. \tag{5.4}$$

The product of $a'$ and $P$ is the rate of prey capture as a function of prey abundance. The $a'PN$ term describes the reduction in the prey population and is proportional to the product of predator and prey abundances.

Eqs. (5.2) and (5.4) describe predator and prey population dynamics in the presence of one another, and together make up the Lotka–Volterra predator–prey model.

Depending on the parameters, the model predicts a cyclical relationship between predator and prey numbers (Fig. 5.3). With available prey ($N$), the number of predators ($P$) and the consumption rate ($a'PN$) increases. An increase in the consumption rate, however, will reduce the number of prey ($N$),

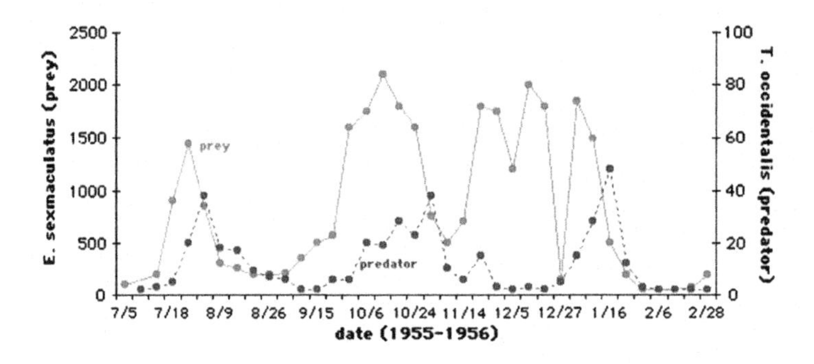


**FIGURE 5.4**

Huffaker et al. (1958) reared two species of mites to demonstrate these coupled oscillations of predator and prey densities in the laboratory. Using *Typhlodromus occidentalis* as the predator and the six-spotted mite (*Eotetranychus sexmaculatus*) as the prey, Huffaker constructed environments composed of varying numbers of oranges (fed on by the prey) and rubber balls on trays. The oranges were partially covered with wax to control the amount of feeding area available to *E. sexmaculatus*, and dispersed among the rubber balls. The results displayed are from one of the many experimental permutations. Note that the prey and predator population axis have different scales (Source: Huffaker et al., 1958, Figure 18).

which causes $P$ (and therefore $a'PN$) to decrease. As $a'PN$ decreases the prey population is released from predation pressures and $N$ increases. With more prey, $P$ can increase, and the cycle begins again.

These models were tested using mites in the late 1950s. After developing an experiment with some spatial complexity, Huffaker was able to replicate the model predictions (Fig. 5.4).

Although these models are mathematically tractable, the Lotka–Volterra model rely on unrealistic assumptions. For example, prey populations are limited by food resources and not just by predation, and no predator can consume infinite quantities of prey. Although cyclical relationships between predator and prey populations have been demonstrated in the laboratory or observed in specific natural settings, models based on the Logistic equation may fit data better because it effectively describes predator's consumption rate changes as prey densities change.

Nevertheless, predator–prey dynamics are an important selective pressure. The capacity of a prey species to help to avoid detection and attack has a clear adaptive advantage. Camouflage and developing hard shells add protection and avoid becoming prey. Meanwhile, fish, birds, and amniotes have developed various structures to "find" and extract prey from protective structures, e.g., powerful jaws, teeth, and beaks.

## PACIFIC LAMPREY: ANADROMOUS PARASITES

Similar to predation, parasitism can be advantageous for some plants and animals. For example, the Pacific lamprey, *Entosphenus tridentatus*, parasitizes fish. Inhabiting the coastal streams that drain into the Pacific Ocean, the Pacific lamprey are also an important ceremonial food for Native American tribes in the Columbia River basin.

Ammocoetes, lamprey larvae, live in freshwater for several years, usually 3–7 years. Ammocoetes are filter feeders and draw overlying water into burrows to obtain food. After the larval period the am-

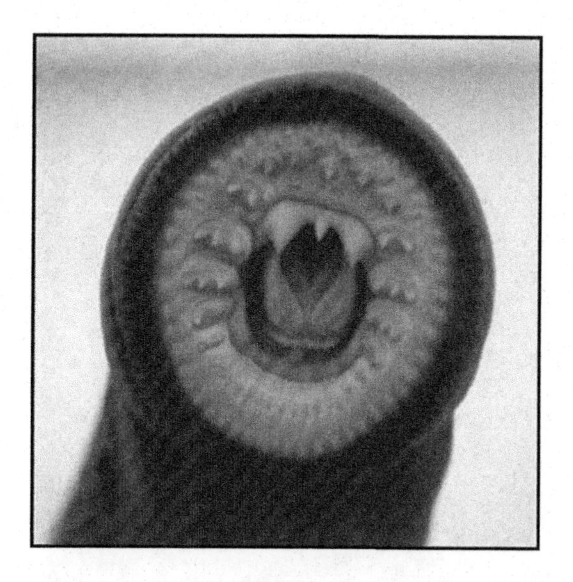

**FIGURE 5.5**

Pacific lamprey mouth part (Source: USGS, Western Fisheries Research Center).

mocoetes undergo metamorphosis and equipped with a juvenile/adult body morphology, they have a sucker-like mouth that allows them to parasitize fish (Fig. 5.5). The anadromous juvenile swims downstream to the ocean. In the marine environment, matured adults can parasitize fish, scavenge, or act as predators. The adults live at least one to two years in the ocean and then return to freshwater to spawn. The trip from freshwater to seawater and back requires a major physiological shift to compensate for change in osmotic pressure in fresh versus salt waters that probably evolved over 300 mya. Little information exists on lamprey use of the ocean; hence, changes in ocean upwelling, acidification, coastal runoff, food availability, and temperature may influence lamprey survival. Reductions in the availability of their hosts, Pacific salmon (*Salmo* spp.), and some other marine fishes, affect adult lamprey survival and growth and can be modeled conceptually as a predator–prey model. Thus based on Eq. (5.2), and $ca'PN$, the decline in Pacific lamprey might be explained by the decline of hosts ($N$).

## BACTERIA AND BACTERIOPHAGE INFECTIONS AND COEVOLUTION

Another antagonistic relationship is between bacteria and bacteriophage (viruses) (Fig. 5.6). For example, cyanophages can infect a broad range of Cyanobacteria, e.g., *Anabaena* spp., *Microcystis* spp. and *Planktothrix* spp., where cyanophages can dramatically reduce blooms. Nevertheless, just as the models in predator–prey dynamics, bacteriophage-bacteria can exhibit a time lag antagonistic pattern (Fig. 5.3).

Although antagonistic interactions influence populations dynamics, antagonistic coevolution may also operate at the gene level. The evidence to support this idea had generally been observed in agricultural fields and human directed selection for pathogenic resistance. While "coevolving" communities

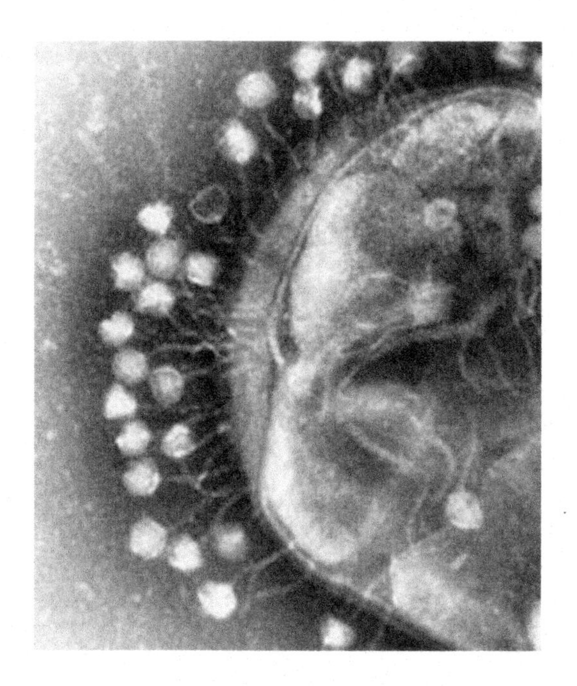

**FIGURE 5.6**

Bacteriophage attached to and infecting a bacterium (Source: Graham Beards).

have been created in the laboratory, they are often sustained for only a few generations, thus stretching the concept of coevolution.

In the case of viruses, for example, the cynophages can upregulate certain genes to generate energy needed for the phage to reproduce. But even beyond gene regulation, cyanophages can carry and insert genes to increase host photosynthetic efficiency. With external sources of DNA and genes for photosynthesis, the cyanophage will infect the hosts and promote an increase of photosynthetic rates before the cells are killed.

## PARASITE MEDIATED SEXUAL AND ASEXUAL HOST REPRODUCTION

Like predation and virus infections, parasitism can play an important role in population dynamics via reproductive mechanisms. At least eleven species of Trematoda (Platyhelminthes, see page 115) parasitize the New Zealand mudsnail. What controls the population dynamics of host and parasites? At low host densities, the spread of parasites might be limited, whereas at high host densities, an epidemic of parasitism can occur, where the parasite reduces the survival or fecundity of the host. Clearly, the relationship is antagonistic, but some interesting subtleties may also play out. (See Fig. 5.7.)

The New Zealand mudsnail populations are composed of sexual and asexual populations. Since asexual lineages tend to have higher per individual growth rates, we might expect asexual populations to replace sexual ones. However, as asexual genotype populations become more common they are also

**FIGURE 5.7**

Infected New Zealand mudsnail, *Potamopyrgus antipodarum* (Source: Gabe Hart).

more vulnerable to trematoide infections. These observations suggest that sexual and clonal individuals respond to different selective pressures (biotic: infection vs. abiotic: resource availability).

## RED QUEEN AND COURT JESTER HYPOTHESES

Coevolution in species with antagonistic interactions, such as in predator–prey, host-parasite or exploiter-victim systems, is also an important process to create and maintain species diversity. In 1973, Van Valen proposed the Red Queen Hypothesis, which emphasizes the primacy of biotic conflict over abiotic forces in driving selection. Species must continually evolve to survive in the face of their evolving enemies, yet—on average—their fitness remains unchanged. (See Fig. 5.8.)

But there are more subtle aspects of the relationship. Pathogenic parasites can drive host resistance; resistant hosts select against ineffective parasites, thus parasite success relies on its capacity to overcome the resistance. Thus the parasite-host relationship has direct bearing on the coevolving characteristics for both the parasite and host. Therefore we might hypothesize that continual adaptation is required for a species to maintain its reproductive success. This hypothesis is referred to as the Red Queen's Hypothesis, taken from Lewis Carroll's *Through the Looking-Glass* and has been used to explain 1) the advantage of sexual reproduction, and 2) the constant evolutionary arms race between interacting species. Alternatively, the Court Jester Hypothesis proposes that macroevolution is driven mostly by abiotic events and forces, which we will begin to address in earnest in Chapters 6 and 7.

In the case of the New Zealand mudsnail, the Red Queen Hypothesis postulates that high infection rates in the common asexual clones could periodically favor the genetically diverse sexual individuals and promote the short-term coexistence of sexual and asexual populations. There is on-going work to test alternative hypotheses.

## YEAST INFECTIONS IN DAPHNIA

The fungi species, *Metschnikowia bicuspidata,* infects *Daphnia dentifera* and with infections rates that sometimes exceed 60%. But *Daphnia* can also develop resistance, although resistant lines are less fecund. In a study comparing lakes with differing levels of epidemics, the proportion of susceptible

**FIGURE 5.8**

The Red Queen and Alice, taken from Lewis Carrol's sequel to Alice and Wonderland, *Through the Looking Glass* (Source: Sir John Tenniel).

*Daphnia* increased in some lakes and declined in other lakes (Fig. 5.9A). It seems counterintuitive that a host might evolve greater susceptibility to their virulent parasites during epidemics.

Yeast infections are not the only biotic interaction with *Daphnia*, they are also subject to predation. When predation is high as indicated when body sizes are smaller, the epidemic size decreases (Fig. 5.9B). Invertebrate predation may depress yeast epidemics because fish selectively cull infected *Daphnia*.

The epidemics may also depend on the nutrient levels and productivity, where higher nitrogen and possibly phosphorous concentrations are associated with large epidemics (Fig. 5.10). With higher lake productivity, *Daphnia* increased their resistance in spite of the costs on fecundity. Meanwhile, in lakes with low productivity, *Daphnia* had an increase in susceptibility.

When there are ecological constraints on the epidemic size (lower nutrient levels, higher predation rates), *Daphnia* became more susceptible to infection. Thus host susceptibility may depend on the ecological context of an epidemic.

## ECO-EVO EFFECTS: GUPPIES IN TRINIDAD

Guppies have been studied since the 1940s and the results are spectacular. In fact, guppies (*Poecilia reticulata*) are one of the best vertebrate species for studying the relationship between ecology and evo-

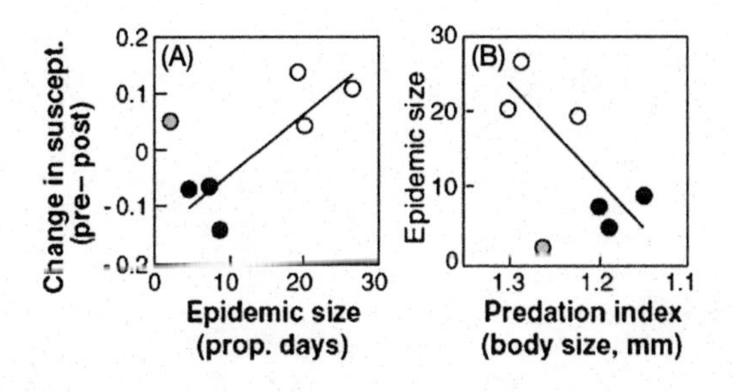


**FIGURE 5.9**

Relationships between epidemic size, predation, productivity, and evolutionary outcomes (Source: Duffy et al., 2012). Top panels show links with epidemic size (quantified as the area under the infection prevalence curve through time). (A) Epidemic size versus change in mean susceptibility (mean proportion infected of post-epidemic genotypes subtracted from mean of pre-epidemic genotypes). (B) Epidemic size is on the y axis, plotted versus predation intensity (where smaller body size indicates higher predation). Open symbols denote lake populations that evolved increased resistance, black symbols indicate increased susceptibility, and gray symbols indicate no significant evolutionary change.

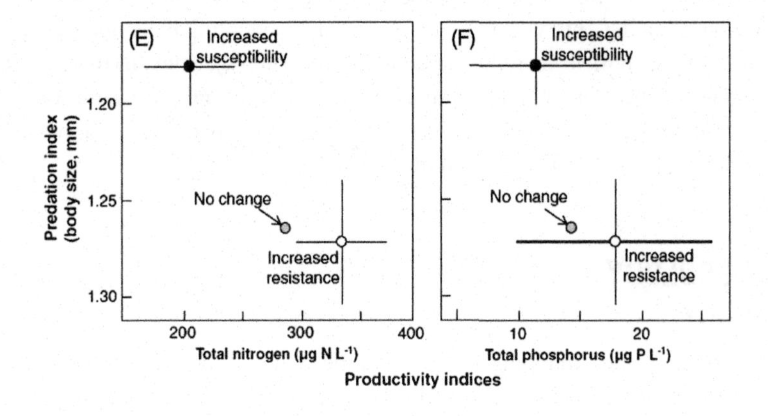


**FIGURE 5.10**

Evolutionary outcomes mapped onto predation-productivity space (Source: Duffy et al., 2012). Note that the y axis scales with increasing predation intensity, so small body size (high predation) is at the top. Points in (E) and (F) are lake means for nitrogen and phosphorous, respectively (±1 SE).

lution in the wild, especially on the island of Trinidad. Because guppies have short generation span and live in streams, where waterfalls restrict migration, there is variation in behavior, life history, physiology, and appearance among populations that can be attributed to variation in the predator community. Ongoing research continues to unravel the eco-evo relationships using guppies as model organisms.

Below waterfalls guppies are subject to predation by large vision-oriented predators, so male guppies have cryptic coloration to avoid being predated. But female selection relies on coloration; the more ornate and colorful, the more attractive the male and likelihood of mating and predation (Fig. 5.11)!

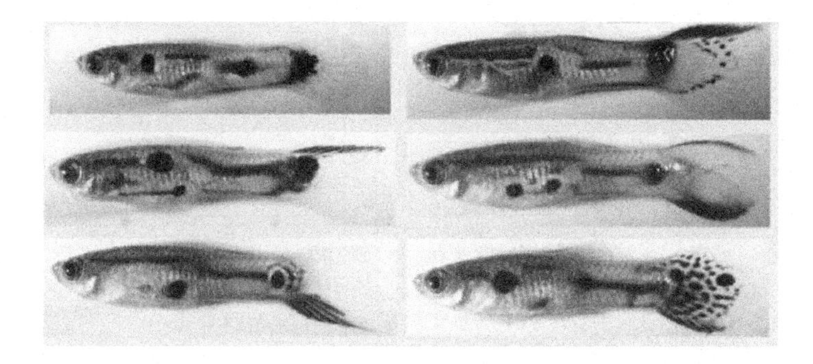

**FIGURE 5.11**

Image of male guppies. Uniquely ornate males are more likely to be selected by females to mate (Source: Hughes et al., 2013).

Above waterfalls the fish community is simpler and predation pressure is much reduced; the male fish in these streams are more colorful, while below the falls they are cryptically colored.

Cryptic coloration is not the only way to avoid predation—schooling behavior, evasive tactics, mating activity, and foraging activities can reduce predation vulnerability. Some of these might be learned, some might be under genetic control, and some associated with geographic variation as antipredator responses, or based on predator behavior. Based on the research being done on guppies, many of these confounding factors may be isolated and tested individually.

## COMMUNITY ECOLOGY

### DYNAMIC ECOLOGY

In contrast to plant catalogs and mapping, the development of ecology in North America was based on how communities (plant assemblage changed over time). Composed of populations of interacting species that occupy the same geographical area, communities seemed to follow predictable patterns of changes or succession. Whether in sand dunes, recently exposed sediments following glacial retreats, or fires, communities often had successional stages (seres) before reaching what might be called a stable "climax". "Dynamic ecology" was coined for some time to emphasize the importance of environmental changes in response to disturbance (secondary succession) or colonization of newly exposed substrates (primary succession).

That communities "developed" into a climax community, referring to the abiotic drivers, i.e., climate, has been a central tenet in community ecology. Although the field is plagued with ontological categories, community ecologists have made peace with these categories by removing some of the anthropomorphic references, e.g., communities develop from "immature to mature" states, like individuals. In addition, after the 1940s a community was best described as a set of organisms that individualistically contribute to what we perceive as a community and not as a coevolving, codeveloping organism.

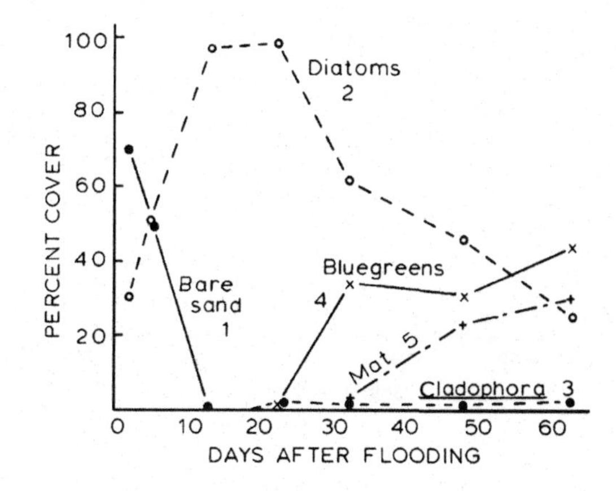


**FIGURE 5.12**

Algal recovery after a flood in Sycamore Creek (Source: Fisher et al., 1982).

## DISTURBANCE AND RECOVERY

Since the early 20th Century, ecologists have appreciated the role of disturbance in structuring communities. First noted in old-fields, disturbance was understood as a process that creates a succession of vegetation patterns. And just like disturbances in forest, such as fires, rivers experience disturbances, e.g., floods, which can be measured in terms of their magnitude and frequency.

In the simplest case, floods might remove the algae, invertebrates, and fish and carry them downstream. For example, a flash flood in Sycamore Creek, Arizona removed all of the algae and reduced invertebrate standing crop by 98%. Physical and morphometric conditions typical of the pre-flood period were restored within 2 days and the algae community recovered in 2–3 weeks. Algal communities responded rapidly and achieved a standing crop of nearly 100 g $m^{-2}$ in 2 weeks. But the algae community composition recover included a series of stages, where diatoms dominated the stream bottom early in the succession, followed by filamentous greens, and finally with Cyanobacteria (Fig. 5.12).

Macroinvertebrates also recolonized denuded substrates rapidly as flying aquatic insects laid eggs in or near the stream. Mayfly and dipteran taxa grew rapidly and completed several generations to become dominant invertebrates during the 1st month after the flood.

Disturbance looks different in lakes. Glaciation is, perhaps, the most common and widespread form of disturbance that affected nearly every high-elevation region in the Sierra Nevada Mountains and Rockies and, of course, much of the higher latitudes of North America. The Great Lakes and most of the lakes of the mid-west of the United States have been formed by glaciation. Lakes have been "developing" every since the glaciers have retreated. In many cases, lakes become more productive and become filled with sediment or vegetation.

A disturbance in the watershed can have dramatic impacts in streams, but these effects are also felt by the receiving waters, such as lakes. For example, fires are known to cause different changes in soil chemistry in different forest types. Furthermore, the frequency and magnitude of fires and the successional patterns of forest after fires can impact stream and lake chemistry. As a record of lake chemistry, diatoms buried in the sediments can record chemistry changes over long periods. In some cases, diatom records demonstrate the role of fire in lake chemistry, but in other cases, the signal is too

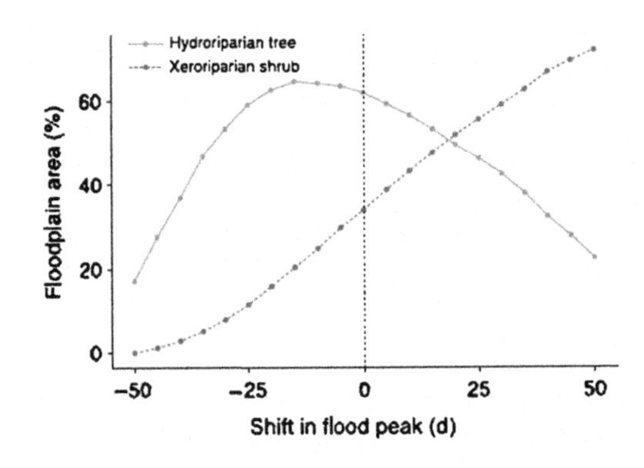


**FIGURE 5.13**

Effect of flood timing on hydroriparian tree and xeroriparian shrub dominance on the floodplain. Negative numbers indicate a shift to earlier flood peaks, positive numbers to later flood peaks. Hydroriparian tree was more abundant with slightly earlier flood peaks, but all other shifts in flood timing favored xeroriparian shrub (Source: Lytle et al., 2017).

small to detect because of forest characteristics. For example, sediments in Kettle Lake, Canada, suggest that fire-induced changes in lake chemistry was muted in catchments dominated by black spruce (*Picea mariana*), where nutrients and major ions losses were minimized because of the humus layers.

## BEHAVIORAL GUILDS

We can distinguish guilds based on habitat selection. For example, guild may be distributed along nutrient and flow-disturbance gradients. For example, in high-disturbance habitats, nutrient-poor waters, algae may present a physically low profile guild, while at the locations with low flow disturbance, high-nutrient waters may present a high-profile. Some algae are motile, and their relative abundance can increase with nutrient concentrations and decrease along the disturbance gradient. (See Fig. 5.13.)

These patterns demonstrate how subtle difference in diatom growth morphologies, environmental conditions, and even behavior can be subject to specific species-environment interactions, and may even be used to evaluate anthropogenic impacts on ecosystems.

## LIFE-HISTORY STAGES AND VEGETATION DYNAMICS IN WETLANDS

Many wetlands have pronounced seasonal vegetation dynamics and animal population patterns. A useful way to understand plant successional patterns relied on grouping plants in groups with similar life-history traits. Using examples from pothole prairie wetlands dominated by annual forbs, there is a succession of species that germinate, grow, and set seed. (See Fig. 5.14.) In simple terms, we can look at some plants using the following characteristics:

- get in and out early,
- get in at the end and thrive quickly and set seed, or
- have a flooded stage and a non-flooded life-history stage.

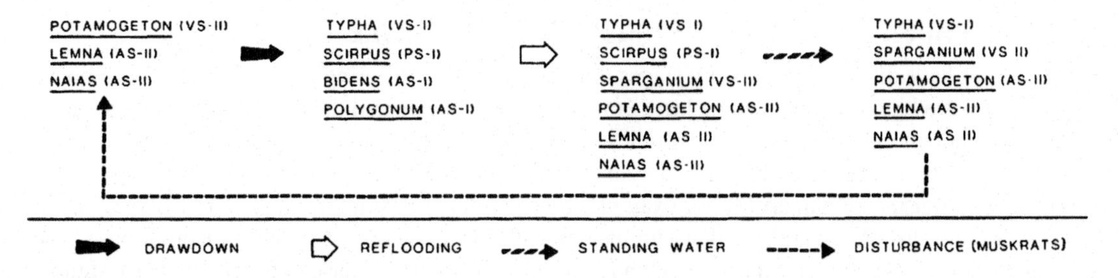


**FIGURE 5.14**

Model of wetland succession (Source: van der Valk, 1981).

**FIGURE 5.15**

Image of vernal pool in San Diego County (Source: San Diego Reader, Daniel Powell, March 23, 2016).

Adding another component, we might map history characters within a wetland. For example, in wetland, drawdown plants can be distributed in concentric rings of showy wildflower patterns. (See Fig. 5.15.)

In another example, Southern California vernal pools occur on hummocky coastal terraces, where compacted clay and cemented hardpans layers block water infiltration, which allows the water to pool.

When there is a large enough rain event, the depressions experience four phases:

***wetting phase*** where detritus releases nutrients for bacterial, cyanobacterial, and algae growth. Seeds germinate and dormant invertebrates will hatch and begin developing;

***aquatic phase*** is characterized by saturated soils, anoxic waters, and the systems are generally phosphorous-limited and dominated by aquatic plants and crustaceans grazing on bacteria and algae. Birds, amphibians, and other aquatic animals colonize the pools;

***drying phase*** as water evaporates, the salinity increases and water-tolerant plants dominate, flower, and set seed. Invertebrates lay eggs, which enter dormant cyst stages;

***drought phase*** the dry depressions are dominated by grasses, most other vegetation is dormant and may be habitat for terrestrial animals.

## METAMORPHOSIS LIFE HISTORY STAGES: AQUATIC LARVAE

Many species have different life-history stages that are adapted to differing conditions. The most common strategy is a biphasic cycle with a larva emerging from the egg and then transform from larvae to juvenile via metamorphosis. This strategy relies on dramatic morphological, physiological, behavioral, and ecological transformations between the larvae and the juvenile. The advantage of such biphasic life-history strategies are numerous. For example, for lotic macroinvertebrates, the larval stages allow for downstream dispersal via drift, dietary changes with instar stages, or habitat selection based on available resources. Although much is known of life-history strategies based on this relatively simple system—larva, metamorphosis and juvenile—very little is known about how it originated, but provides information regarding critical options for organisms with changing conditions.

## LINKING SUCCESSION TO POPULATION DYNAMICS

But one question that ecologist ask is, "Why do some species dominate early after the disturbance and others dominate in later stages?" Based on observations, such as the those of Sycamore Creek, ecologists have categorized taxa between "r-selected" and "K-selected" species. Using the Verhulst model (Chapter 4, Eq. (4.7)) parameters

$$\frac{dN}{dt} = rN\left(1 - \frac{N}{K}\right),$$

where r-selected species are those that emphasize high growth rates, whereas K-selected species are those that have low reproductive rates and are competitive when the population is near the carrying capacity.

In unstable or unpredictable environments, r-selection allows for quick reproduction. Thus, putting energy to better compete against other species is lost, because these species will find new places (in time or space) to exploit. r-selected species tend to have high fecundity, small body size, early maturity onset, short generation time, and the ability to disperse offspring widely. Organisms that exhibit r-selected traits can range from bacteria to diatoms, aquatic insects to crustaceans.

In contrast, K-selected species compete successfully for limited resources and their population size is relatively stable, approaching the carrying capacity. These strong competitors may have the following characteristics: large body size, long life expectancy, and the production of fewer offspring, which often associated with extensive parental care until they mature. With fewer offsprings, parental investment is high, which increases the probability of survival. Organisms with K-selected traits include large organisms, such as dolphins, riparian trees, and even green algae.

## R/K SELECTION AND AGE-SPECIFIC MORTALITY

Although r/K selection theory found good use in 1970s, gaps in the theory and ambiguities in how to interpret empirical data limited the concept's utility. For example, willow trees produce a multitude of offsprings and capitalize on newly opened riparian habitats (r-selected). But willow thickets are often monotypic and other species are shaded out and can persist for decades (K-selected). This example demonstrates the judgment calls needed to categorize species, which makes its use as a predictive tool quite limited.

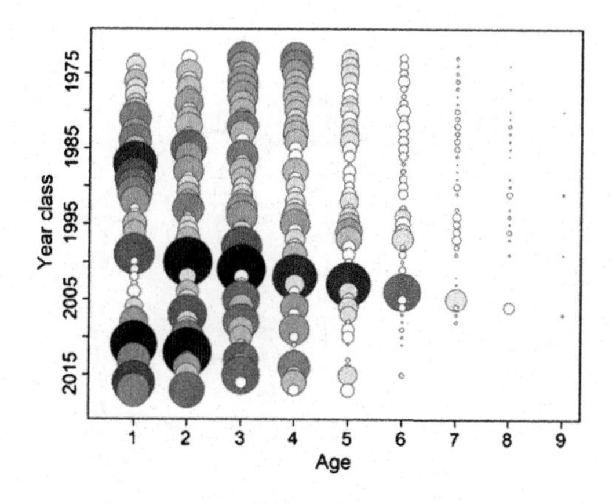


**FIGURE 5.16**

Bubble plot showing the proportion of the annual trawl survey catches attributed to each class, through time (1973–2017). The size and color of the circles represent the proportion of each age class relative to total abundance in a given year. Larger darker circles represent larger proportions, whereas smaller lighter circles represent smaller proportions. (Source: Vidal et al., 2019).

As more quantitative analyses were conducted, ecologists noted that the density-dependent selection could depend on organisms' life-histories. By providing a mechanistic causative link between an environment and an optimal life-history, ecologists used age-structured models (e.g., Leslie Matrix models, page 138) as a framework to incorporate many of the themes important to the r/K selection concepts. (See Fig. 5.16.)

For example, in the case of alewife (*Alosa pseudoharengus*), the life stages have been truncated by mortality—in this case with the introduction of chinook salmon (*Oncorhynchus tshawytscha*). In Lake Michigan, alewife biomass has generally been declining through time, although infrequently strong year classes have been observed, substantially boosting biomass at times. In recent decades, the age composition reflects fewer old fish.

Evidence from exploited fish populations suggests that age truncation can lead to greater variability in spawner biomass, which has been correlated with greater sensitivity to the environment

# ECOSYSTEM ECOLOGY

## LINDEMAN AND ECOSYSTEM OF A LAKE

Whereas community ecologists stress the temporal changes in communities, ecosystem ecologists focus on energy flux and nutrient flow within a system. Like community ecology, ecosystem ecology includes the community, but tends to place a strong emphasis on abiotic factors (e.g., precipitation, temperature, etc.) and trophic transfers. The work of Raymond Lindeman is often cited as a model for ecosystem ecology. He conducted research in several small lakes or ponds and measured the biomass and energy transfer from one trophic level to the next. (See Fig. 5.17.)

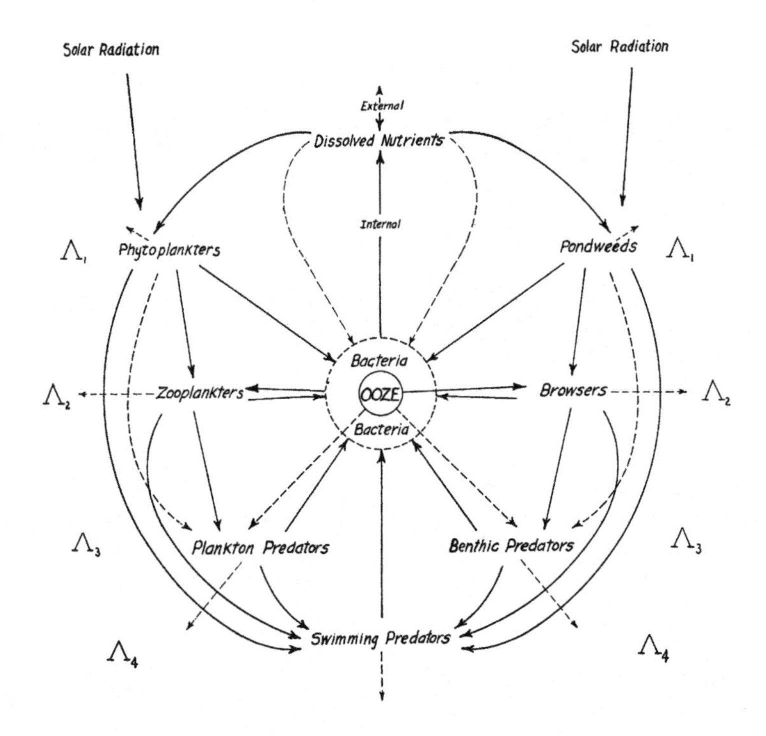


**FIGURE 5.17**

The original illustration by Lindeman that identifies trophic relations in terms of energy leads, which he then used to formulate a concept of biological efficiency (Source: Lindeman, R.L., 1942, The trophic-dynamic aspect of ecology, Ecology, 23, 399–417 (Lindeman, 1942)). Note the $\Lambda_{1...4}$ symbols that are used to identify trophic efficiency, i.e., the proportion of energy transferred from one trophic level to the next.

By defining a functional integration of organic matter as a source of energy and inorganic nutrient cycles, Lindeman defines an ecosystem as "the system composed of physical-chemical-biological processes active within a space-time unit of any magnitude, that is, the biotic community plus its abiotic environment." (Lindeman, 1942, p. 400).

Although the calculations of standing biomass and productivity is key to Lindeman's paper (Fig. 5.18), he framed these "biomass pools" within the context of successional development—thus subsuming community ecology concepts into ecosystem ecology. With an emphasis on rate-controlling processes and the ecological efficiency of energy transfer, he reduced trophic relations to a common denominator, energy, and created a theoretical structure that would provide predictions that future ecologists could test.

## TROPHIC FUNCTIONAL GROUPS

The concept of trophic functional groups is a cornerstone of ecosystem ecology, where organisms are categorized according to how they obtain energy and what they ingest. Ecologists characterizations of trophic groups:

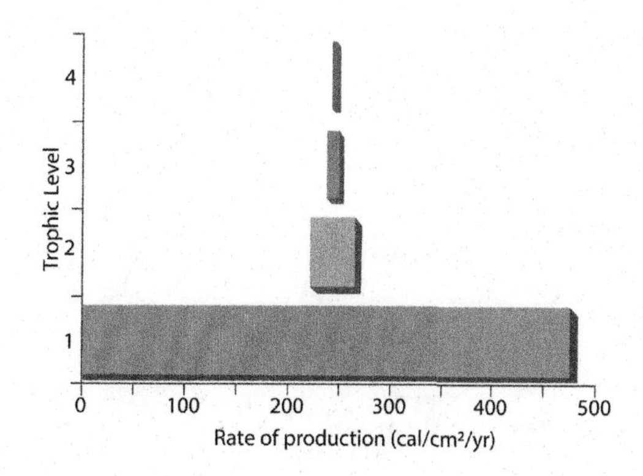


**FIGURE 5.18**

The trophic pyramid based on Lindeman's research (Source: Lindeman, 1942).

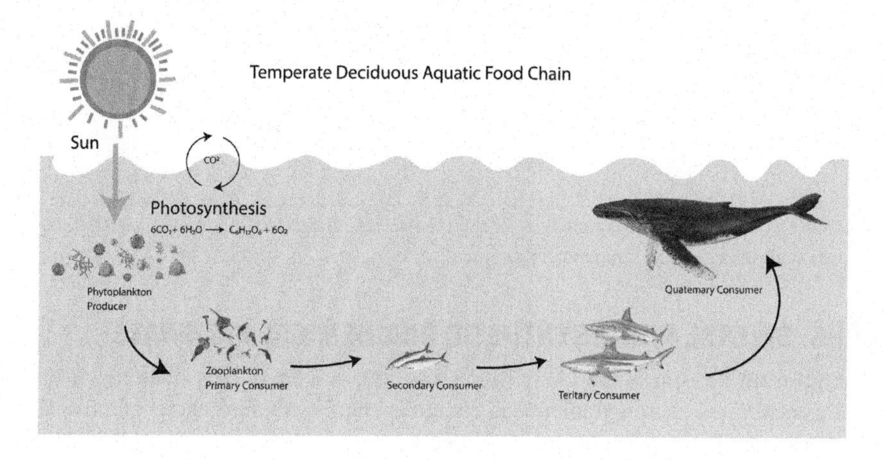


**FIGURE 5.19**

Image of a simple food chain.

- primary producers, autotrophs, organisms that fix carbon and often obtain their energy from the sun;
- primary consumers, which are heterotrophs that eat primary producers;
- secondary consumers or predator eat primary consumers, followed by
- tertiary and higher level consumers or top predators.

In a simple system, food chain diagrams are popular renditions of how these groups interact (Fig. 5.19). Unfortunately, few systems are simple. And not all organisms are cooperative enough to fit into these nice categories. For example, some species are mixotrophic: both autrophic and heterotrophic. Even Lindeman recognized the microbial component, which he referred to as ooze—saphrophytic but consume organic matter from every level.

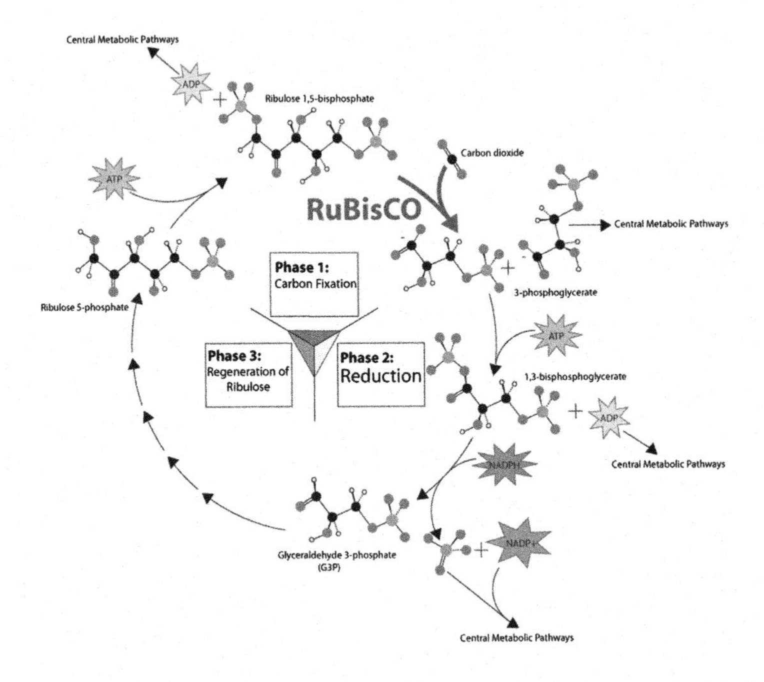


**FIGURE 5.20**

Image of the Calvin Cycle (Source: Mike Jones), which is sometimes referred to as the dark reaction, because it can continue to operate without a solar radiation energy source.

## AUTOTROPHS: DIVERSE PHOTOSYNTHETIC BIOCHEMICAL PATHWAYS

Capturing energy from the sun is the "base" for most aquatic ecosystems. Photoautotrophs convert electromagnetic energy into proton gradient and chemical energy, whereas chemoautotrophs capitalize on redox reactions from inorganic molecules to generate energy. We will explore these topics further in Chapter 11.

Autotrophs fix $CO_2$ to make sugars. For many algae and plants, the actual fixation of $CO_2$ relies on the enzyme RuBisCO to combine $CO_2$ to a five-carbon sugar molecule to produce two three-carbon sugars (Fig. 5.20). These three-carbon sugars are further processed to produce glucose. After fixing $CO_2$, the Calvin Cycle recycles the 3-carbon sugars into a 5-carbon sugar to fix another $CO_2$ molecule.

The conversion from solar energy to "biochemical" energy begins with the light reaction, where solar energy is used to create a $H^+$ gradient and NADPH, which are then used to drive the dark reaction, e.g., Calvin Cycle. The conversion efficiency of photosynthesis varies, where algae can approach 3%, terrestrial crops are ~1%, producing biomass.

The amount of carbon biomass produced in a given area and time period is called primary productivity. Primary productivity is often reported as net primary productivity (NPP), which is gross primary productivity (GPP) minus the respiration and herbivory. Wetlands are known to have some of the highest proclivities, but they can also be quite variable. In general, the values in nonforested wetlands between 500–2000 g C $m^{-2}$ $yr^{-1}$, where the global terrestrial average is approximately 450 g C $m^{-2}$ $yr^{-1}$.

Various physiology mechanisms can increase the conversion efficiency of light into energy, especially in terms of water relations. For example, plants with the C4 pathway have anatomical and physiological adaptations to concentrate $CO_2$ (Fig. 5.21). As a result C4 plants, can fix $CO_2$ more efficiently per molecule of $H_2O$ water lost to evapotranspiration. In contrast to terrestrials plants, where water is limited, aquatic plants also have several mechanisms to concentrate $CO_2$ because of limited $CO_2$ concentrations.

Though the concentration in the atmospheric $CO_2$ is only ~410 ppm (as of 2019), it is in relatively good supply compared with water. In water the diffusion rates of $CO_2$ are lower than in air. In addition, $CO_2$ is in equilibrium with carbonate and bicarbonate, which can further reduce carbon availability for photosynthesis.

To capture limited $CO_2$ amounts, numerous aquatic photoautotrophs have $CO_2$-concentrating mechanisms. These mechanisms rely on active inorganic C influx across a membrane (or membranes) that are widespread in aquatic photoautotrophs, e.g., C4 and CAM physiology mechanisms; they will be discussed in more detail in Chapter 11.

The biomass produced by autotrophs and its annual production represents what is available to primary consumers as the base of the foodweb for many ecosystems.

## PRIMARY CONSUMERS: PRODUCTION AND BIOMASS

Primary consumers or secondary producers are organisms that feed on basal food sources (i.e., plants or detritus) and form the first feeding link in community foodwebs and trophic structures. In the water column, examples include crustaceans such as *Daphnia* spp. and *copepods* (Fig. 5.22). On stream channel bottoms, insect larvae might graze algae attached to rocks (periphyton), whereas snails are important consumers of algae growing on plants (epiphyton). Unfortunately, the terminology has confounded several trophic patterns. For example, do primary consumers eat living autotrophs or does this include dead autrophic material, such as leaves. Often these taxa are called detrivores, which can lead to some ambiguities in the terminology and requires one to pay attention to how species are grouped.

Nevertheless, invertebrates that consume (allochthonous) organic matter provide an important linkage between the leaves that fall and enter the water column and the foodweb dynamics in the stream.

For example, the New Zealand mud snail (*Potamopyrgus antipodarum*) can process a great deal of inorganic biomass and become a dominant consumer. In fact, *Potamopyrgus antipodarum* in Polecat Creek, Georgia had one of the highest production rates ever reported for a stream invertebrate (194 $g/m^2/yr$). In contrast, native invertebrate production ranged from 4.4 to 51 $gm^2/yr$. It is not rare that invasive species dominate aquatic systems, and in this case, *Potamopyrgus* might be responsible for over 65% of the invertebrate productivity.

However, besides species composition, water chemistry may influence secondary production. For example, in streams on the Allegheny Plateau of West Virginia, crayfish (*Cambarus bartoni*) may constitute >55% of shredder biomass in West Virginia streams, while contributing significantly less (<20%) to the annual shredder production. However, annual production of the other common shredder species differ among streams as a function of pH and alkalinity (Fig. 5.24), where crustaea, *Gammarus minus* (Fig. 5.23) and the giant crane fly (Diptera: *Tipula abdominalis*) dominate alkaline waters; caddisflies (Tricoptera: *Pycnopsyche* sp.), and stoneflies (Plecoptera *Peltoperla arcuata*) dominate in intermediate waters, and different set of stoneflies (Plecopter: *Amphinemura* sp.; *Leuctra* spp.; and *Paracapnia angulata*) dominate acidic waters.

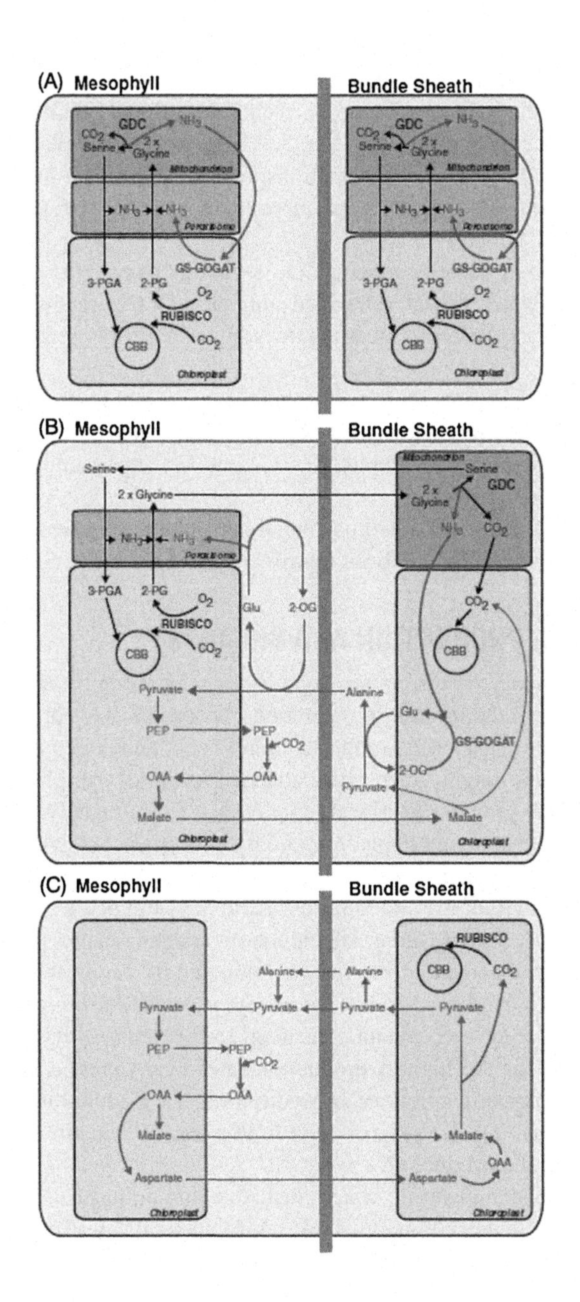


**FIGURE 5.21**

Compared to C3 plants (A), plants using C4 (C) biochemical and anatomical adaptions can be considered $CO_2$-concentrating mechanisms. The C3 terminology references the 3-carbon sugar produced (3-phosphogluterate) with $CO_2$ fixation in the Calvin Cycle, whereas C3 plants fix $CO_2$ into malate, a four carbon sugar. Some species have intermediary (B) pathways that developed a photorespiratory pump (Source: Schulze et al., 2016).

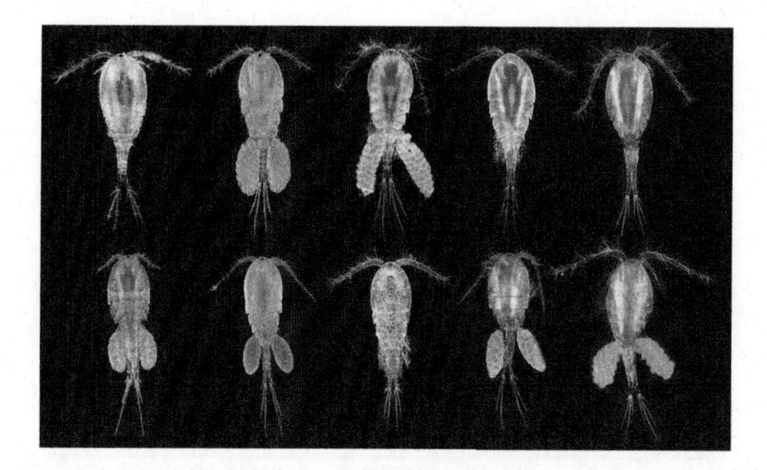

**FIGURE 5.22**

Copepods of different species (Source: Wiki Commons), using a combination of microscopy methods.

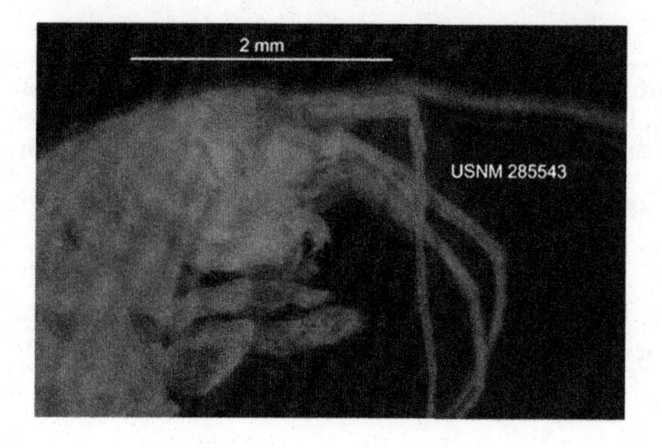


**FIGURE 5.23**

*Gammarus minus* (Source: Michal Manås). Note: *Gammarus minus* has larger eyes when populations share habitat with high density of predatory fish (Glazier and Deptola, 2011). In addition, this species has been found in caves with specialized adaptions.

The differences in annual production as a function of stream pH and alkalinity demonstrates how sensitive secondary consumers can be to stream chemistry (Table 5.2). However, whereas the annual production of the shredders in these streams depend on water chemistry (i.e., pH and alkalinity), other abiotic and biotic factors can also have profound impacts on community species and development, e.g., riparian vegetation; temperature; substrate, predators, etc.

Measurement of biomass is usually a straightforward matter of quantitative sampling. To assess relative importance of primary consumers and foodweb structure and function, ecologists measure both production and biomass, which can then be used to estimate annual turnover rates (annual production divided by mean annual biomass (P/B)). Turnover rates for fast-growing organisms are generally cal-

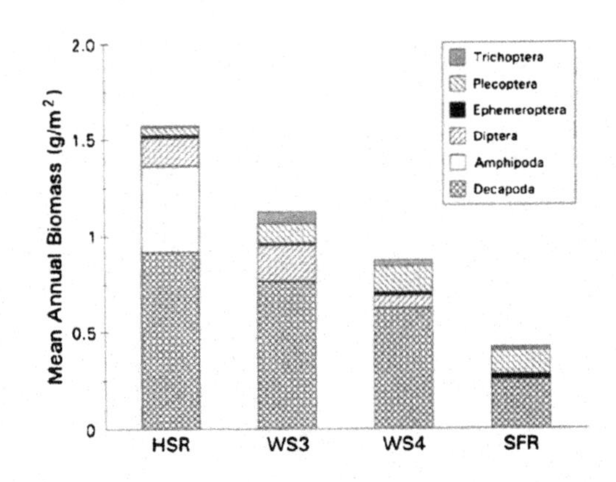


**FIGURE 5.24**

Production by species in four headwater streams (Source: Griffith et al., 1994). This work demonstrates how geology and water chemistry influences consumer production (DM = dry matter) on the Allegheny Plateau of West Virginia (see Table 5.2 for stream chemistry parameters).

**Table 5.2 pH, alkalinity, and productivity in four headwater streams (Source: Griffith et al., 1994). This work demonstrates how geology and water chemistry influence consumer production on the Allegheny Plateau of West Virginia.**

| ID | Description | pH | Alkalinity ($mg \cdot L^{-1}$) | Production (g $DM/m^2$) |
|---|---|---|---|---|
| HSR | Slightly Alkaline | 7.5 | 40.8 | 3.77 |
| WS3 | Acidic | 6.1 | 0.9 | 2.06 |
| WS4 | Acidic | 6.0 | 0.7 | 1.56 |
| SFR | Very Acid | 4.2 | 0.0 | 1.19 |

culated by day. With an estimate of biomass turnover rate, consumer production measurements can be used to quantify foodweb links.

In fact, evaluating biomass turnover rates can lead to counterintuitive results. For example, the biomass of chironomids is often low in comparison to many other primary consumers (e.g., mayflies). However, their contributions to foodwebs may be underestimated, because they may have high turnover rates. In the Ogeechee River (Georgia), woody snags or large woody debris (LGM) provide important habitat for chironomids, but these diptera might have low standing biomass, but are major prey for caddisflies. Yet, the biomass for the caddisfly might be 10x the amount of biomass of the chironomids. How can the caddisflies be supported by such a low biomass prey?

In general, estimates of P/B is usually below 10, but some estimates are between 50/60. This range suggests that our understanding of these measures require additional work. In the case of chironomids, their small size and low biomass annual P/B have produced extraordinary estimates of 65–100. Thus they are able to provide a significant food source for the caddisflies.

## SECONDARY CONSUMERS: PREDATORS AND DIEL LAKE MIGRATION PATTERNS

Secondary consumers, such as planktivorous fish or predaceous invertebrates, eat zooplankton. Whereas photosynthesis limits plant growth in deeper part of a lake, consumers may inhabit a wide range of depths, although hypoxia may restrict their distribution in stratified lakes. Secondary consumers are also subject to predation, and in many cases, they migrate to lower portions of a lake during the day to avoid detection from predators. This diel migration can reduce predations rates, but can also reduce secondary consumer feeding time and reproductive capacity and release algae from grazing controls. Thus a combination of migration and selective feeding can reenforce spatial and temporal patterns in lakes that might determine primary and secondary production and predator activities.

## FEEDING GUILDS

Feeding guilds allow ecologists to interpret the structure of foodwebs—where taxa might have specialized feeding behavior or feeding apparati, e.g., fish gap size, aquatic insect jaw and tooth design. Whereas no universal definitions exist, ecologists attempt to categorize organisms based on their feeding preferences. For example, shredders tear large pieces of microbially conditioned leaf detritus with specialized mouth parts, whereas scrapers feed on attached algae or "biofilms" of bacteria and algae (Fig. 5.25).

As species scrape and shred coarse plant litter in the process of obtaining food, they convert coarse litter into fine particulates. Collectors filter suspended organic particulates from flowing waters or from small, water-filled spaces within the sediments. Although these functional classifications provide valuable insight, they can obscure important foodweb dynamics that result from differences among individual species and changes in feeding behavior under specific conditions.

## OMNIVORY: AS A PROBLEMATIC REALITY

The concept of trophic levels is one of the oldest in ecology and informs our understanding of energy flow and population dynamics, but its use has been criticized for ignoring omnivory. In reality, ecologists have recognized omnivory for decades, but incorporating omnivory in models to test is quite challenging.

For example, if trophic levels are defined as integers, where 0 represents plants, 1 represents herbivores, and 2 represents secondary consumers, we can test how many organisms might sort into these categories. When ecologists test this structured approach, a large proportion of taxa (~65%) occupy integer trophic positions, suggesting that discrete trophic levels exist. However, the majority of those trophic positions were aggregated as either plants and herbivores. Meanwhile, secondary consumers are much harder to categorize. Thus above herbivore trophic level, foodwebs should be thought of as a tangled web of omnivores.

However, the preponderance of omnivory seems to vary with ecosystem type, where it is most common in marine systems, rare in streams, and intermediate in lakes and terrestrial foodwebs. Although several hypotheses attempt to explain this pattern, the level of disturbance in streams compared to other systems may be an important factor.

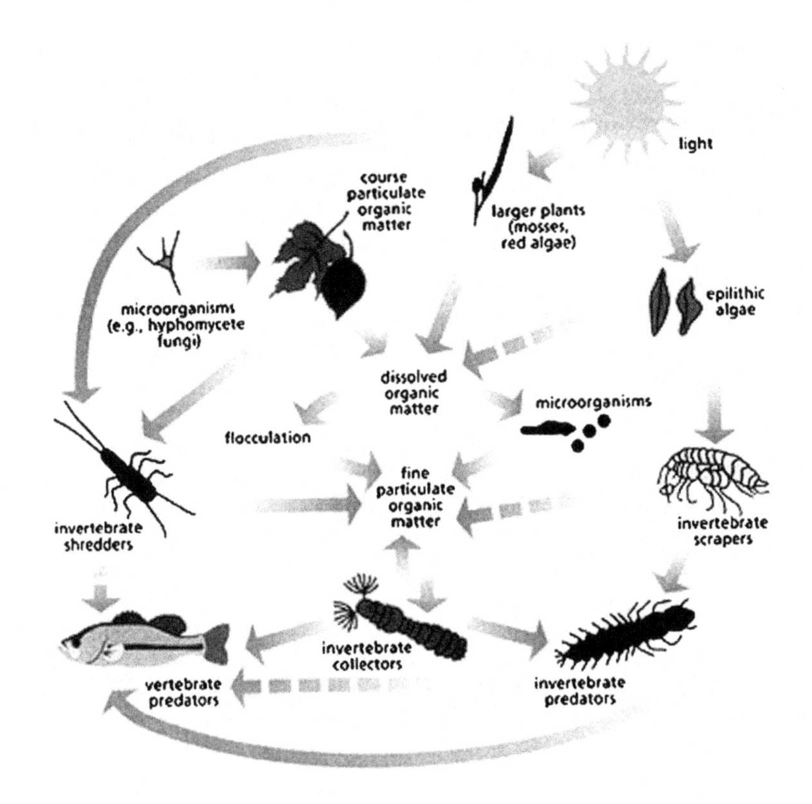


**FIGURE 5.25**

Image of feeding guilds (Source: Department of Environmental Protection, West Virginia).

## MICROBIAL LOOP

The matter cycling in aquatic foodwebs depend on the linkages between heterotrophic bacteria, fungi, and protists in the so-called "microbial loop" (Fig. 5.26). In spite of their small size, the biomass of heterotrophic bacteria can make up the majority of the total heterotrophic biomass in aquatic systems.

With this relative importance, it should be no surprise that microbial communities are key players in organic matter degradation and the subsequent release of nutrients and use by primary production.

## ORGANIC MATTER DEGRADATION

As a source of energy, organic matter can be classified into three degradation pathways: predation, particle feeding, and dissolved organic matter (DOM) uptake. In all cases, the breakdown relies on a complex set enzymatic processes. For example, hydrolysis reactions of organic substrates are ubiquitous (a dominant pathway), which is mediated by extracellular enzymes in the intestines of animals and by the enzymatic activity of bacteria. But different organisms might differ in their specialties. For

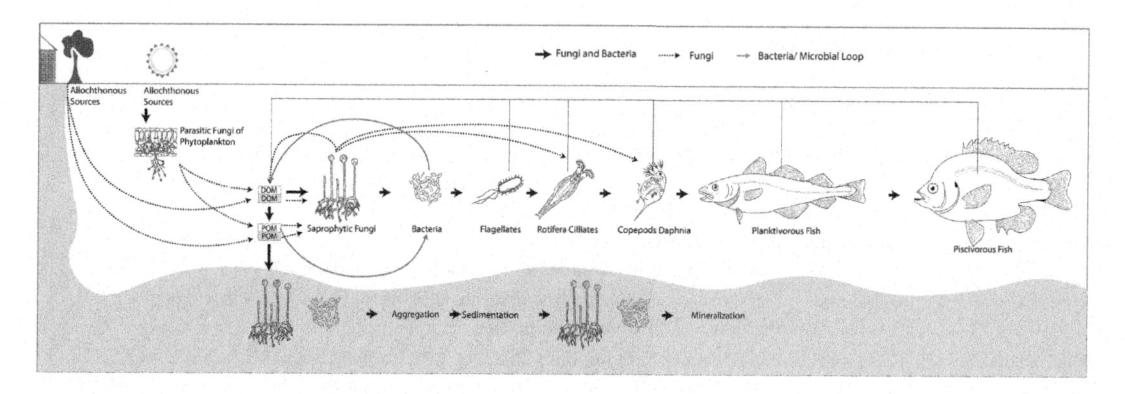


**FIGURE 5.26**

Image of the microbial loop (redrawn from Grossart and Rojas-Jimenez, 2016).

example, while bacteria and particle feeders might hydrolyze nonliving particles, the uptake of small organic molecules is almost exclusively done by bacteria.

The decomposition of dissolved organic macromolecules is mediated mainly by the enzymes of free-living bacteria, which subsequently incorporate the small molecules. Since extracellular hydrolysis is a relatively slow process in comparison to the uptake of low-molecular-weight organic matter and depends on the compound, bacterial activity has a strong influence on the concentration and speciation of dissolved organic molecules in water.

Finally, the efficiency of animals feeding on particles can be dynamic and varies considerably. For example, the slow but continuous microbial component is in contrast to pulse-feeding activities of animals. Added to these dynamics, some particles are fast-sinking large particles and others are nonsinking small particles; these differences will lead to very different degradation processes.

## SIZE GUILDS: PLANKTON

Another way to evaluate water column processes relies on categories of different plankton sizes in water column. Although the cutoff between categories can depend on the worker (Table 5.3), the general categories can have important ecological implications. Larger organisms have lower surface area to volume ratios compared to small organisms. Thus smaller plankton can often access resources that are unavailable to larger taxa.

The existence and importance of nano- and even smaller plankton was only discovered during the 1980s, though they are thought to make up the largest proportion of all plankton in number and diversity. Algae cells are in the Pico- and Nanoplankton size classes and vary in size across two orders of magnitude.

Small plankton come from diverse evolutionary lineages and associated biochemical differences. For example, the content of polyunsaturated fatty acids in the genus *Nannochloropsis* (Eustigmatophyceae: SAR Supergroup) had 50–100 times higher content of Eicosapentaenioc acid than other fresh-water picoplankton, such as *Choricystis* and *Psuideodictyospehaerium*. Although Eicosapentaenioc acid has been implicated as a good "fatty acid" for human diets, the ecological significance

**Table 5.3 Size classes of plankton.**

| Category | Size | Examples |
|---|---|---|
| Femtoplankton | < 0.2 μm | viruses |
| Picoplankton | 0.2–2 μm | small eukaryotic protists; bacteria; Chrysophyta |
| Nanoplankton | 2–20 μm | small eukaryotic protists; small diatoms; small flagellates; Pyrrophyta; Chrysophyta; Chlorophyta; Xanthophyta |
| Microplankton | 20–200 μm | large eukaryotic protists; most phytoplankton; Protozoa (Foraminifera); ciliates; Rotifera; juvenile metazoans—Crustacea (copepod nauplii) |
| Mesoplankton | 0.2–2 mm | metazoans, e.g., copepods; Medusae; Cladocera; Ostracoda; Chaetognaths; Pteropods; Tunicata; Heteropoda |
| Macroplankton | 2–20 mm | metazoans, e.g., Pteropods; Chaetognaths; Euphausiacea (krill); Medusae; ctenophores; salps, doliolids and pyrosomes (pelagic Tunicata); Cephalopoda |
| Megaplankton | > 20 mm | metazoans, e.g., jellyfish; ctenophores; salps and pyrosomes (pelagic Tunicata); Cephalopoda |

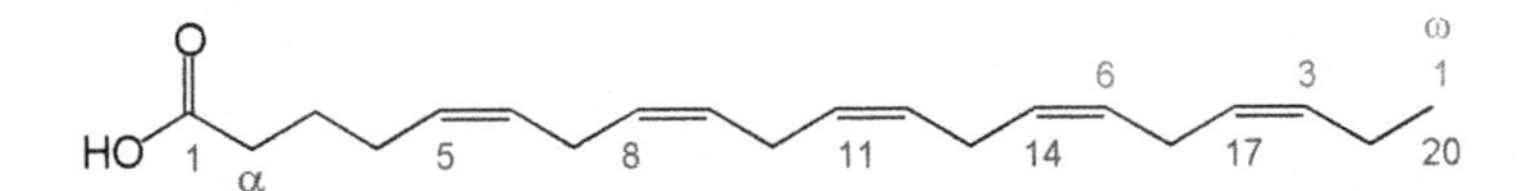


**FIGURE 5.27**

Eicosapentaenioc acid, which is a precursor to prostaglandine synthesis and subsequent precursors for other tissue hormones.

of this precursor to various hormones for *Nannochloropsis* is unknown, but deserves interest. (See Fig. 5.27.)

## ECOSYSTEM ENGINEERS: LINKING SPECIES TO ECOSYSTEM PROCESSES

Beginning in the early 1600s, fur trappers captured and killed millions of beavers (*Castor canadensis*), so the pelts could be made into hats in Europe. After decimating the populations, their population has rebounded.

Their presence in riverine systems has a wide range of implications because they "engineer" habitats for themselves and modify stream geomorphology and provide a range of ecosystem services. For example, beaver dams can broaden streams, submerge meadows, and raise water tables. In the Acadia National Park wetlands doubled between 1944 and 1997 after beavers were reintroduced. Their ponds can store water and prevent downstream flooding, increase groundwater recharge, improve fish habitat, provide hunting areas for birds, and can even improve water quality.

In this case, the Beaver provides a valuable asset for a stream ecosystem, but landowners are not impressed if the beaver floods portions of their productive lands or homes. Thus the relationship between the ecosystem services and localized impacts on private landowners need creative management policies to navigate this perceived conflict.

# EARTH SYSTEMS SCIENCE: BIOSPHERE AND PLANETARY SCIENCES

## ATMOSPHERIC AND ECOSYSTEM CONNECTIONS

The biosphere includes the atmosphere, oceans, soils, and the physical and biological cycles that affect them. Using satellite imagery, NASA coined the term "earth systems science" in 1990s to study the linkages between the biosphere and planetary system. The term and methods have been applied to Earth, but also our planetary and nonplanetary neighbors. Within Earth Systems Science, Climate Science is probably the most recognizable sub-category.

By 1896 scientists had appreciated the sensitivity of the Earth's climate on the composition of the atmosphere. In spite of the controversies, the areas of uncertainty are quite well constrained. Now, we know $CO_2$, $CH_4$, $N_2O$, and other trace gases absorb infrared light, effectively trapping it in the atmosphere. Known as the greenhouse effect, and these gases known as greenhouse gases, the planet has been warming due to the anthropogenic increase of these gases.

## WATER TEMPERATURE, SALMON, AND INDIGENOUS PEOPLES

Climate change has challenged and will continue to challenge peoples throughout the world. For example, river and stream warming will impact indigenous populations in the Pacific Northwest. Salmon play a particularly important role in the diet, culture, religion, and economy of Native Americans in this region, but salmon are extremely sensitive to water temperatures, which have been increasing. Average August water temperatures at selected sites in the Snake River increased 0.77°C between 1960 and 2015. Although several mechanisms can lead to elevated water temperatures (e.g., stream shading, impoundments), the prospect of climate-change-generated stream temperatures limits regional restoration options to protect fisheries.

However, climate change in general will have a dramatic impact on all inland waters with increasing probabilities of extreme weather, on the ability of the infrastructure to provide safe and reliable waters, and lead to increased competition between human and environmental uses of water.

## PALEOCLIMATES, ENDORHEIC BASINS, AND CLIMATE SCENARIOS

Lower to mid-Permian deposits in central United States record a significant and long shift from more humid environment, characterized by coal and organic shale to semiarid environment. These changes demonstrate the ephemeral nature of surface waters, where their disappearance is a function of increasing aridity and seasonality.

Lakes are particularly vulnerable to changes in climate parameters. Variations in air temperature, precipitation, and other meteorological components directly cause changes in evaporation, water balance, lake level, ice events, hydrochemical and hydrobiological regimes, and the entire lake ecosystem. Under some climatic conditions, lakes may disappear entirely. Paleoclimate research may provide important clues regarding what we might expect with climate change, especially in terminal lakes that are sensitive to water withdrawals from the watershed.

The largest endorheic lakes include the Caspian and Aral Seas, Lake Balkash, Lake Chad, Lake Titicaca, and the Great Salt Lake. Inflow changes have profound implications for these lakes. Whereas the Aral Sea, for example, has been significantly reduced by increased water use for irrigation, the Great Salt Lake has increased in size because the precipitation has increased in its watershed.

In addition, currently exorheic lakes could become endorheic with drying trends. For example, Lake Winnipeg has shifted from an exorheic to endorheic basin during a warm dry period in the mid-Holocene. However, over the 5000 years, the drying trend reversed itself and a moist climate allowed the lake to become exorheic. Climate change scenarios indicate that the lake could become endorheic again.

Climate change is generating complex responses in both natural and human ecosystems that vary in their geographic distribution, magnitude, and timing across the global landscape. One of the major issues that scientists and policy-makers now confront is how to assess such massive changes over multiple scales of space and time. Lakes and reservoirs comprise a geographically distributed network of the lowest points in the surrounding landscape that make them important sentinels of climate change. Their physical, chemical, and biological responses to climate provide a variety of information-rich signals. Their sediments archive and integrate these signals, enabling paleolimnologists to document changes over years to millennia. Lakes are also hotspots of carbon cycling in the landscape and as such are important regulators of climate change, processing terrestrial and atmospheric, as well as aquatic carbon.

## NESTED HIERARCHY AS AN ORGANIZING PRINCIPLE

Ecologists have tried to organize how organisms interact with their environment using various organization principles. One of the most popular relies on a hierarchy model and levels of organization (Fig. 5.28). Although these categories provide value to communicate how organisms interact in the environment, the categories breakdown quickly as we try to begin to appreciate how these patterns and processes actually function.

Beginning at the individual's level, the physiology of an organism defines environmental limits for development and reproduction in the environment. For example, the physiology of fish constrain water temperature preferences. The interacting individuals within a species form a population. The control of populations include birth, death, and migration. The study of how individuals and their populations interact with the environment has often been referred to as autecology and define the "lower levels" of ecological hierarchy.

In contrast to autecology, the term "synecology" refers to study of interacting species in the environment, which includes how species compete for resources, or trophic patterns of herbivory and predation. These types of analyses include community ecology and ecosystem ecology. As we will evaluate further below, these analyses imply some type of scale, often for the convenience of the ecologist. For example, a forest ecologist might look at a group of trees, a patch, or forest stand, or a stream ecologist might evaluate a reach, set of reaches, or portion of a river. At larger scales, we might begin to approach landscape ecology that might push the boundaries of the term by, for example, including an analysis of many rivers or lakes or an entire forest using new tools, such as remote sensing using satellite images.

Finally, as we have developed a more sophisticated understanding of the globe's processes, the entire biosphere might be of interest. Although the term "biosphere ecology" is not used much, many ecologists work at this scale, especially when climate change is the focus of study.

We have been using the word environment without providing definition; the term is related to the common use of "everything around us". Ecologists use the word with a bit more precision, where

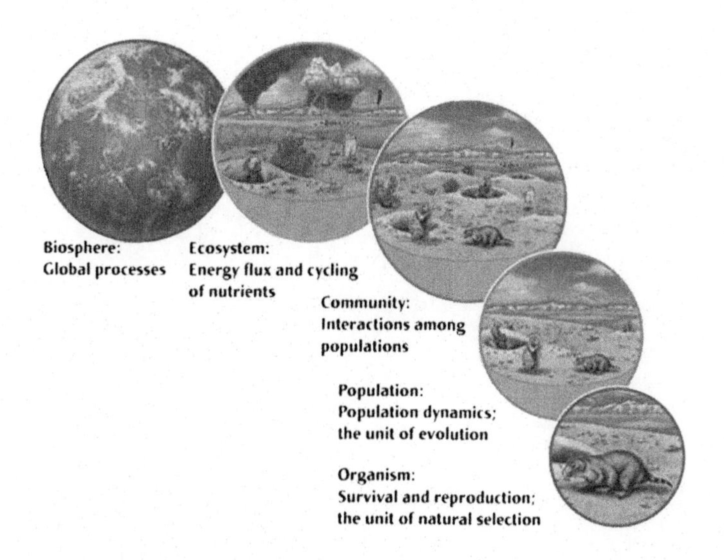


**FIGURE 5.28**

Levels of organization. (Source: Raven, Peter H., Hassenzahl, David M., Hager, Mary Catherine, Gift, Nancy Y., Environment, 9th Edition (page 42), Wiley, Kindle Edition.)

the environment includes physical and chemical factors, which are referred to as abiotic factors, and interacting organisms or biotic factors. The combination of the biotic and abiotic factors are what ecologists try to evaluate to better understand how individuals, populations, communities, ecosystems, landscapes, and the biosphere function.

Without doubt, it is a challenge, but as we have learned in the last century, the sustainability of our planet depends on our understanding and effective management of these systems.

## NEXT STEPS

### CHAPTER STUDY QUESTIONS

1. Describe how predator–prey equations can be used to evaluate the role of trophic interactions to predict population dynamics.
2. Summarize how sexual selection can influence population dynamics, e.g., predator/prey and infection vector and host.
3. Describe succession and how communities change with time and how these can be evaluated.
4. Diagram energy transfer and nutrient flux in lotic and lentic systems.

CHAPTER

# THE MATRIX: THE PHYSICAL AND CHEMICAL PROPERTIES OF WATER

# 6

*There are these two young fish swimming along, and they happen to meet an older fish swimming the other way, who nods at them and says, "Morning, boys, how's the water?" And the two young fish swim on for a bit, and then eventually one of them looks over at the other and goes, "What the hell is water?"*

**David Foster Wallace (1962–2015)**

## CONTENTS



Ecology and Management of Inland Waters. https://doi.org/10.1016/B978-0-12-814266-0.00019-2

Inland waters are wildly different than marine waters, where organisms face high velocity river currents, and temperature can often vary more than 5 or 10°C over a 24 hour period. Organisms may contend with UV radiation in shallow waters or lose all sense of sight with high water turbidity. Perhaps, the most challenging aspect of inland waters is the variation in water chemistry—in particular the changes in salt concentrations.

Anadromous salmon and steelhead migrate from marine habitats into freshwaters to reproduce. To accomplish this, young fish born in freshwaters must acclimate to marine waters with a dramatic physiological shift to avoid the osmotic pressure changes that will "suck" the water from the fish in the marine environment. And when they return to their natal waters a physiological shift must occur so water does not burst their cells and organs. Of course, the timing of these migrations coincide with a peak in flow and available spawning ground habitats. Thus these fish navigate a wide range of the physical characteristics of water.

Understanding the physical properties of water relies on a wide range of topics, such as hyrdrology, geomorphology, fluid dynamics, optics, temperature, and chemistry. Far from the deep-end of the pool, we will wade in gently into these topics with the hope that the reader become deeply appreciative of the value of these topics relevant to understanding the nature and ecology of inland waters.

After reading this chapter, you should be able to

1. Describe the physical and chemical characteristics of inland waters and how various species have adapted to these habitats.
2. Understand how water movement, fluid dynamics, light, temperature, and chemicals structure biotic communities.

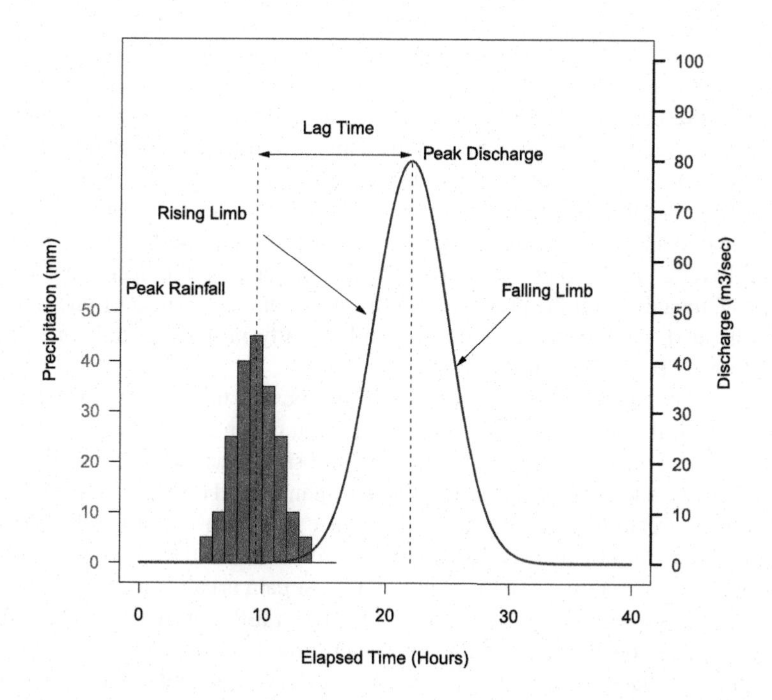


**FIGURE 6.1**

Idealized hydrograph demonstrating the lag between peak rainfall and peak discharge. The response on a stream will depend on the watershed characteristics, e.g., geology, size, and ground water-surface water relationships, and precipitation events, e.g., previous events, rainfall intensity.

# MOVING WATER AND MOVING IN WATER

## DYNAMICS OF FLOW: THE HYDROGRAPH

Water flows downhill. However, the velocity $v = L/t$ (length per time, such as meters per second) and volume of flow or discharge $q = V/t$ (volume per time such as cubic meters per second (cms)) varies with time and space. With rainfall, a stream may fill with water from surface runoff and subsurface seepage to that might generate peak flow. The flow dynamics can be graphed with a hydrograph to display various stages of streamflow (Fig. 6.1).

A hydrograph plots time versus discharge on a graph, and several characteristics of streamflow can be characterized. For example, the part of the graph where discharge rapidly increases is called the rising limb. The maximum flow is the peak flow, followed by the falling limb. The relatively flat discharge before rising limb and after the falling limb is called the baseflow.

## SPATIAL PATTERNS OF DISCHARGE

The water velocity of a headwater stream might be 200 cm/s and a steep elevation gradient or a slow meandering river may be running slow (1 cm/s), but carrying thousands of $m^3/s$ of water. Whereas the headwater stream might have a catchment or watershed area of 500 ha, the slow-moving river might be capturing a million ha of runoff.

Over the decades the United States Geologic Survey (USGS) has measured discharge from thousands of stream locations and made these records available online. Using these records, we can begin to identify patterns of streamflow. For example, by reanalyzing the three streams we encountered in Chapter 2 (Fig. 2.18), we plotted the data by Julian Date, i.e., day of the year (Fig. 6.2, left panel). Using another method to analyze the autocorrection of discharge, we found a narrow range of autocorrelation in the Eel River and even narrower in the Mojave. In contrast, the Sacramento River has very high autocorrelation, meaning that the predictability from one day to the next or one week to the next, or one month to the next, is high, whereas the streamflow early in the year tells one almost nothing about the streamflow in 90 days later in the Mojave.

There are many more details regarding the hydrograph, but for now this is enough for us to make some observations that link the hydrograph to ecological patterns and processes.

Stream discharge is a combination of surface runoff and subsurface interactions. Stream discharge gain or lose water into the surrounding substrates, via the channel bed or banks. When the groundwater elevation adjacent to a stream is above the stream's surface elevation, water can flow into the stream from the groundwater and the stream is called a gaining stream. When the adjacent watertable is lower than the streams, then water can flow into the banks and the stream is called a "losing stream." Over the course of a watercourse streams may be gaining and losing in different reaches. In addition, streams segments my vary with season, at times gaining and other times losing.

When the volume of water exceeds the capacity of a river's channel, water overtops banks and floods adjacent areas, i.e., the flood plain.

## LENTIC CURRENTS

The dominant direction of stream currents follow the topographic gradient in the stream channel. However, back currents, eddies, and side channels make the streamflow a bit more complex. When water flow around corners, over rocks or through obstructions, the velocities change. Waters with differing velocities are shear against each other, creating eddies, turbulence, and chaotic flows. To measure the volume of water flowing in a stream, hydrologists need to capture the variation in flow velocities and may need to take 10–20 velocity measures across a stream.

In terms of the ecology of the stream habitat, this creates a wide range of habitat types, e.g., pools, riffles, and runs, where algae, invertebrates, and fish taxa may specialize in different habitat types.

## SEDIMENT TRANSPORT AND GEOMORPHOLOGY

River channels are subject to three main geomorphological processes, which contribute to their transitory state: erosion, transport, and deposition of material. Water naturally follows a path that changes from high to low elevation as a function of gravity; the difference between the starting and ending elevations or potential energy drives sediment and channel morphology. Excess energy dissipates through turbulent reaches, channel bed roughness forms, development of meander bends, and sediment transport.

The capacity of a stream to transport sediment depends on the water depth, velocity, and the size and amount of material moved through a system. Velocity and depth are influenced by slope, pattern, dimensions, discharge, and roughness of a channel.

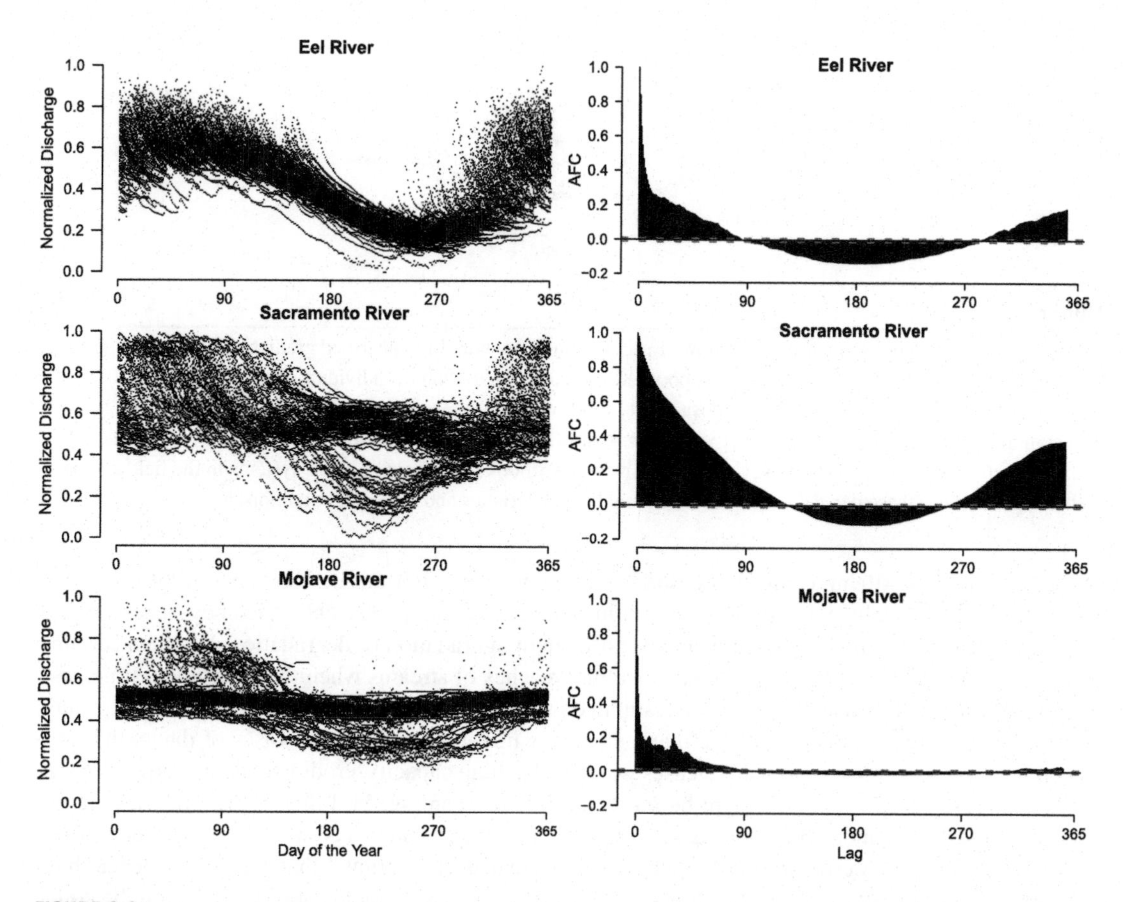


**FIGURE 6.2**

Discharge patterns for three contrasting lotic systems (Source: USGS Discharge Data). The flow patterns from each of the three rivers vary based on their hydrological characteristics, some of which have been modified by humans, i.e., hydromodification. For example, the Eel River in northern California demonstrated high flow in the early and late part of the year, whereas flows in the fall are quite low. Meanwhile the Sacramento River's summer flows have two patterns—low flows before the Shasta Dam was built, and then moderate, regulated flows to provide drinking and irrigation water during the regular summer drought conditions of California's Mediterranean-type climate. Finally, the Mojave River has consistent flow with very high flow, but uncommon high flow rates in the early winter and low rates in the summer. However, one wonders if there is a human component to the summer flows that seem out of place in the desert.

A watershed's geologic history and climate influence the amount and type of sediment that enters the river system. Sediment load is made up of material dissolved in solution, suspended as fine particles in the water column or as gravel, cobbles, and boulders in the bed of the channel.

When the capacity to transport sediment load is high, the sediments might be eroded. When hydraulic conditions shifts so that the capacity to carry sediments is low, the sediments might be deposited

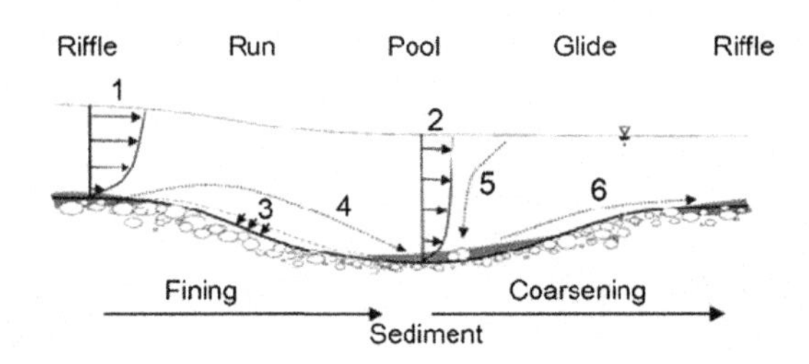


**FIGURE 6.3**

Schematic of a riffle-pool sequence at low flow: 1 accelerating flow with high near-bed gradients; 2 decelerating flow above a pool; 3 regressive run erosion; 4 pool refilling; 5 accumulation of individual coarser grains due to bank erosion; 6 pool scouring and riffle deposition at high flows (velocity reversal) (Source: Garcia et al., 2012). The combination of stream gradient, water velocities, sediment dynamics, and hydrology can create repeated sequences of stream flow characteristics. Usually idealized in textbooks, recognizing these systems in the field can be a challenge, because these depend on particular regional characteristics and temporal conditions.

in the streambed. A stream channel is described as being in equilibrium if, over a period of time, the hydraulic variables delicately and mutually adjust to provide the power and efficiency necessary to transport the load supplied by the drainage basin without aggradation or degradation. Within a certain time scale, aggradation or degradation occurs along reaches of streams when the insufficient time has elapsed for the interactions of variables to result in equilibrium conditions. Depending on the time scale and careful analysis, we might assume that a stream is in equilibrium. In the cases of dams, the impoundment traps sediments and then releases water with a high capacity, eroding channels downstream of the dam. However, examples might be more subtle. For example, Mill dams built 200 years ago, forgotten, and buried about 100 years ago continue to influence steams in New England (see page 46). Tectonic uplift of the Sierra Nevada Mountains, have created 'nick points' that provide evidence that these streams are out of equilibrium. Thus environmental scientists should carefully question the concept of equilibrium by evaluating geomorphological processes across temporal and spatial scales.

## GEOMORPHOLOGY OF THE BENTHOS: SUBSTRATE

Instream channels show a great deal of variation depending on stream velocity, sediment load, etc. The velocity patterns of stream directly influence the channel habitat structure. Muds are found in low-energy systems, sands are found in higher-energy systems; pebbles, rocks, boulders cover the bottom in high-energy systems. (See Fig. 6.3.)

In a similar way, lakes bottoms tend to be composed of relatively deep muds, where current velocities are very low. Meanwhile the shoreline might experience constant wave action, thus characterized as a higher-energy system and be composed of sand or rocks. Even where a stream enters a lake might also have courser substrates than other parts of the lake.

These substrate patterns structure stream and lake communities, where aquatic species might specialize in different habitat types. For example, along the shoreline of a lake that include types of substrate, various organisms might use these habitats for specialized purposes, e.g., to attach, graze, hunt, and hide. Table 6.1 summarizes some of the commonly used habits that organisms might associate with various types of substrate.

**Table 6.1 Aquatic habit terminology.**

| Habitat | Description |
|---|---|
| Benthic | Ecological region at the lowest level of a body of water, such as an ocean or a lake, including the sediment surface and some subsurface layers |
| Emergent | Aquatic plant with leaves and flowers that appear above the water surface |
| Endolithic | Growing within cavities of rock |
| Endopelic | Growing within mud (sediment) |
| Endosammic | Growing within sand (sediment) |
| Epigean | Growing on the benthic surface |
| Epilithic | Growing attached to rock surfaces |
| Epipelic | Growing on mud (sediment) |
| Epiphytic | Growing attached to other plants |
| Epipsammic | Growing on sand |
| Epizoic | Growing attached to animals |
| Neustonic | Inhabit the region on or just below the surface of a body of water |
| Periphytic | Inhabit the surface of submerged plants and other underwater objects |
| Stygophilic | Inhabit both surface and subterranean aquatic environments |
| Stygobitic | Inhabit groundwater systems or aquifers, such as caves, fissures, and cavities |

In many cases, the growth habits can be predictive—where a certain species might often be found on one type of substrate. For example, the diatom, *Bacillaria paxillifer*, can slide along the substrate, which might be the mud or other plants.

## CURRENTS IN LAKES

Lake currents are associated with localized inputs, outlets, and currents generated by the wind. Wind-driven currents travel in horizontal and vertical directions. Wind adds kinetic energy to the lake and can create the potential energy. Wind affects the lake through the shear it imparts on the water surface. This shear drags the water in the downwind direction, adding kinetic energy and causing surface currents, surface waves, and a so-called surface set up: the mean lake surface downwind is tilted upward compared to the upwind side of the lake (Fig. 6.4). This set up results in a basin-scale circulation: bottom-water return currents compliment the surface-water motion. Although the surface set up may only be a few millimeters or centimeters, this results in a much greater (sometimes on the order of meters) tilting of the internal isopicnals, lines of constant density. Thus the set up results in a reorganization of the potential energy in the lake: downwind regions have increased potential energy. When the wind stops, this potential energy is released, causing basin-scale waves called seiches.

These currents mix the water column that might bring plankton, that might otherwise sink beyond the photic zone back to the surface, where photosynthesis can occur. Similarly, nutrients having moved up to the surface might become limited with stratification. These are important themes that will come up on page 205.

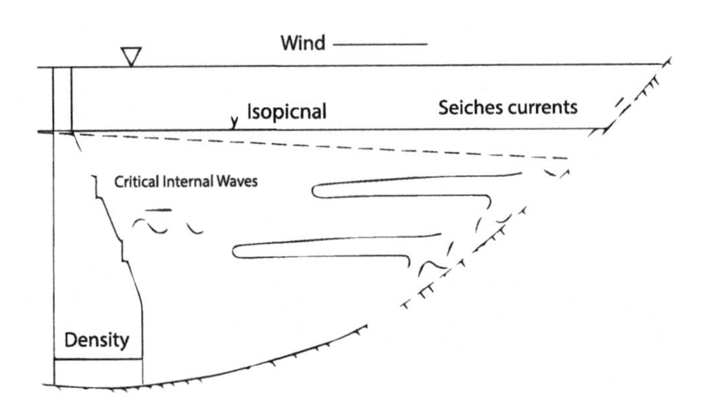


**FIGURE 6.4**

Schematic of physical processes leading to boundary mixing and intrusions in a lake. Wind can create seiches and critical internal waves. Turbulence from seiching currents or wave breaking can mix fluid in the bottom boundary layer, and gravitational readjustment of this mixed layer can lead to intrusions of mixed fluid propagating into the interior (Source: Wain and Rehmann, 2010).

## ADAPTING TO CURRENTS IN STREAMS AND LAKES

Capitalizing on physical properties of water, aquatic taxa maintain their position in the water column using a range of active (swimming, swim bladders, etc.) and passive activities (body shape, density, etc.). In some cases, animals force water movements to capture prey, e.g., bivalves circulate water across gills to filter out zooplankton from the water column.

For many organisms in streams, water currents are a mechanism to migrate, predominantly downstream. Insects mitigate drift according to the colonization cycle hypothesis. This hypothesis predicts that upstream-directed aquatic insect flight behavior compensates for the gradual downstream movement of insect larvae. However, in some cases, invertebrates can hitch a ride with upstream migrating fishes as well.

In streams, water velocities vary across a channel and with depth. These variations provides a wide range of habitats for many aquatic organisms. The lowest water velocities are in pools or along the stream-channel edges. Water velocities can create geomorphological patterns, e.g., sequences of pools, riffles, and runs. Pools are usually one to a few meters deep, and where water velocities approach zero. Deeper parts of a pool might be cooler, provide hideouts for to avoid predation and areas to rest. Runs are fast-moving waters that do not have much turbulence and often are relatively quiet. Riffles may also have relatively high velocities, but because they also have turbulence, they might have eddy currents that generate a complex mix of water velocities. With trapped air bubbles, riffles usually generate sound that might vary from a pleasant fountain-like sound to a roar approaching the volume of jet engines.

Algae species and invertebrates can often be found within specific habitat-types. In fact, debates remain active about how to best characterize stream habitat because species distributions are far from homogeneous. Thus some argue that targeted sampling, e.g., riffle and pool sites, whereas others use stream reach segments that are undifferentiated in the sampling design. The selection between the

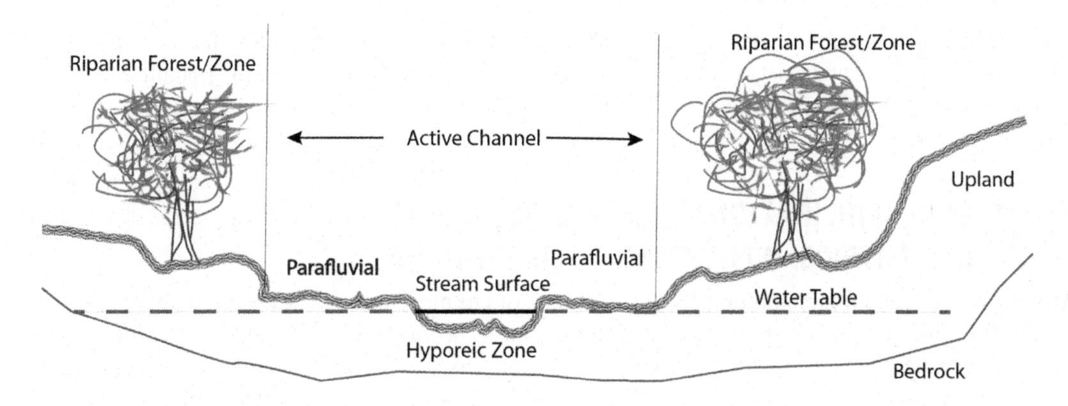


**FIGURE 6.5**

Four interaction zones (Source: Levick et al., 2008).

methods may have as much to do with researcher bias and project goals as the resources available for the project.

Currents in lakes might be more subtle, but just as important. In general, lakes will experience currents propagated from where water inputs enter the lakes or be generated in situ by wind.

## GROUNDWATER AND SURFACE WATER EXCHANGE

Streams and lakes are in hydrologic connectivity with the groundwater. In fact, stream flow contributes to aquifer recharge in many basins. A losing stream is characterized by water seeping into the channel bed or banks and moves into the regional groundwater table. If, however, the groundwater head potential is higher than the surface of the stream, water can flow into what is called a gaining stream.

Moreover, water is regularly exchanged with the bed and banks. This inflow and outflow is called hyporheic exchange. Water in the hyporheic zone acts as a reservoir of water, a function that is referred to as transient storage. The relative volume of water in transient storage compared to stream can be highly variable (between 3 and 460 percent). Thus the transient storage capacity can play a major role in modulating the height of a stream. Across all scales, the functional significance of the hyporheic zone relates to its activity and connection with the surface stream and can extend beyond into riparian and upland areas. Conceptualized on a catchment scale, the hyporheic corridor describes gradients extending to alluvial aquifers that might extend 10s of kilometers from the main channel and might include several interacting types of systems (Fig. 6.5).

At the stream-reach scale, hydrological exchange and water residence time influence several patterns and processes of hyporheic zone. For example, interstitial invertebrate communities may represent a significant biomass, specializing on hyporheic zone of a stream, and can respond to seasonal changes in stream flow. For example, in gaining streams, subsurface water supplies stream organisms with nutrients, whereas losing streams water provide dissolved oxygen and organic matter to microbes and invertebrates in the hyporheic zone. As a function of discharge and bed topography and porosity, gradients in the hyporheic zone exist at all scales and vary temporally. For example, redox potential gradients will influence biogeochemical processes on particle surfaces.

# ECOLOGICAL IMPLICATIONS OF FLUID DYNAMICS

## NEWTONIAN PHYSICS: INERTIA AND VISCOSITY IN WATER

Moving water exerts a force on objects in the water. And to move through water, an object must exert a force. A force is a push or pull upon an object resulting from the object's interaction with another object. Whenever there is an interaction between two objects, there is a force upon each of the objects and can be specified with the following units: mass · length · time$^{-2}$, usually expressed as the SI units of Newtons (kg·m·s$^{-2}$).

Inertial force $F_i$, as the name implies, is the force due to the momentum of the fluid. Inertia is the resistance to change its state of motion (including a change in direction). In other words, it is the tendency of objects to keep moving in a straight line at constant linear velocity, $v$ (length per time, such as m·s$^{-1}$).

In fluid dynamics, this concept is usually described as a resistance for an object to accelerate or decelerate within fluids and is usually expressed as momentum equation (Mass · velocity) times acceleration ($a$): $F_i = M \cdot v \cdot a$.

But in fluids, the expression is converted to the mass/volume of fluids, i.e., their density, $\rho$ $(M/V)$:

$$F_i = M \cdot v \cdot a = (\rho \cdot v) \cdot v. \tag{6.1}$$

Using a dimensional analysis, the units are mass/volume · length/time · length/time, which can simplify to mass · length · time$^{-2}$, which are the dimensions of force (kg·m·s$^{-2}$). By specifying the momentum (inertia) in terms of density, we can appreciate how an increase in water density and velocity can increase the force of moving waters.

Viscosity (represented by $\mu$) measures a fluid's resistance to flow. Viscosity is the resistance to change in form or a sort of internal friction. Said in another way, viscosity measures how easily a fluid moves across a surface. In everyday terms (and for liquids only), viscosity is often referred to a fluid's "thickness." Thus water is thin, having a lower viscosity, whereas vegetable oil is thick having a higher viscosity. But compared to air, water is viscous. The viscosity of water is roughly 18 times greater than air. Because air flows easily, tidal ventilation, i.e., moving the respiratory medium in and out of a respiratory organ, is possible in lungs. In water, it is difficult and very rare.

Fluids resist the relative motion of objects through them, but also the motion of layers of fluids with differing velocities. This resistance can be referred as viscous forces or $F_v$, but is related to the surface area of the object:

$$F_v = \mu A \frac{\Delta v_x}{\Delta z}, \tag{6.2}$$

where the Greek letter $\mu$ defines viscosity and is measured as a Pascal second ((kg m$^{-1}$ s$^{-2}$) · s). At first the equation appears quite different than the generic momentum equation for force, but looking at these units carefully, we see aspects of mass and velocity.

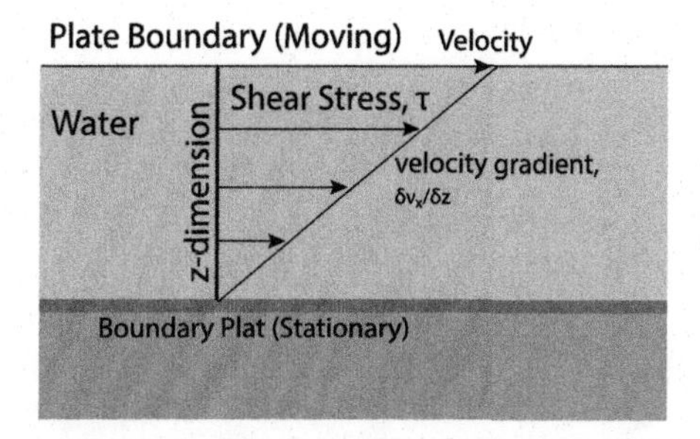


**FIGURE 6.6**

Fluids moving across a stationary plate create shear stress in the direction of flow. The force depends on distance from the stationary plate ($z$) or in our case the stream bottom.

**Table 6.2 Dynamic viscosity of selected fluids.**

| Fluid | Dynamic Viscosity (Pa · s) |
|---|---|
| Air (15°C) | $1.81 \times 10^{-5}$ |
| Water (20°C) | $10.02 \times 10^{-4}$ |
| Blood (37°C) | $3\text{–}4 \times 10^{-3}$ |
| Motor oil SAE 40 (20°C) | 0.32 |
| Motor oil SAE 10 (20°C) | 0.65 |
| Corn syrup | 1.38 |
| Honey | 2–10 |
| Peanut butter | ~250 |

The ratio $\frac{\Delta v_x}{\Delta z}$ is a key aspect of this force—referred to as to the velocity gradient in a fluid. The velocity gradient can be used to describe how water velocity and force vary as one gets farther from the object, i.e., as a function of $z$.

Thus the dynamic (shear) viscosity of a fluid expresses its resistance to shearing flows, where adjacent layers move parallel to each other with different speeds (Fig. 6.6). If a fluid with a viscosity of 1 Pa·s is placed between two plates, and one plate is pushed sideways with a shear stress of one pascal, it moves a distance equal to the thickness of the layer between the plates in one second:

$$\tau = \nu \frac{\Delta v_x}{\Delta z}, \tag{6.3}$$

where $\tau$ is the shear force, which depends on the velocity gradient and the dynamic viscosity $\nu$ (Table 6.2).

The viscosity of water changes with temperature (Fig. 6.7), but is the change ecologically significant? As water temperatures decrease, viscosity increases. In other words, water thickens and for some taxa, this can make their swimming a bit sluggish. Alternatively, fish in warmer waters swim in less viscous waters, and thus experience less drag as they swim.

Besides shear stress, fluid dynamics can be applied to airplane and insect wings to understand flight, and to fish and invertebrates to understand swimming and foraging behavior. Thus fluid dynamics is

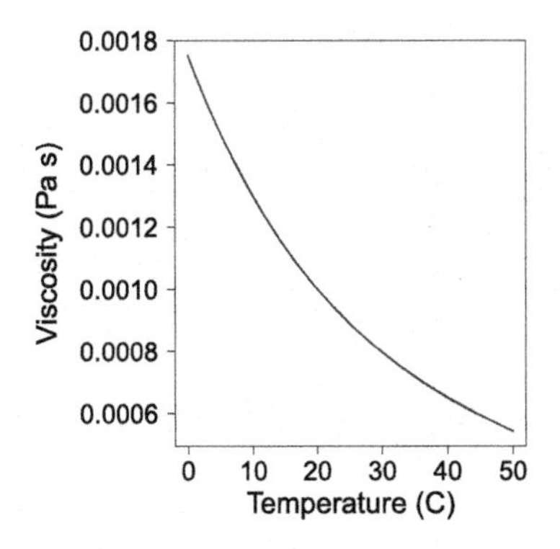


**FIGURE 6.7**

Dynamic viscosity of water as a function of temperature, which can be modeled using the following equation: $\mu_T = A \cdot 10^{B/(T-C)}$, where $A = 2.414 \cdot 10^{-5}$ Pa·s; $B = 247.8$ K; and $C = 140$ K.

**FIGURE 6.8**

Bonytail chub (Gila elegansus) is an example of a fish with a morphological characteristics that might improve their hydrodynamics. Certainly, the streamlined body is clearly selected to reduce drag forces. But what about the hump? Note: Breeding males have red fin bases (Source: Brian Gratwicke).

a powerful tool to understand how organisms move and behave in aquatic systems and even help us understand the evolutionary constraints of body shape and swimming in aquatic systems. We will explore these themes next.

## LIFT AND DRAG

The bonytail chub (*Gila elegansus*) is native to the Colorado River and can grow to become over 60 cm in length. Like many other desert fish, their coloring tends to be darker above and lighter below; these markings may serve as camouflage. They have streamlined bodies that allow them to prey on other fish. Some bonytail chub bodies arch into a smooth, predorsal hump (in adults) (Fig. 6.8). The enlarged hump may help them stay on the bottom of the river—but how?

As early as the 1980s, ecologists noted that various fish and aquatic macroinvertebrates have in-stream distribution patterns correlated to the variation in stream velocity. Fluid dynamics provide a

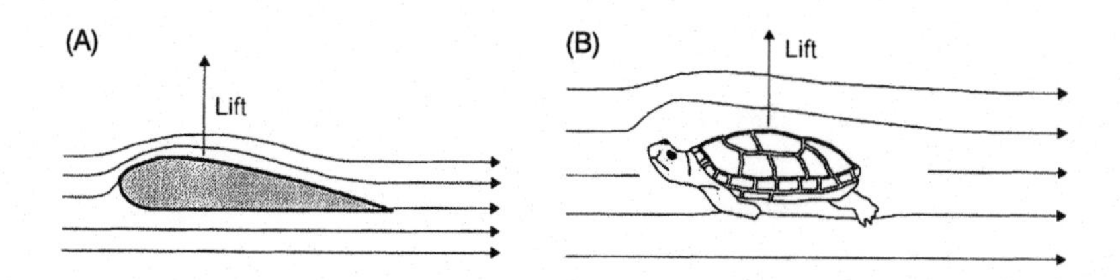


**FIGURE 6.9**

Streamlines and lift: (A)—an airfoil, designed to create lift when moving through the air, and (B)—a turtle body may generate lift when swimming (Source: Vogel, 1994).

good explanation for these patterns. For example, the forces of lift and drag have been measured for various taxa and used to explain how they negotiate moving waters.

Lift is a force that is perpendicular to the oncoming flow direction. In contrast to lift, drag is parallel to the flow direction. Drag or fluid resistance are forces acting opposite to the relative motion of any object moving with respect to a surrounding fluid and depends on velocity.

Lift allows objects moving in a fluid to control their depth or height in a fluid. For example, the shapes of a wing or a turtle can both produce upward lift to counter the force of gravity (Fig. 6.9).

Hydrologists can estimate the drag of an object when they know the velocity of the object or fluid, density of the fluid, effective area of the object, and drag coefficient.

Drag coefficients vary depending on the object's shape (Fig. 6.10). When an object is more streamlined, the drag coefficient can be dramatically reduced by reducing the turbulence behind the fish. In fact, we can often make accurate predication about swimming speeds based on the body and head shapes.

The turbulence is a function of Reynolds Number, which is a ratio of the viscosity forces ($F_v$) and inertia forces ($F_i$). Before we delve into these forces, we need to distinguish between laminar and turbulent flow and the way waters flow across solids.

## LAMINAR AND TURBULENT FLOW

The movement of fluids can be described as either laminar or turbulent flow. At very slow velocities, the fluid tends to flow without lateral mixing. In other words, there are no cross currents, nor eddies or swirls of flow, perpendicular to the direction of flow. Laminar flow depends on viscosity forces preventing mixing. As the velocity increases, the inertial forces exceed the forces of viscosity, and the ordered pattern of laminar flow breaks down. As eddies form, flow is defined as turbulent. These eddies occur at various scales, where eddies can become nested within eddies to create complex flow patterns. The transition between laminar and turbulent flow can roughly be estimated by the ratio of $F_v$ and $F_i$, which is referred to as Reynolds Number (Fig. 6.11).

Reynolds Ratio is the ratio of inertia force over viscous forces and can be simplified as follows:

$$R_e = \frac{F_i}{F_v} = \frac{\rho S v}{\mu}, \tag{6.4}$$

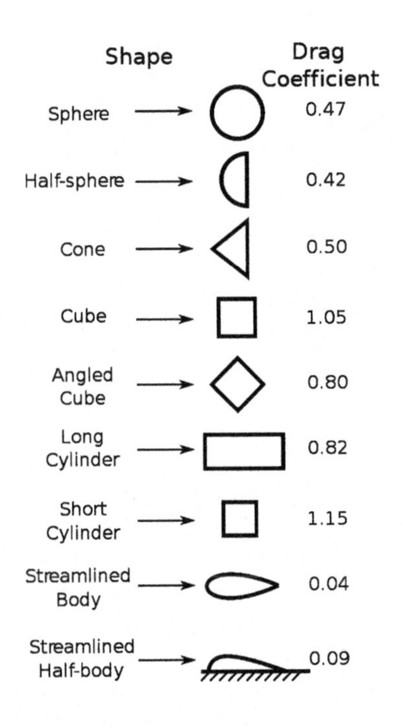


**FIGURE 6.10**

Measured drag coefficients for a range of shapes (Source: Public Domain).

where $\rho$ is the density of the water, $S$ is the surface area of the submerged object, $v$ is the velocity, and $\mu$ is the dynamic viscosity. In other words, the ratio is composed of the driving force over the resisting force.

With low velocities, fluid laminar is dominant and viscous force dominate to define low Reynolds Numbers. When the inertial forces are dominant (high flow) and Reynolds Numbers are high, then flow patterns are chaotic and flow is turbulent.

Turbulent flow is characterized by a range of velocities and vectors that define a mechanical solution. Whereas laminar flow can be described using simple equations (e.g., Newton's law of viscosity), turbulent flow can only be described in a statistical fashion, where the velocity is a mean of the velocity vectors. In fact, whereas the mean streamflow might be fast, fish can remain stationary by capitalizing on eddies and expending far less energy than would be needed in a laminar flow context.

Reynolds Number can be used to predict the abundance, diversity, and community composition of invertebrates streams. Alternatively, microscope organisms might experience nutrient limitations, because diffusion rates across the laminar flow is exceedingly slow. Thus turbulence might provide better access to a well mixed water column and greater access to resources.

Streams have complex flows since they are three-dimensional, have temporal variations, and friction, which creates ripples downstream. Reynolds Number drives body anatomy and locomotion in water (Fig. 6.12). For lotic macroinvertebrates the range for Reynolds Numbers are lower in range, since drag forces overcome the lift forces. Small individuals usually have a lower Reynolds Number,

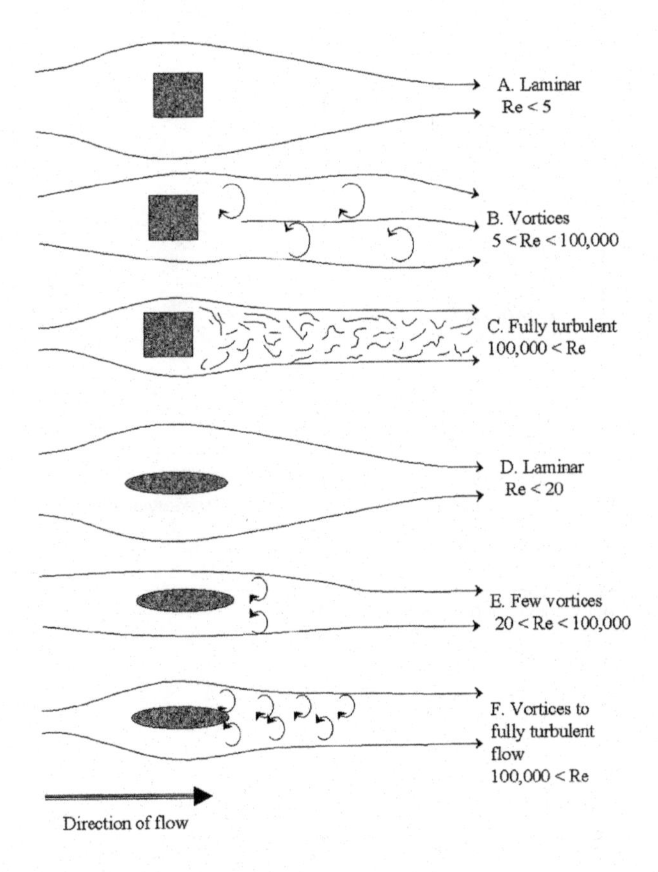


**FIGURE 6.11**

The distinction between laminar and turbulent flow can depend on fluid velocity and the fluid-boundary characteristics, such as the stream bottom roughness. In general, slow-moving waters are more likely to show laminar (parallel) flow patterns, whereas increasing velocities increase turbulence (Source: Dodds, 2002).

since their bodies experience the friction drag, because they stay in the sublayer, whereas older individuals have higher Reynolds Numbers and experience more of the drag forces. Usually drag creates so much stress on the benthic macroinvertebrate that it plays a key role in the energy budget of that animal. Hydraulics affect distribution patterns of lotic macroinvertebrates.

## NAVIGATING SHEAR STRESS: FRESHWATER MUSSELS

The spatial distribution benthic mussels in streams can be described as patchy. Some sections of a stream have high densities, whereas in other areas mussels are missing or at very low abundances. These distributions seem to be related to stream velocity and sediment characteristics, but can we use the principles of fluid dynamics to understand the mechanisms that influence mussel abundance? In particular, the shear stress imparted onto the benthos may explain the patchy distribution of mussels.

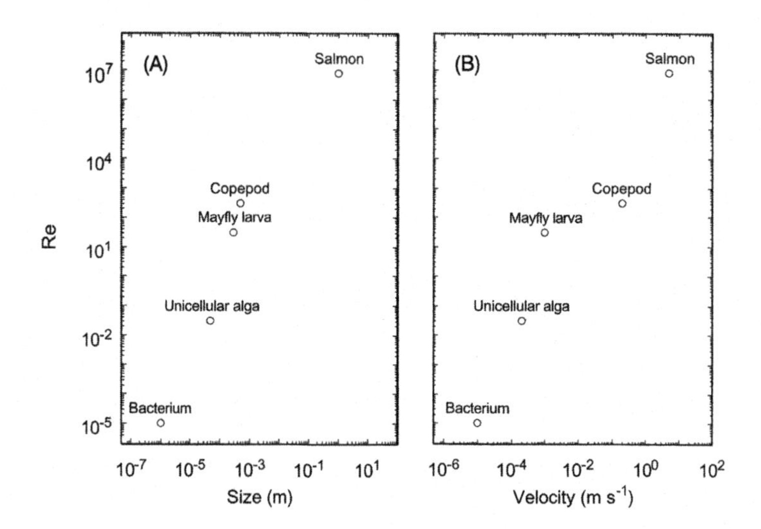


**FIGURE 6.12**

Reynolds Number ($R_e$) and body size (Source: Dodds, 2002).

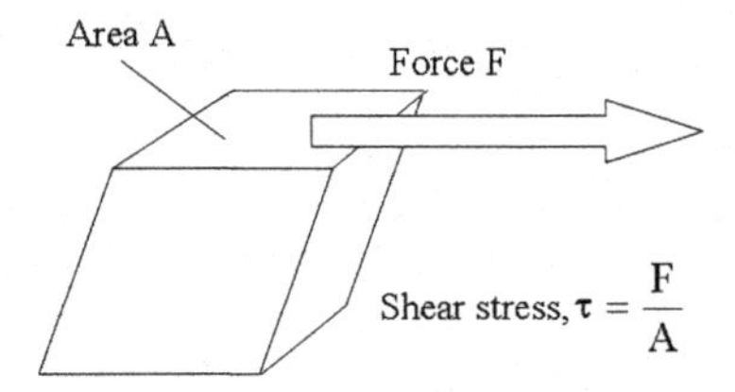


**FIGURE 6.13**

Shear stress is proportional to the force and area of the object experiencing the force.

As described in the previous section, fluids (liquids and gases) moving along solid boundary will induce a shear stress on that boundary. As shear stress is imparted onto the boundary, the fluid loses velocity.

Shear stress, denoted as $\tau$, is defined as the component of stress coplanar with a material cross-section and is often measured in Newtons (N). Shear stress arises from the force vector component parallel to the cross-section (Fig. 6.13).

The relationship between the shear stress and the velocity gradient was studied by Isaac Newton. He proposed that the shear stresses are directly proportional to the velocity gradient:

$$\tau_{(s)} = -\mu \frac{\partial v_x}{\partial z}, \tag{6.5}$$

where $\mu$ is the dynamic viscosity of the fluid; $v_x$ is the velocity of the fluid along the boundary; $z$ is the height above the boundary.

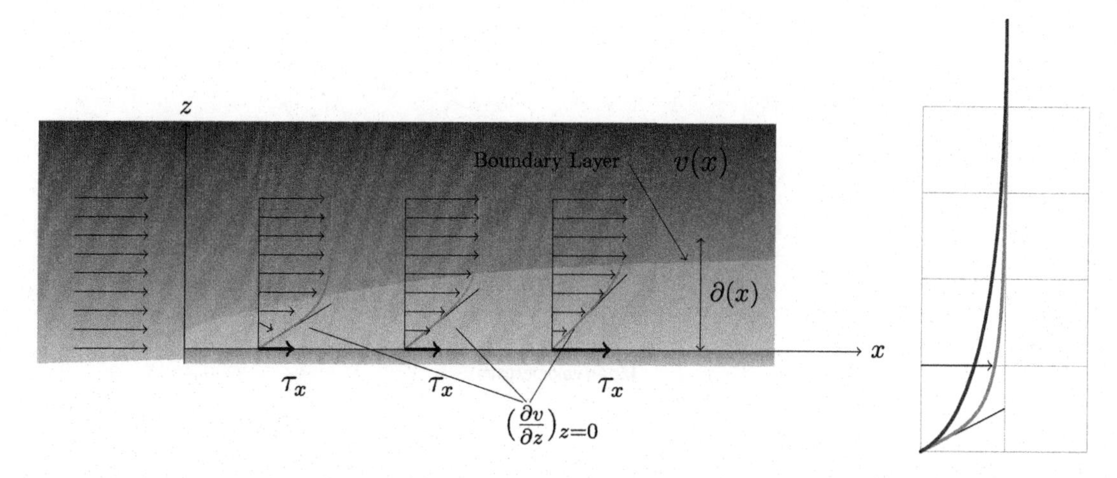


**FIGURE 6.14**

The boundary layer increases in size and shear stress increases as fluids move across a surface. Stream velocity is near the maximum beyond the boundary layer.

At a surface element parallel to a flat plate, at the point $x$, specifically, the wall shear stress, is defined as:

$$\tau_w \equiv \tau(z=0) = -\mu \left.\frac{\partial v_x}{\partial z}\right|_{z=0}. \tag{6.6}$$

One key aspect of the velocity gradient is that the velocity at the boundary is zero, which is usually referenced as the no-slip condition. The no-slip condition dictates that the speed of the fluid at the boundary (relative to the boundary) is zero, but at some height from the boundary, the flow velocity must equal that of the fluid. Recall, the region between these two points is aptly named the boundary layer (Fig. 6.14).

Returning to the strong habitat preferences observed by mussels, we can explore how shear stress might impact their habitat preferences. For example, in the South Fork of the Eel River (California), mussels avoided riffles, where shear stress can be as high as 80 N/m$^2$. Mussel aggregations were usually found in pools when shear stress was ~5 N/m$^2$ in the winter. Pool selection seems to be highly related to where channel gradient was low, yet not where shear stress and velocity were necessarily lowest (Fig. 6.15).

In contrast, mussels in the Salmon River (Idaho) are restricted to runs and are absent in pools or riffles. However, this behavior may be a relatively recent change, because pools have been experiencing seasonal sediment filling; thus given the sediment loads, the mussels select a less preferable habitat. But with access to seston transport, while avoiding the seasonal sediment filling and scouring, mussels avoid the pools.

Thus we find that mussels may depend on a combination of factors, such as sediment load and dynamics and shear stress. Using these tools, we can better appreciate and manage our streams to maintain healthy populations and even use these taxa as indicators of water quality.

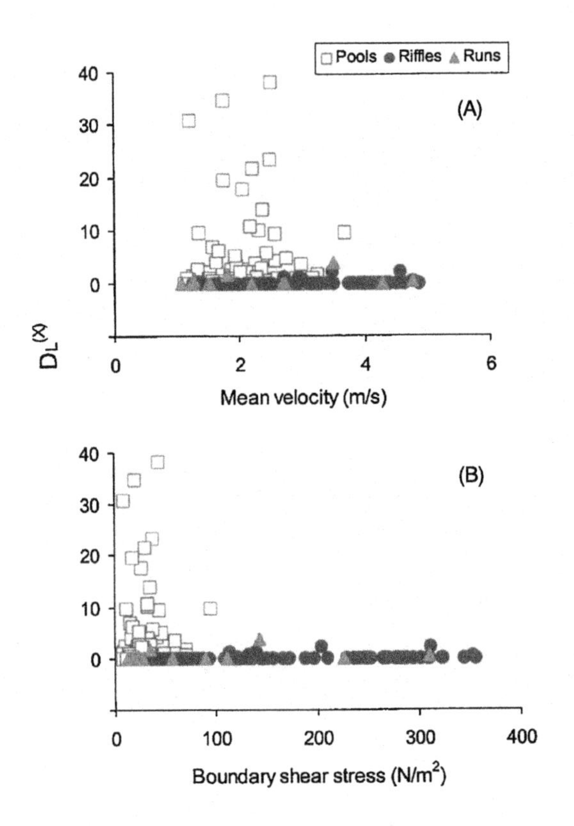


**FIGURE 6.15**

Abundance of mussels (as measured in $D_L(x)$) as a function of velocity and shear stress in the South Fork of the Eel River. Note that the mussels seem to aggregate at low shear stress values, but areas with appreciable velocities, where they will have better access to seston (Source: Howard and Cuffey, 2003).

# LIGHT

## ELECTROMAGNETISM AND WAVES

Light is electromagnetic radiation, whose quality is defined by the wavelength. Visible light is measured with a wavelength in the range of 400–700 (nm or me$^{-9}$), between the infrared (with longer wavelengths) and the ultraviolet (with shorter wavelengths). (See Fig. 6.16.)

The main source of light on Earth is the Sun. Sunlight provides the energy for photosynthesis that provides virtually all the energy used by living things. The primary properties of visible light are intensity, propagation direction, frequency or wavelength spectrum, and polarization. Its speed in a vacuum, 299,792,458 m/s, is one of the fundamental constants of nature, *c*. This is the same c in Einstein's famous equation, $e = mc^2$.

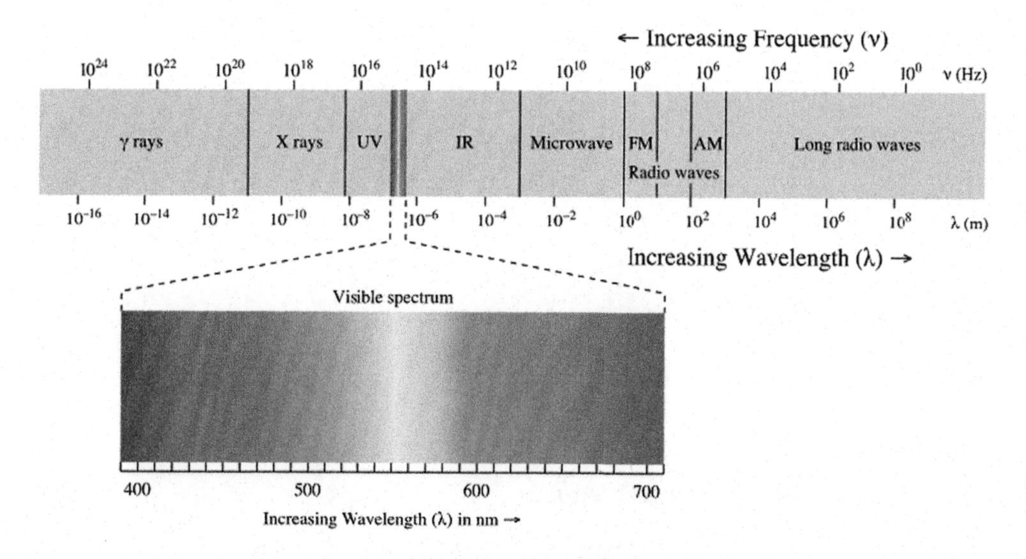


**FIGURE 6.16**

The light spectrum demonstrates the relationship between color and light wavelengths (λ) (Source: Nikita Patel (UCD), Kevin Vo (UCD), Mateo Hernandez (UCD)).

## LIGHT ATTENUATION

Light disappears quickly as one descends into the water column. More accurately, light is reflected, scattered, and absorbed by water and/or the particles in the water. The combination of these processes attenuates light with depth, i.e., there is less light with depth. The lack of light at depth prevents animals that rely on sight to find and capture prey. It also means that photosynthetic bacteria, algae, and plants are unable to fix carbon. Thus maintaining a position in the euphotic zone, where light is available can be an important strategy for many organisms. And as a corollary, organisms might avoid being eaten by visual hunters with behaviors or characteristics that prevent being seen, such as remaining in the dark or having translucent bodies.

The percent absorption is measured as a function of the light at the surface and at depth, which can be calculated as

$$\%A = \frac{100(I_0 + I_z)}{I_0}, \tag{6.7}$$

where $A$ is absorption, $I_0$ is the intensity at the surface, and $I_z$ is the intensity at a certain depth, $z$. The decline of light with depth is called attenuation, which can be described mathematically with an attenuation coefficient ($k$),

$$I_z = I_0 e^{-kz}. \tag{6.8}$$

The larger the extinction coefficient, the greater the attenuation of light or lower intensity with depth (Fig. 6.17).

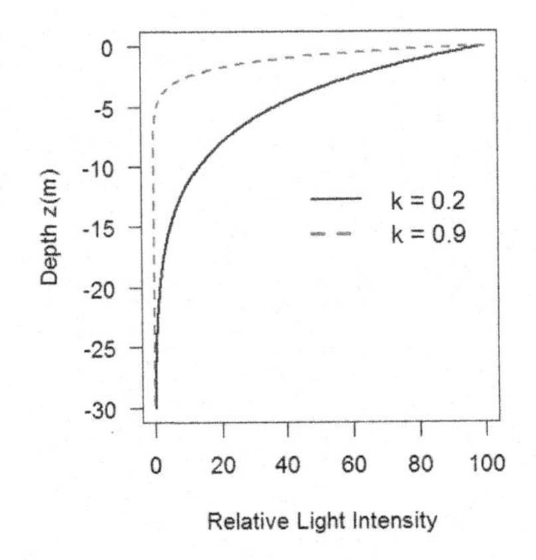


**FIGURE 6.17**

Light extinction with depth where $k = 0.2$ and $k = 0.9$.

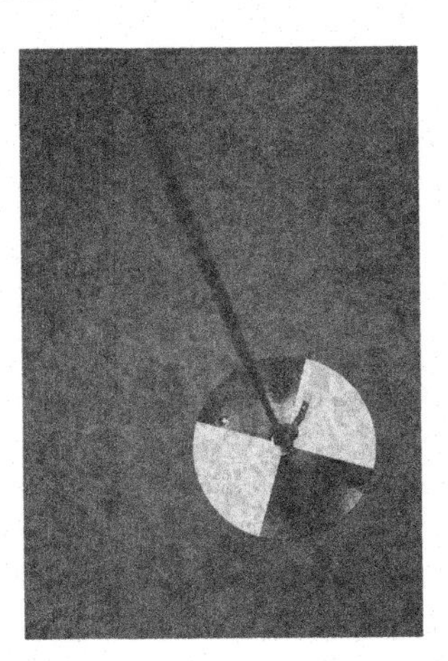

**FIGURE 6.18**

Secchi Disk.

Light extinction coefficients vary between lakes (Table 6.3). For example, Crater Lake is an exceptionally clear, sky-blue lake. In comparison, Lake Tahoe is also quite transparent, but its clarity has been decreasing since the 1960s due to development around the lake.

One simple way to measure water clarity relies on the Secchi disk, which was invented by Angelo Secchi in 1865. Now using a black and white pattern (Fig. 6.18), the Secchi depth is defined as the average depth when the disk disappears and appears in the water column. Because of its simplicity of use the Secchi Disk has generated a long record of Secchi Depths in thousands of lakes worldwide (Table 6.3).

Secchi Transparency is a function of the absorption properties of the water as a function of the dissolved organic carbon and particulate matter in the water column. As mentioned earlier, Crater

**Table 6.3 Light attenuation in selected lakes (Sources: † Larson et al., 2007, * Tahoe Environmental Research Center). Because Secchi Depths are relatively simple, it is easier to find these data than extinction coefficients and depth of the Euphotic Zone.**

| Lake | $k$ | Secchi depth (m) | Euphotic Zone |
|---|---|---|---|
| Crater Lake (OR)† | 0.06–0.12 | 35–45 | >120 |
| Lake Tahoe (CA/NV) | 0.12 | 40 | 90–136 |
| Lake Tahoe (1968)* | – | 31.2 | – |
| Lake Tahoe (2014)* | – | 23.7 | – |
| Lake Baikal, Siberia | 0.2 | 5-40 | 15–75 |
| Lake Erie | 0.2–1.2 | 2–10 (1970–1990) | – |
| | – | >10 (1993–1995) | 12–26 |

Lake is known for its clarity has a Secchi Depth between 35–45 m. In contrast, Lake Tahoe has had a reduction in Secchi Depth of 10 m between 1968 and 1998 with an average decline in transparency of 0.37 m each year. By measuring lake clarity on a regular basis, environmental scientists can evaluate how lakes might be changing over time.

Light attenuation depends on more than just the depth of water. There are also temporal and spatial patterns of light attenuation. For example, the Secchi Depth in Lake Tahoe has been as deep as 40 m, usually in winter months, whereas a low of 8.25 m was recorded in the summer of 1983.

A robust estimate of the light attenuation coefficient is a function of depth, properties of water, absorption properties of the water quality and back-scattering (reflection):

$$k(z) = k_{water}(z) + k_{DOM}(z) + k_{PM}(z) + b_{downward}(z) - b_{upward}(z)R(z), \tag{6.9}$$

where $k(z)$ is calculated for various ranges of depth; $k_{water}$ is for distilled water; $k_{DOM}$ is for dissolved organic matter; $k_{PM}$ is for particulate matter; $b_{downward}$ is back-scatter from below; $b_{upward}$ is back-scatter from above combined with reflectance—$R(z)$.

Although Eq. (6.9) is more difficult to quantify and requires specific sensors, it allows researchers to distinguish between different components of the water column to understand the mechanisms that attenuate light.

## CONTRASTS IN LAKE CLARITY

A lake with clarity means there is relatively low suspended sediments and/or phytoplankton biomass. The factors that control lake clarity are numerous, including direct-controlling factors. Some examples include suspended sediments; plankton populations, and DOC and DOM concentrations. In addition, other drivers influence these factors to increase or decrease clarity, thus creating a dynamic system. For example, temperature; nutrient concentrations; species composition and density of bacteriophyton; phytoplankton, zooplankton, and predators-prey population dynamics can play an important role in controlling lake clarity.

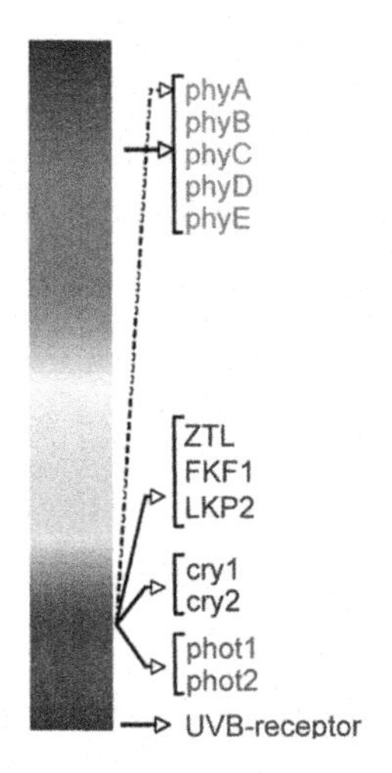


**FIGURE 6.19**

Plant response to light quality (Source: adapted from Frankhauser).

**FIGURE 6.20**

A LICOR PAR sensor that is mounted on a frame and lowered in the water on a line to measure PAR with depth. These sensors are most effective in lakes and ponds.

## LIGHT QUALITY AND PAR

Only a subset of the light spectrum is used biologically. In particular, a suite of wavelengths between 400 and 800 nm ($m^{-9}$) are known as photosynthetically active radiation. Although other wavelengths can influence biochemical and behavioral physiological processes, PAR drives photosynthesis (Fig. 6.19).

How is PAR measured? In the water column, PAR is measured using a sensor that is calibrated to measure the energy of light (in Einsteins) (Fig. 6.20). PAR sensors measure direct and scattered light from nearly every direction in a similar way that phytoplankton might absorb light.

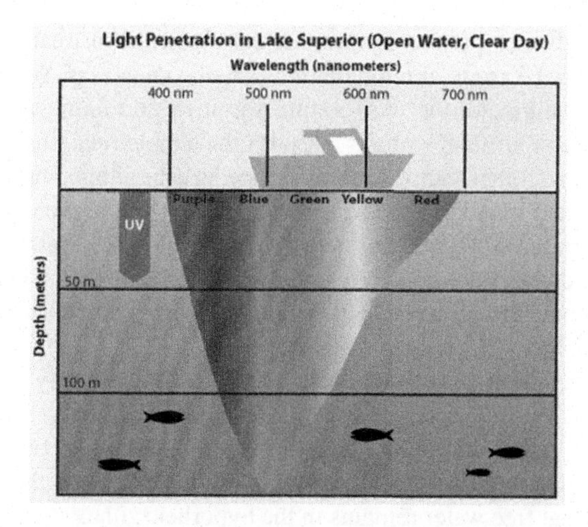


**FIGURE 6.21**

Visible light penetrates differently into the surface waters. Longer wavelengths (e.g., red) are absorbed at a shallower depth than shorter wavelengths (e.g., blue), which penetrate deeper into the water. (Source: The credit for the design (not the data) is "Minnesota Sea Grant." The credit for the data is "Jay Austin, Thomas Hrabik, Trevor Keyler, University of Minnesota Duluth.")

In addition, particulates and macroscopic taxa absorb light with varying levels of influence on the water column. Besides the light absorption properties of an intact organism, a range of DOC compounds also absorb specific wavelengths that can change light quality in the water column.

## LIGHT AND COMMUNITY STRUCTURE

Because of the strong gradient of light that changes with depth, various species might favor different habitat-types, which will vary with depth and function of the extinction coefficient ($k_Z$).

Light attenuation differs by various light wavelength (Fig. 6.21); thus the quality and quantity of light contribute to how various taxa might use or avoid water depth. For example, ultraviolet energy reaching the lake bottom can predict primary production in small Antarctic lakes.

# TEMPERATURE

## LOTIC WATER TEMPERATURES

The temperatures of rivers and streams vary much more rapidly than those of lakes. Depending upon size and origin of streams, their seasonal and diurnal temperatures follow atmospheric temperatures more closely than do those of lakes. The source of the stream water and, in some instances, the nature of the drainage pattern determines the thermal properties of a stream. For instance, water temperature varies along the lengths of the valleys. Large river and stream temperature generally reflect the mean monthly air temperature of the region.

The temperature of headwater streams is variable, but as the downstream water volume increases and becomes more constant, the range of temperature variation decreases. With respect to stream size, smaller streams tend to have greater the temperature variation and more rapid response to environmental fluctuations of ambient air temperatures. Besides the simple relationship between atmospheric temperature with stream size, other factors can influence stream temperature regimes. For instance, during spring, snowmelt water may keep the temperature below that of the air. Also, sunny days after a precipitation events can raise water temperatures rapidly as warm soil water flows into a stream.

In the summer, the headwaters are relatively cool, either because they contain spring water that is at the annual mean soil temperature or because they come from high elevations and are only slowly warmed by the air, the sun, and conduction from the ground.

In winter the reverse may apply to spring-fed streams that start relatively warm and become cooler. Also, during cold weather, ice may form on the water's surface although it does so less readily on running rather than on still water. Once ice has been formed, snow can accumulate on it and, together, they form an excellent insulator against further heat loss. However, even if the water is frozen down to the streambed, it is likely that free water remains in the hyporheic zone.

Besides seasonal temperature changes, diurnal cycles drive temperature as well. In general, maximum temperatures occur in the late afternoon, and minimum temperature just before dawn. This is caused primarily by the radiation into and out of the water. The range of daily variation of water temperature is largest when there is the greatest differential between mean diurnal air temperature and mean water temperature. However, these changes depend on upstream conditions, volume of water, and turbidity. For example, with increased volume and turbidity, the range of variations is reduced. The difference between day and night may exceed 6°C in small streams in the summer, whereas the difference might be undetectable in large rivers. Even in small streams, the deeper the water, the less the daily variation. Shallow streams a few meters wide or small pools, especially if unshaded, can experience extremely high temperature variations. Finally, although these fluctuations exist throughout the year, they are minimized beneath the ice cover.

Because of turbulence and the shallow nature of most streams, thermal stratification is not generally an attribute of streams. When streamwaters do stratify, the process usually takes place in pools along the stream course. This may be caused by sunshine on the shallows, inflowing groundwater or the water from a tributary hugging the bank on which it had entered.

## TEMPERATURE AND LAKE STRATIFICATION

Because light is attenuated with depth (Fig. 6.21), especially infrared, the surface of water is preferentially warmed as sunlight is converted to heat. As the surface warms, the water itself becomes less dense—thus can float on cooler waters below. This layering of the water column by temperature, which is called thermal stratification, can be a permanent feature of some aquatic systems; it can also vary seasonally or even shorter time frames. Limnologists have identified three different layers within a lake (Fig. 6.22): epilimnion (uppermost layer), metalimnion (middle layer), and hypolimnion (bottom layer). The thermocline is an imaginary plane within the lake, where the rate of temperature decrease is greatest. The thermocline or metalimnion is a zone, often defined as a region, where temperature gradient is at least 1°C decrease with every 1 m increase in depth.

The thermal stratification is maintained by differences in water density, where less dense water floats on the higher-density water. Water density can be influenced by both temperature and salinity.

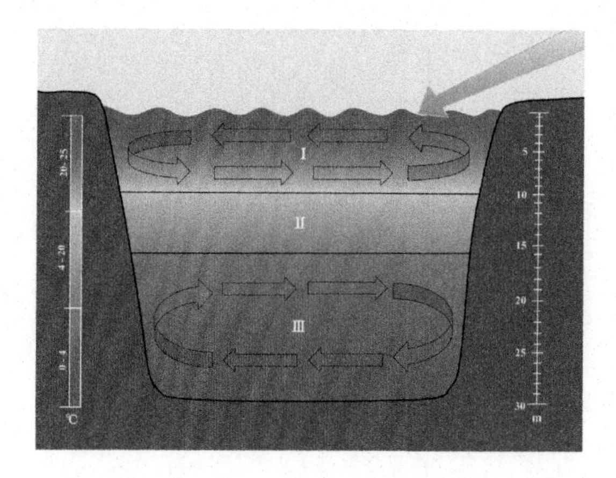


**FIGURE 6.22**

Lakes are stratified into three separate layers: the epilimnion (I), metalimnion (II), and (III) hypolimnion. The scales are used to associate each section of the stratification to their corresponding depths and temperatures. The arrow is used to show the movement of wind over the surface of the water, which initiates the turnover in the epilimnion and hypolimnion (Source: Wikicommons).

Water density reaches its maximum at 3.94°C (Fig. 6.23). Surface waters, warmed by solar radiation, are less dense than deeper colder and more dense water. In temperate regions, where lake water warms up and cools through the seasons, a cyclical pattern of overturn occurs that is repeated from year to year as the cold dense water at the top of the lake sinks.

Temperature profiles are common ways to show how temperature changes with depth. Using the $x$-axis as temperature and $y$-axis as depth, we can identify the epilimnion, metalimnion (thermocline), and hypolimnion (Fig. 6.22).

The temperature profile depth will change seasonally, which can be displayed in a more complex figure (Fig. 6.24). The $y$-axis is still depth, but the $x$-axis is now time, and the contours are interpolated isopleths of temperature.

## LAKE MIXING

Although many lakes are stratified during specific times of the year, lakes can also experience mixing events. The thermocline can breakdown as a result of strong wind events or drastic temperature changes. When this occurs water from the epilimnion will mix with the hypolimnion and the loss of a temperature gradient. The climate and lake physiology contribute to define the mixing regime. For example, in dimictic lakes, the lake water turns over during the spring and the fall. These regimes will be described in more detail Chapter 7.

The mixing of lakes relies on wind to generate enough shear stress to break down the stratified layers. In fact, in spite of their placid look, lakes are in constant motion. Although currents may not be visible, they are not only present, but can control fundamental biological processes within the water column. For example, currents can resuspend phytoplankton into the euphotic zone.

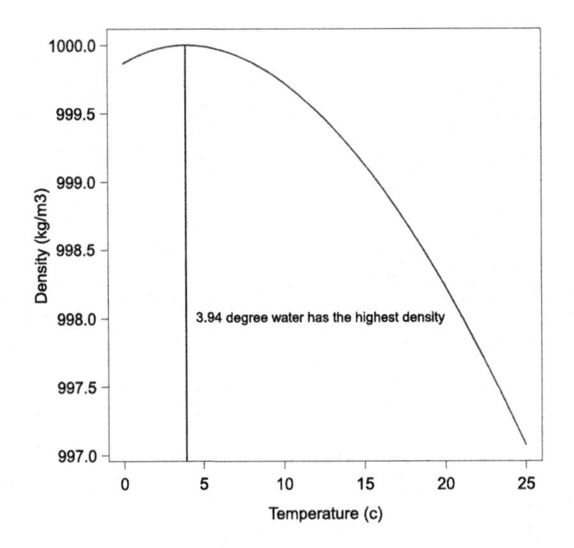


**FIGURE 6.23**

Water density is a function of temperature, where the maximum density is 3.94°C. What is missing? Dissolved solids, salts, also determines water density. We will address this as we discuss terminal lakes.

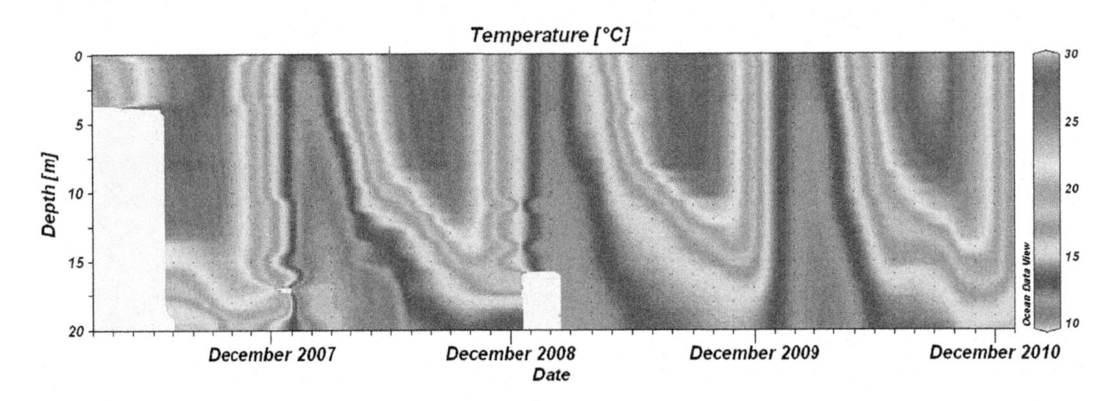


**FIGURE 6.24**

Temperature profile Roosevelt Lake, Arizona (Source: Susanne Neuer). Roosevelt Lake mixes once per year, usually around February and March each year. This mixing pattern is referred to as monomictic.

Thermal stratification effectively separates lake waters vertically, which allow differing habitats between the layers. Whereas lake mixing might allow limited nutrients to be evenly distributed in the water column, stratified lakes might favor different biogeochemical process in each layer, thus have differing nutrient concentrations in the epilimnion and hypolimnion. For example, phosphorous might be limited by photoplankton uptake in the epilimnion, where phosphorous concentrations are high in the hypolimnion because of phosphorous release from anaerobic lake sediments.

**Table 6.4 Major compounds of freshwaters. Note the distinction between particulate and dissolved is arbitrary and operational. Filters of various pore sizes (which is even a problematic concept) are used to separate dissolved from particulates, whereas there is a continuum of size factions within the water column.**

| Dissolved Compounds | |
|---|---|
| **Inorganic Ions (mM)** | **Organic Compounds** |
| $Ca^{+2}$ (0.3–3.0) | C, H, O, N, P, S (1 μM–1.0 M) |
| $Na^{+}$ (0.3–4.0) | - proteins (10%) |
| $Mg_2^{+}$ (0.1–1.0) | - polysaccharides (20%) |
| $HCO_3^{-}$ (0.5–5.0) | - humics (70%) |
| $SO_4^{2-}$ (0.01–0.5) | |
| $Cl^{-}$ (0.01–5.0) | |
| $Si_2^{-}$ (0.1–6.0) | |
| $O_2$ (0.0–.6) | |
| **Particulate Compounds** | |
| **Inorganic Compounds** | **Organic Compounds** |
| Formed in Soils | |
| Clay, silicates | Debris from organisms |
| FeIII Iron (III) oxides | Humic compounds |
| Mn(III–IV) oxides | Organo-mineral particles |
| Formed in Waters | |
| (Ca, Mg)-$CO_3$ | Cellular Debris |
| $SiO_2$ | Fecal pellets |
| FeOOH, $Mn_x$ | Organo-mineral particles |

## CHEMICAL COMPOSITIONS

### MAJOR CHEMICAL CONSTITUENTS

The composition of waters include inorganic and organic constituents. Aquatic chemists analyze and characterize aquatic systems by using observable (pragmatic) categories, e.g., dissolved and particulate compounds, where dissolved substances pass through a 0.45 μm filter and particulates are trapped on this filter paper (Table 6.4).

Dissolved compounds might be inorganic or organic (biologically or synthetically produced). When the concentrations of these dissolved compounds decrease, based on dilution or increase due to evaporation, we refer to this as a conservative behavior. When compound concentrations, usually nutrients, change due to uptake or release by organisms, we refer to this as non-conservative behavior.

The source of these compounds might be from terrestrial sources in the watershed—based on characteristics of the geology, soils, vegetation, or other biota in the watershed. These are allochthonous sources of material, i.e., external to the water column. Alternatively, materials created in the water column are referred to as autochthonous materials, e.g., carbohydrates created by periphyton. In general, we expect allochthonous materials to be the major soure of inorganic compounds listed in Table 6.4, while autocthanous and allocthanous sources might contribute organic compounds.

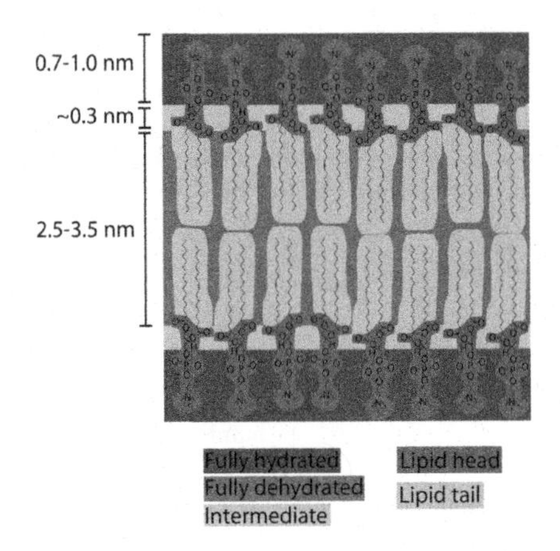


**FIGURE 6.25**

Illustration of lipid bilayer (Source: https://commons.wikimedia.org/wiki/File:Bilayer_hydration_profile.svg). The fatty acid chains in the middle are hydrophobic, whereas the ester external are hydrophilic, creating a boundary that limits water to move across the member without aquaporins.

Every stream, lake, and wetland usually has a characteristic chemical signature. For anadromous fish, this signature might allow adults to navigate back to their natal waters to reproduce, or to drive physiological patterns that might promote biological diversity. For example, in the highlands of Scotland, differences in calcium carbonate concentrations ([$CaCO_3$]) drive the composition of *Daphnia* species.

## MAINTAINING CONCENTRATION GRADIENTS

The concentration of chemicals inside organisms differ from their environment. Maintaining this concentration gradient is indicative that the organism is alive.

At the cellular level, all organisms must maintain some type of concentration gradient—as a combination of electrical and chemical gradients between and across a membrane. At the scale of organelles and cells, organisms use active and passive mechanisms to control the concentration of solutes inside the membrane. Maintaining an appropriate mixture of salts inside these membranes is required for the proper protein development (e.g., folding) and function.

Some key fats and oils make up the lipid bilayer that maintains solute concentrations that differ from inside to the outside of these membrane-bound spaces (Fig. 6.25). Membranes serve a number of purposes, but for our current purpose, you should appreciate the hydrophobic region within the bilayer that prevents many solutes from crossing cell membranes.

Water-soluble substances typically do not cross lipid membranes easily, unless specific transport mechanisms are present.

The concentration of these solutes, then in turn, influences the amount of water inside the membrane-bound spaces. The water pressure on the cellular membrane is called osmotic pressure. By understanding the concept of osmosis and how cellular mechanism maintain solute concentrations

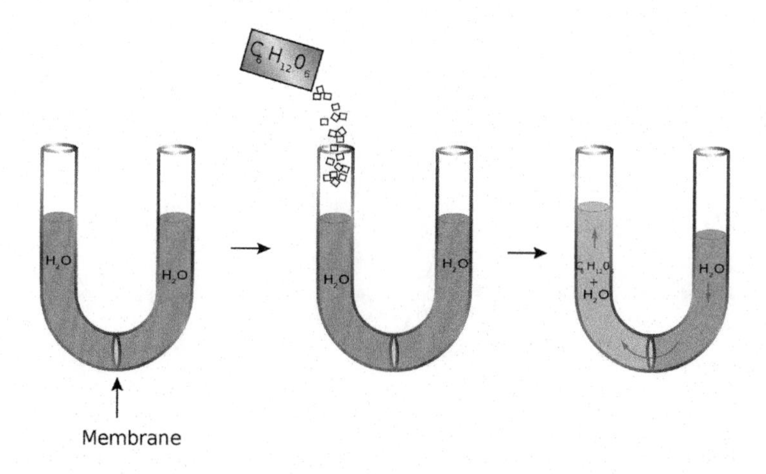


**FIGURE 6.26**

Diagram of osmosis in a U-shaped tube (Source: Wiki Commons). Osmosis occurs when two solutions, containing different concentration of solute, are separated by a selectively permeable membrane. Solvent molecules, such as water, pass preferentially through the membrane from the low-concentration solution to the solution with higher solute concentration. The transfer of solvent molecules will continue until equilibrium is attained.

relative to the extra-cellular environment, we will appreciate a great deal of how organisms function in the liquid environment and the capacity to invade inland waters.

Osmotic pressure is a measure of the tendency of water to move into one solution from another. The higher the osmotic pressure of a solution, the more water wants to move into the solution (Fig. 6.26). A hypertonic (or hyperosmotic) solution has higher solute concentrations on the outside of the semipermeable membrane:

**Hypotonic** or (hypoosmotic) solutions have a lower concentration on the outside of the semipermeable membrane.

**Hypertonic** solutions have a higher osmotic pressure than a particular fluid, typically a body fluid or intracellular fluid.

Pressure must be exerted on the hypertonic side of a selectively permeable membrane to prevent diffusion of water by osmosis from the side containing pure water.

If this idea sounds counter intuitive, then consider it in terms of water concentration: The higher the solute concentration, the lower the water concentration, molecule for molecule. Net movement therefore occurs from regions of higher water concentration to regions of lower water concentration, exactly what one would expect.

Movement of water across a semipermeable membrane is called osmosis. Without an opposing force, water tends to flow across semipermeable membranes. Water will flow until there is an equilibrium, i.e., the osmotic pressure and solute concentration is equal on both sides of the membrane. However, living organisms maintain an opposing force, so an equilibrium is never reached—as along as the organism is alive.

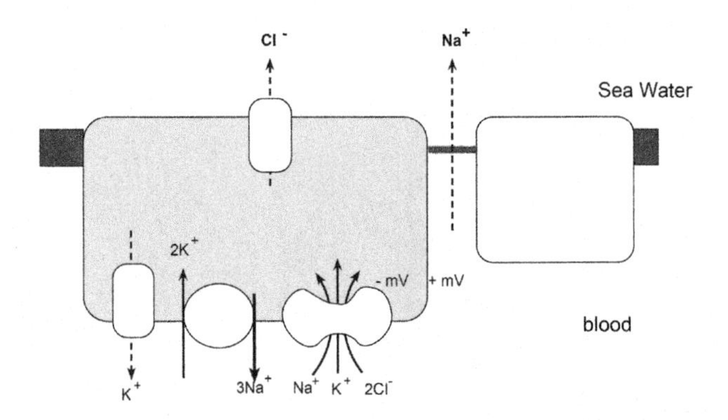


**FIGURE 6.27**

Model of teolist osmoregulation cell transport, where ionic charge and osomotic pressure are simultaneously controlled.

## OSMOREGULATION: FROM ORGANELLES TO BODIES

The movement of water across cell membranes is essential for cellular integrity, but can cause problems. A small difference in solute concentration results in a very large osmotic pressure gradient across the cell membrane. The cell membranes of animal cells cannot withstand any appreciable pressure gradient. Water movement can eliminate differences in osmolality across the cell membrane, but this alone is itself a problem as it leads to alteration in cell volume. Consequently regulation of intracellular solute concentration is essential for control of cell volume. (See Fig. 6.27.)

Taxa are sensitive to the concentration of salts. Saline and freshwaters are only one chemical gradient in aquatic systems. Organisms in both aquatic and terrestrial environments must maintain the right concentration of solutes and amount of water in their body fluids; this involves excretion (getting rid of metabolic wastes and other substances, such as hormones that would be toxic if allowed to accumulate in the blood) via organs, such as the skin and the kidneys. Keeping the amount of water and dissolved solutes in balance is referred to as osmoregulation.

Osmoregulation is the active regulation of the osmotic pressure of an organism's fluids to maintain the homeostasis of the organism's water content; that is, it keeps the organism's fluids from becoming too diluted or too concentrated. Organisms, according to osmoregulation, are put into two major categories: osmoconformers and osmoregulators.

Osmoconformers match their body osmolarity to their environment. It can be either active or passive. Most marine invertebrates are osmoconformers, although their ionic composition may be different from that of seawater.

Osmoregulators tightly regulate their body osmolarity, which always stays constant, and are more common in the metazoa. Osmoregulators actively control salt concentrations despite the salt concentrations in the environment.

The blood of freshwater fishes is typically more salty than the water in which they live. Osmotic pressure, the force that tends to equalize differences in salt concentrations, causes water to diffuse, or enter, into the fish's body, primarily through the gills, mouth membranes, and intestine. The gills actively uptake salt from the environment by the use of mitochondria-rich cells. Water diffuses into the fish, so it excretes a very hypotonic (dilute) urine to expel all the excess water. A marine fish has an

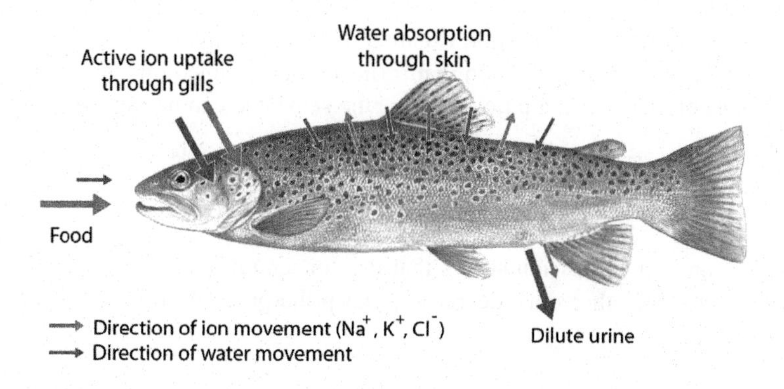


**FIGURE 6.28**

Osmoregulation in freshwater fishes (Source: Wiki Commons).

internal osmotic concentration lower than that of the surrounding seawater, so it tends to lose water and gain salt. It actively excretes salt out from the gills. To eliminate excess water, freshwater fishes produce a large amount of very dilute urine. Lampreys, for example, may daily produce an amount of urine equal to as much as 36% of their total body weight; bony fishes commonly produce amounts of urine equaling from 5 to 12% of their body weight per day. As these fishes are gaining water, they are losing salts. Salts contained in their foods are insufficient to maintain the proper salt balance. Freshwater fishes have therefore developed the capacity to absorb salts from water by means of their gills. (See Fig. 6.28.)

As in teleost fish, the gills, gut, and kidney are thought to be the main effector organs involved in the active regulation of salt and water balance, although firm data is scarce.

Most fish are stenohaline, which means they are restricted to either salt or freshwater and cannot survive in water with a different salt concentration than they are adapted to. However, some fish show a tremendous ability to effectively osmoregulate across a broad range of salinities; fish with this ability are known as euryhaline species. An example is salmon.

# PH AND INLAND WATERS

## DISASSOCIATION OF WATER

Chemically, water is amphoteric: It can act as either an acid or a base in chemical reactions. Pure water ($H_2O$) disassociates into hydrogen cations and hydroxyl anions easily, although the amount of cations and anions are relatively small:

$$H_2O(l) \rightleftharpoons H^+(aq) + OH^-(aq)\cdot \quad (6.10)$$

The reaction is reversible and the relative concentrations between the left and right side of the equation are in equilibrium:

$$K_w = [H^+][OH^-] \quad (6.11)$$

The equilibrium constant for this reaction is called the water disassociation constant ($K_w$); equals $1.01 \times 10^{-14}$ at 25°C. Because every $H^+$ ion that forms is accompanied by the formation of an $OH^-$ ion, the concentrations of these ions in pure water are the same and can be calculated from $K_w$:

$$K_w = [H^+][OH^-] = (x)(x) = 1.01 \times 10^{-14};$$

$$x = [H^+] = [OH^-] = 1.01 \times 10^{-7}\ M.$$

The equilibrium constant expression shows that the concentrations of $H^+$ and $OH^-$ in water are linked. As one increases, the other must decrease to keep the product of the concentrations equal to $1.01 \times 10^{-14}$ (at 25°C).

## ACIDS AND BASES

In addition to $H_2O$, lake and river water includes salts, acids, and bases which contribute to this mixture of $H^+$ and $OH^-$ within the water. When one ion is added to the water, the amount of other ions shifts, because they are in an equilibrium, and pH can drive some of the most important shifts in aquatic systems.

According to the Arrhenius theory of acids and bases, when an acid is added to water, it donates a proton to water to form $H_3O^+$ (often represented by $H^+$ as above).

Acids tend to have a higher concentration of $H^+$ and are characterized as easily losing their protons in water, whereas bases have a higher concentration of $OH^-$.

Acidity refers to water having a pH less than 7. We can measure this through the pH, known as the logarithm of the reciprocal of the concentration of free hydrogen ions, i.e., $pH = -\log_{10}[H^+]$.

## PH IN CHEMICAL EQUILIBRIUM

Just as the disassociation of water is reversible, many aquatic reactions are can proceed in either direction, and pH can alter the equilibrium.

The higher the concentration of $H_3O^+$ (or $H^+$) in a solution, the more acidic the solution is. An Arrhenius base is a substance that generates hydroxide ions, $OH^-$, in water. The higher the concentration of $OH^-$ in a solution, the more basic the solution is.

If an acid, like hydrochloric acid, is added to water, the concentration of the $H^+$ goes up, and the concentration of the $OH^-$ goes down, but the product of those concentrations remains the same. An acidic solution can be defined as a solution in which the $[H^+] > [OH^-]$. The example below illustrates this relationship between the concentrations of $H^+$ and $OH^-$ in an acidic solution.

If rainwater has a concentration of 0.025 M HCl at 25°C, then there will be more $[H^+]$ than $[OH^-]$. Furthermore, by using $K_w$, we can calculate the concentration of each ion and the pH.

First, we can assume that hydrochloric acid, HCl(aq), like all strong acids, is completely ionized in water: $[H^+] = [Cl^-] >> [HCl]$.

Thus, $[H^+]$ is equal to the HCl concentration, or 0.025 M. We can calculate the $[OH^-]$ by rearranging the water dissociation constant expression (Eq. (6.11)), and solve for $[OH^-]$:

$$[OH^-] = 1.01 \times 10^{-14}/0.025 = 4.04 \times 10^{-13}.$$

Note that the $[OH^-]$ is not zero, even in a dilute acid solution with a pH of 1.6.

If a base, such as sodium hydroxide, is added to water, the concentration of hydroxide goes up, and the concentration of hydronium ion goes down. A basic solution can be defined as a solution in which the $[OH^-] > [H^+]$.

## PH OF RAIN WATER

Pure water has a pH of 7.0 (neutral); however, natural, unpolluted rainwater actually has a pH of about 5.6 (acidic). The acidity of rainwater comes from the natural presence of three substances ($CO_2$, NO, and $SO_2$) found in the troposphere (the lowest layer of the atmosphere). Since carbon dioxide ($CO_2$) is present in the greatest concentration, it therefore contributes the most to the natural acidity of rainwater.

Sulfur dioxide dissolves in water and then, like carbon dioxide, hydrolyzes in a series of equilibrium reactions. In the gas phase sulfur dioxide is oxidized by reaction with the hydroxyl radical via an intermolecular reaction:

$$SO_2 + OH \longrightarrow HOSO_2^{\cdot} \tag{6.12}$$

which is followed by

$$HOSO_2^{\cdot} + O_2 \longrightarrow HO_2^{\cdot} + SO_3 \cdot \tag{6.13}$$

In the presence of water, sulfur trioxide ($SO_3$) is converted rapidly to sulfuric acid:

$$SO_3(g) + H_2O(l) \longrightarrow H_2SO_4(aq) \cdot \tag{6.14}$$

After creating sulfuric acid, the $H_2SO_4(aq)$ will disassociate into protons to make the pH decline:

$$SO_2 \cdot H_2O \rightleftharpoons H^+ + HSO_3^- \tag{6.15}$$

$$HSO_3^- \rightleftharpoons H^+ + SO_3^{2-} \cdot \tag{6.16}$$

Nitrogen dioxide reacts with OH to form nitric acid. For example,

$$NO_2 + OH^{\cdot} \longrightarrow HNO_3 \tag{6.17}$$

which can disassociate just as above.

Acid-producing gasses are created by anthropogenic sources, but also biological processes that occur on the land, in wetlands, in the oceans, and in volcanoes. Thus, for example, fumaroles from the Laguna Caliente (Costa Rica) crater of Poás Volcano create extremely high amounts of acid rain and fog with acidity as high as a pH of 2. The acid kills the surrounding vegetation.

Unfortunately, human industrial activity produces additional acid-forming compounds in far greater quantities than the natural sources of acidity described above. In some areas of the United States, the pH of rainwater can be 3.0 or lower, approximately 1000 times more acidic than normal rainwater. In 1982 the pH of a fog on the West Coast of the United States was measured at 1.8! When rainwater is too acidic, it can cause problems, ranging from killing freshwater fish and damaging crops, to eroding buildings and monuments.

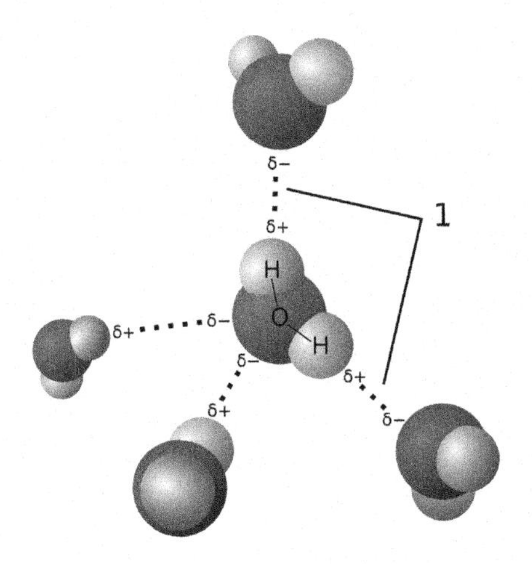


**FIGURE 6.29**

3D model of hydrogen bonds between molecules of water (Source: Wiki Commons).

# UNIONIZED DISSOLVED GASES IN AQUATIC SYSTEMS

## UNIONIZED GASES AND WATER'S CRYSTALLINE STRUCTURE

As we have discussed above, atmospheric $CO_2(g)$ can become dissolved in the water column as $CO_2(aq)$, which then disassociates to $HCO_3^-$ and $CO_3^{2-}$. However, there are other gases that can enter the water column as unionized forms with differing ecological affects.

Composed of hydrogen and oxygen, water acts as a semicrystalline structure because of the hydrogen bonding. This bonding allows water molecules to form clusters that create a nonrandom structure in water (Fig. 6.29). As the temperature of the water increases, the clusters are less stable and the crystalline structure breaks down. However, this crystalline structure allows a wide range of gases to be stabilized (i.e., dissolved) in the crystalline structure, such as $CO_2$.

Using Henry's Law, which models a linear relationship between the gas phase activity and aqueous phase activity of gases in water

$$P_a = K' C_A, \tag{6.18}$$

where $P_a$ is the partial pressure of the gas, $C_A$ is the liquid-phase concentration for specie $A$, and $K'$ is Henry's Law Constant. For gases that do not disassociate into ions in water, such as $O_2$, $N_2$, and $CH_4$ this relationship is quite powerful. For others that can disassociate, such as $CO_2$ and $NH_3$, the liquid-phase concentration is a bit more complex and often depends on pH.

An important application of this issue includes the issues associated with $NH_3$, which is toxic to fish because, as an unionized form, it can pass across the gill structure of fishes and enter the blood stream. However, the reaction

$$NH_3(aq) + H_2O \leftrightharpoons NH_4^+OH^- \tag{6.19}$$

depends on the pH. When the total $NH_3$ is measured, the toxicity at a minimum depends on the pH and temperature—where lower pH reduces the toxicity.

# GEOCHEMISTRY OF INORGANIC CARBON

## CARBON DIOXIDE, CARBONATE, AND BICARBONATE

Carbon dioxide from the atmosphere is a significant source of terrestrial carbon. Aquatic macrophytes, including emergent and floating vegetation, obtain a substantial amount of carbon from atmospheric sources. However, for submerged photoautrophics, carbon dioxide must be obtained from the water column, where the concentration is 10,000 times lower than that atmosphere. Inorganic carbon occurs in the water as dissolved and unionized form ($CO_2$(aq) or $H_2CO_3$(aq) (carbonic acid)) or in a dissolved, ionized state ($HCO_3^-$ (bicarbonate) and $CO_3^{-2}$ (carbonate)). The sum of the concentration of these compounds is referred to as DIC, Dissolved Inorganic Carbon. The relative concentration of each of these is partially controlled by the water's pH. But concentrations can also vary with stream size, sunlight, and groundwater contribution. For example, the amount of $CO_2$ dissolved in the Eel River depends on groundwater inputs, algae uptake, and pH.

Carbonic acid generation occurs through a series of reactions, which include the following representations:

$$CO_2(g) + H_2O \rightleftharpoons CO_2(aq) + H_2O \rightleftharpoons H_2CO_3(aq) \tag{6.20}$$

$$H_2CO_3 \rightleftharpoons HCO_3^- + H^+ \tag{6.21}$$

$$HCO_3^- \rightleftharpoons CO_3^{-2} + H^+ \tag{6.22}$$

Note that with the [$H^+$] as a component of the reactants means that with decreasing pH (increase in [$H^+$]) will drive reactions (6.21) and (6.22) to the right and possibly drive $CO_2$ from the water column, making $CO_2$ a limiting resource for plant photosynthesis. For example, some taxa, e.g., red algae and bryophytes, preferentially use $CO_2$ over $HCO_3^-$ because less energy is required for respiration. In addition, when pH favors DIC in the form of $CO_2$ these taxa may better compete for inorganic carbon.

## BIOGEOCHEMISTRY OF DISSOLVED INORGANIC CARBON

DIC chemistry is rather complicated, but for our purposes, we can begin by appreciating that the form of DIC depends, in part, on the pH of the water (Fig. 6.30). Thus $CO_2$(aq) can partially drive pH in water, but the overall geochemistry of DIC depends on pH in the water.

The summation of DIC (or $\Sigma CO_2$) budget within a watershed can be simplified into five internal mass balance essential components: photosynthesis, respiration, gas exchange, groundwater discharge, and geochemical dissolution or precipitation. Each control may exert variable influence within the DIC budget based on productivity, allochthonous supply, and watershed geology. These processes play a critical role in forming linkages between terrestrial ecosystem reactions and surface water transport of DIC.

The general ability of a system to neutralize acidity resulting from a variety of atmospheric, geochemical, or biological reactions is associated with inorganic carbon buffering action. Generally the pH of a system is largely controlled by the interaction of $H^+$ ions produced within the hydrolysis of bicarbonate derived from geochemical reactions.

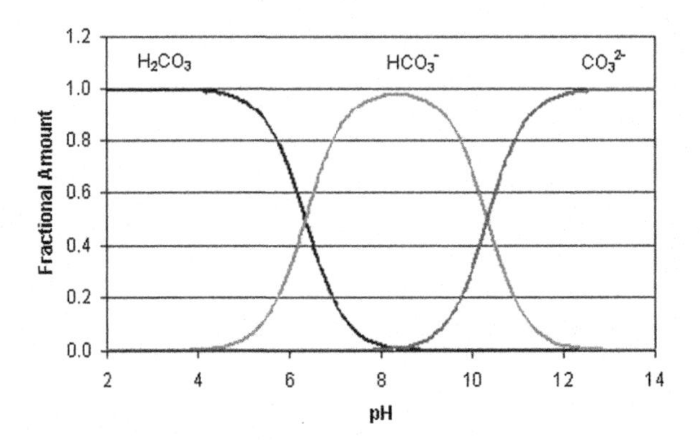


**FIGURE 6.30**

Dissolved Inorganic Carbon speciation and pH, where carbonate, bicarbonate, and carbon dioxide are in equilibrium.

## FIXING CARBON FROM DIC

In water carbon dioxide is easily dissolved. However, the diffusion rate is around ten thousand times slower. Because of the low levels of $CO_2$ in water, plants have adapted morphological structures (thin tissues, reduced cuticles, and dense chloroplasts) to capture available carbon dioxide. Additionally, intercellular spaces in leaves and stems allow for the rapid internal diffusion of $CO_2$ within the plant.

Different species of submerged vegetation obtain inorganic carbon as carbon dioxide ($CO_2$) and bicarbonates ($HCO_3$). Some plants take up $HCO_3^-$ and then metabolize to $CO_2$ before it can be fixed to sugars. These species are able to survive in hard water environments. Plants that are not able to metabolize bicarbonates are restricted to soft waters with low pH or streams with high concentrations of $CO_2$.

## DISSOLVED OXYGEN CONCENTRATION AND SATURATION

The concentration of oxygen is approximately 30x lower in aquatic systems, and yet aquatic organisms seem to obtain enough to meet their physiological requirement.

As we described above, $O_2$ is one of these compounds that does not disassociate in water. Instead oxygen is trapped in the lattice structure of the water. At warmer temperatures, the crystalline structure breaks down, thus water will hold less oxygen. This is one of the key issues in stream temperatures—waters of higher temperature have less capacity to hold oxygen, which can then become limiting to aquatic organisms.

For example, we can use Henry's Law to estimate the concentration in water at 25°C (Eq. (6.18)). Since $O_2$ in the atmosphere is about 20.95% (v/v) and $K_{\mathrm{H,pc}} = 769.2$, we calculate $[O_2(aq)]$ as follows:

$$O_2 = K_{\mathrm{H,pc}} \cdot O_2(conc) = 769.2 \cdot 20.95\% = 5.16\ \mathrm{mg/L}, \tag{6.23}$$

usually reported mg/L or ppm. This is the concentration at equilibrium, and in water quality discourse, this is referred to as 100% saturation, because that is all the water can "hold". However, this concentration will vary with temperature. In fact, the concentration of oxygen dissolved in water, simply called Dissolved Oxygen or DO, can vary with many factors:

- temperature,
- salinity,
- water turbulence and atmospheric mixing, and
- water column production (photosynthesis) and consumption (respiration).

The relationship between oxygen saturation and temperature is nonlinear, but there are various methods to calculate percent saturation—the equations are more complex (i.e., include the water vapor pressure, atmospheric pressure, and salinity) and are generally more accurate.

The DO concentration for 100% air-saturated water at sea level is 8.6 mg $O_2$/L at 25°C and increases to 14.6 mg $O_2$/L at 0°C.

## OXYGEN SUPERSATURATION AND HYPOXIA

When the rate of oxygen consumption exceeds the reaeration rates, the measured dissolved oxygen will be less than that predicted by Henry's coefficient. In the water column, photosynthesis may increase and respiration may reduce oxygen concentrations. The relative rates for these to process will determine oxygen concentrations that are above or below the equilibrium predicted by Henry's coefficient. The biological demand for or production of oxygen means that $[O_2]$ exhibits non-conservative behavior. When there is no oxygen demand, then dissolved oxygen concentrations show conservative behavior. However, the nonconservative behavior can be masked as reflected by the following:

$$\frac{dO_2(demand)}{dt} = \frac{O_2(production)}{dt} \tag{6.24}$$

and

$$\frac{dO_2(demand)}{dt} = \frac{O_2(production)}{dt} > \frac{dO_2(aeration)}{dt} \tag{6.25}$$

Of course, since photosynthesis varies with light, dissolved oxygen might vary with the time of the day or clarity of the water. The dissolved oxygen might vary with depth. During spring or fall turnover, the lake might have the same oxygen concentration, whereas during stratified conditions, oxygen may be near saturation at the surface, where phytoplankton are active during the day and low at the benthos, where respiration dominates and hypolimnion does not mix with the epilimnion. (See Fig. 6.31.)

During at 24-hour (diel) cycle, dramatic oscillations in dissolved oxygen concentrations are common. During the night, when photosynthesis cannot counterbalance the loss of oxygen through respiration and decomposition, DO concentration may steadily decline. It is lowest just before the sun rises, at which time photosynthesis resumes. These oscillations will influence species distributions and even drive daily migrations.

If the DO is chronically low, either in a constant fashion or in a diel cycle, fish and invertebrates respiration capacity might increase mortality. For salmon, these DO dynamics might affect different life stages in different ways. For example, midsummer, when strong thermal stratification develops in a lake, may be a very hard time for fish. Water near the surface of the lake, epilimnion, is too warm for them, whereas the water near the bottom, hypolimnion, has too little oxygen. Low DO forces the fish to spend more time higher in the water column, where the warmer water is suboptimal for them. This may also expose them to higher predation, particularly when they are younger and smaller. In lotic systems, salmon rely on cold, well-oxygenated water to exchange through the sediments, where their

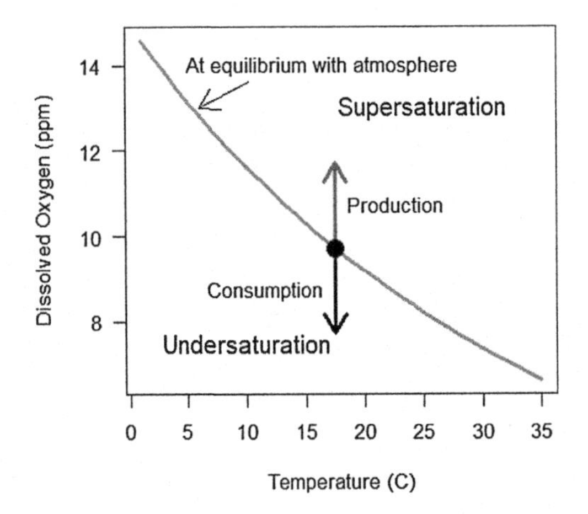


**FIGURE 6.31**

Relationship of Dissolved Oxygen, temperature, and water column processes (Source: MLH). Dissolved Oxygen or DO saturation depends on temperature, but the concentration is further influenced by water column processes. As temperature increases, the saturation concentration declines—although we still refer to the water column being 100% saturated, which was plotted here using Eq. (6.18). However, with photosynthesis the DO can be over the saturation concentration, i.e., over 100%—oversaturation. If on the other hand, heterotrophic respiration dominates the water column, the oxygen can be consumed and the concentration can be less than 100%—undersaturation. In each of these cases, we measure DO in terms of $mg{\cdot}L^{-1}$ and % saturation, because they both have important implications to aquatic organisms.

eggs are buried. As waters warm, lower DO concentrations can stress larval development and reduce their fitness.

In contrast to animals that are sensitive to low DO, some can actually survive hypoxic conditions using a range of behavioral and physiological adaptations. For example, eggs of the tropical killifish can be buried in the mud and survive without oxygen for months.

## MECHANISMS TO EXTRACT $O_2$ FROM THE WATER COLUMN

Like terrestrial animals, fish and other aquatic organisms need $O_2$ to live; in addition, photoautotrophs use $CO_2$ for fixation. However, the diffusion of gases through water is slow, which limit $O_2$ and $CO_2$ concentrations.

Even with some variability, the high concentrations are still quite limiting for most organisms. So how do these organisms get oxygen?

In the water, $O_2$ is in the dissolved form, but dissolving oxygen in water is not easy, so the concentrations are low. Where air and water meet, this tremendous difference in concentration causes oxygen molecules in the air to dissolve into the water. More oxygen dissolves into water when wind stirs the water; as the waves create more surface area, more diffusion can occur.

To live, fish must extract oxygen from the water and transfer it to their bloodstream. This is done by gills, lungs, specialized chambers, or skin, any of which must be richly supplied with blood vessels to act as a respiratory organ. Extracting oxygen from water is more difficult and requires a greater

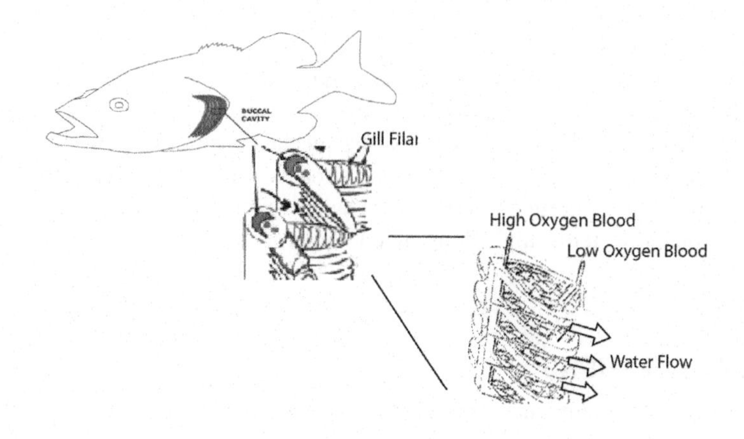


**FIGURE 6.32**

Fish gill structure, showing countercurrent exchange of blood and water (Source: Pearson Education).

expenditure of energy than does extracting oxygen from air. Fishes have necessarily evolved very efficient systems for extracting oxygen from water; some fishes are able to extract as much as 85% of the oxygen contained in the water passing over the gills, whereas humans can extract only about 25% of the oxygen from the air taken into the lungs.

As water moves past their gills (or other breathing apparatus), microscopic bubbles of oxygen gas in the water, called DO, are transferred from the water to their blood. Like any other gas diffusion process, the transfer is efficient only above certain concentrations. In addition, the following are points worth noting regarding water-to-blood oxygen transfer in fish:

- Gills have high surface area and may provide 10 to 60 times more oxygen than that of the whole body surface;
- Short diffusion path for oxygen reduces the time to incorporate oxygen into the blood stream;
- A countercurrent maximizes the concentration gradient between the water and blood (Fig. 6.32);
- High gill-water contact, so a high proportion of the water interacts with gills; and
- Water flows through but does not require a reversal of flow (in and out like in mammal lungs), which takes more time and energy.

However, $O_2$ can be too low to sustain a range of vertebrates and invertebrates. Thus DO is an important measure of water quality and habitat value in streams, lakes, and wetlands. In addition, $O_2$ also is needed by virtually all algae and all macrophytes, and for many chemical reactions that are important to lake functioning. Thus it is important to note that even when producing $O_2$, these taxa or processes may also be sensitive to $O_2$ concentrations, especially at night.

## SOUND

Water does not conduct sound in the same way as air. Thus it is not surprising that most aquatic organisms perceive and respond to sound in different ways that many terrestrial taxa. We should note, however, that aquatic mammals, such as whales and dolphins, because of their shared ancestry with terrestrial mammals respond to and vocalize in ways that we recognize.

Approximately, two thirds of all freshwater fish have adaptations that enhance auditory sensitivity and broaden frequency ranges. These so-called hearing specialists may have evolved in quiet habitats, such as lakes, backwaters of rivers or slowly flowing streams. However, when exposed to areas with noise (streams and river noise), these fish may not be able to capitalize on these adaptations. Thus the value of hearing in aquatic systems might be limited to quiet waters.

## NEXT STEPS

### CHAPTER STUDY QUESTIONS

1. Describe the physical and chemical characteristics of inland waters and how 10 species have adapted to these habitats.
2. Create an image that describes how water movement and fluid dynamics structure biotic communities in lotic systems.
3. Create an image that describes how light, temperature, and nutrients structure biotic communities.

### PROBLEM SETS

1. Secchi Depth measurements have been recorded from Lake Tahoe since 1968. Using the values from Table 6.5 answer the following questions:

**Table 6.5 Annual average Secchi depth (m) for Lake Tahoe.**

| Year | Depth (m) | Year | Depth (m) | Year | Depth (m) |
|---|---|---|---|---|---|
| 1968 | 31.2 | 1985 | 24.2 | 2002 | 23.8 |
| 1969 | 28.6 | 1986 | 24.1 | 2003 | 21.6 |
| 1970 | 30.2 | 1987 | 24.6 | 2004 | 22.4 |
| 1971 | 28.7 | 1988 | 24.7 | 2005 | 22 |
| 1972 | 27.4 | 1989 | 23.6 | 2006 | 20.6 |
| 1973 | 26.1 | 1990 | 23.6 | 2007 | 21.4 |
| 1974 | 27.2 | 1991 | 22.4 | 2008 | 21.2 |
| 1975 | 26.1 | 1992 | 23.9 | 2009 | 20.8 |
| 1976 | 27.4 | 1993 | 21.5 | 2010 | 19.6 |
| 1977 | 27.8 | 1994 | 22.6 | 2011 | 21 |
| 1978 | 25.9 | 1995 | 21.5 | 2012 | 22.9 |
| 1979 | 26.7 | 1996 | 23.4 | 2013 | 21.4 |
| 1980 | 24.8 | 1997 | 19.5 | 2014 | 23.7 |
| 1981 | 27.4 | 1998 | 20.1 | 2015 | 22.3 |
| 1982 | 24.3 | 1999 | 21 | 2016 | 21.1 |
| 1983 | 22.4 | 2000 | 20.5 | 2017 | 18.4 |
| 1984 | 22.8 | 2001 | 22.4 | 2018 | 21.6 |

- What is the clarity trend in the lake?
- Is a linear model the best fit? Use an AIC selection model approach and evaluate 5 different models and determine how well they satisfy the statistical assumptions.

## LITERATURE RESEARCH ACTIVITIES

1. Search for online Secchi Depths for ~10 long-term records. Evaluate the trends and determine if the slopes are different from each other. Evaluate and suggest causes for each result.
2. Select 5 streams and determine the mean concentration for each stream and how much variation takes place between the dry and rainy season. Discuss how these results vary by stream.

CHAPTER

# THE STAGE: TYPOLOGIES OF AQUATIC SYSTEMS

7

*The marsh, to him who enters it in a receptive mood, holds, besides mosquitoes and stagnation, melody, the mystery of unknown waters, and the sweetness of Nature undisturbed by man.*
**Charles William Beebe, Log of the Sun**

*Observe what happens when sunbeams are admitted into a building and shed light on its shadowy places. You will see a multitude of tiny particles mingling in a multitude of ways... their dancing is an actual indication of underlying movements of matter that are hidden from our sight.*
**Lucretius's scientific poem, On the Nature of Things (c. 60 BC)**

*Only a fool tests the depth of the water with both feet.*
**African Proverb**

## CONTENTS



Ecology and Management of Inland Waters. https://doi.org/10.1016/B978-0-12-814266-0.00020-9

The Pantanal is a region encompassing the world's largest tropical wetland. The name "Pantanal" comes from the Portuguese word pântano, meaning wetland, bog, swamp, quagmire, or marsh. Whereas most of the Pantanal is located in the Brazilian state of Mato Grosso do Sul, and includes portions of Bolivia and Paraguay. In its entity the Pantanal encompasses an area between 140,000 and 195,000 km$^2$. Roughly 80% of the Pantanal floodplain is submerged during the rainy seasons nurturing an astonishingly diverse collection of aquatic plants and supporting a dense array of animal species (Fig. 7.1).

**FIGURE 7.1**

The Pantanal (Brazil, Bolivia and Paraguay) is a mix of habitat-types that allow one of the highest diversity in a wetland system (Source: Wikicommons).

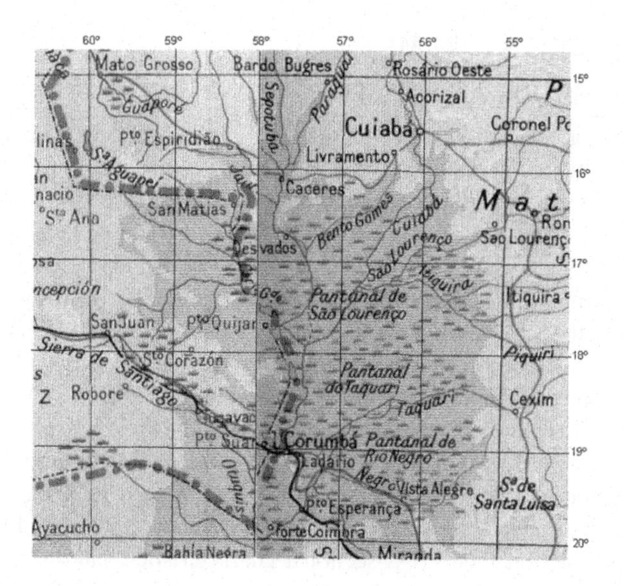


**FIGURE 7.2**

Map of the Pantanal.

Using distinct hydrological, geological, and ecological characteristics, sixteen habitat-types have been identified within the Pantanal, such as grassland, cerrado woodland, cerrado (bush savanna), marshes, semideciduous forest, gallery forest, and floating mats (Fig. 7.2). The patchy nature of the upland and wetlands allows for a dynamic habitat diversity and complexity that depends on the extent of seasonal flooding that can be between 2–5 meters each year. However, as humans modify the hydrology of the region, these ecosystems types are being transformed; we should view these as dynamic communities instead of static categories.

Nevertheless, humans use classifications schemes to help us organize information, and classifying inland waters provides a cultural anchor to identify and analyze various ecosystems. Even the use of regional words based on local languages have become incorporated into standardized vocabulary in the scientific literature. In fact, a dizzying plethora of formal and informal terms categorize various aquatic ecosystem types. For example, wetlands might be known as wet meadows, bogs, fens, swamps, vernal pools, potholes, bottomlands, mire, moor, musket, peal, tundra, playa lake, slough, ponds, tidal wetlands, salt marshes, riparian forests, and mangrove forests.

By relying on the patterns and processes in aquatic system, this chapter outlines the nomenclature to classify and describe inland waters.

After reading this chapter, you will be able to

1. Discuss how patterns and processes are used to classify aquatic systems.

## CLASSIFYING AQUATIC SYSTEMS

### STRUCTURE AND FUNCTION AS A BASIS

Developing a classification scheme helps us to organize various patterns and processes. No perfect classification scheme exists, because the number of ambiguities and exceptions are impossible to avoid completely. There have been numerous attempts at classifying aquatic systems. The following constitute examples common vernacular in regards to the same: rivers, streams, creeks, lakes, ponds, reservoirs, wetlands, swamps, marshes. But in many cases, there are ecological transition zones where the distinction might be difficult to make, e.g., what is the difference between a pond and wetland if the wetland dries out during the dry season, such as a vernal pool? Additionally, is there a way to decide when a stream becomes a river? And most important for our purposes, what might be ecologically useful categories that help us understand aquatic systems.

As we will see, there have been a wide array of attempts to develop a systematic way to classify aquatic systems. For example, "The Classification of Wetlands and Deepwater Habitats of the United States" has been developed for a robust regulatory purpose. Originally published in 1976 and updated in 1979 and 2013, aquatic systems are categorized into 5 systems: Estuarine, Lacustrine, Palustrine, Marine, and Riverine, based on their cross-sectional characteristics, landscape position, vegetation cover, and hydrologic regime.

***Estuarine Systems*** are usually semienclosed by land, but have open, partly obstructed, or sporadic access to the open ocean. Estuarine ecosystems are composed of deepwater tidal habitats and adjacent tidal wetlands, in which ocean water is at least occasionally diluted by freshwater runoff from the land. Because of the marine influence, thus not an inland system, estuaries are outside the scope of this text.

***Lacustrine Systems*** are lake habitats that include wetlands and deepwater habitats in topographic depression or a dammed river channel; do not have substantial emergent habitat, and are at least 8 ha in size if there is an active wave-formed or bedrock shoreline.

The Lacustrine System includes permanently flooded lakes and reservoirs, intermittent lakes (e.g., playa lakes), and tidal lakes with ocean-derived salinities below 0.5 ppt. Typically, deep water habitats are extensive and considerable wave action exist. Islands of Palustrine wetlands may lie within the boundaries of the Lacustrine System. These systems are often divided between the Limnetic System (deep water, > 2.5 m) and Littoral System (from the shoreline to the depth of 2.5 m) (Fig. 7.3).

***Marine Systems*** consist of the open ocean overlying the continental shelf and its associated high-energy coastline. Because these systems are beyond the scope of this text, we will not consider marine systems any further.

***Palustrine Systems*** are generally wetland systems with emergent vegetation, i.e., trees, shrubs, persistent emergents, emergent mosses or lichens, and all such wetlands, or small shallow waters with active wave-form or bedrock shorelines. These systems can be intermixed with wide range of other wetland systems (Fig. 7.4).

***Riverine Systems*** include all wetlands and deepwater habitats contained within a channel, with two exceptions: (1) wetlands dominated by trees, shrubs, persistent emergents, emergent mosses, or lichens, (Palustrine systems), and (2) habitats with water containing ocean-derived salts of 0.5 ppt or greater (Estuarine systems). Channels can be naturally or artificially created (e.g., canal)

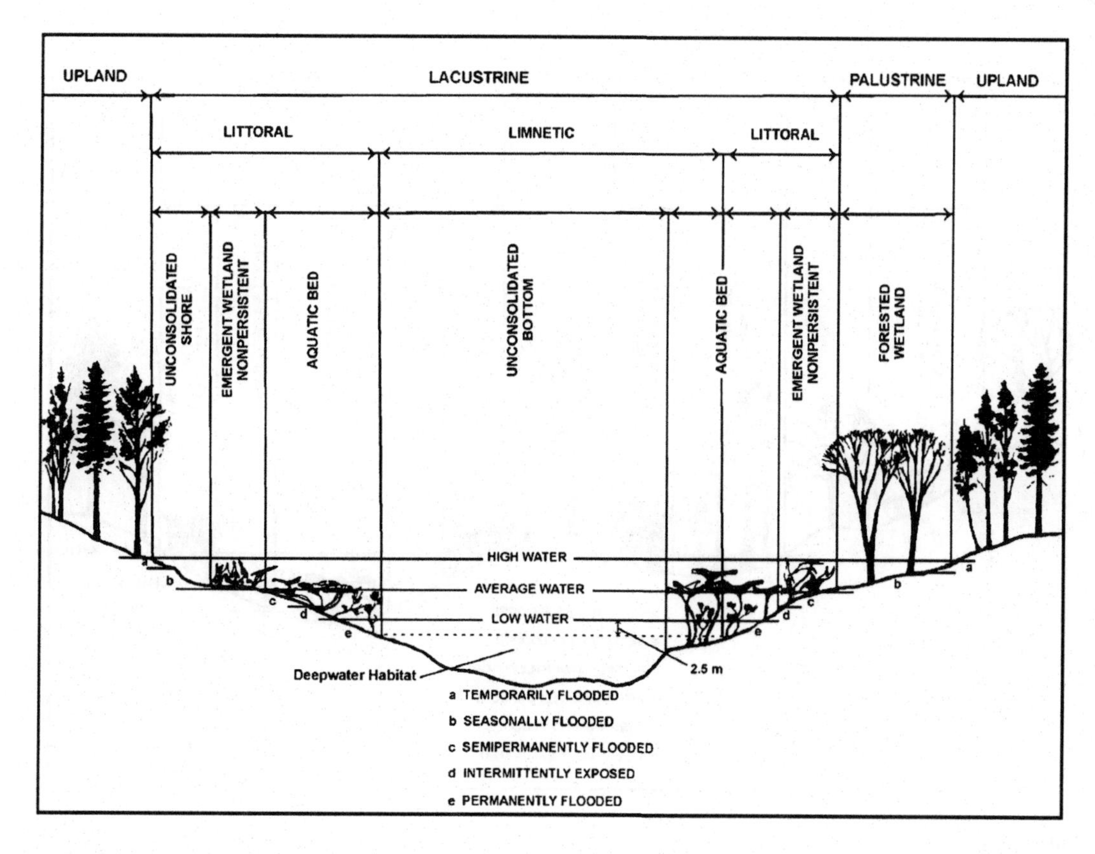


**FIGURE 7.3**

Lacustrine System classification (Source: Committee, 2013).

that periodically or continuously contains moving water, or form a connecting link between two bodies of standing water.

The riverine systems can be categorized as tidal, perennial, and intermittent and further defined in terms of water permanence, gradient, substrate, and the extent of floodplain development (Fig. 7.5).

There are two considerations when using this system. First, it is based on structural characteristics and few ecological functions. For example, in lake ecosystems, the Littoral Zone is usually defined by the Photic Zone—where PAR reaches the benthos, which allows submerged vegetation to thrive. This might be shallower or deeper than the 2.5 m criterion used in the classification system. Second, by focusing on cross-sectional characteristics, the system does not capture how some systems change along a hydrology gradient, e.g., from upstream to downstream locations. Thus for the remainder of the text, we will focus on how landscape position, hydrology, geomorphology, and habitat help us define aquatic systems.

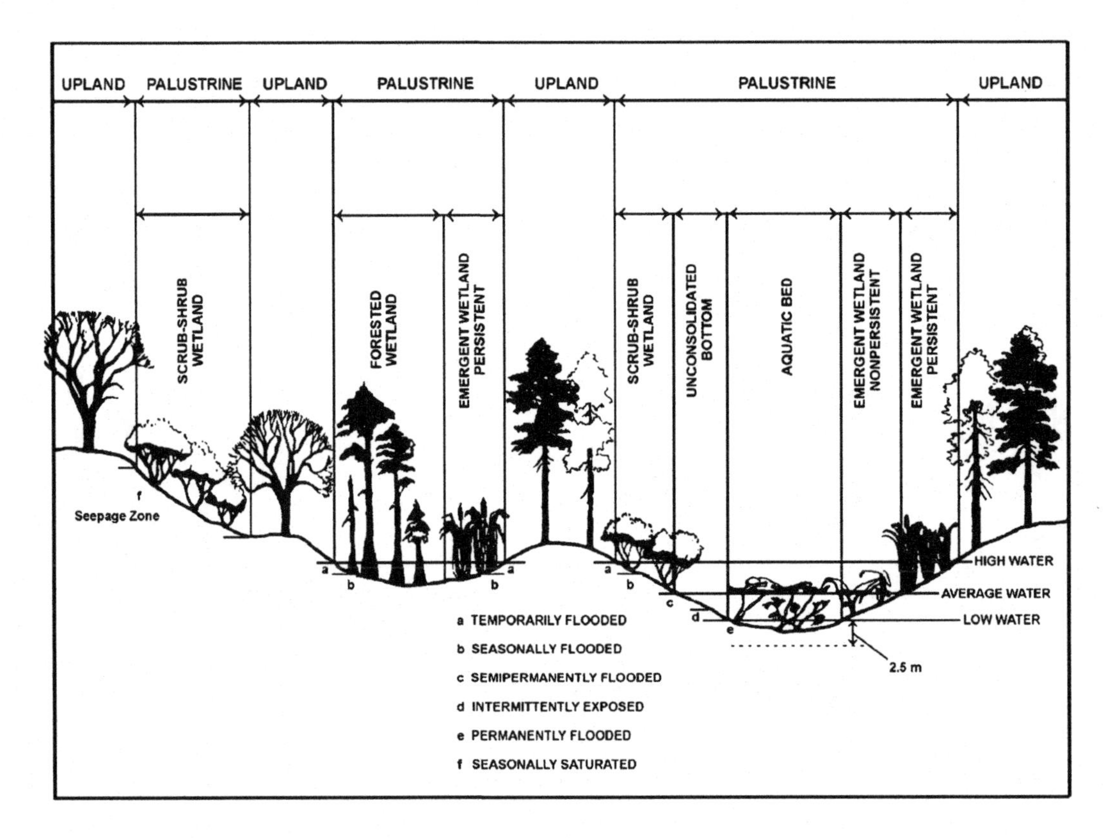


**FIGURE 7.4**

Palustrine ecosystem classification system (Source: Committee, 2013).

## CATCHMENTS AND FLOODPLAINS

### WATERSHEDS AND TRIBUTARIES

As precipitation falls on the landscape, the waters flow down-gradient until a stream is formed. A stream is a body of water that flows across the Earth's surface via a current and is contained within a narrow channel and banks. Streams join with other streams, which combine to form larger streams or rivers. The streams that join rivers are call called tributaries. The river and its tributaries is known as the river basin. The whole area around the basin is called the watershed, where all precipitation that falls within the watershed flows through a common point and out of the basin (exorheic basin), or into a terminal lake (endorheic basin).

### HEADWATERS AND FLOODPLAINS

The source or headwaters of a river or stream is the furthest place in that river or stream from its estuary or confluence with another river, as measured along the course of the river. Headwaters are up-gradient of the outlet, and this often at higher elevations. But the headwaters of the everglades, which acts like

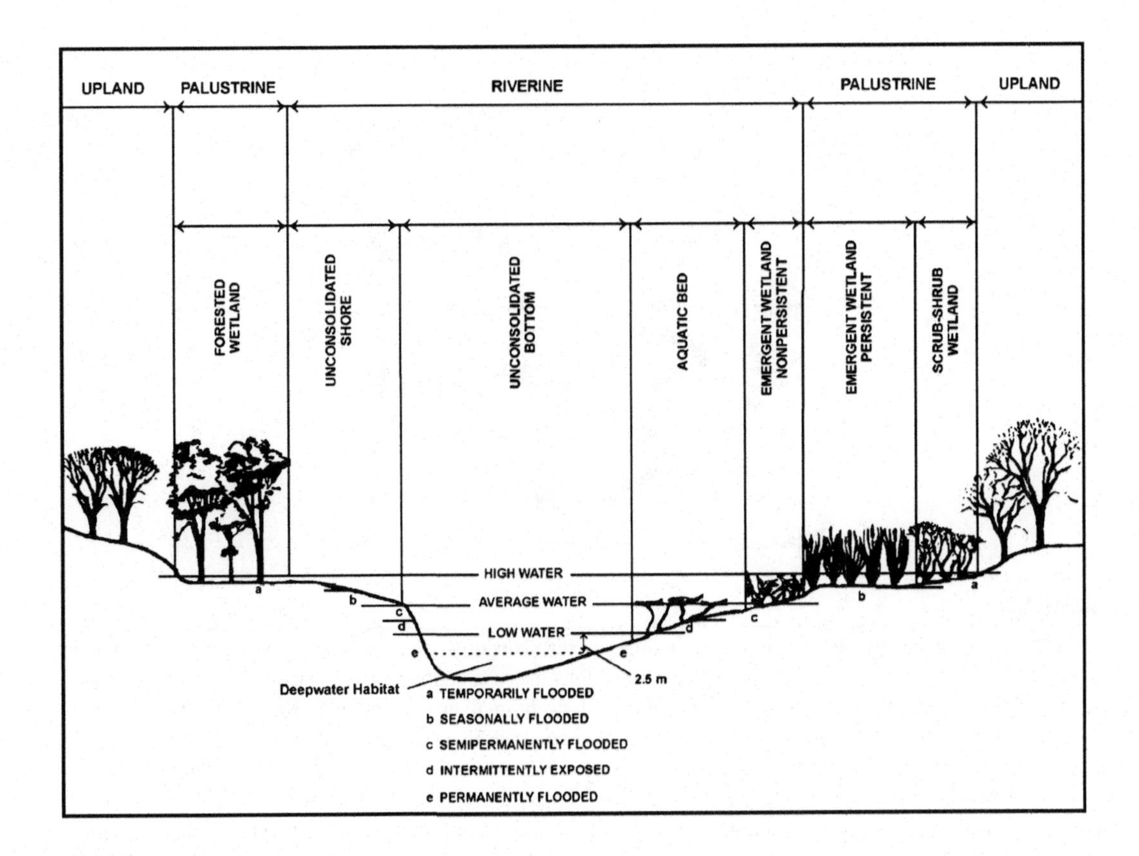


**FIGURE 7.5**

Riverine system classification (Source: Committee, 2013).

a really slow and wide river, may be less than a few meters higher than the estuarine outlets. Over half of the stream miles are headwaters.

Flow in a headwater may be year-round, seasonal, or rain-dependent. Approximately, 60% percent of stream miles in the continental US only flow seasonally or after storms. For example, a headwater stream may flow briefly when snow melts or after rain. But during dry period, the stream may stop flowing and become a network of pools filled with water. Desert headwater streams can be fed by a spring and only flow above ground for 10s to 100s of meters before infiltrating into the ground. Some spring-fed headwaters are characterized by clear water with a steady flow and constant temperatures, whereas headwaters that are fed from marshes or swamps might be brown- or black-colored waters.

All rivers depend on a network of unknown, unnamed, and underappreciated streams. Because small streams and streams that flow for only part of the year are the source of fresh waters, changes that harm these headwaters affect streams, lakes, and rivers downstream.

As headwaters join other streams, the volume of water increases and they begin flow through larger and larger valleys, where much of the valley floor is the floodplain. A floodplain is the land adjacent, which is flooded when discharge is high. The plain is composed of sediment deposited by floodwaters.

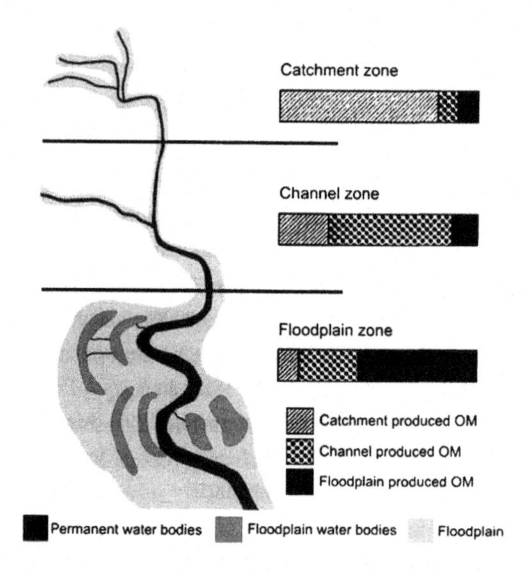


**FIGURE 7.6**

Schematic interplay of variable carbon sources in different segments of a river. Owing to the topographical variation in the landscape, the sequence of the segments can vary along the river course. Top: In catchment areas with strong aquatic-terrestrial interfaces, the floodplain extension is relatively small and inputs of terrestrially produced organic matter are high. Middle: In natural (mountainous, steep-bordered) or human-made (channelized) segments, the extension of the floodplains is restricted, and carbon fixation occurs largely by riverine plankton and aquatic macrophytes. Below: In floodplain areas, carbon can contribute to river carbon budget via waterflow from the floodplain to the mainstream, or via feeding migration of fish and other aquatic animals between the floodplain and the main channel (Source: Wantzen et al., 2008).

Some floodplains have a more complex geology, influenced by previous high sea levels, ancient rivers, or tectonic activity that lifted or created specific geologic conditions.

As water move from headwaters, through channel zones, and into floodplains, we see that the sources of organic matter will shift from terrestrial (catchment) sources to channel sources, and finally to floodplain sources (Fig. 7.6).

## STREAM ORDER NUMEROLOGY

Geographers, geologists, and hydrologists use stream order to qualitatively characterize the size of the waterways. When using stream order to classify a stream, the orders range from a first-order stream all the way to the largest, a 12th-order stream (Fig. 7.7). A first-order stream is the smallest of streams and consists of small tributaries, and may be referred to as creeks or brooks. These streams may be intermittent and flow into and "feed" larger streams. In addition, first- and second-order streams generally form on steep slopes and flow quickly until they slow down and meet the next order waterway. Large waterways (at the highest level the stream order) are called rivers and exist as a combination of many tributary streams.

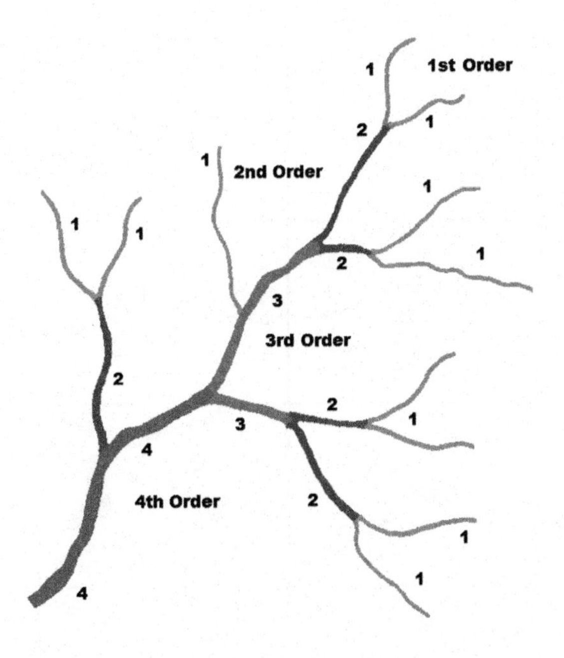


**FIGURE 7.7**

Illustration of the river numerology and stream order (Source: Dickson Despommier).

Stream order increases when two streams combine with the same number—otherwise the higher-order stream carries forward downstream. Stream order is tricky to implement, especially with ephemeral streams.

## THE DENDRITIC LANDSCAPE

River systems form distinct landscape drainage patterns. These patterns are the result of fluvial dissection of a landscape and depends on patterns of geological formations. For example, dendritic patterns, which resemble a tree branch, are common where rocks or sediments are flat-lying and preferential zones of structural weakness are minimal. Four distinct patterns have been recognized: dendritic, radial, trellis, and annular (Fig. 7.8). Even the patterns can vary with climate, where the patterns of rainfall, vegetation, and geology will produce different patterns as well (Fig. 7.9).

## VALLEY SEGMENTS, STEAM REACHES, AND CHANNEL UNITS

Independent of landscape position, streams can be thought of as three hierarchically nested subdivisions of the drainage network. For example, a stream is composed of valley segments, stream reaches, and channel units. Channel units are taken as point locations, where specific measurements might be made. A reach is a length of a stream or river, usually suggesting a level, uninterrupted stretch. The beginning and ending points may be selected for geographic, historical or other reasons, such as landmarks that include gauging stations, river miles, natural features, and topography. A valley segment

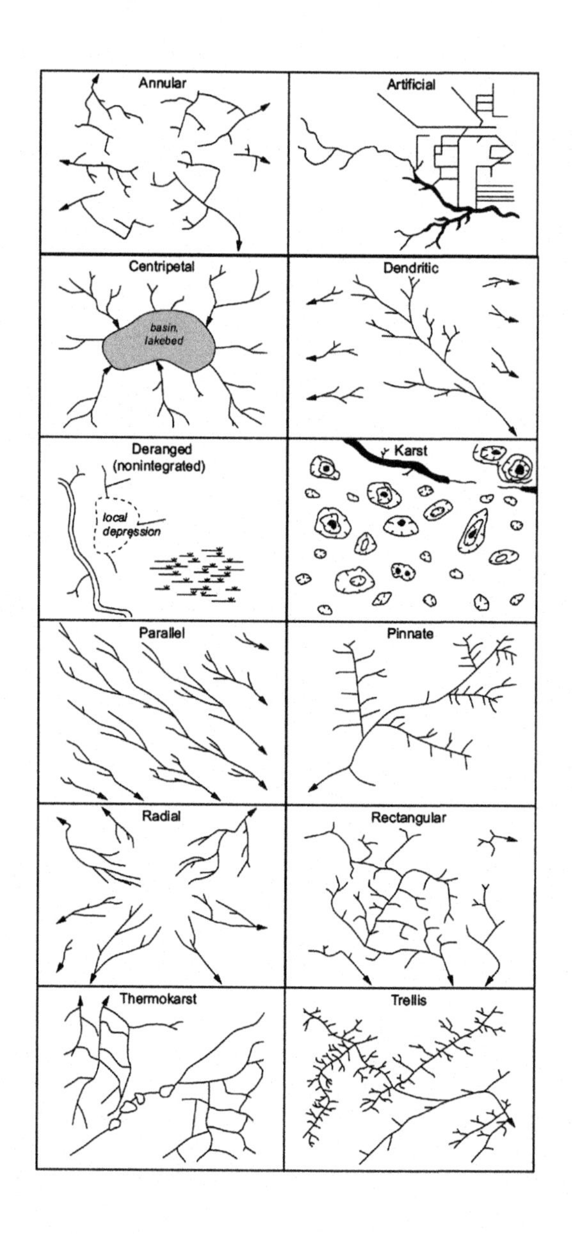


**FIGURE 7.8**

Drainage patterns at the landscape level (Source: NRCS).

includes a number of similar stream reaches that are similar in size and scale within valley without major confluences.

Within this hierarchy of spatial scales, valley segments, stream reaches, and channel units can facilitate accurate, repeatable descriptions, and convey information about biophysical processes that

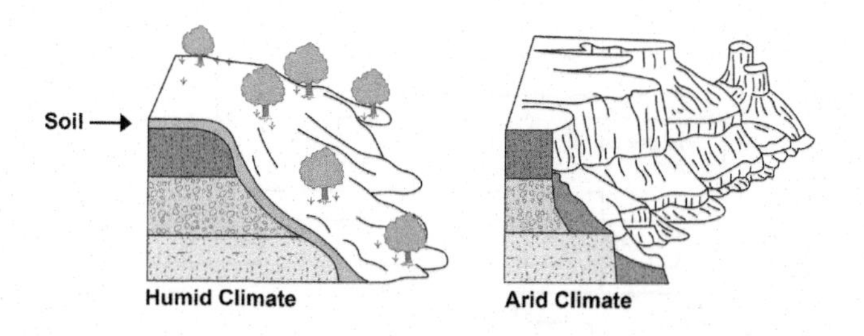


**FIGURE 7.9**

Topographic patterns in humid and arid climates (drawn by Luyi Huang.

**Table 7.1 Hierarchical classification scheme based on channel characteristics (modified from Bisson et al., 2017).**

| Classification level | | Spatial scale | Temporal scale (years) |
|---|---|---|---|
| Channel/Habitat Units | | 1–10 m$^2$ | < 1–100 |
| | Fast Water | | |
| | Slow Water | | |
| | Bars | | |
| Channel Reaches | | 10–1000 m$^2$ | 1–1000 |
| | Colluvial Reaches | | |
| | Bedrock Reaches | | |
| | Free-formed Alluvial Reaches | | |
| | Force Alluvial Reaches | | |
| Valley Segment | | 100–10,000 m$^2$ | 1,000–10,000 |
| | Colluvial Reaches | | |
| | Bedrock Reaches | | |
| | Free-formed Alluvial Reaches | | |
| Watershed | | 50–500 km$^2$ | 1,000–10,000 |
| Geomorphic Province | | 1,000 km$^2$ | 1,000–10,000 |

influence channel geomorphology and habitat quality. These subdivisions have spatial and temporal meanings. Thus it is important to acknowledge ambiguities—where an area might be easily categorized another might be impossible to distinguish with precision. Nevertheless, a range of environmental scientists rely on these categories (Table 7.1).

Alternatively, some rely geomorphological characteristics to classify channel reach characteristics (Fig. 7.10).

## ECOREGIONS

Ecoregions and large catchments can be used to classify stream assemblages with the idea that ecoregion boundaries are "natural" breaks between community assemblages, hydrology, and geomorphic patterns. The development of a classification scheme assumes that the river corridors concerned can be subdivided into units by clearly defined boundaries. Whereas geography can account for some variation

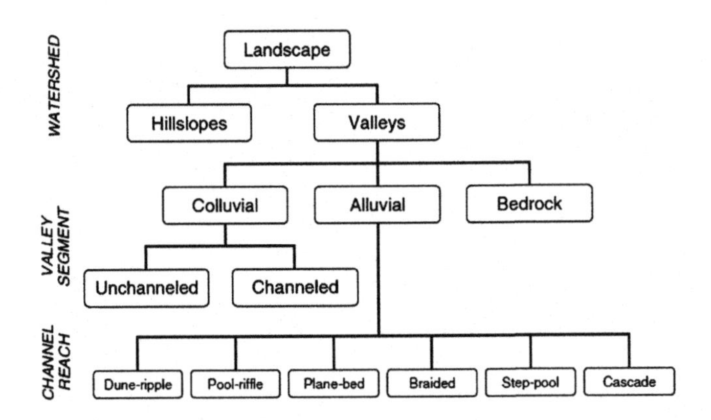


**FIGURE 7.10**

Hierarchy classification of streams into valley segments and stream reaches (Source: Bisson et al., 2017).

seen in stream assemblages across a large region, developing boundaries is problematic. Because of the gradually varying nature of river environments, both along and across their corridors, identification of clear boundaries is nearly impossible.

# THE RIVER CONTINUUM CONCEPT

## GEOMORPHOLOGY AND THE RIVER CONTINUUM

The terms river morphology and its synonym fluvial geomorphology describe the shapes of river channels and how they change in shape and direction over time. The morphology of a river channel is a function of a number of factors:

- composition and erodibility of the bed and banks (e.g., sand, clay, bedrock);
- power and consistency of the current;
- vegetation and the rate of plant growth;
- availability, size, and composition of sediment moving through the channel;
- the rate of sediment transport through the channel, and the rate of deposition on the floodplain, banks, bars, and bed, in addition to
- regional aggradation or degradation due to subsidence or uplift.

A river regime is a dynamic equilibrium system, which is a way of classifying rivers into different categories. For example, river regimes can be categorized by the sinuosity and channel characteristics (Fig. 7.11). Stream gradient, sediment load, and hydrograph all contribute to the type of patterns we see in a river system. River morphology can also be affected by human interaction, which is a way the river responds to a new factor in how the river can change its course.

These in-stream processes also drive processes downstream, e.g., eroded sediment upstream is desposited downstream. Thus connecting the geomorphology to the continuum of waterflow can link the structure and function of stream to create classification structure, but also define an entire landscape

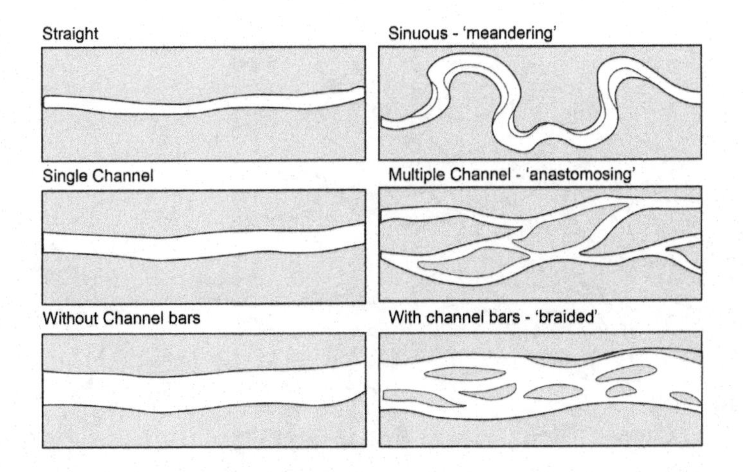


**FIGURE 7.11**

Channel-type based on sinuousity (drawn by Luyi Huang).

morphology, where downstream patterns and processes depend on upstream patterns and processes (Fig. 7.12).

## CHANGING APPRECIATION OF THE STREAMS

Thus far we have been exposed to a wide range of competing ways to classify Lotic Systems. Over time, environmental scientists began to characterize reaches by creating generalizable and predictive categories based on the concepts we have discussed to propose the River Continuum Concept (Fig. 7.13).

As a centralized classification scheme for streams and rivers, "River Continuum Concept" can be thought of as one of the most powerful because of its capacity to create testable hypotheses. Formally introduced by Vannote and colleagues, the RCC has been tacitly accepted, or criticized for it is robustness. Taking the cues from other ecological frameworks, Robin Vannote and others hypothesized that rivers follow a predictable gradient of characteristics from headwaters to floodplain rivers, thereby proposing the River Continuum Concept, which emphasizes the longitudinal connectivity in stream systems.

As the physical gradient changes from source to mouth, chemical systems and biological communities shift and change in response. The River Continuum Concept can be applied to this linear cycling of nutrients, continuum of habitats, influx of organic materials, and dissipation of energy.

The theory assumes that hydrologic characteristics, geomorphological processes, and biological factors are poised in a dynamic equilibrium. By identifying stream reaches within the continuum and dynamic equilibrium, the biological properties of the Lotic System become predictable, or at least allows for predictions to be explicit and tested.

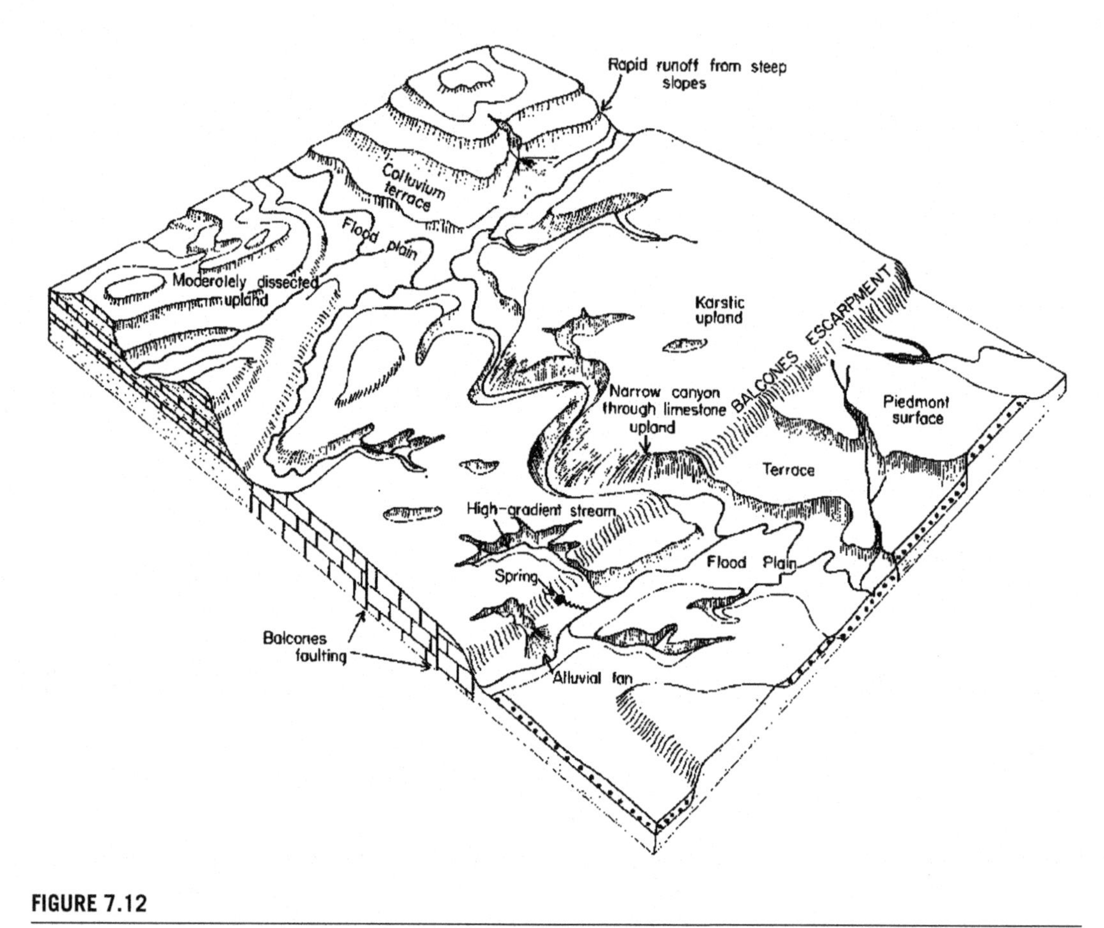


FIGURE 7.12

The escartment of the Balcones is a powerful image of the role of water to create a landscape.

## RIVER CONTINUUM CONCEPT: FASCINATION, DISREPUTE, & MILD EMBRACED

Based on the observations made in temperate zone waters, the RCC predicts that nonalpine headwaters are dominated by leaf litter as a carbon source and that invertebrate shredders are dominant in the water column. The ratio of in-stream photosynthesis is relatively low, and the foodweb is relatively simple with few secondary consumers or top predators. Further downstream, shredded leaf matter and moderately processed leaves drive collector invertebrate populations. These wider streams allow for in-stream photosynthesis to occur, thus the ratio of photosynthesis to respiration increases. As more tributaries combine to form larger rivers, the amount of terrestrial carbon declines, in-stream photosynthesis accounts for most of the new carbon, the ratio between photosynthesis to respiration is now low, and the food web is much more complex.

On the basis of the use of the concept to organize research results, a range of 'exceptions' mounted. In other words, researchers found that various stream systems did not fit the model, so the model was far from universal. For example, the model falls apart with alpine headwaters, recognizing that many

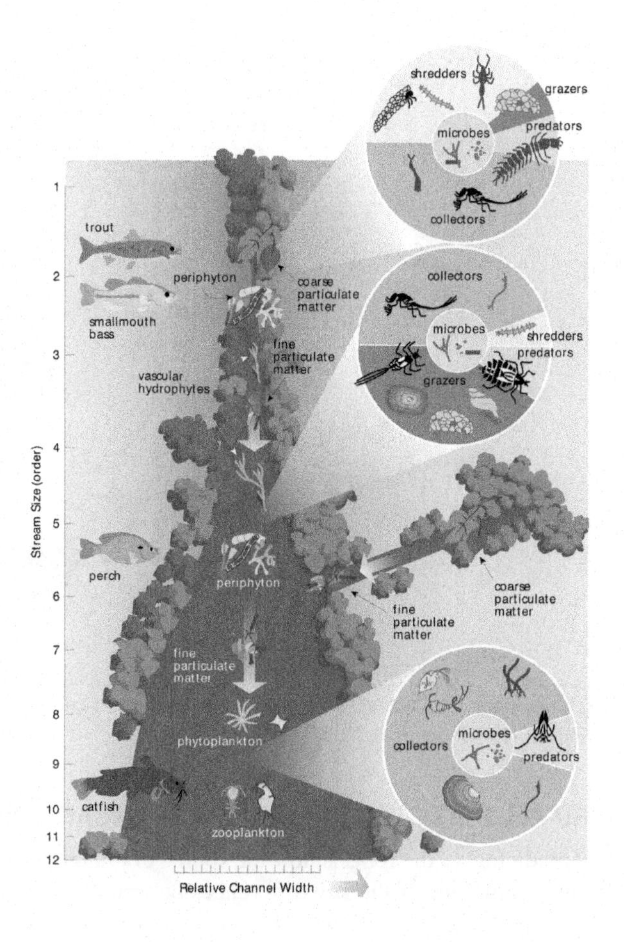


**FIGURE 7.13**

River Continuum Concept (Source: The Federal Interagency Stream Restoration Working Group).

predators in headwaters may not be aquatic but terrestrial, and fails to adequately capture the role of the floodplain as a central component of the Lotic System. In addition, the model's predictions do not do well in desert or tropical systems. Finally, with the introduction of dams, the RCC was modified as to be a River Discontinuum, with a new set of predictions.

Thankfully, the ecologists have worked hard to use these weaknesses to propose alternative models and continue developing generalizations that can be used to better understand Lotic System ecologies (Fig. 7.14). Although the River Continuum Concept lacks a qualitative physical model to represent downstream trends in habitat, stream ecologists use the concept as heuristic tool, or as a way to frame hypotheses to test. For this reason, the RCC remains an important way to characterize and classify streams.

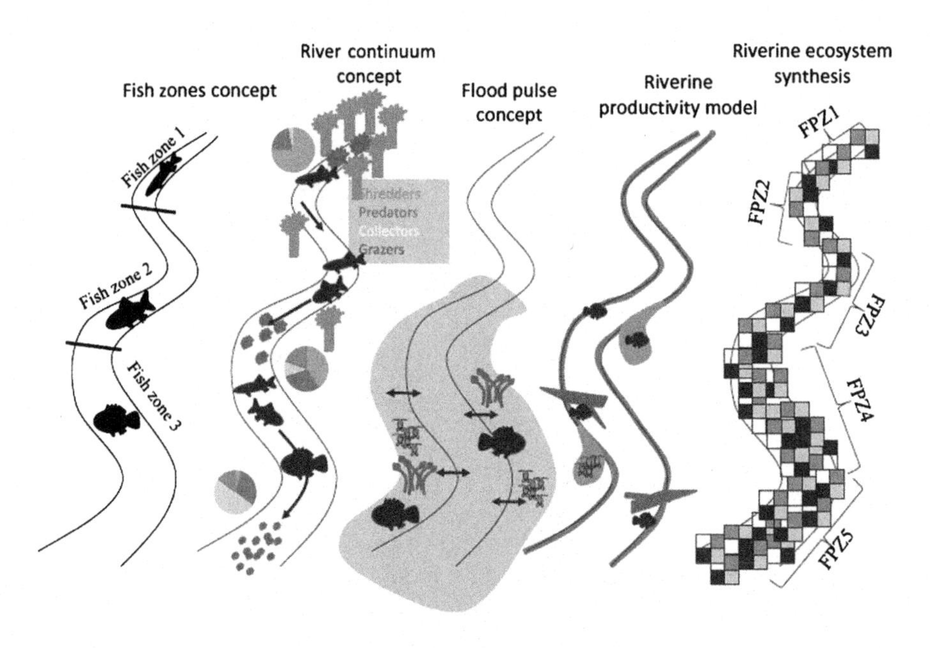


FIGURE 7.14

The development of the River Continnum Concept is itself a part of a continuum of conceptual changes in how we think of streams and rivers (Source: Humphries et al., 2014).

## AUTOCHTHONOUS AND ALLOCHTHONOUS SOURCES OF ORGANIC MATTER

One of the most important predictors of the RCC is how the source of organic matter and carbon inputs changes along the river continuum. One of the key differences between headwater (primary streams) and floodplain rivers is the source of organic matter. Organic matter in Aquatic Systems can be from terrestrial sources or produced within the water column. Allochthonous inputs are of terrestrial origin, whereas autochthonous organic matter is produced by submerged plants and algae.

# LENTIC SYSTEMS: GEOLOGICAL AND GEOMORPHOLOGICAL ORIGINS

## ENDORHEIC AND EXORHEIC BASINS

Tectonic processes drive mountain building, whereas water erodes the mountains and carries sediments downstream. As we discussed before, on landscape scales, basins can be classified as endorheic and exorheic. In the development of lakes, this distinction has profound implications, where exorheic basins have relatively short residence times, depending on the volume of inputs to the volume of storage to keep the salt balance relatively stable. For endorheic basins, a long-term increase in salt concentrations exist, which drives physiological adaptations to these changes or species extinctions.

Although this classification scheme is central to our understanding of lake ecology, another key factor is the age of the lake. A young endorheic lake might behave identical to a nearby exorheic lake, thus we will begin our classification scheme in the context of lake age.

**FIGURE 7.15**

Global ice sheets (Source: National Park Service).

## PROCESSES THAT CREATE YOUNG AND OLD LAKES

In general, lakes are young—most are less than 10,000 years. In the higher latitude > 40°, the vast majority of the lakes are products of the last Ice Age. As a result of glacial moraines that have created impoundments, these lakes may be in higher elevations, or where massive ice sheets existed, e.g., the northern portions of North America (Fig. 7.15).

Lakes that are older are usually the result of tectonic or volcanic activity or both. These lakes can be categorized by how they were formed.

***Lava Dam Lakes*** Approximately 10 mya, a river flowed through the Lake Tahoe valley floor with headwaters at the south and an outlet at the north until Mt. Pluto, now an extinct volcano north of Lake Tahoe, produced a lava flow that connected the Carson and Crystal ranges and blocked the outlet of the river, creating Tahoe Lake. Over time, the valley filled with water until the rim was full. The upper the Truckee River is one of many streams to flow into the lake and the lower Truckee River drains the lake at the north end, and flows through Reno into a terminal lake, Pyramid Lake, Nevada.

In contrast, the Owyhee River, along the eastern Oregon and Idaho boarder, has had a history of lava dams that have been stable for periods of > 1000 years and were sufficiently impermeable for lakes to form, and subsequently fill with lacustrine sediment to elevations approaching the dam crests. These lakes have since eroded the face of the dams allowing the river to flow unimpeded.

***Glacial Lakes*** are scattered across the northern latitudes of the Asia, Europe, and the United States. In the midwest, they are associated with the ice sheets that covered the northern latitudes (Fig. 7.15). These lakes are formed by glacial runoff being captured in valleys or areas dammed by moraines (rocks and sediments carried and deposited by a glacier). Retreating glaciers carve away at the

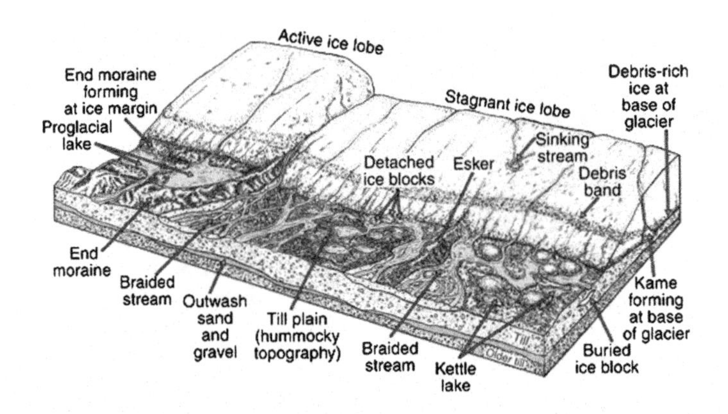


**FIGURE 7.16**

The development of the landscape behind retreating glaciers allows for the development of a complex terrain (Source: Illinois State Geological Survey).

mountains leaving behind steep topography, and, in some cases, nearly vertical basins typical geomorphology (Fig. 7.16).

Due to the extreme topography alpine lakes commonly do not have much soil, basins will instead be largely made up of rock. This lack of soil provides little opportunity for vegetation to grow in the areas surrounding the lake. In addition to affecting the soil and vegetation, the steep sides of the basin reduce the lakes' exposure to direct sunlight.

***Rift Zone*** Rift zones are areas where the crust is being ripped apart and can create deep and slowly deepening lakes. In active fault areas, the deepening rate can exceed sedimentation rates. In these cases, life of the lake can be quite long. In fact, some of the oldest lakes in the world are in rift zones. The lakes in the African Rift Zone are some of the most well known (Fig. 7.17).

***Sag Lakes*** are fresh water bodies in the lowest parts of a depression formed between two sides of an active strike-slip, transtensional or normal fault zone. Lake Elizabeth is an example of a sag lake.

***Oxbow Lakes*** are U-shaped lakes that forms when a wide meander of a river is cut off, creating a free-standing body of water. This landform is so named for its distinctive curved shape, which resembles the bow pin of an oxbow. In Australia, an oxbow lake is called a billabong, from the indigenous Wiradjuri language. In south Texas, oxbows left by the Rio Grande are called resacas.

An oxbow lake forms when a river creates a meander, due to the river's eroding the bank. After a long period of time, the meander becomes very curved, and eventually the neck of the meander becomes narrower and the river cuts through the neck during a flood, cutting off the meander and forming an oxbow lake (Fig. 7.18).

***Mudslide, Landslide, Rockslide*** are types of natural damming of a river by some kind of landslides, such as debris flows and rock avalanches. If the damming landslides are caused by an earthquake, it may also be called a quake lake. Some landslide dams are as high as the largest existing artificial dam.

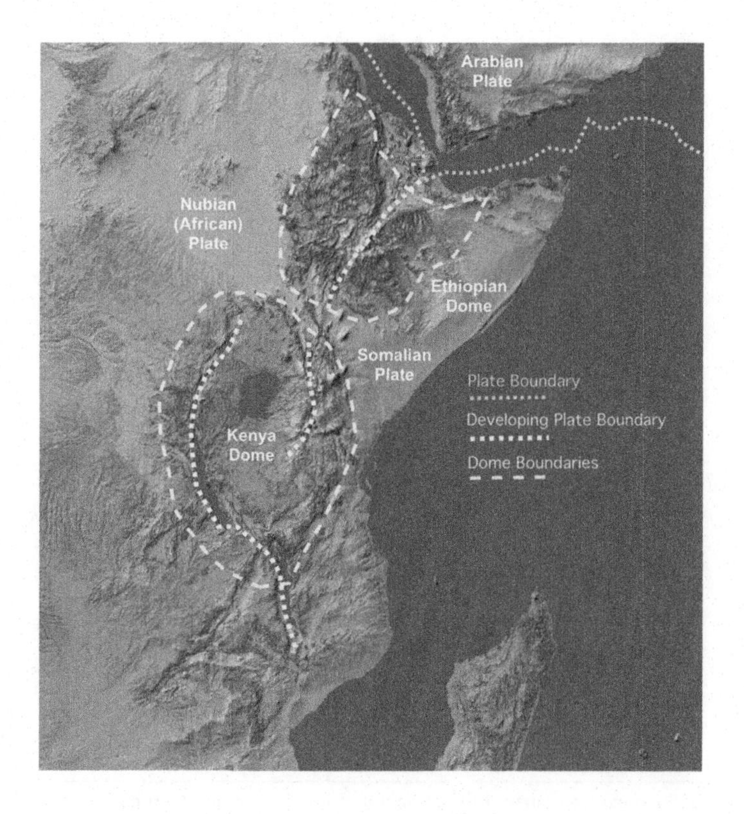


**FIGURE 7.17**

East Africa.

The major causes for landslide dams investigated by 1986 are landslides from excessive precipitation and earthquakes, which account for 84%. Volcanic eruptions account for a further 7% of dams. Other causes of landslides account for the remaining 9%.

***Caldera Lakes*** is a lake that forms in a volcanic crater or caldera, such as a maar. Crater lakes covering active (fumarolic) volcanic vents are sometimes known as Volcanic Lakes, and the water within them is often acidic, saturated with volcanic gases, and cloudy with a strong greenish color.

Lakes located in dormant or extinct volcanoes tend to have fresh water, and the water clarity in such lakes can be exceptional due to the lack of inflowing streams and sediment.

***Dissolution Lakes*** The action of the groundwater continually dissolves permeable bedrock, such as limestone and dolomite, creating vast cave systems, but can also create lakes. Formed by dissolution of soluble rock (often limestone) by percolating water, for example,

$$CaCO_3 + CO_2 + H_2O \longleftrightarrow Ca^{2+} + 2\,HCO_3^- \tag{7.1}$$

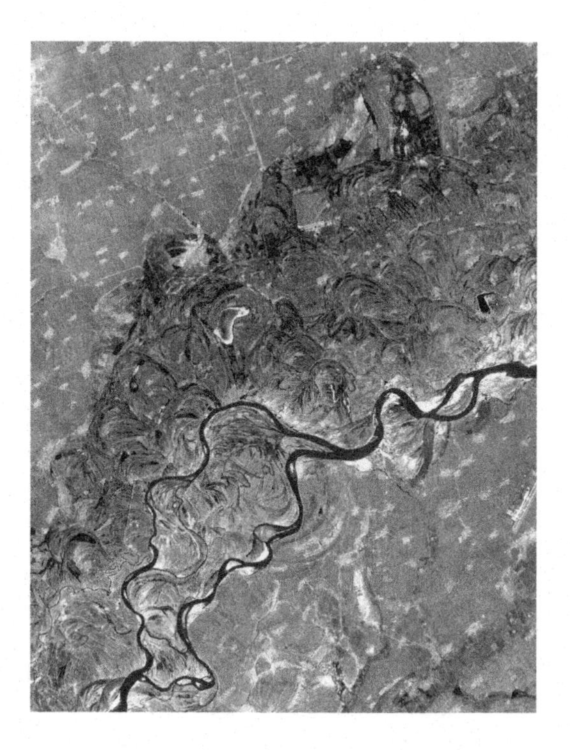

**FIGURE 7.18**

The swirls and curves of the previous courses of the Songhua River in Northeast China can easily be seen in this satellite photo (Source: NASA/GSFC/METI/ERSDAC/JAROS, and the U.S./Japan ASTER Science Team).

Areas with numerous solution lakes are known as "Karst topography," but these may also be areas where sink holes can form quickly and be short-lived. Finally, cave ponds and mound springs are important in karst areas with very unique organisms.

***Reservoirs*** are artificial impoundments. Although these are sometimes also called lakes, it seems prudent to refer to these as reservoirs to acknowledge the anthropocentric construction, but they are often built with different purposes:

- Tributary-storage reservoirs are constructed near the headwaters of the river, and are typically very deep and used for flood control.
- Mainstem storage reservoirs are built in river floodplains, using the land on either side of the river. Mainstem storage reservoirs are used to control floods and generate hydroelectric power.
- Off-stream reservoirs are not built near a stream; they are used to store water delivered through aqueducts.

Spillways are constructed at different heights in dams to control release in different scenarios; we will cover the same in more detail in Chapter 9.

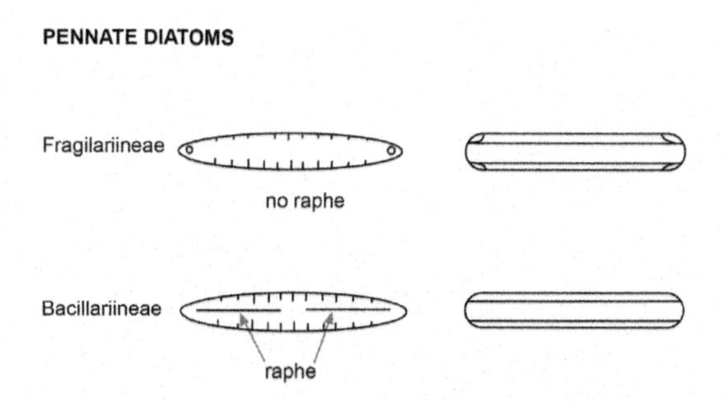


**FIGURE 7.19**

Schematic of pennate diatoms (drawn by Luyi Huang). Distinction using characteristic with high-powered magnification.

## PALEOLIMNOLOGY AND PROXIES

Paleoclimatology (or the study of past climates) uses proxy data to relate elements collected in modern-day samples. Sediment can preserve a range of proxies that might include fossils, biological materials, human artifacts, geochemical markers, and isotopic data. Because of the continual accumulation of sediment, lake sediment cores can offer a comprehensive analysis of an area. Proxies can then be used as indirect metrics correlated with climate conditions:

***Pollen and spores*** can be analyzed within a lab setting to determine the family, genus, or species which the pollen grains belong. The overall taxonomic assemblage of pollen-types reflects the surrounding vegetation of the lake. Pollen analysis is a major field in paleoecology. A range of statistical and qualitative treatment of pollen records can be used to reconstruct past vegetation and climate.

***Diatoms*** The assemblage of diatoms reflect many aspects of the lake temperature, chemical, and nutrient environment. Because the frustule (especially centric shaped ones) can be preserved in sediments (Fig. 7.19). Using morphological characters, diatomists can relate the taxa to water conditions, which can be extrapolated in the regional climate and watershed characteristics.

***Charcoal*** from woody material can age and can be used to infer fire history in the watershed.

***Organic matter and Isotopic Analyses*** combined can provide information regarding the organic matter dynamics in terms of 1) the amount (accumulation rate, reflecting productivity), 2) the origin (aquatic or terrestrial determined from the C:N ratio), 3) the carbon isotopic composition (reflecting plant photosynthetic pathways of C3 vs C4 plants, and reflect drought and/or aquatic productivity), 4) the nitrogen isotopic composition (reflecting sources of nitrogen and nitrogen cycling).

***Chironomids*** are in the Diptera is one of the best indicators of climate change. Chironomids have great ecological importance due to their diverse habitat in the aquatic ecosystem, feeding habitat, their being and important component in the foodweb, and above all their usage as a plaeolimnological studies because their head capsule and feeding structures are fossilized in the lake sediments. Chironomids complete their larval stage in the water and its lifecycle continues up to several years, whereas their adult lifespan is very short. They have an important role in the

**FIGURE 7.20**

Image of reservoir showing the dendritic pattern along the perimeter of the reservoir (Source: Rens van der Sluijs). When water levels rise, some rivers reveal a dendritic pattern similar to lightning. This image shows the La Serena Reservoir, established in the Zújar River, Badajoz, Estremadura, Spain.

degradation of material in the aquatic ecosystem and serve as an important part of the foodweb during their whole lifecycle.

This bottom-dwelling family of flies respond to fluctuations in the surrounding environment. Thus they are good indicators of salinity, water depths, stream flow, aquatic productivity, oxygen level pH, pollution, temperature, and overall ecosystem health. Since the assemblage of Chironomids depends on several environmental factors, they can easily be related to changing environmental factors.

## PHYSIOGRAPHY OF LAKES

### LAKE AND RESERVOIR GEOMETRY AND BATHYMETRY

The physical shape and geometry of lakes and reservoirs have important implications for Limnetic Systems. For example, the perimeter of the shoreline relative to the surface area will vary between young lakes and reservoirs and older, well developed lakes. Young Limnetic Systems tend to have a dendritic (tree-like) pattern with more lake edge than older, rounder lakes (Fig. 7.20).

The bathymetry of a lake plays an important role in lake ecology. For example, where macrophytes can obtain enough PAR, they can become rooted in the lake sediments. The zone is called the Littoral Zone. Beyond the littoral zone is the pelagic zone. The depth that light penetrates (usually as PAR), is called the Glsphotic Zone and beyond that is referred to as the aphotic zone (Fig. 7.21).

The depth of lakes and reservoirs often vary in predictable ways. Lakes tend to be deepest in the center, unless relatively recently formed, whereas reservoirs are almost always deepest where the impoundment (dam) is located. Of course, there are several ecological implications of this—where in lakes the littoral and photic zone might be well-distributed around the basin, in reservoirs the deepest part is on one edge with no Littoral Zone to speak of. In addition, natural lakes tend to have developed shorelines with gentle slopes and abundant vegetation, emergent and submerged. Thus fish or invertebrates that depend on macrophyotes may find higher quality habitat in lakes compared to reservoirs.

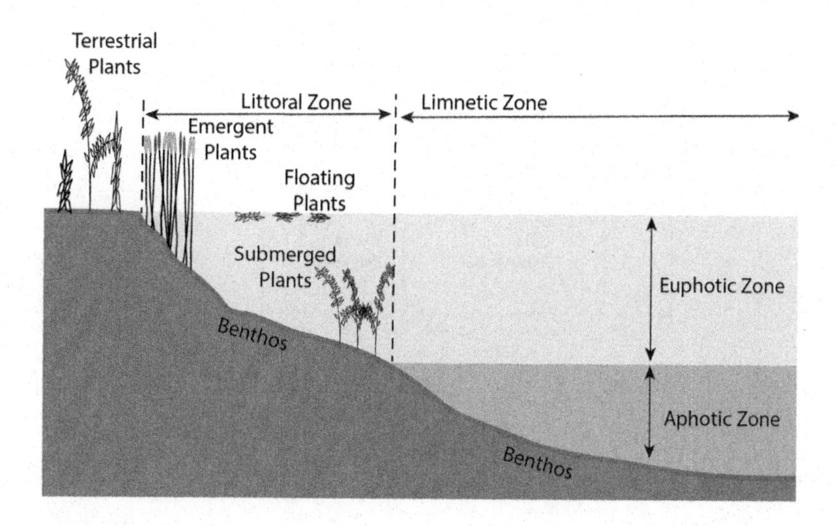


**FIGURE 7.21**

Horizontal lake structure.

Furthermore, reservoirs with steep slopes and variable stages provide little opportunity for vegetation development along the waters edges, giving an appearance of bathtub rings, where shoreline erosion cuts small shelves into the shore.

The average depth is calculated using a simple equation:

$$z_{mean} = V/A, \tag{7.2}$$

where $z$ is the depth, $V$ is volume, and $A$ is area. The volume of a lake or reservoir is determined by the bottom topography or bathymetry and depth. But with some modifications, we could calculate the area or volume of the littoral, pelagic, or photic zones or the volume of the Aphotic Zone.

Using simple equations we can calculate volume at various water depths (stage) in a relatively straightforward fashion. We might modify the equation above so it is a function of stage. The area will also vary by the stage of the water—more water implies a larger area:

$$z_{mean,s} = V_s/A_s, \tag{7.3}$$

where $s$ is stage. With a reasonable bathymetric map, we can also rearrange the equation, $V_s = z_{mean,s} \cdot A_s$, to estimate the volume of water stored in a lake at a given stage. Furthermore as lakes or more likely, reservoirs are drawndown, we might predict change to the Littoral, Pelagic, Photic and Aphotic Zones.

## STRATIFICATION

As described in Chapter 6, water density is a function of temperature and salinity. Furthermore, dense water sinks and less dense water floats, a principle that leads to lake stratification. When lakes stratify we can identify three layers: epilimnion, metalimnion (or thermocline), and hypolimnion.

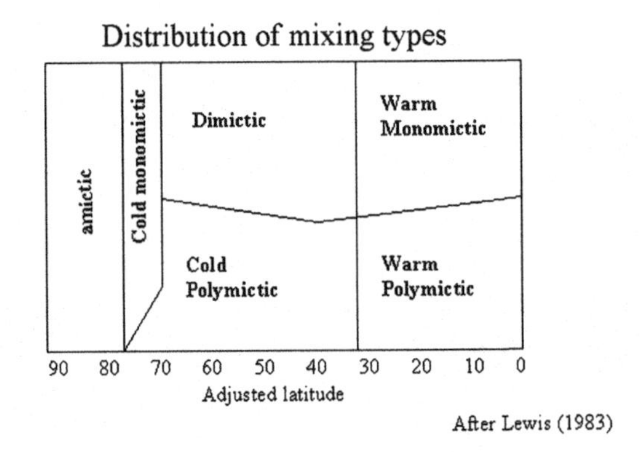


**FIGURE 7.22**

Types of Lake mixing types.

Stratification relies on a density gradient between the epilimnion and hypolimnion, but as that gradient declines, a stratified lake will mix. Usually wind will impart shear stress on the surface and generate water currents strong enough to mix the lake. Lakes can be classified by their mixing patterns: meromictic (never mixes), monomictic (mixes once per year), dimictic (mixes twice per year), or polymictic (mixes multiple times per year) (Fig. 7.22). In addition, amictic lakes are perennially sealed off by ice from most of the annual seasonal variations in temperature. Conversely, for most of the time, the relatively shallower lakes are unstratified; that is, the entire water column can be considered the epilimnion.

## TROPHIC CLASSIFICATIONS

### TROPHIC TERMINOLOGY: OLIGO-, MESO-, AND EUTROPHIC

The trophic status of lakes are often the most studied and discussed topics in Limnology. And for this section we will follow suite. However, let us begin with the historical problem of the definitions and try to define lake population dynamics from a somewhat unique position.

Limnologists cite two important scientists (Naumann and Thienemann) in the development of the terms eutrophic and oligotrophic, which refer to high and low productivity, respectively. It spite of decades of about the value and use of the terms, they remain central in how lakes productivity is classified and for what purpose. However, in the original papers, Naumann and Thienemann included another category: hetertrophic, which is hardly used. Is there a reason for its disuse?

Traditionally, these three main categories were defined as:

***Oligotrophic*** refers to deep lakes, the depth being 18 meters or more; have a U-shaped basin and scanty littoral vegetation. They are less productive and their vegetation consists mainly of Chlorophyceae.

***Eutrophic*** are generally shallow lakes with larger Littoral Shelf and usually dense vegetation . The water is rich in calcium, nitrate and phosphate. They are highly productive and their algal population consists mainly of diatoms and Cyanobacteria.

***Heterotrophic*** references are not clearly defined lakes but their waters are supposed to contain very high percentage of dissolved organic matter.

By focusing on the relative productivity of lakes, Limnologists were subject to some awkward conflated ideas where productivity was often correlated with better fisheries, where in some cases this was true and in other cases, this was not.

## LAKE SUCCESSION

Most early paleolimnological studies focused especially on the biological productivity of lakes, and the role of internal lake processes in directing lake development. Although Naumann had speculated that the productivity of lakes should gradually decrease due to deposition of catchment soils, Thienemann suggested that the reverse process likely occurred. Early sediment core documenting midge populations changes seem to support Thienemann's view.

Through the 1950s Limnologists suggested that lakes would progressively develop through oligotrophic, mesotrophic, and eutrophic stages, before senescing to a dystrophic peaty stage and filling completely with sediment. In-filling would accelerate under eutrophic conditions, when the hypolimnion would experience oxygen depletion that would release of iron-bound phosphorus to the overlying water. This process of internal fertilization would stimulate biological productivity, thus increasing in-filling.

These views were criticized as Palaeolimnologists identified a host of external factors that are equally or more important as regulators of lake development and productivity. For example, late-glacial climatic oscillations (e.g., the Younger-Dryas) seem to accompany changes in productivity. Thus the assumption that lake development is unidirectional process is too simplistic and climatic change have a profound effect on lake communities.

Nevertheless, the terms oligotrophic, mesotrophic, and eutrophic continue to be used as categories of productivity, without the progressive development implications.

## STRATIFICATION AND OXYGEN IN LAKES

Dissolved oxygen concentrations may change dramatically with lake depth. Oxygen production occurs in the top portion of a lake, where sunlight drives the engines of photosynthesis. Oxygen consumption is greatest near the bottom of a lake, where sunken organic matter accumulates and decomposes. In deeper, stratified, lakes, this difference may be dramatic—plenty of oxygen near the top, but practically none near the bottom. If the lake is shallow and easily mixed by wind, the DO concentration may be fairly consistent throughout the water column as long as it is windy. When calm, a pronounced decline with depth may be observed.

Eutrophication exacerbates this condition by adding organic matter to the system, which accelerates the rate of oxygen depletion in the hypolimnion. Urban and other forms of runoff can also add to this problem very suddenly and dramatically by causing fish kills after excess soils and road hydrocarbons are washed in from intense rainstorms. Conditions may become especially serious during a stretch of hot, calm weather, resulting in the loss of many fish. You may have heard about summertime fish kills in local lakes that likely results from this problem.

In eutrophic and hypereutrophic lakes, summertime fish kills can happen most easily during periods with high temperatures, little wind and high cloud cover. The clouds reduce daytime photosynthesis with its oxygen production, and so the DO in the mixed layer. Even the entire water column of a shallow unstratified lake can become critical for fish and other aquatic organisms.

The same basic phenomenon can occur in winter (winterkill) when ice cover removes reaeration from the atmosphere and snowcover can light-limit algal and macrophyte photosynthesis under the ice. Many lakes in the upper midwest are mechanically reaerated or injected with air, oxygen, or even liquid oxygen to keep ice off of some of the lake and to add oxygen directly to prevent winterkills.

The development of anoxia in lakes is most pronounced in thermally stratified systems in summer and under the ice in winter when the water mass is cut-off from the atmosphere. Besides the direct effects on aerobic organisms, anoxia can lead to increased release of phosphorus from sediments that can fuel algal blooms when mixed into the upper Euphotic (sunlit) Zone. It also leads to the buildup of chemically reduced compounds, such as ammonium and hydrogen sulfide (H2S, rotten egg gas), which can be toxic to bottom-dwelling organisms. In extreme cases, sudden mixing of H2S into the upper water column can cause fish kills.

## WETLANDS: SPRINGS, MARSHES, POOLS, RIPARIA, AND SWAMPS

### LEGAL, GEOMORPHOLOGICAL, AND BIOLOGICAL DEFINITIONS

A wetland is a land area that is saturated with water, either permanently or seasonally. As a distinct ecosystem, wetlands include plants that are adapted to the unique hydric soil. The water found in wetlands can be freshwater, brackish, or saltwater. The main wetland types include swamps, marshes, bogs, and fens, and subtypes include mangrove, carr, pocosin, and varzea. Wetlands occur naturally on every continent, except Antarctica, the largest includes the Amazon River basin, the West Siberian Plain, and the Pantanal in South America.

Many definitions for wetlands have been proposed and utilized over the years. Among the most widely accepted definitions is that of Cowardin et al. (1979), which was adopted by the Unites States Fish and Wildlife Service. This definition includes three aspects—water, soil, and plants and provides the ability to recognize, delineate, and describe wetland environments. The three basic components of wetlands are the following:

***Hydrology*** where flooding and water (e.g., water table or zone of saturation) is at the surface or within the soil root zone during all or part of the growing season.

***Hydric soils*** are characterized by frequent, prolonged saturation and low oxygen content, which lead to anaerobic chemical environments, where reduced iron is present.

***Wetland vegetation*** adapted for growing in standing water or saturated soils, e.g., sedges, rushes, rices, mangroves.

Wetlands are present in all climatic and landscape settings. Wetlands are common in tropical and temperate lowlands. They can occur in alpine regions and even some deserts have wetlands supported by groundwater or infrequent storm runoff. Much of the Arctic Region turns to wetland during the brief period of summer melting. The rich diversity of wetland environments requires a flexible definition.

**Table 7.2 Categories of springs by streamflow.**

| Magnitude | Flow (cfs, gal/min, pint/min) | Flow (L/s) |
|---|---|---|
| 1st magnitude | > 100 cfs | 2800 L/s |
| 2nd magnitude | 10 to 100 cfs | 280 to 2800 L/s |
| 3rd magnitude | 1 to 10 cfs | 28 to 280 L/s |
| 4th magnitude | 100 US gal/min to 1 cfs | 6.3 to 28 L/s |
| 5th magnitude | 10 to 100 gal/min | 0.63 to 6.3 L/s |
| 6th magnitude | 1 to 10 gal/min | 63 to 630 mL/s |
| 7th magnitude | 2 pint to 1 gal/min | 8 to 63 mL/s |
| 8th magnitude | Less than 1 pint/min | 8 mL/s |
| 0 magnitude | No flow (sites of past/historic flow) | |

## SPRINGS, SEEPS AND WET MEADOWS

A spring is any natural situation, where water flows from an aquifer to the surface. For example, the source of a spring may be waters traveling through karst topography. Crenobionts are organisms that live in springs. The presence of endemic crenobionts and rare taxa highlights the importance of these habitats in maintaining high levels of biodiversity, as well as their contribution to a better understanding of biodiversity patterns in freshwaters. Spring habitats often include disjunct and relict distribution. A comparison with other geographic areas suggests that springs contain a significant fraction of the total number of species found in freshwater habitats and may contribute almost one third of regional freshwater biodiversity.

Springs can be categorized by their discharge (Table 7.2). However, few researchers use these terms and instead report the discharge.

The forcing of the spring to the surface can be the result of a confined aquifer, in which the recharge area of the spring water table rests at a higher elevation than that of the outlet. Spring water forced to the surface by elevated sources are artesian wells. Nonartesian springs may simply flow from a higher elevation through the earth to a lower elevation and exit in the form of a spring, using the ground like a drainage pipe. Still other springs are the result of pressure from an underground source in the earth, in the form of volcanic activity. The result can be water at elevated temperature, such as a hot spring.

A seep is a moist or wet place where water, usually groundwater, reaches the surface from an underground aquifer. Seeps are usually not of sufficient volume to be flowing beyond their above-ground location. Like a higher volume spring, the water is only from underground sources. Seeps often form a puddle, and are important for small wildlife, bird, and butterfly habitat and moisture needs. Some butterflies (Lepidoptera) species, including some endemic species, rely on seeps and obtain nutrients, such as salts and amino acids.

A wet meadow is a type of wetland with soils that are saturated for part or all of the growing season. Debate exists whether a wet meadow is a type of marsh or a completely separate type of wetland. Wet prairies and wet savannas are hydrologically similar. Wet meadows may occur because of restricted drainage or the receipt of large amounts of water from rain or melted snow. They may also occur in Riparian Zones and around the shores of large lakes.

Unlike a marsh or swamp, a wet meadow does not have standing water present except for brief to moderate periods during the growing season. Instead, the ground in a wet meadow fluctuates between brief periods of inundation and longer periods of saturation. Wet meadows often have large numbers of wetland plant species, which frequently survive as buried seeds during dry periods, and then regenerate after flooding. Wet meadows therefore do not usually support aquatic life, such as fish. They typically have a high diversity of plant species, and may attract large numbers of birds, small mammals, and insects (including butterflies).

Wet meadows were once common in wetland-types around the world. They remain an important community-type in wet savannas and flatwoods. They also survive along rivers and lakeshores, where water levels are allowed to change within and among years. But their area has been dramatically reduced. In some areas, wet meadows are partially drained and farmed, and therefore lack the biodiversity described here. In other cases, the construction of dams has interfered with the natural fluctuation of water levels that generates wet meadows. Like other wetlands, the vegetation in a wet meadow usually includes a wide variety of herbaceous species, including sedges, rushes, grasses, and a wide diversity of other plant species.

## BOGS AND FENS

A bog is a wetland that accumulates peat, a deposit of dead plant material and if often composed of mosses usually or "peat moss" in the *Sphagnum* genus. Sphagnuma includes over 300 species and can store over 20 times as much water as their dry weight. As *Sphagnum* moss grows, it can spread into drier areas that form large bogs. Other names for bogs include mire, quagmire, and muskeg; alkaline mires are called fens.

Bogs occur where the water at the ground surface is acidic and low in nutrients. When the water is derived entirely from precipitation, the bog is ombrotrophic (rain-fed). Because of the low fertility and cool climate, bogs' plants grow slowly. Nevertheless, growth is faster than decomposition, so organic matter accumulates as peat, which can be meters thick (Fig. 7.23).

Fens receive mineral-rich surface water or groundwater and called minerotrophic peatlands. They usually have a neutral or alkaline pH. Fens frequently have a high diversity of plant species, including carnivorous plants. They may also occur along large lakes and rivers, where seasonal changes in water level maintain wet soils with few woody plants. The distribution of individual species of fen plants is often closely connected to water regimes and nutrient concentrations.

Fens are distinguished from bogs, which are acidic, low in minerals, and usually dominated by sedges and shrubs, along with abundant mosses in the genus Sphagnum. Bogs also tend to exist on dome-shaped landmasses, where they receive almost all of their usually abundant moisture from rainfall, whereas fens appear on slopes, flats, or depressions, and are fed by surface and underground water in addition to rain.

## VERNAL POOLS

Vernal pools are rain-fed systems and they have no appreciable surface-groundwater loss or gain of water. In some cases, they can form a network of pools, that are connected during rainfall events or while surface runoff flows across them, but soon the pools become independent of each other, thereby hydrologically isolated.

The term, vernal pools, appeared in early 20th Century, where W.L. Jepson described "empherally wet depression". Over the years the definition has become a bit more restricted, "episaturated seasonal wetlands that are characterized by a unique assemblage of vegetation and soils." It is generally agreed

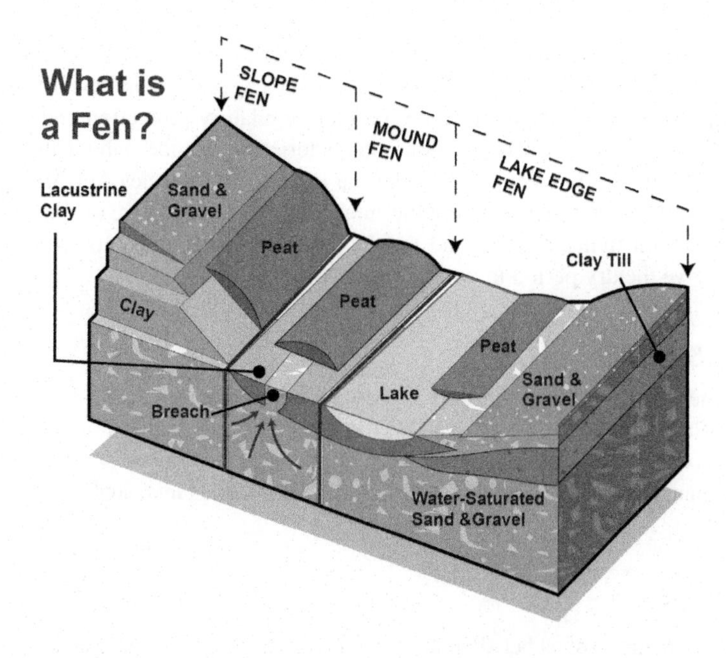


**FIGURE 7.23**

Diagram of a Fen (drawn by Luyi Huang). Fens are peat-forming, groundwater fed wetlands. Fens differ from bogs in that they are less acidic, have higher nutrient concentrations and have a more diverse plant and animal community. They often are dominated by grasses, sedges, rushes and wildflowers.

**FIGURE 7.24**

A California vernal pool (CNPS 2000).

upon that vernal pools are "seasonal wetlands that form in shallow basins and alternate on an annual basis between a stage of standing water and extreme drying conditions". They exhibit four cyclical stages: (i) a wetting phase, (ii) an aquatic or inundation phase, (iii) a waterlogged-terrestrial phase (Fig. 7.24), and (iv) the drought phase.

In California vernal pools are covered in water from winter to spring, but transition to a dry climate over the summer and fall months. Over 90% of California's vernal pools have disappeared. Vernal pools became of special interest when they were granted regulatory protection under the Endangered Species Act and Clean Water Act. Their unique characteristics provide habitat to many rare plants and animals, some of which have been listed under the Endangered Species Act. Vernal pools include opportunistic species, those that reside during the inundated phase and leave during dry periods (such as birds, amphibians, and flying insects) and resident species (such as plants, crustaceans or beetle species), which survive the dry period in dormant stages.

## MARSHES AND SWAMPS

Swamps and marshes are easily interchanged as they are both areas of vegetation that are susceptible to flooding. In North America they are defined a little differently. A swamp refers to ecological areas, where woody plants or trees are dominant in the ecosystem. For example, woody plants may be cypress trees, willows, or mangroves. A marsh is composed of herbacious plants, such as Spartina, rushes, or sedges.

## RIPARIAN FORESTS

Riparian zones are transition zones between an upland terrestrial environment and an aquatic environment. Organisms found in this zone are adapted to periodic flooding. Many not only tolerate flooding, but require it to maintain health and complete their lifestyles. A riparian forest or riparian woodland is a forested land adjacent to a body of water. Riparian forests are subject to frequent inundation, and usually depend on surface and groundwater as opposed to rainfall.

Riparian forests can trap sediment from terrestrial sources, reduce the damaging effects of flooding and aid in stabilizing stream banks.

## TIDAL MARSHES AND MANGROVE FORESTS

The tidal zone along coastal environments are some of the most productive ecosystems. *Spartina alterniflora*, smooth cordgrass is a perennial deciduous grass, which is found in intertidal wetlands. Mangrove forests only grow at tropical and subtropical latitudes near the equator, because they cannot withstand freezing temperatures. Many mangrove forests can be recognized by their dense tangle of prop roots that make the trees appear to be standing on stilts above the water. This tangle of roots allows the trees to handle the daily rise and fall of tides, which means that most mangroves get flooded at least twice per day. The roots also slow the movement of tidal waters, causing sediments to settle out of the water and build up the muddy bottom. Although the estuarine system is beyond the scope of this text, these systems are also in connection with Tidal-freshwater Systems, which can extend over 100 miles from the ocean.

Freshwater tidal wetlands occur in the Intertidal Zone along coastal rivers, upstream from the Estuarine System, where salinity ranges from 0.5 ppt to 0 ppt. The tidal cycle is similar to that on the coast, but timing is delayed and the amplitude of tidal surge attenuates further upstream. Freshwater tidal wetlands can be herbaceous, shrubby, or forested, and like salt marshes are hydrated by meandering tidal creeks. Plant species diversity in these wetlands is extremely rich, and many of thousands of acres have been diked, drained, and farmed.

**FIGURE 7.25**

Hidden genetic variation in typical Astyanax mexicanus (top) may quickly lead to eye loss in cavefish populations (bottom).

# CAVES

## CAVE AS ECOSYSTEMS

Cave environments have been classically separated into three separate zones. These zones are the twilight zone near the entrance, a middle zone with complete darkness but variable temperature, and the deep interior which bodes complete darkness with constant temperatures. As green plants cannot live in areas with permanent darkness, troglobites (i.e., cave-dwelling organisms) must find other sources of food.

## ADAPTATIONS TO THE DARK

Cave dwelling organisms are called troglobites. One of these is the Mexican tetra (*Astyanax mexicanus Astyanax*, which is a genus of freshwater fish in the family Characidae and are known as tetras. The blind and colorless cave tetra of Mexico is a famous member of the genus, but its taxonomic position is disputed: Some recognize it as part of the Mexican tetra (*Astyanax mexicanus*), but there have been some calls to move it to its own species.

Growing to a maximum overall length of 12 cm, the Mexican tetra is has unremarkable, drab coloration. Its blind cave form, however, is notable for having no eyes or pigment; it has a pinkish-white color to its body (resembling an albino). The fish swims in midlevel water above the rocky and sandy bottoms of pools and backwaters of creeks and rivers of its native environment. The fish feeds on crustaceans, insects, and annelids, although in captivity it is omnivorous. Instead of hunting with visual cues, the Mexican tetra use their nasal cavity to detect food and their sensitive lateral line system can feel vibrations in the water to locate prey (Fig. 7.25).

## CAVE TROPHIC PATTERNS

Within dark portions of cave ecosystems, primary production is done by chemosynthetic autotrophic bacteria, which utilize iron and sulfur as electron donors. As the amount of energy made available through primary production is minute, we can make the generalization that cave communities are decomposer communities.

The detritus-based, or decomposer food chain is characterized by a series of trophic levels, namely detritus, decomposers, and detritovores. Decomposers include both fungi and bacteria capable of breaking down plant detritus to extract nutrients and organic carbon. Detritovores encompass all those organisms that feed on these decomposers.

Food enters cave ecosystems in three main ways through both biological and physical agents, either continuously or in pulses, and in a range of spatial patterns. For example, receding waters may leave a layer of plant detritus behind them that becomes food for both aquatic and terrestrial communities within the Cave Ecosystem. Percolating water through the limestone rock contains dissolved organic matter, bacteria, and protozoa, and the fecal matter of animals regularly entering and leaving the cave is rich in organics. Areas of a cave, where mud layers rich in organic matter left by retreating flood waters have a rich fauna, whereas caves that have been subject to severe or rapid flooding are void of this rich mud layer, and as a consequence have distinct fauna that are usually not typically found in Cave Ecosystems. In stone-bottomed cave streams, most of the food is plant detritus, with microbes only present in low numbers. Many microorganisms are supported by plant detritus, and are a part of the diet of many terrestrial cave invertebrates. This is support for the presence of decomposer food chains in cave ecosystems.

There is also support that microfungi, decomposers, are an important trophic level in the decomposer food chain.

As a substantial amount of energy available to the fauna within cave ecosystems are dependent not on primary production, but on those mechanisms outlined above, this provides evidence that Cave Ecosystems are a Detritus-based Trophic System.

# NEXT STEPS

## CHAPTER STUDY QUESTIONS

1. Create a cross-section that includes each of the following water classification categories: Lacustrine, Palustrine, and Riverine. Describe the processes that might allow each of these to exist adjacent to each other as you have diagrammed.

## LITERATURE RESEARCH ACTIVITIES

1. Review ten recent articles that classify Aquatic Systems and describe what processes are used for these classifications. Do the type of taxa studied that influence these Classification Systems.
2. Describe how key studies and their methods have undermined the universality of the River Continuum Concept.

CHAPTER

# THE PLOT: COMMUNITY DYNAMICS AND TROPHIC INTERACTIONS

*Shall we surrender to our surroundings or shall we make our peace with nature and begin to make reparations for the damage we have done to our air, to our land and to our water?*
**Richard Nixon (1913–1994), 37th U.S. President, State of the Union Message, 22 Jan 1970**

*Many estuaries produce more harvestable human food per acre than the best midwestern farmland.*
**Stanly A. Cain, Ecologists testimony, U.S. House of Representatives, Merchant Marine and Fisheries subcommittee, March 1967**

## CONTENTS



Ecology and Management of Inland Waters. https://doi.org/10.1016/B978-0-12-814266-0.00021-0

The *threespine stickleback* (*Gasterosteus aculeatus*) occur in marine, estuarine, and freshwaters in the Northern Hemisphere (circumpolar), including California. Following glacial melting 15,000–20,000 years ago, threespine stickleback have repeatedly and independently invaded freshwaters. Their success in freshwater depended on a wide range of morphological, physiological, and behavioral changes, some of which we can link directly to genetic changes.

Freshwater threespine sticklebacks look different than their marine varieties. Marine populations have robust pelvic structures, whereas many freshwater invaders have reduced pelvic structure (Fig. 8.1). Because these morphological changes occurred across European, North American, and Japanese freshwater populations, i.e., parallel evolution, there must be some adaptive advantage. For example, reduced (~9) armor plates in freshwater populations, possibly an adaption to low calcium levels, allows for more rapid growth, increases maneuverability and swimming performance, and improves survival in different predation regimes. On the other hand, complete (30–36) armor plates may facilitate escape from predation or reduce ingestibility in marine waters. Whatever the mechanism or combination of mechanisms, the Ectodysplasin (EDA) signaling pathway plays a key role in evolutionary change in natural populations and of parallel evolution of stickleback low-plated phenotypes in freshwaters around the globe (Fig. 8.2).

The EDA signaling pathway is a complex set of genes that regulate the expression of genes that control animal morphology and the *eda* gene in particular is associated with stickleback armor development. The reduced body armor allele is found, rarely, in marine populations, but increases in estuarine or euryhaline environments and occurs in nearly 100% of the individuals in the freshwaters. In addition to morphological changes, *eda* influences several phenotypic characters, including behavior, e.g., willingness to colonize new habitats. Besides *eda*, there are dozens of genes that differ between marine and freshwater stickleback that control food digestibility, osmoregulation, dorsal spines, growth, thermal response, maturation, pigmentation, etc.

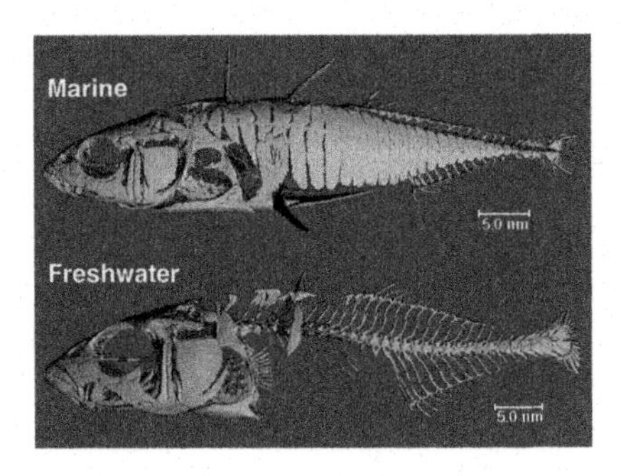


**FIGURE 8.1**

Freshwater and marine stickleback skeleton (Source: Kingsley, Howard Hughes Medical Institute).

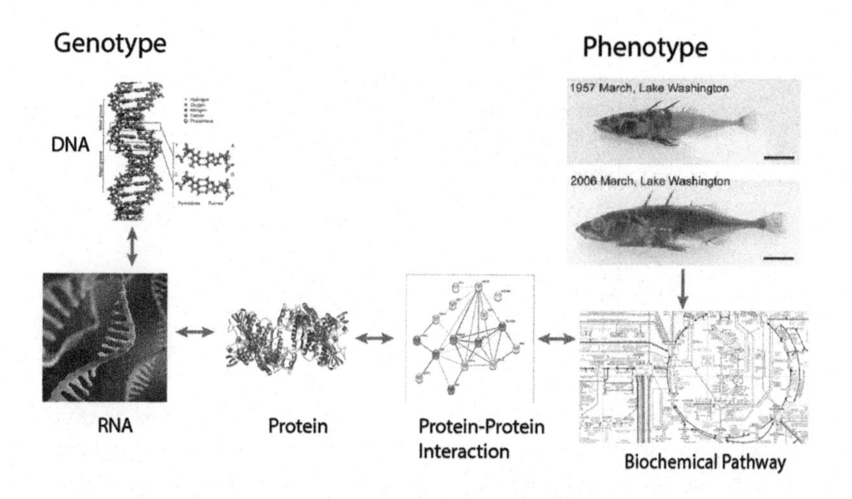


**FIGURE 8.2**

DNA to phenotypes.

Since the glacial retreat about 18,000 years ago, higher-latitude landscapes have changed dramatically. With warmer conditions and Ice-free Aquatic Ecosystems, a wide range of plants, invertebrates, amphibians, reptiles, and mammals have been invading new regions. And as exemplified by the stickleback, marine taxa invaded freshwaters. Each population arrived in these habitats with a history—a genetic record of selection and in a context of competing species and dynamic trophic relationships. These interacting factors define the processes in Aquatic Systems and with the increasing role of humans on a global scale, the patterns and processes in Aquatic Systems are changing even more rapidly since the last glaciation.

The combination of community dynamics and trophic interactions define Aquatic Ecosystems, but individuals negotiate these processes within genetically controlled physiological tolerances, morphological characters, and behavioral responses. After reading this chapter, you should be able to

1. Summarize how individual tolerances, reproductive strategies, and trophic dynamics influence Aquatic Ecosystems;
2. Describe the contrasts between top-down and bottom-up controls in Aquatic Systems.

## ABIOTIC TOLERANCES, PHENOTYPIC EXPRESSIONS, AND EVOLUTION

### GENETIC BASIS FOR TOLERANCES

Phenotype is the set of observable characteristics of an individual resulting from the interaction of its genotype with the environment. Phenotypic plasticity is the capacity for the genetic code to respond to environmental conditions, e.g., changes in the expression of certain genes often via regulatory genes. This plasticity can improve individual survival and even allow populations to invade new habitats, e.g.,

**Table 8.1 Frequency of *fads2* copy number in freshwater and marine ray-finned fish (Source: Ishikawa et al., 2019).**

| *fads2* CN | Marine | Freshwater |
|---|---|---|
| 0 | 1 | 1 |
| 1 | 12 | 17 |
| 2 | 1 | 7 |
| 3 | 0 | 15 |
| 4 | 0 | 3 |

freshwater habitats. Thus we might expect plasticity to be higher where organisms experience greater abiotic variation, such as in varying salinity concentrations, or changes in food availability.

Marine *Gasterosteus aculeatus* display poor growth and health when fed freshwater prey. Meanwhile freshwater sticklebacks seem to do much better with this freshwater diet; moreover female sticklebacks do better than males. The differences seem to be the result of the number of copies of the *fads2* gene in the stickleback genome. Because one copy is normally found on the X chromosome, female fish do better than male fish with a freshwater diet. *fads2* encodes a fatty acid desaturase, a key enzyme catalyzing desaturation and Docosahexaenoic Acid (DHA) biosynthesis. Without DHA fish do not develop properly and are generally unhealthy, and this seems to be a common story for freshwater ray-finned fishes (Table 8.1).

The duplication is associated with a transposon ("jumping gene"), where additional (4 or more) copies can be found on another chromosome. This example demonstrates how phenotypic tolerances are built on genetic diversity of the taxon. Furthermore, as more stressors, e.g., climate change, are put in freshwater biota, maintaining this genetic diversity may be key to maintaining aquatic communities and ecosystem services.

## SMOLTIFICATION AND CHINOOK SALMON

Smoltification is a genetically controlled process by which freshwater-adapted anadromous salmonids alter morphology, physiology, and behavior in preparation for their out-migration to the ocean. Individual fish undergo a series of morphological changes, resulting in a streamlined body shape and silvery coloration (Fig. 8.3). The parr–smolt transformation also includes a decrease in fat storage as energy is consumed. Fish show behavioral changes as well. For example, freshwater parr, territorial bottom dwellers, maintain their orientation to face the water currents, i.e., positive rheotaxis and position in the river, they transform. However, smolts that are less territorial, surface oriented, are less prone to holding position against the current.

A change in salt tolerance is a key physiological factor in adapting to the freshwater-seawater transition for chinook salmon. Researchers commonly measure the activity of the $Na^{+}K^{+}$-ATPase enzyme (NKA) in the gill tissue as an indication of migratory preparedness, because this enzyme is required for extrusion of salt to maintain osmotic balance in seawater (Fig. 8.4). Additionally, increases in NKA activity have been repeatedly linked to migration behavior and survival in salmonids. But, how does NKA activity vary among populations or parental lineages? In one study, little population difference in salinity tolerance was found, but there was significant intra-family variation. Thus variation exists

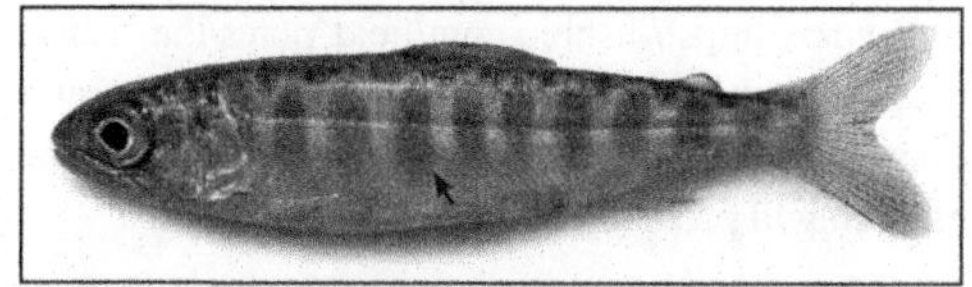


**FIGURE 8.3**

Marking differences in par smolt (Source: Johnson et al., 2015).

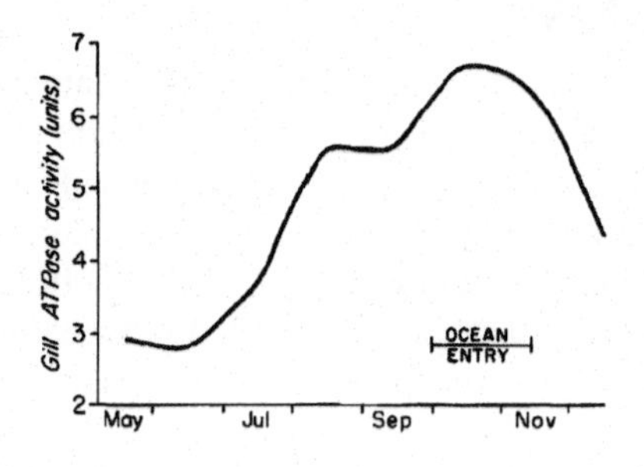


**FIGURE 8.4**

ATPase activity (Source: Wedemeyer et al., 1980).

within a clutch, which can spread the risk of mortality along a family line. These results remind us to be careful when considering the role of populations-level and cohort-level variation, recognizing that selection might be operating on multiple scales, but in different ways.

## REPRODUCTIVE STRATEGIES

### RISK AVERSION AND REPRODUCTIVE SUCCESS

Fish and other taxa develop a wide range of reproductive strategies to minimize risk and maximize reproductive success. Fish egg size, number of eggs, competition for mates, and level of parental

**FIGURE 8.5**

Water hyacinth in California canal (Source: https://fishbio.com/field-notes/the-fish-report/hyacinth-woes-gift-curse). In 2014, an outbreak of water hyacinth occurred in California. Drought conditions and warm weather facilitated its growth, though the plant has been a pest in California for decades. The control and impact of the plant is poorly understood, but ongoing research will probably begin to fill these gaps.

investment are examples of how reproductive strategies can vary for reproductive success in inland waters.

## FITNESS AND AQUATIC PLANTS

Between 1–2% of the vascular plants are aquatic, but the invasion into the waters has occurred between 50–100 times by a wide range of unrelated taxa, e.g., 78 angiosperm families. The majority of aquatic plants reproduce by sexual and clonal asexual means. The preponderance of asexual reproduction in aquatic plants allow them to spread without relying solely on completing the sexual reproductive cycle of flowering, pollination, seed maturation, and successful dispersal. Instead, bulbils, rhizomes, runners, stolons, and turions can break-off the plant and reestablish downstream.

The trade-offs between sexual and asexual production are poorly understood, but some plants, such as common water hyacinth (*Eichhornia crassipes*), asexual reproduction has allowed the species to spread from lowlands in South America to over 50 countries and five continents. The water hyacinth reproduces primarily by way of runners or stolons, which eventually form plantlets. The common water hyacinth are vigorous growers and mats can double in size in two weeks, becoming a major weed, clogging open channels (Fig. 8.5).

The plant reproductive strategies in Aquatic Systems are as varied as the ecosystems themselves and depend on many physical (e.g., water temperature, day length, hydrology) and biological (e.g., pollinator dynamics) factors and readers are encouraged to learn more about these patterns for specific ecosystems of interest.

## PARENTAL CARE VS CANNIBALISM: REPRODUCTIVE STRATEGIES

Parental care is costly for animal parents, who may desert, abort, or kill their offspring and sometimes even eat them. How can eating offspring provide any adaptive advantage? Is this a way to limit raising unrelated offspring ("cuckold"), recapture nutrients and resources when they are limited, reduce investing in offspring with low chances of survival?

**FIGURE 8.6**

Image of Bluegill Sunfish (*Lepomis macrochirus*) (Source: Scott Harden—Own work, CC BY-SA 4.0, https://commons.wikimedia.org/w/index.php?curid=76318632).

Some fish will selectively eat eggs when they are not developing as quickly as the others, or will eat their eggs when they suspect they may not actually be the father. Some fish will eat their eggs when the clutches are too small, and they want to start over—and as it turns out, by eating the eggs the fish can maintain hormone levels that allow them to remain reproductive.

In the case of bluegill sunfish (*Lepomis macrochirus*, Fig. 8.6), parental care is costly, but individuals can adapt their parental investment. In addition, bluegill males can follow parental or cuckolder life histories. For example, bluegill sunfish display both total and partial filial cannibalism. Their behavior, to protect or consume their offspring, can depend on a wide range of factors, but their need to replenish their energy may be an important predictor.

## LINKAGES BETWEEN TERRESTRIAL AND AQUATIC SYSTEMS

The Mangrove Swallow (*Tachycineta albilinea*) inhabits areas around water, including mangroves, marshes, wet meadows, and other open waters in Mexico and Central America. Just like the stickleback, they require long-chain omega-3 polyunsaturated fats, but these insectivores do not obtain these fats from terrestrial insects. Instead they rely on adult aquatic insects for these fats. With this understanding, we begin to appreciate how the management of the Mangrove Swallow depends on maintaining a healthy insect macroinvertebrate populations and waters that support them and land use to successfully maintain terrestrial and aquatic habitats.

The relationship between forests and salmon in Alaska and Canada demonstrates another linkage between Aquatic and Terrestrial Systems. In this case, new forests expand into valleys of retreating glaciers, but soils have low organic matter and nitrogen content, thus limiting forest growth. However, forest productivity increases with ocean-derived nitrogen brought in by spawning salmon. After migrating upstream and spawning, the majority of the salmon die in these streams, becoming food for bears or their decomposing carcasses becoming a source of nitrogen in headwaters streams, adjacent trees, or deposition areas of bear excrement. And since these trees provide shade that reduce algae growth and keep streams cool and oxygenated, rapid forest growth provides a benefit for the salmon's developing eggs and next generation of spawners.

Similar to the Alaska and Canada example, spring run Chinook salmon in California are a keystone species that link aquatic and terrestrial ecosystems and support directly or indirectly many ecosystem services. This relationship has another twist in California watersheds, where delivery of marine de-

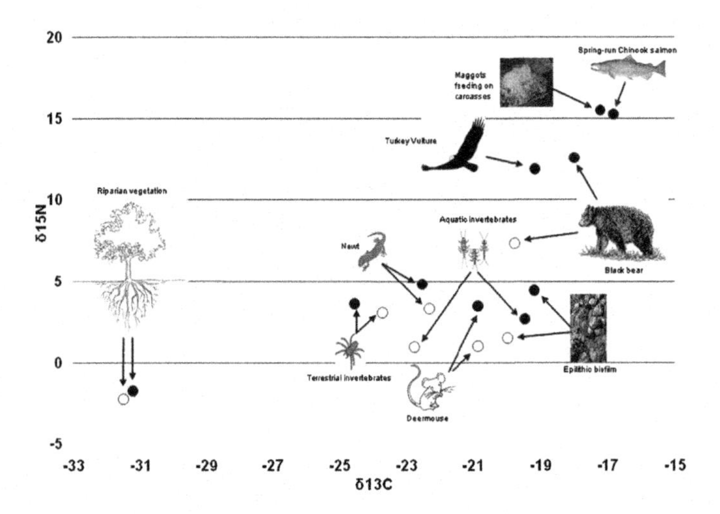


**FIGURE 8.7**

Biplot of $\delta^{13}C$ and $\delta^{15}N$ values in tissues of sampled organisms taken from sites above and below a barrier to salmon migration. Closed circles represent mean values from tissues at spawning sites. Open circles are mean values from sites located above a waterfall barrier to salmon migration. In most cases, open circles plot lower than closed circles, indicating that organisms from non-salmon sites are less enriched in marine-derived nitrogen than similar taxa at spawning sites (Source: Moyle et al., 2011).

rived nitrogen can be interrupted by migration barriers. Using stable isotopes associated with aquatic and terrestrial food webs in Butte Creek, California, researchers at the University of California, Davis found a clear signal of marine derived nitrogen ($\delta^{15}N$) and carbon ($\delta^{13}C$) in both aquatic and terrestrial food webs in areas where Chinook salmon spawn below a dam and waterfall barrier to migration (Fig. 8.7). However, upstream of the dam, marine derived N and C was much reduced. Without marine derived nutrients, we would expect cascading effects on riparian ecosystems, including lower ecosystem productivity. Moreover, the annual run of spring run Chinook is estimated to have declined by 90% over the past century, which suggests that overall declines of marine derived nutrients in Sierra Nevada mountain ecosystems has been profound.

These examples demonstrate linkages between Aquatic and Terrestrial Systems, a conclusion further developed in the flood pulse concept (see page 285 for further discussion on this topic).

## ENVIRONMENTAL FILTERS

### FISH ASSEMBLY RULES

Fish diversity is not evenly distributed around the world. Fish diversity varies by region, where environmental filters reduce the number of taxa that occur in inland waters. For example, there are only 67 inland native fish species in California (~0.6% of the total number of freshwater fish species on the planet), with an additional 51 introduced species. Is this number of fish typical of inland waters, or are circumstances that reduce the number species in these inland waters unique to California?

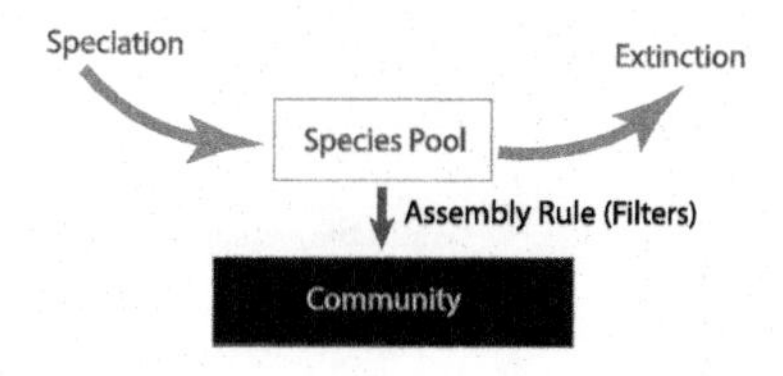


**FIGURE 8.8**

Conceptual model for how assembly rules (purple arrow, dark gray in print version) might be used to describe community composition.

Some Ecologists have tried to apply "assembly rules" to predict community assemblages based on the pool of available species (Fig. 8.8). Assembly rules are sets of environmental filters that facilitate selection from a pool of potential taxa: those that pass through the filter become members of the community. For our purposes, we will use these filters as post hoc explanation to show how these "assembly rules" might be applied.

The first filters that select freshwater fish communities of California are the dramatic extinction events through in the geologic record. Fish species in California come from a pool of successfully evolving taxa. Surviving extinctions allow the ancestors of contemporary fish to invade California's inland waters. The earliest fish emerge in the Earth's fossil record in the Cambrian Period, but their radiation takes place in the Silurian and Devonian Periods—historically referred to as the "Age of Fish." When Pangaea broke up in the middle of the Jurassic Period, most of the major fish orders already existed in fossil record, thus representatives of these orders are nearly globally distributed. But these species pools change with time as a result of the interplay between speciation and extinction.

Historically, the original fish ancestors were marine taxa, although some argue that fish evolved in freshwaters. As fish species followed available food into freshwaters, they overcame a wide range of ecological tolerances (e.g., osmoregulation) to successfully invade freshwaters (Fig. 8.9). Nevertheless, as these animals move into freshwaters, they bring with them some level of evolutionary "capacity," i.e., genotypic and phenotypic plasticity, and inertia that limit future adaptive capacity to changing environmental conditions and subsequent environmental filters. For example, *Gasterosteus aculeatus* is famous for its evolutionary flexibility, having evolved from fresh, to brackish, and marine environments, and back.

The Limnologist, G.E. Hutchinson, identified four potential evolutionary pathways by which organisms might invade freshwater:

1. Active invasion from the sea and adaptation to lowered salinities;
2. passive invasion from the sea, via waters that gradually freshened over time;
3. invasion from terrestrially evolved taxa (e.g., higher plants and possibly insects) as secondary invasions, e.g., aquatic mammals; or
4. adaptive radiation within inland waters.

In the case of fish, we can immediately exclude #3, since there are no fish that have left aquatic systems to return later as fish, while amniotes, a derived form of fish, have done so. We cover two groups of amniotes elsewhere: aquatic tetrapods and birds in later sections.

Since most marine fishes are associated with continental margins, we can surmise that ecosystem productivity was higher in these regions. Even the fossil records demonstrates high level of diversi-

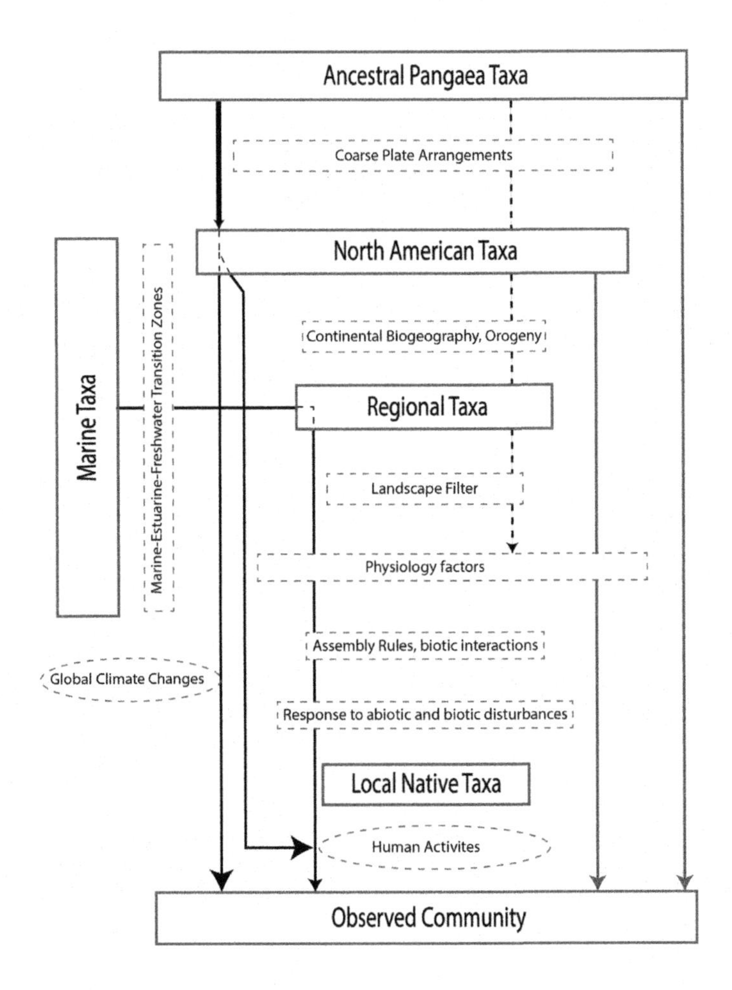


**FIGURE 8.9**

Environmental filters for fishes that define the observed inland fish community based on Moyle (2002).

fication at the coastal margins. Therefore, associated with the transition zones between marine and freshwater, is the possibility of explaining the diversity in freshwater to develop deep ancestry lines similar to the marine taxa.

Moving to freshwaters was probably a way for fish to take advantage of available food and avoid predation. For example, well-developed algal mats and periphyton, algae growing on rock surfaces, were probably common in the Cambrian Period; herbivores took advantage of these food sources, and predators were quick to follow. With the advent and dominance of vascular plants in the Carboniferous Period, the terrestrial contribution of organic matter dramatically increased. Thus Paleoecologists cite the explosion of land plants as a source of food for fish and helped draw fish into freshwaters. In fact, the Ordovician Period coincides with both a prolific increase in terrestrial vegetation and a dramatic radiation of fish taxa.

Being associated with the Coastal Zones allow us to also conceptualize how coasts might confer reproductive isolation and regionally defined biodiversity. Compared to the eastern shore of Laurentia, the coastal margin of California is going to have a different pool of species as potential invaders.

The foregoing demonstrates that Environmental Filters are tested and provide a conceptual framework to test hypotheses (Fig. 8.8).

Landscape Filters might include migration barriers. For example, vernal pool taxa in a terrestrial landscape matrix can move with migratory birds, thereby allowing dispersal barriers to be avoided. Biotic factors can also be seen as filters, where a predator might exclude specific preys or prey from a community.

## BEYOND PRESENCE AND ABSENCE

Environmental Filters provide a conceptual framework that defines presence and absence within communities, but they are not designed, as of yet, to make predictions about the processes in inland waters. To predict population dynamics, density-independent and density-dependent models were tested with less than satisfactory results (Chapter 4).

Phytoplankton growth could be modeled as a function of reproductive rates and resource limitations, such as a nutrient (e.g., bottom-up controls). When competing species were grown together, one species would disappear. Competition Models could be parameterized to model unstable populations, where one species could exclude another. Finally, Predator–Prey Models added another complexity and modeling efforts demonstrated how prey dynamics might be predator-controlled, but when tested in laboratory settings, where model predictions fell short. For example, by testing Lotka–Volterra Equations using paramecium populations in small test tubes, we learned that stable populations required habitat heterogeneity. In the end, translating these population models to complex Aquatic Ecosystems is well beyond capacity of Mathematical Models.

However, these models are tacitly based on two modes of population control: resources limitations (bottom-up) and predator–prey dynamics (top-down). The tension between these to forces (in terms of population dynamics and controversy in the literature) provides framework to evaluate the structure and function of inland waters.

# TOP-DOWN VERSUS BOTTOM-UP CONTROLS

## WHY IS THE WORLD GREEN?

In 1960 three Ecologists wrote a paper describing the relationship between global patterns of plant biomass, herbivory, and predator controls. The paper, referred to as HSS or "The World is Green," suggested that plant biomass is not mowed down, because herbivore populations are held in check by predators. But how can we reconcile this top-down hypothesis with the observations that link primary productivity with nutrient additions, e.g., phosphorus, or as a bottom-up control.

The HSS paper outlines one of the most important questions about how ecosystems function, how is productivity controlled, and what does it mean when these controls are modified or altered?

## BOTTOM-UP CONTROLS

### LIMITING NUTRIENTS AND STOICHIOMETRY

Ecological Stoichiometry links the energy balance with chemical elements that compose groups of organisms. Similar to Chemical Stoichiometry, Ecological Stoichiometry is based on mass balance, i.e., chemical inputs = chemical outputs within selected taxa or functional groups (e.g., phytoplankton, consumers, and secondary consumers). For example, one might ask how do specious interactions depend on the balance of energy or ratio of elements available?

To illustrate this question, we can compare the Stoichiometry Ratio for phytoplankton and zooplankton. Phytoplankton generally have a high C:N:P Ratio, e.g., the Redfield Ratio 106:22:1 for oceans (Note: recent surveys report a higher average ratio, i.e., 163:22:1). In contrast to phytoplankton, zooplankton tissues are composed of less C relative to the N and P, thus they may require higher consumption rates of high C:N:P phytoplankton to obtain enough N and P. This ratio "imbalance" or mismatch can limit the growth of individual or population growth. In any case, differences in stoichiometric homeostasis between plants and animals can lead to large and variable elemental imbalances between consumers and resources.

The roots of Ecological Stoichiometry are based on observations made by Liebig, Lotka, and Redfield, but now we know photoautotrophs can exhibit a very wide range of physiological plasticity in elemental composition, and thus have relatively weak Stoichiometric Homeostasis. In contrast, other organisms, such as multicellular animals, have relatively strict homeostasis and have distinct chemical composition. For example, carbon to phosphorus ratios in the suspended organic matter in lakes can vary between 100 and 1000 whereas C:P ratios of Daphnia, a crustacean zooplankton, remain nearly constant at 80:1.

## NOTE: MARINE AND LAKE CONTROLS OF PHYTOPLANKTON

Net primary production are positively related to nutrient availability in both the sea and lakes. Oceanographers have observed that the highest rates of primary production by marine phytoplankton are generally concentrated in areas with higher levels of nutrient availability. Similar to lakes, vertical mixing can be an important source of nutrients and for marine systems upwelling brings water with higher nutrient concentrations from the depths to the surface. These upwelling areas are concentrated along the west coasts of continents and around the continent of Antarctica, which is especially true along the coast of California, and are associated with high productivity.

In contrast to many freshwaters, marine systems are thought to be limited by nitrogen. In 1990, Edna Graneli added nutrients (nitrate, phosphorous, and control) to filtered seawater from the Baltic Sea. Incubated flasks with nitrate additions increased chlorophyll *a*, whereas flasks with phosphate additions and the controls had similar chlorophyll *a*. Thus the rate of primary production in the Baltic Sea is limited by nitrate. However, open ocean primary productivity may be iron-limited.

In another study, researchers altered the nutrient inputs and concentrations in Himmerfjard, Sweden, a brackish water coastal inlet of the Baltic Sea. The results indicated that a nitrogen limitation can shift to phosphorus limitation by altering nitrogen:phosphorus ratios. Increasing additions of phosphorus to Himmerfjard reinforced nitrogen limitation, whereas decreasing phosphorus additions and increasing nitrogen additions led to increased phosphorus limitation.

**FIGURE 8.10**

In the Siskiyou Mountains (Klamath Range), Castle Lake is 1645 m above sea level and is part of the headwaters of the Sacramento River. The lake or tarn, formed in a glacial cirque, is the largest (by volume) of the 25 Alpine and Sub-Alpine Lakes within the larger Upper Sacramento River Watershed. Dr. Chandra and student during late winter sampling (2008). Currently, the University of Nevada, Reno is managing the long-term monitoring of the lake (Source: Department of Biology, University of Nevada Reno).

## PRODUCTIVITY IN CASTLE LAKE: LEIBIG LAW OF THE MINIMUM

Since Charles Goldman began monitoring Castle Lake in 1959, a small glacial lake in the Sacramento River watershed has been an important research site (Fig. 8.10). For instance, using radioactive $^{14}C$ isotopes, Goldman evaluated the limitations of primary productivity. He determined that primary productivity increased with molybdenum, a cofactor for nitrogenase, a key element for N-fixation. Later he determined that areas with alder trees, which have a N-fixer symbiont, were associated with higher primary productivity as well. By identifying Mo and alder-mediated N-fixation as mechanisms that limit primary productivity, Goldman's conclusions conform to Carl Sprengel's "Theorem of minimum" (known as Law of the Minimum). This "Theorem" states that plant growth is not determined by the total resources available, but by the scarcest available resource. Thus the growth of an autotroph is limited by the one essential mineral, e.g., Mo, that is in the relatively shortest supply.

Further work in Castle Lake later noted that bacteriophyton was correlated with primary productivity, which suggests that the nutrient regeneration process could limit primary productivity, and static nutrient concentrations might be an overly simplistic measure as a limitation to primary productivity (Fig. 8.11). We will revisit Castle Lake later in the chapter and turn our attention to the importance of the "microbial loop."

Finally, we note the value of Castle Lake as a research and training facility. Goldman established the Castle Lake Environmental Research and Education Program (CLEREP) to facilitate collaborative research and education on a specific ecosystem-type in California (Fig. 8.10). CLEREP is now the longest running, mountain lake research and training program in the Americas.

## THE BLUE LAKE TAHOE

Lake Tahoe is known for its blue color, which is a function of its clarity. However, the transparency has been declining, with average decline in transparency of 0.37 m each year between 1959 and 1998.

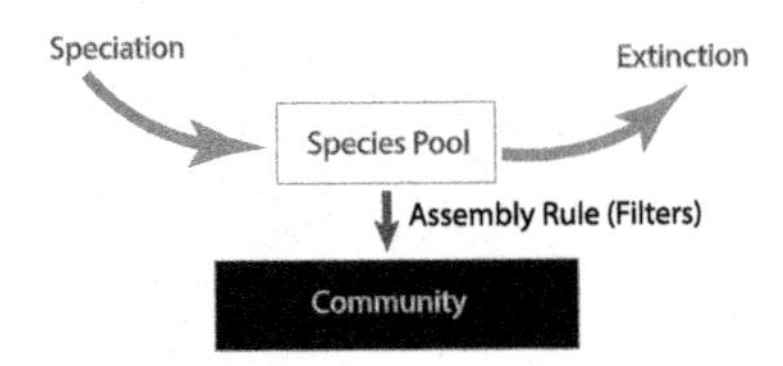


**FIGURE 8.11**

Relationship between bacteriophyton and productivity as measured by chlorophyll, Castle Lake, CA (Source: Brett et al., 1999).

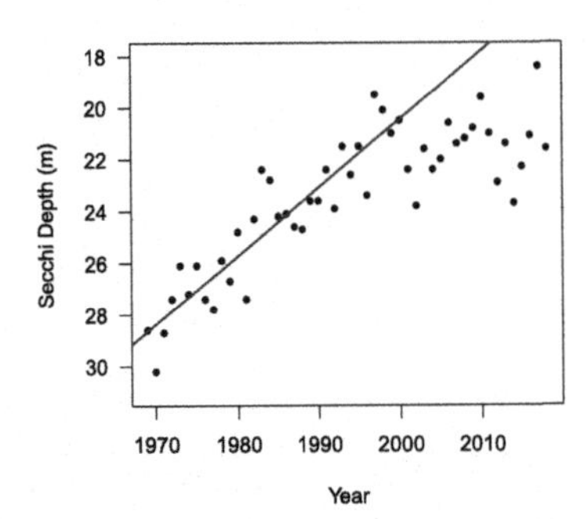


**FIGURE 8.12**

Tahoe Secchi Depth time series (Source: https://tahoe.ucdavis.edu/secchi). Best-fit line includes the years 1968–1998, after which the Secchi Depths may level off.

Annual productivity doubled in Lake Tahoe between 1959 and 1981, with an average increase of 5.6% between 1960 and 1988 (Fig. 8.12).

This increase corresponds to include a 10-fold increase in human population between these years. But the mere presence of people does not provide Ecologists with a mechanism to know how the clarity has declined. The increased nutrient additions combined with the lake's lengthy retention time has been gradually increasing the productivity (and eutrophication) of the lake. Lake Tahoe has a large volume of water, ~26 million $m^3$. And with a water residence time off 650 years, any nutrient loading added by humans is going to play a role in the lake for a long time, effectively forever.

To address this issue, strict regulations have been developed to reduce nutrient inputs. For example, all human waste is exported from the basin, erosion from roadways and during construction must be carefully managed, and wetland restoration to filter nutrients has been an important priority. However, atmospheric deposition (dust from roadways) and in and ex situ exhaust may be important sources of P and N into the basin (Fig. 8.13).

## APPLYING BOTTOM UP RESOURCE MODELS

As described in Chapter 4, the Monod and Droop Equations (Eqs. (4.20) and (4.21)) have been used to model for the growth of microorganisms based on nutrient concentrations. For example, these equations have been applied to phytoplankton growth and predicted how various population growth rates

**FIGURE 8.13**

It is with irony that the "Keep Tahoe Blue" decal (Source: League to Save Lake Tahoe) is usually seen on bumper stickers near the exhaust pipes on vehicles that burn fossil fuels. The exhaust from these fossil fuels may be an important source of N deposition in the lake that promotes eutrophic conditions. Keeping Tahoe blue could be associated with a saying "Don't go there".

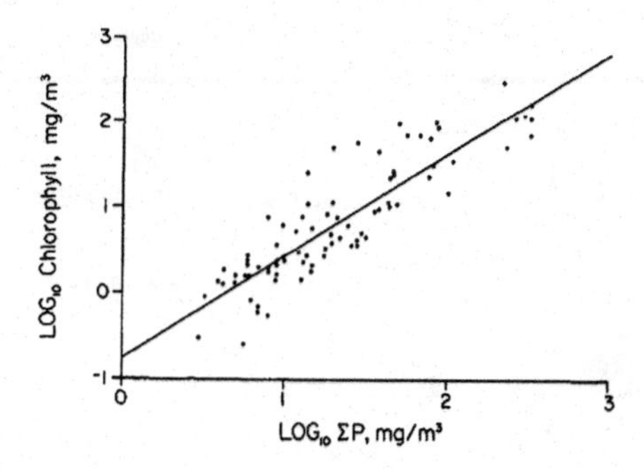


**FIGURE 8.14**

Relationship between mean annual chlorophyll concentration and mean annual concentration of total phosphorus (Source: Schindler, 1978). $\log[Chl] = 1.213\log[\sum P] - 0.848$, $r = 0.88$.

can vary based on how observed differences in their modeled coefficients. However, the assumptions needed to use these equations are often violated in nonlaboratory conditions. Thus models that describe bottom-up population dynamics remain compelling on a theoretical level, but their predictive ability in Aquatic Systems remains out of reach.

## BOTTOM-UP MECHANISM FALL SHORT

The relationship between P and Chlorophyll *a* is pretty compelling on the face of it (Fig. 8.14) and even quite good for an ecological study. However, the x- and y-axes scales are log-transformed. Thus lakes with similar phosphorous concentrations might vary in primary productivity by two or three orders of magnitude. Combined with these observations, the practical limitations of nutrient modeling to predict primary production, and the knowledge that predator–prey relationships can determine population dynamics, it is no wonder that Limnologists would evaluate additional mechanisms, e.g., top-down controls, as a control of primary productivity.

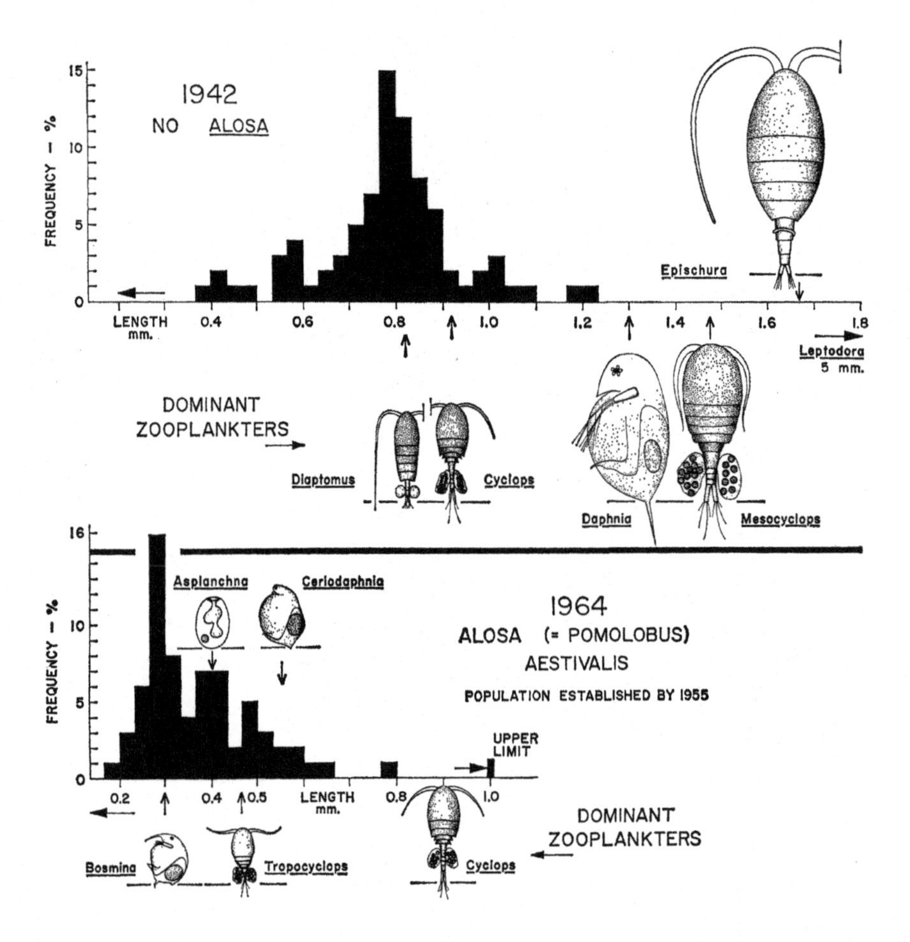


**FIGURE 8.15**

The composition of zooplankton of Crystal Lake, Connecticut (Source: Brooks and Dodson, 1965). The dramatic shift in zooplankton size classes after the introduction of *Alosa aestivalis* demonstrates the impact of predators on zooplankton composition and size.

# TOP-DOWN CONTROLS

## TROPHIC CONTROLS AS A THEORETICAL CONCEPT

When considering the trophic interactions in lakes, primary production might be controlled by population reductions (by consumption, grazing, predation, etc.) or released from growth constraints (by nutrient remobilization, reduction of competition, etc.) whereas total nutrient amounts may be constant.

By assuming that top trophic levels exert control for the next level below, we can make a prediction of the entire trophic structure. For example, with the introduction of the blueback herring (*Alosa aestivalis*) in 1955, the zooplankton size-classes dramatically shifted in Crystal Lake (Fig. 8.15). With smaller zooplankton, there are likely to be further changes in grazing and phytoplankton biomass.

**FIGURE 8.16**

Image of mesocosms (Source: Wageningen University & Research). Mesocosm generally have water circulation systems. Water quality, i.e., nutrient concentrations, temperature, pH, etc., can be monitored and controlled. Fish and/or various zooplankton and phytoplankton can be introduced at various densities according to the hypotheses being tested.

Although the concept relies on several key observations (Fig. 8.15), it can be traced to the work of Hairston, Smith, and Slobodkin and their Green World Hypothesis and bringing attention to the role of top-down forces (e.g., predation) and indirect effects in shaping ecological communities.

## EVIDENCE FOR TROPHIC CASCADES

In the late 1970s, Limnologists designed experiments to determine the role of trophic controls. To reduce the sources of variation, a few studies were conducted using mesocosms. Mesocosms are large tanks of water that can be carefully monitored and controlled, and can be replicated to conduct experiments (Fig. 8.16).

Mesocosm studies suggested that an increase zooplankton biomass can stimulate phytoplankton productivity; in addition, intermediate levels of zooplankton biomass is associated with maximum algae production. When zooplanktivorous fish had limited access to zooplankton, zooplankton biomass was limited, but phytoplankton growth was maximized. However, the results are almost impossible to generalize to whole-lake environments, because mesocosms do not have the habitat or species diversity of lakes.

The Trophic Cascade Concept was more fully developed in 1985, when Stephen Carpenter and colleagues proposed that while nutrient inputs determine the potential rate of primary production in a lake, piscivorous and planktivorous fish can cause significant deviations from potential primary production. In support of their hypothesis, they cited a negative correlation between zooplankton size, an indication of grazing intensity, and primary production (Fig. 8.17). The basic Trophic Cascade Theory is that nonadjacent levels in a food web vary together (Fig. 8.18).

## WHOLE LAKE EXPERIMENTS

The number of whole-lake studies to evaluate trophic cascades are limited, but in some cases the response is dramatic. One type of experiment relies on removal of a trophic control and other attempts to introduce a trophic control. Neither approach yields the perfect design, but we have learned something about trophic controls in each case.

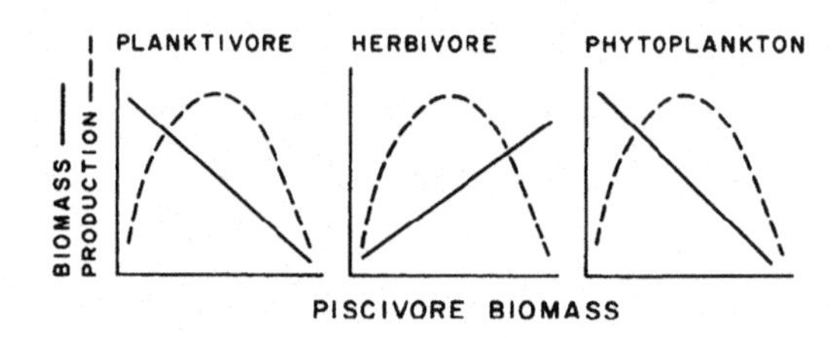


**FIGURE 8.17**

The hypothesized role of lentic piscivores on biomass (Source: Carpenter et al., 1985). Piscivore biomass in relations to biomass (solid line) and production (dashed line) of vertebrate planktivores, large herbivore and phytoplankton.

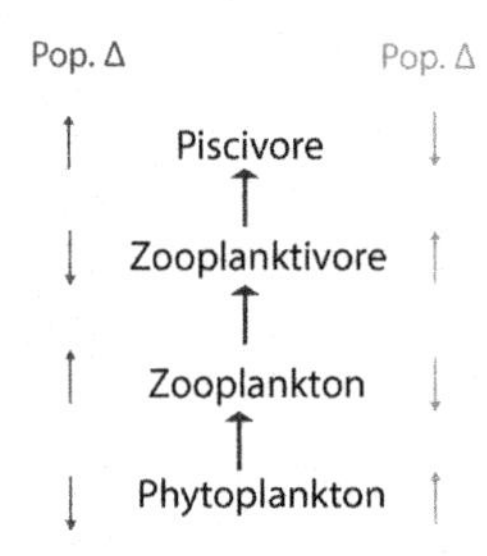


**FIGURE 8.18**

Schematic of the basic trophic cascade theory, where non-adjacent levels in a food web are positively correlated.

In the mid 1950s, Poltruba Lake, Czech Republic was poisoned with rotenone, which killed the fish zooplantivores. As a result the zooplankton size increased and there was a dramatic decline in phytoplankton biomass. This study indicated that fish can control the zooplankton and change the amount of phytoplankton grazing.

Another approach relied on the introduction of a fish predator. For example, Carpenter tested their trophic cascade model by manipulating the fish communities in two lakes and using a third lake as a control. Two of the lakes contained substantial populations of largemouth bass (*Micropterus salmoides*). A third lake had no bass, due to occasional winterkill, but contained an abundance of planktivorous minnows, an assemblage of species, e.g., redbelly dace (*Phoxinus eos*), finescale dace (*Phoxinus neogaeus*), and central mudminnows (*Umbra limit*). They removed 90% of the bass from one lake and put them into the other. They simultaneously removed 90% of the planktivorous minnows from the second lake and introduced them to the first. They left the third lake unmanipulated as a control.

Adding planktivorous minnows produced a complex ecological response. Increasing the planktivorous fish population led to increased rates of primary production. However, though the researchers increased the population of planktivorous fish in this experimental lake, they did so in an unintended way. Despite their best efforts, a few bass remained. So, by introducing a large number of minnows, prey for bass, they fed the remaining bass. An increased food supply combined with reduced population density induced a strong numerical response by the bass population. The manipulation increased the reproductive rate of the remaining bass 50-fold, producing an abundance of young bass that fed voraciously on zooplankton.

In summary, the piscivores, such as bass, feed on planktivorous fish and invertebrates. Because of their influence on planktivorous fish, bass indirectly affect populations of zooplankton. By reducing populations of planktivorous fish, bass reduce predation rates on zooplankton. Large-bodied zooplankton, the preferred prey of size-selective planktivorous fish dominated the zooplankton community. The large zooplankton reduced phytoplankton biomass and the rate of primary production. This interpretation was consistent with the negative correlation between zooplankton body size and primary production reported by Carpenter and his research team.

## TROPHIC CASCADE EXAMPLES

Trophic Cascades have been proposed as a way to explain top-down, indirect effects of food web populations dynamics. Because of the concept's simplicity, as well as its possible application in controlling unwanted environmental outcomes, it has gained popularity as a theory and a potential management strategy.

The following are selected examples of Trophic Cascades:

- Piscivorous fish can dramatically reduce populations of zooplanktivorous fish, zooplanktivorous fish can dramatically alter freshwater zooplankton communities, and zooplankton grazing can in turn have large impacts on phytoplankton communities. Removal of piscivorous fish can change lake water from clear to green by allowing phytoplankton to flourish.
- In the Eel River, in Northern California, fish (steelhead and roach) consume fish larvae and predatory insects. These smaller predators prey on midge larvae, which feed on algae. Removal of the larger fish increases the abundance of algae.
- In the Pacific kelp forests, sea otters feed on sea urchins. In areas where sea otters have been hunted to extinction, sea urchins increase in abundance and decimate kelp.
- Wolves in Yellowstone predate deer, but the deer respond by staying in the riparian zone. By being constrained to the riparian zone, the deer browse a narrow band of vegetation and dramatically improve the quality of the stream.

## ECOLOGICAL APPLICATIONS

With the success of the reintroduction of the wolf, some might argue that the model to introduce a top predator might be a good strategy to reduce harmful algal blooms. For example, what if phytoplankton blooms in eutrophic nutrient-rich lakes be more easily controlled by adding trophic levels rather than controlling the source of the nutrient addition? Controlling eutrophication with addition of predatory fish is seductively cheaper compared to the costs to reduce nutrient loading. Although some have seen potential to apply Trophic Cascades, where they can be sustained, the introduction of a new predator is unlikely to have simple trophic impacts that reduce algal blooms. Because of the complexity of lake food webs and risk of unintentional consequences, any recommendation to introduce new species as a way to improve water quality should be viewed with suspicion.

## COMPENSATING AND CONFOUNDING FACTORS

Dave Wetzel was not impressed by the terminology or concept of Trophic Cascades and joined the criticisms that the theory may be an oversimplification or just plain wrong. For example, Wetzel (2001, pp. 464–466, emphasis added) criticized the concept and complained that

> *a considerable number of studies drew conclusions that were unsupported and proposed quite imaginary coupling, all of which persists as an embarrassment to limnology. Among the worst were repeated studies that claimed, without data, that differential predations (unmeasured) was inducing differential grazing (unmeasured) that was causing major nutrient regeneration (unmeasured) and inducing major alteration of photosynthetic productivity and phytoplankton community structure (unmeasured).*

Much depends on what is being measured and how. In this next section, we will explore some of these issues.

Wetzel cited two main issues with the Trophic Cascade Model. First, there are numerous other factors that control productivity in Lentic Systems. Second, to maintain these controls, the systems need to be continuously disturbed.

***Omnivory–Modeling Nightmare*** Trophic feeding giving rise to group fidelity is probably more of a figment our imagination than a static entity. Categories, such as primary producer, primary consumer, plantivore and piscivores help us categorized taxa, but these groups are quite heterogeneous. The utility of the trophic-level concept breaks down when food web interactions change with life-history stage or body size (Fig. 8.19). Since more than 80% of the earth's animal taxa undergo metamorphosis and over 75% of all vertebrate taxa are size-structured, it seems that the generalizability of trophic levels is questionable.

The presence of size structure means that the intensity of specific processes (competition, predation, etc.) changes over time due to the size-dependence of these processes. For example, a competitive interaction could shift to a predatory-prey with size class changes. Ontogenetic shifts in habitat use over the lifecycle also constitute a major reason behind habitat couplings, which can also dramatically shift the species-species interactions.

***Consumer-Mediated Nutrient Regeneration*** Early research using plastic bag filled with filtered lake water and various combination zooplankton and fish suggested that Trophic Cascades might be caused by nutrient dynamics. For example, using nutrient-permeable chambers in the enclosures (Fig. 8.20), phytoplankton productivity was mediated by consumers, regardless of the grazing pressure.

The importance of nutrient recycling by fish and zooplankton is likely to vary greatly among lakes. Although these results support Trophic Cascades, the mechanism suggested here is indirect and via bottom-up nutrient dynamics, and undermining the dichotomy between top-down and bottom-up control.

***Lake Morphometry*** Another factor complicating the trophic theory is the influence of lake morphometry. For example, when lakes are studied that have different depths, stratification may influence the outcome. A thermocline provides refuge for the zooplankton when fish are restricted to the epilimnion. By forcing the fish to the epilimnion, a higher ratio of zooplankton to fish biomass in the deeper lake occurs, creating a weaker positive indirect effect between photoplankton to fish biomass (Fig. 8.21).

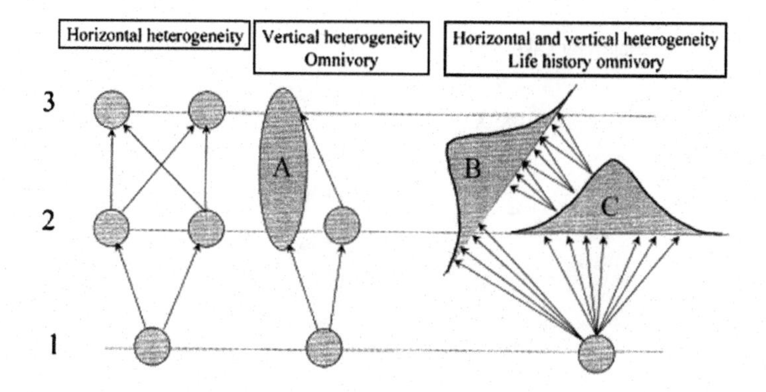


**FIGURE 8.19**

Persson (1999) proposed three types of heterogeneities in interaction food webs that will give rise to community patterns not expected from a trophic-level approach. In the left case, several species compete at each trophic level and may confound clear identification of the controlling species. In the middle case, omnivory prevents a clear trophic control. In the right case, size variation may confound trophic prey sources and influence changes in omnivory with life history (Source: Persson, 1999).

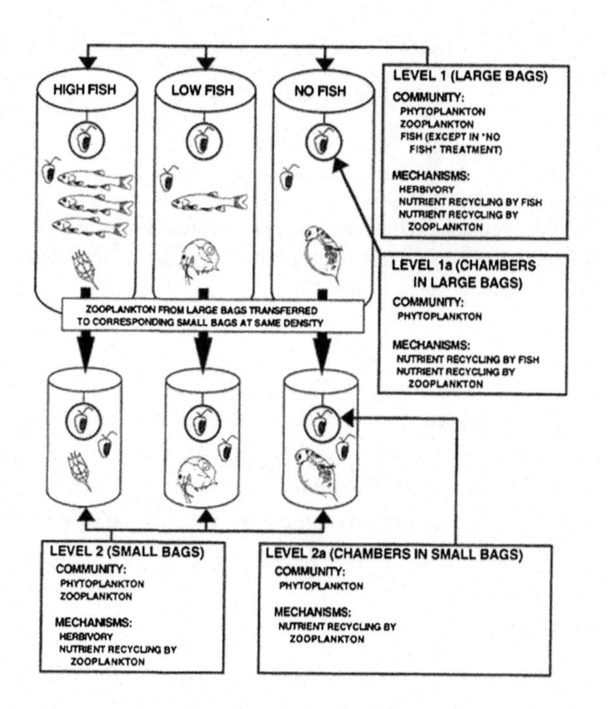


**FIGURE 8.20**

Schematic of in-lake enclosure experiment designed to test trophic cascades (Source: Vanni and Layne, 1997). Container volumes were 2450 L for large bags, 60 L for small bags, and 60 mL for chambers.

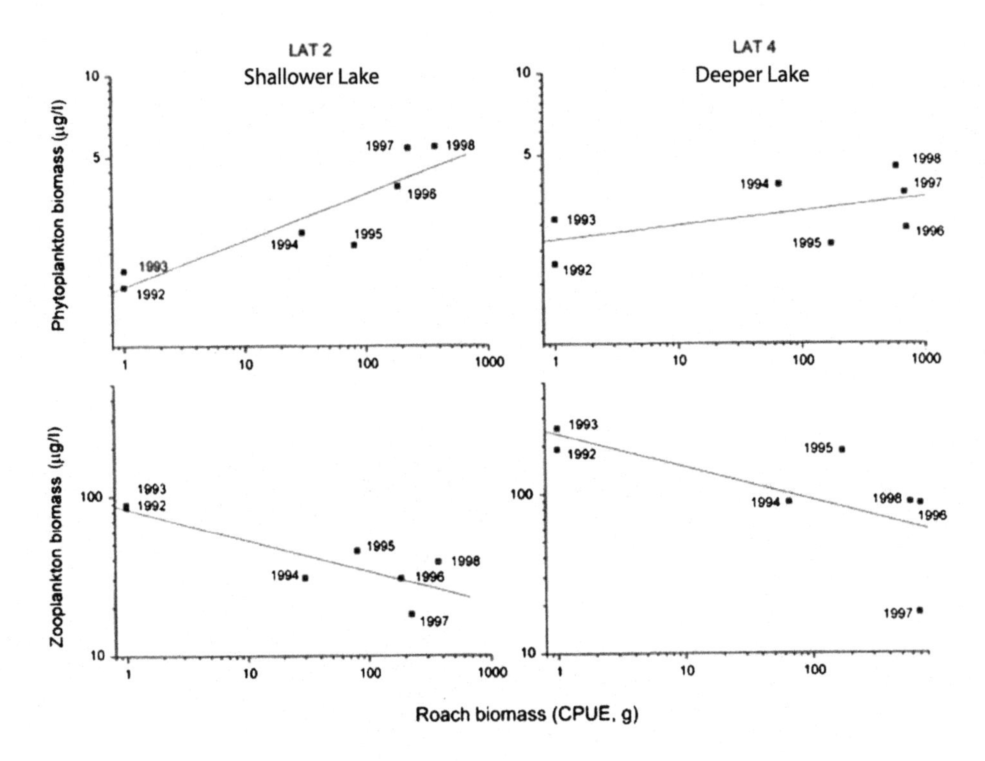


**FIGURE 8.21**

Morphometric differences between treatment of lakes resulted in a different strength of relationship between roach (*Rutilus rutilus*) biomass and both zooplankton biomass. Zooplankton found refuge from both roach and perch (*Perca fluviatilis*) YOY predation in the cooler water of the deeper lake (Lat 4), thus exerting a weaker trophic control (Source: Persson et al., 2004).

***Hydrology*** In many lake systems, the retention time (water flushing rate) has been an important predictor of eutrophication in lakes. Nutrient limitations, flow rates, depth, and biodiversity can interact to play a role in lake phytoplankton blooms, especially as loading varies temporally.

***Role of Macrophytes*** Macrophytes provide habitat and refugia for grazers. With protection from secondary consumers, grazers can be effective in controlling primary productivity, thus limiting the strength of the interactions between trophic levels.

## TROPHIC CASCADE LEGACY

As early as 1992 Ecologists were making strong arguments criticizing Trophic Cascades, including subtleties in the model, and limiting its scope. By 1996, some argued that over half of the peer reviewed articles testing the Trophic Cascades obtained results that did not support the concept.

As we complete the discussion about Trophic Cascades, we need to return to the role of microbes, which seem to mediate between the top-down and bottom-up forces.

## MICROBIAL LOOP AND HETEROTROPHIC METABOLISM

The Microbial Loop describes a trophic pathway where DOC is returned to higher trophic levels via its incorporation into bacterial biomass, and then coupled with phytoplankton-zooplankton productivity. Although the term was proposed for marine systems, the process has also been found in inland waters, e.g., Castle Lake.

In general, DOC is released from bacterial lysis, the leakage or exudation of fixed carbon from phytoplankton (e.g., mucilaginous exopolymer from diatoms), sudden cell senescence, sloppy feeding by zooplankton, the excretion of waste products by aquatic animals, or the breakdown or dissolution of organic particles from terrestrial plants and soils. Bacteria in the Microbial Loop decompose this particulate detritus to utilize this energy-rich matter for growth. As bacteria introduce organic carbon into the food web, additional energy becomes available to higher trophic levels. In addition, based on the results from Castle Lake, bacterioplankton can be stimulated by P, as opposed to labile DOC. If the bacterial demand for P is greater than the algal demand for then bacterioplankton may provide a key control on primary productivity.

In summary, the patterns and processes of inland waters depend on a wide range of dynamic controls that include nutrient availability, top-down controls, and microbial processes. It is probably not the primacy of any one of these that determines aquatic productivity, but a combination of each that vary with time and space.

## NEXT STEPS

### CHAPTER STUDY QUESTIONS

1. Describe how individual tolerances might influence Lotic Systems.
2. How do reproductive strategies define community dynamics in vernal pools?
3. Summarize how trophic dynamics influence lentic community dynamics.
4. Summarize the differences between top-down versus bottom-up controls in Aquatic Systems.

PART

# 3

# WATER IN THE ANTHROPOCENE

CHAPTER

# DEVELOPING AND APPROPRIATING WATER

*When the well is dry, we know the worth of water.*
**Benjamin Franklin (1706–1790), Poor Richard's Almanac, 1746**

## CONTENTS



Ecology and Management of Inland Waters. https://doi.org/10.1016/B978-0-12-814266-0.00023-4

The Salton Sea was created accidentally by zealous developers and overly confident engineers, who failed to appreciate the hydrology and sediment load of the Colorado River. After one attempt and bankruptcy, the California Development Company was incorporated in 1896 to divert the Colorado River water into the Salton Sink, a dry lake bed near the Coachella and Imperial valleys. Hoping to turn the desert green with agricultural fields, the Imperial Canal was completed by 1901, but filled with silt at alarming rates, especially near the canal's start at the Colorado River diversion.

The poorly conceived project became a disaster, when winter flooding in 1905 destroyed the diversion structures (Fig. 9.1). The whole of the Colorado River poured into the Salton Sink, forming the Salton Sea for two years until the canal breach was mended. Without this source of water the size Salton Sea has declined over the year, but remains the largest lake in area in California.

Historically, over millions of years, the lower Colorado River has consistently changed it course building the terrain along the California and Arizona and Mexico boarder region. For thousands of years, the river flowed into and out of the Imperial Valley alternately, creating a freshwater lake, an increasingly saline lake, and a dry desert basin, depending on river flows and the balance between inflow and evaporative loss. The cycle of filling has been about every 400–500 years and has repeated itself many times. The latest natural cycle occurred around 1600–1700 as recounted by Native Americans, while the Colorado River overflowed into the Alamo River in 1884, 1891, 1892, and 1895. Fish traps still exist at many locations, and the Native Americans evidently moved the traps depending upon the cycle.

The idea to bring irrigation water into the Salton Sink could be seen part of the geomorphologic pattern of the region, but with the dams and withdrawals along the Colorado and Gila rivers has turned the riverine outlet into the Sea of Cortez into a desolate landscape of dried channels, salt pans, and degraded habitat. Before we understood the ecological relationship between inland waters and marine waters, water draining into the sea was considered a wasted resource. Thus, diversions to supply irrigation and drinking was an accepted tradition, and now maintaining reliable supplies must be negotiated across economic development agendas, efforts to promote social well-being as various human population sizes approach their transitions, and ecological values and services.

In the context of a highly variable spatial and temporal pattern of rainfall, water is unevenly available. Thus human beings build infrastructure to store and move these waters as part of our development. Therefore, projects to "develop" water have direct links to economic activity and political power. As a result, changes to the water cycle have profound implications to the sustainable use of water that considers a broad range of uses and values. After reading this chapter, you should be able to

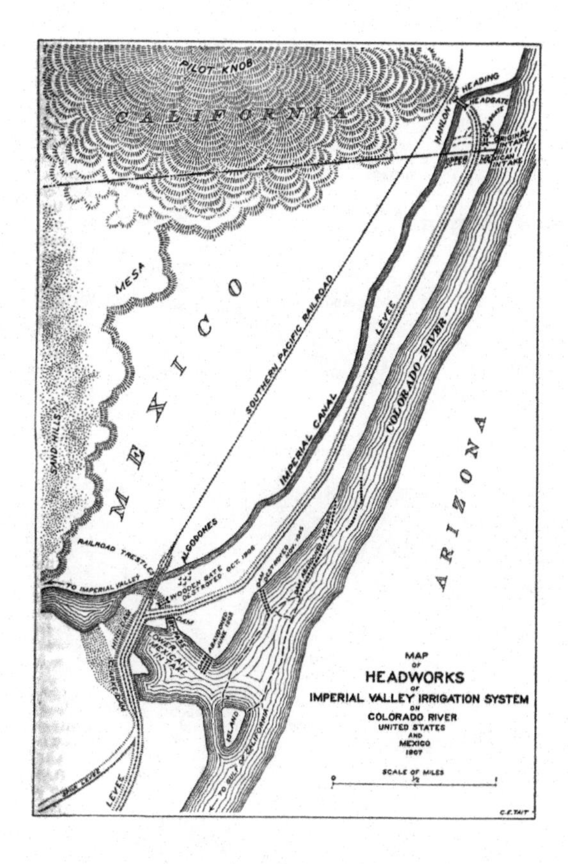


**FIGURE 9.1**

Map of headworks of the Imperial Valley irrigation system, dated 1907 (Source: Bayley, 1995).

1. Describe how water management goals can be consistent with the Flood Pulse Concept;
2. Describe how ground-surface water interactions influence water supplies;
3. Summarize how Aquatic Ecosystems are modified by regulated rivers;
4. Evaluate how environmental flows might be used to improve ecological outcomes, and
5. Apply watershed management concepts to improve water supply management goals.

# FLOOD PULSE CONCEPT

## THE FLOOD PLAIN AS AN AQUATIC ECOSYSTEM

The Flood Pulse Concept stipulates that the annual flood pulse drives productivity, community dynamics, and biodiversity in river's ecosystem. It contrasts with the view that floods are catastrophic events—although they certainly can be. The Flood Pulse Concept describes the dynamic interaction between a river ecosystem and the Transition Zone between water and land (Fig. 9.2). And this Tran-

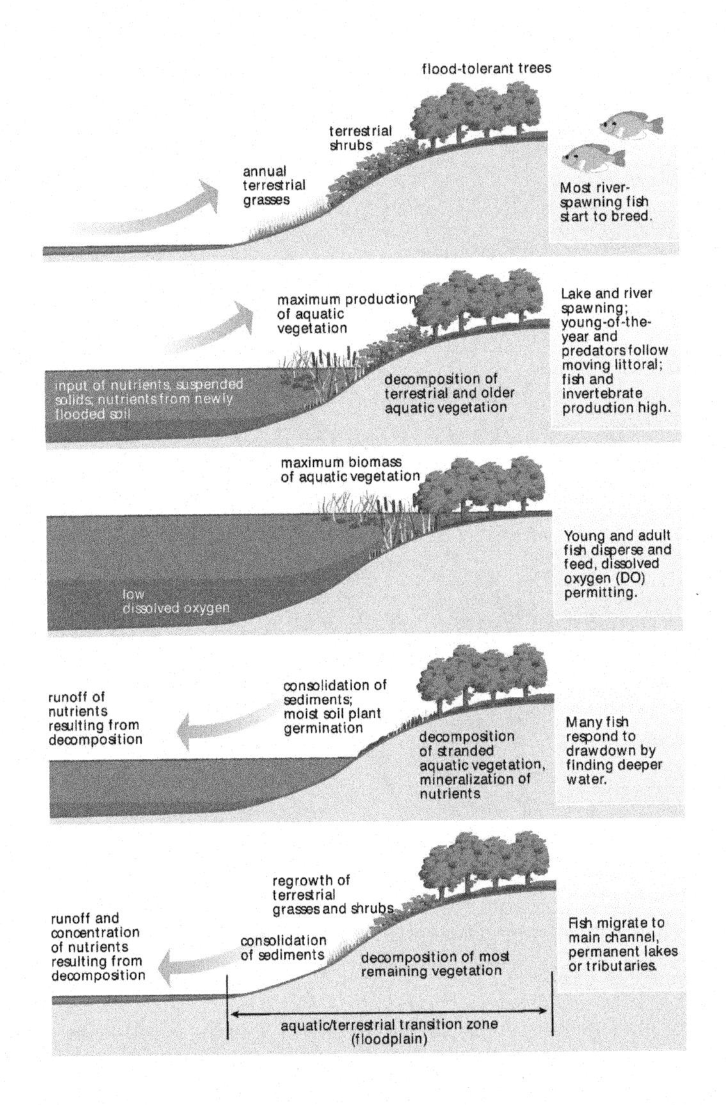


**FIGURE 9.2**

Diagram of the Flood Pulse Concept (Source: Bayley, 1995).

sition Zone is also where human activities are concentrated and in tension with Aquatic Ecosystems. This tension with floods drives humans to channelize waters and build impoundments in an effort to protect themselves and managed water availability.

River Flood Plain Systems consist of an area surrounding a river that is periodically flooded by the overflow of the river as well as by precipitation, called the Aquatic/Terrestrial Transition Zone (ATTZ). The ATTZ is the area covered by water only during the flooding. This flooding in turn is essential habitat for many species. The Flood Pulse Concept is unique because it incorporates the outlying rivers

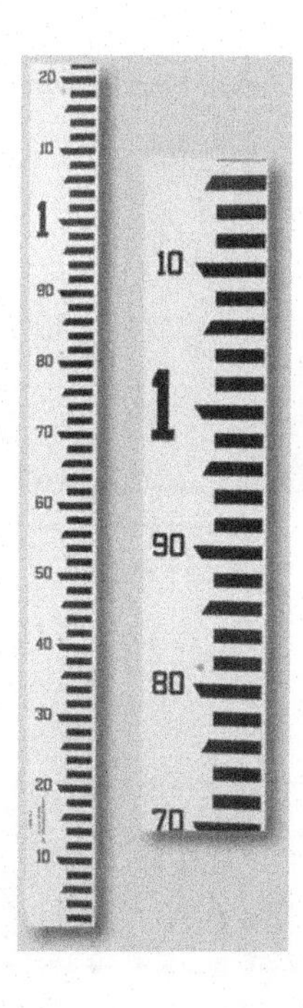


**FIGURE 9.3**

Examples of staff plates used to measure stream stages (height). With manual discharge measurements, hydrologists can create a rating curve that can be used to estimate flow within range of discharges measured.

and streams, which add a lateral aspect to previous concepts. From this lateral perspective, rivers can be seen as a collection of Width-based Water Systems.

Rivers carry water and sediment (eroded soil particles) from upstream to downstream locations. On the island of Borneo, the volume of riverine flow depends on precipitation with a total rainfall between 1.5 and 4.8 meters per year. With a pronounced dry and wet season, rainfall is ~80 mm/month in the dry season and 2000 mm/month in the wet season. Thus the river's discharge can vary dramatically based on the amount of precipitation. When the discharge exceeds the capacity of a river's channel, water overtops banks and floods adjacent areas, i.e., the floodplain. Floods are a natural part of Borneo's ecological functioning, as described by observations and indicated by settlements of indigenous peoples. With access to the floodplain, the overall river velocity is reduced, fish have access to floodplain resources, and floodplain vegetation receive inputs carried by the water.

## THE FLOOD PROCESS AND FLOOD STAGE

Flood stage refers to the height of a river (or any other body of water) above a locally defined elevation. This locally defined elevation is a reference level, often referred to as datum. At gaged locations, we can often find staff plates that might be used for measuring the stage of a stream (Fig. 9.3). Although staff plates are used to estimate discharge with discharge curves and develop hygrographs (Chapter 6,

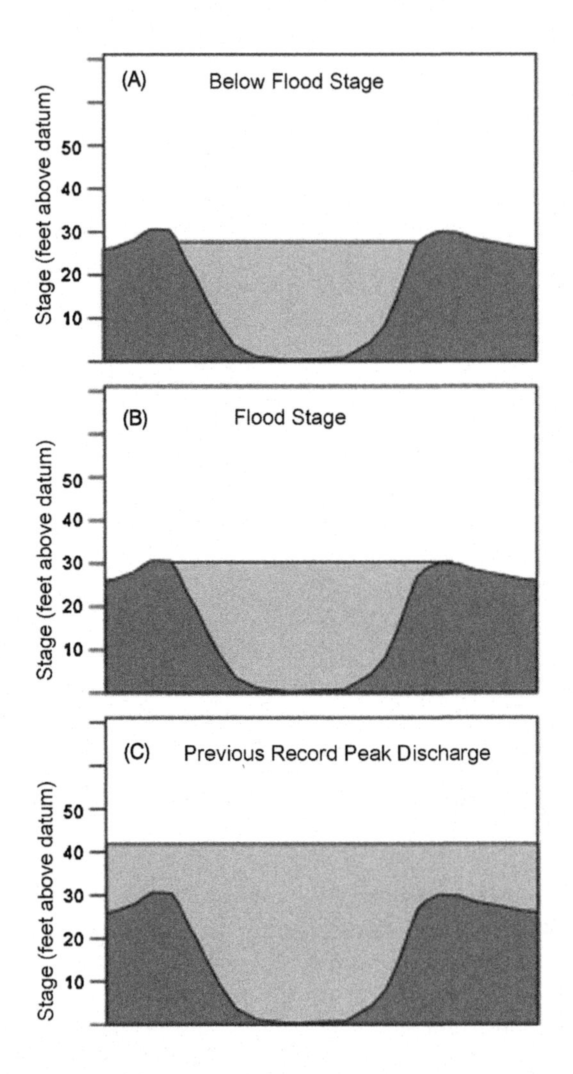


**FIGURE 9.4**

Flood stage and discharge.

section "Moving Water and Moving in Water"), in this case, stage readings indicates if a stream will overtop its banks and cause flooding (Fig. 9.4).

First, at the start of the flooding, nutrients rush in from the area where the flood begins. During flood periods, the most important element is called the moving littoral. As flooding begins and water levels increase nutrients that have been mineralized in the dry phase are suspended with sediments in the floodwaters and main river. The moving littoral consists of the water from the shoreline to a few meters deep in the river. This pulse of water is the primary driver of high productivity and decomposition rates as it moves nutrients in and out of the floodplain and is good breeding ground for many species of

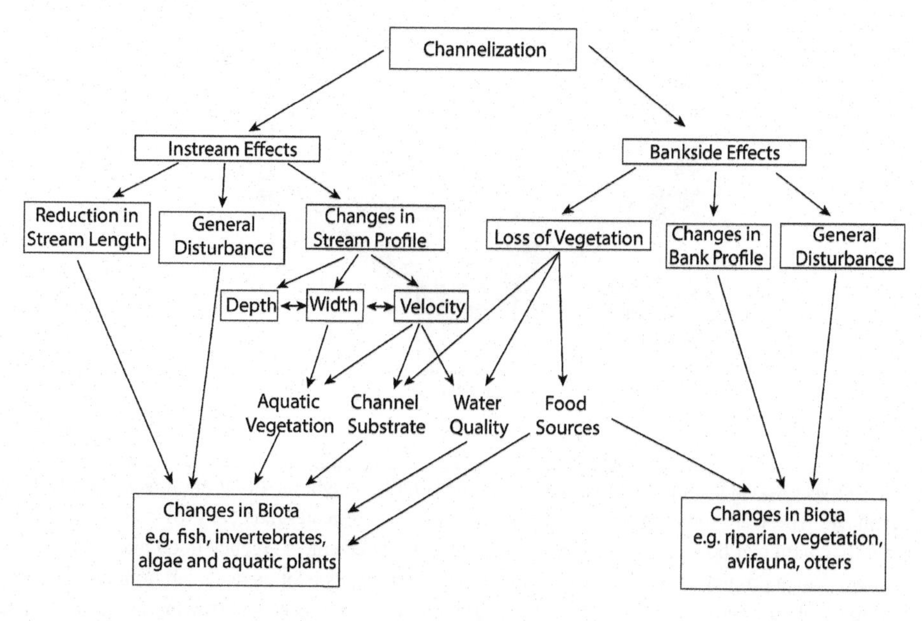


**FIGURE 9.5**

Ecological impacts of channelization (modified from Brooker, 1985).

estuarine organisms. At this point in time production rates exceed decomposition rates. As water levels stabilize, decomposition rates outpace production rates, frequently contributing to dissolved oxygen deficiency. When the water starts receding, the moving littoral reverses, concentrating nutrients and contributing to phytoplankton growth.

## CHANNELIZATION: ATTEMPTS TO MANAGE THE CHANNEL

We can protect floodplain activities by increasing the size of the banks with levies and channeling waters, but these practices have numerous ecological effects (Fig. 9.5). Channel straightening leads to an immediate increase in bed gradient and in a natural deepening and widening of the channel. Losses of river channel habitat resulting from straightening can be substantial— the loss of over 40% in low gradient floodplains are common.

With a goal to increase the flow-carrying capacity of a river channel generally, flow velocity will increase, which has direct ecological implications because many aquatic organism habitats are partially defined by velocity. In addition to local increases in discharge rates from channelization, downstream flooding risks may be increase, thus effectively displacing the area of disturbance downstream.

Natural channels supports higher densities of fish, greater biomass, and more species, often due to the loss of in-stream and riparian cover. Channelization often devastates bankside tree and ground cover; a few objective studies address such effects. Macroinvertebrates communities are dramatically altered, coupled with a severe reduction in habitat value and geomorphical habitat (e.g., riffles, pools, vegetation). Channelization has a direct impact on riverine birds, in particular based on the reduction of areas available to breeding and wintering wetland birds. Finally, as implied throughout this section,

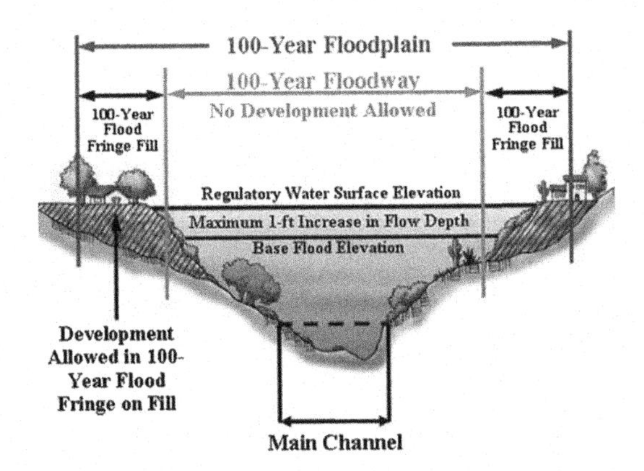


**FIGURE 9.6**

Diagram of a 100-year flood (Source: FEMA). When a home is built, bought or sold it often receives a FEMA flood elevation certificate. These are sometimes known as flood elevation certificates, and are provided to show the risk of flooding for that particular home. Whereas Federal Emergence Management Agency flood maps are used to determine whether or not a property is in a FEMA-designated flood plain, its is often necessary to hire a professional land surveyor to fill out the certificate. These certificates are required if a person is borrowing money to buy a home.

the floodplain is cut off from the natural processes driven by a stream hydrograph. Thus not only has the ecological processes of the stream been altered, but the floodplain as well.

## THE ONE HUNDRED YEAR FLOOD AND ITS DISCONTENTS

Even with all the infrastructure—levees and dams—flooding still seems to be a regular event. Perhaps, you have experiences of flooding in your life time or certainly heard about floods on the news. Often they are described as extreme events, for example, described as a '100-year flood'. When linked to the stage (height of the stream), the implication is that this event only happens once in 100 years.

However, the terminology can be misunderstood.

For Hydrologists, the term "100-year flood" is better described as a flood having a 100-year recurrence interval or a probability of once per 100 years. In other words, a flood of that magnitude has a 1 percent chance of happening in any year.

What is a recurrence interval? Using a statistical technique, a process called Frequency Analysis, the probability of the occurrence of a given discharge event is estimated. The recurrence interval is based on the probability that the given event will be equaled or exceeded in any given year. For example, assume there is a 1 in 100 chance that a streamflow of 900 cubic meters per second ($m^3/s$) will occur during any year at a certain site. Thus a peak flow of 900 $m^3/s$ at the site is said to have a 100-year recurrence interval. Flood recurrence intervals are based on the magnitude of the annual peak flow (Fig. 9.6).

Ten or more years of data are required to perform a Frequency Analysis for the determination of recurrence intervals. Of course, with more records an analysis will improve the estimates and confidence in the estimate.

Recurrence intervals for the annual peak streamflow at a given location change if there are significant changes in the flow patterns at that location, possibly caused by an impoundment or diversion of flow. The effects of development (conversion of land from forested or agricultural uses to commercial, residential, or industrial uses) on peak flows is generally much greater for low-recurrence interval floods than for high-recurrence interval floods, such as 25-, 50-, or 100-year floods. During these larger floods, the soil is saturated and does not have the capacity to absorb additional rainfall. Under these conditions, essentially all of the rain that falls, whether on paved surfaces or on saturated soil, runs off and becomes streamflow.

## FLOODPLAIN DEVELOPMENT

When water supplies and flood risks are reduced as a result of the construction of upstream impoundments, few think to implement growth restrictions in the urbanizing basin. Ironically, flooding regularly plagues these developed areas as most Houston residents have long known even before Hurricane Rita arrived (Fig. 9.7). Even the city of Los Angeles regularly floods during intense rainfall events. Before being covered by impermeable surfaces, the deep alluvial deposits throughout the Los Angeles River encouraged significant runoff infiltration and groundwater recharge. Most of the year, stream channels experienced low flow levels, whereas the stream continued below ground. With long and intense winter precipitation events, these shallow channels are suddenly transformed to torrents. Even more perverse, with the impermeable surfaces, water no longer infiltrates to recharge groundwater levels and instead it becomes a flood risk; this is a ubiquitous characteristic of streams in the west.

Building dams can be used to limit peak flows and store water for later uses. Deciding whether the motivation to build dams as flood control infrastructure is a response to protect or to promote the development of the floodplain is not easy to answer. However, the impact on a river is the same: dams dramatically alter a river's hydrograph; the magnitude and variation of flow is dramatically altered.

In spite of a vast literature that promotes better management of lands subject to floods and the reduction, if not the prevention of flood damages, we continue to observe people practicing the opposite. Annual expected flood damages continue to increase in the United States, despite increased investments in flood damage protection and mitigation. Housing is built in the floodplains along rivers, or on coastal lands bordering seas and oceans. Political acquiescence allows this even at the risk of being flooded. The question is not if these developments will be flooded. Rather it is a question of when, and how severely.

According to the US National Center for Atmospheric Research, between 1970 and 1995 inland floods killed more than 318,000 people and left more than 81 million homeless around the world. Between 1991–95 flood-related damage exceeded $200 billion, which is about 40% of all the economic damage from natural disasters during this period. Over the past 25 years the federal government has spent 140 billion USD to prepare for and recover from natural disasters. Over 70% of these disasters involved flooding (Fig. 9.8). Even more striking is that the cost of annual flood losses in the US has more than doubled since 1900 in inflation adjusted dollars.

## FLOOD CONTROL: MIXED BLESSINGS

As described above, natural floodplains have a close relationship to the flood waters of a stream. Without these floods, nutrient cycling, primary productivity, fish habitat, and biodiversity might all

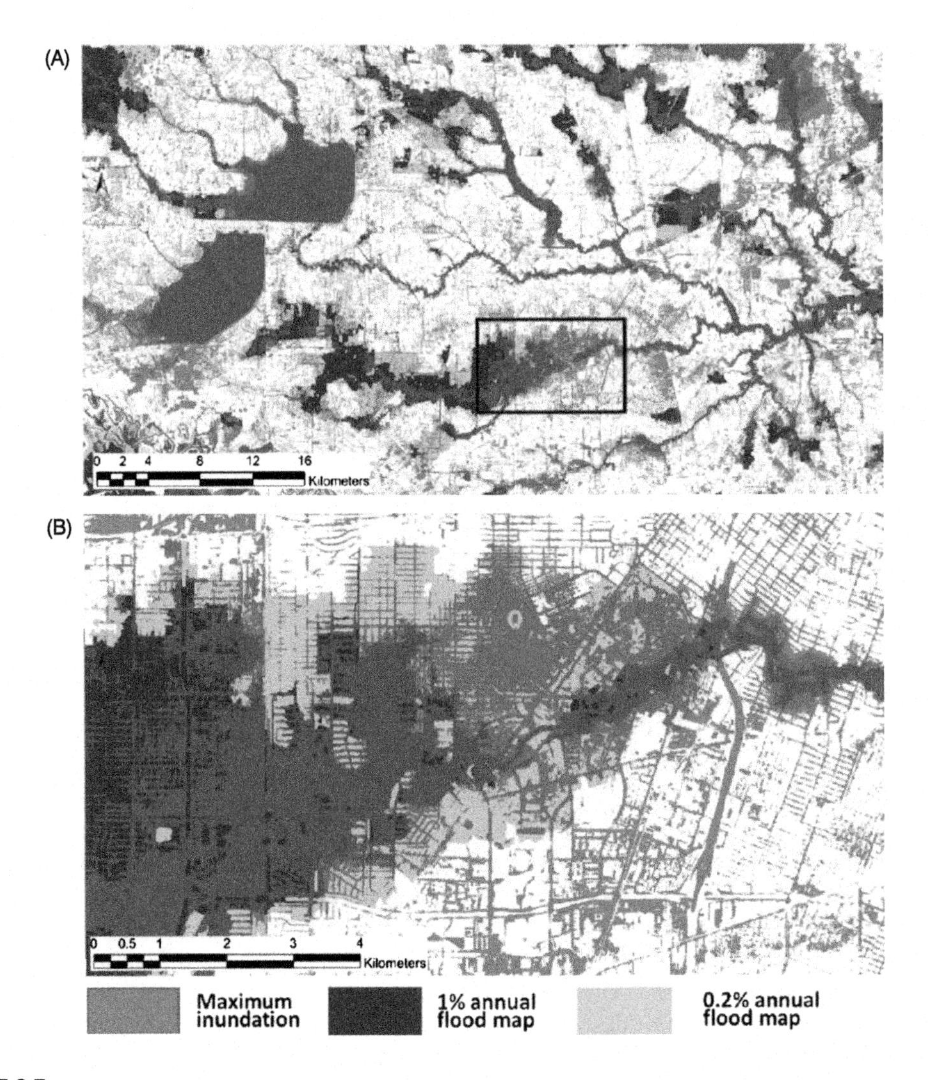


**FIGURE 9.7**

A comparison between the simulated maximum inundation extent (red, gray in print version) and the 1-percent (dark blue, dark gray in print version)/0.2-percent (light blue, light gray in print version) annual-chance flood maps. (A) The entire domain. (B) The enlarged domain marked in (A). (Source: Noh et al., 2019).

be impacted. With flood control channels, the velocity of water changes how aquatic organisms navigate streams—perhaps, finding it difficult to find refuge during flood events. Hyperheic exchange is a key process in streams. Flood control projects limit that exchange, which can be important to habitat and also water quality protection, e.g., temperature and nutrient cycling.

Finally, as a justification for most dams and channelized streams, the flood risk for downstream residents is usually reduced, but it is important to acknowledge that most western societies depend

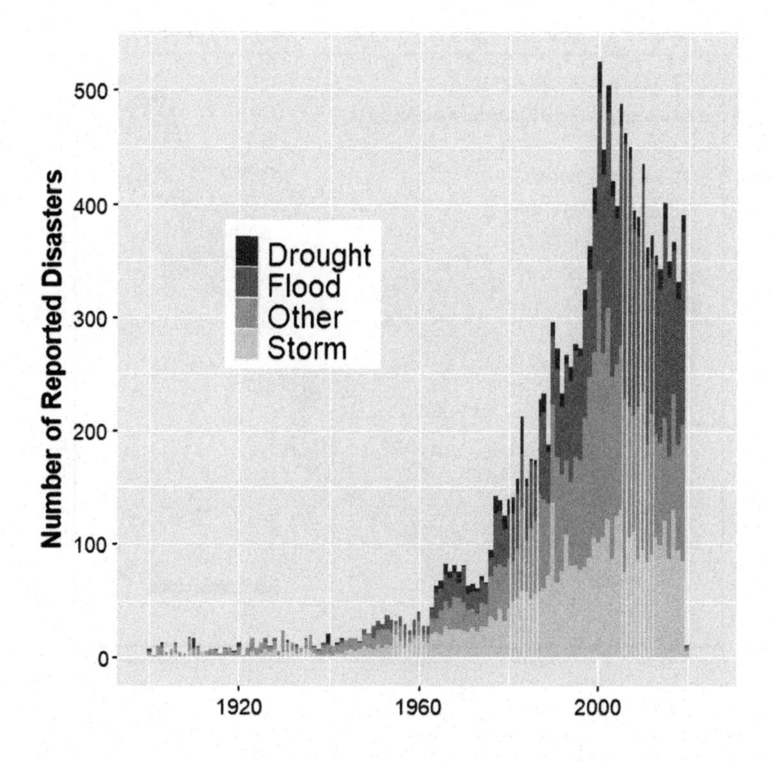


**FIGURE 9.8**

Number of natural disasters from 1900 to 2009 classified in geological (earthquakes and volcanic eruptions), flood, storm, mass movement, drought (including extreme temperature and wild fire), and biological (epidemics, insect infestations) triggers (Source: CRED 2010. EM-DAT – The OFDA/CRED International Disaster Database, Database Advanced Search, Centre for Research on the Epidemiology of Disasters. Université Catholique de Louvain, Brussels, Belgium. http://www.emdat.be/Database/AdvanceSearch/advsearch.php).

on property value to moderate and stabilize economic growth. The initial development of flood-prone land was a transfer of wealth to the landed-elite and subsidies by the government to take on flood risks. Once the floodplain is developed, the loss of property value can decimate the social fabric of a region—thus the Army Corp of Engineers has become the government arm to limit these losses as much as possible.

Of course, the framework ignores the losses of life associated with flooding, so most developed countries respond with investments to prevent repeat outcomes, which seems reasonable. But, as mentioned before, it is accurate—based on experience—to say that by such investments the flood risk is altered.

Once floodplains are no longer subject to flooding, few compelling reasons remain to limit development in these areas. Thus flood-control projects may be growth-inducing into the flood plain, making it nearly impossible to consider (or pay for) substantial restoration should that ever be desired.

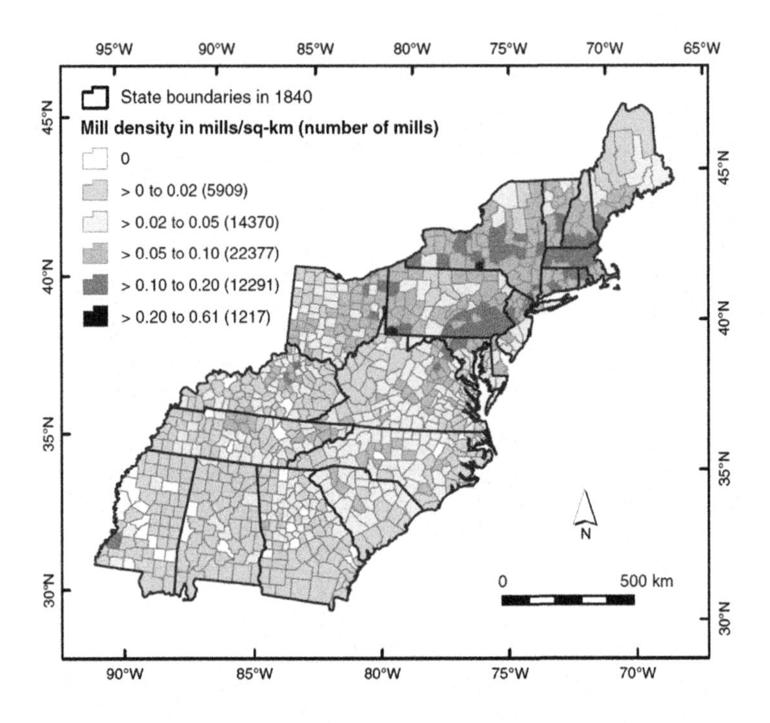


**FIGURE 9.9**

Density of mill ponds along New England (Source: Walter and Merritts, 2008).

## RIVERS AS SOURCES OF TRANSPORTATION AND POWER

Channelization has also been used to constrain a river and make it deeper, thus providing transportation. The Mississippi River is famous for the locks and levees that are used to promote transportation. Designed and built as a way to improve the capacity of farmers in the watershed to export grains and avoid the rail monopolies, this beginning of the system was constructed in the early 20th century. Much of the midwest continues to rely on the system for barge transport of commodities.

Using water as a source of power has been an important part of the industrialization process. Before coal and oil, water's potential energy was often converted to kinetic energy to drive saw and grain mills. For example, along the eastern Piedmont Region of the United States, there was a high density of grain mills to support increasing population within the country (Fig. 9.9).

These mill dams and land use changed New England's streams from a marshy multichannel morphology to today's meandering single-channel form. Compared to the tectonically active streams of the west, these streams tend to have low gradients and did not carry gravel (or even much fine sediment) during the Holocene Period.

However, the formation of gravel bars in the wake of eroding, meandering channels, followed by overbank deposition of fine sediment, was not the natural process of floodplain formation as originally thought. Coupled with deforestation and agricultural practices, an increased sediment supply gradually converted mill ponds to sediment-filled reservoirs. Later, along with dam breaching, has lead to channel incision through postsettlement alluvium and accelerated bank erosion by meandering streams.

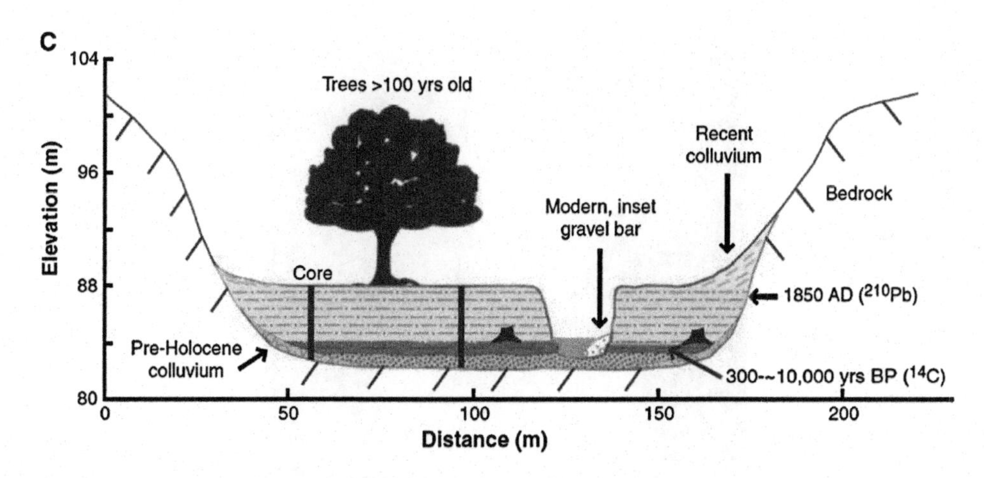


**FIGURE 9.10**

Geomorphology signal of mill ponds (Source: Walter and Merritts, 2008).

Ironically, the history of these disruptions in the Piedmont was not fully appreciated in the interpretation of the stream geomorphology, and some have argued that such a situation has led to dramatic misinterpretations of "natural" stream geomorphology processes, which were then inappropriately used as models for restoration design and goals (Fig. 9.10).

# ECOLOGY OF REGULATED RIVERS AND MAJOR DAMS

## HYDROGRAPHS AND RESERVOIR TYPES

This global impact of dams is hard to understate, given that there are more than 50,000 dams of over 15 meters in height around the world. Whereas dams provide value as a source of national pride or symbols of development, the practical reasons to build dams are quite diverse and include flood control, irrigation, power generation, urban drinking water supplies, recreation or a combination of two or more of these reasons. And whereas some dams are being removed, because their environmental costs are higher than their benefits, other large dams are still planned or under construction.

As one might guess, the hydrograph of the regulated river depends on the dam's purpose. The dams in the lower Colorado River in the western United States release water based on energy demand, which in the Southwest is highest in the summer and afternoons when peak demand powers air conditioning (Fig. 9.11). Other dams store water during the rainy season for use in agricultural irrigation or for drinking water supplies, so water releases might be unseasonably low during the wet season, and relatively high during the dry season.

***Power generating*** Over the past hundred years, successful economic growth relied on 'fungible' power. In other words, an efficient of kinetic energy was to convert it to electrical energy that could be transported and used for many different purposes—in the industry, commerce, and residential sectors. The first hydroelectric dam was built in Appleton, Wisconsin, and began to

**FIGURE 9.11**

Hover Dam (Source: US Bureau of Reclamation).

generate electricity by 1882. The power output was at about 12.5 kW. By 1889 200 hydroelectric power plants have been built in the US.

What are the ecological implications of power generation? As it turns out, many of the reservoirs in the Sierra Nevada Mountains are designed with power generation in mind—if they do not directly generate electricity, they transfer water to power-generating facilities. High-elevation basins in the Sierra Nevada Mountains are responsible for almost 50% of all hydroelectricity generated in California, due largely to their head potential and snow storage. These systems vary in terms of managing utility, storage capacity, conveyance capacity, and altitude.

Two important objectives in the operation of a hydropower system are to generate during periods when demand is high and energy is more valuable, and to minimize unnecessary spilling (water lost without electricity generation). Peak energy demand in the western United States occurs during hot summer afternoon hours, when the demand for air conditioning is high, rather than in the winter. Spills are most recurrent during the reservoir-refilling periods (spring months), but can also occur in winter months.

***Water storage*** Water storage in California is key to ensure water availability for drinking (and power) and irrigation in the summer and fall. By facilitating this high waterflow during the summer drought, effectively adding water in some systems that might otherwise experience low flow often promotes invasive species.

***Flood control*** Floods have plagued the lowlands in California since the first residents made their homes there. In the last one hundred years, levies and impoundments have been constructed to prevent downstream flooding. Of course, in spite of these activities, we still find many areas susceptible to floods each year. Probably the most obvious is the dramatic rise and fall of water levels, and the complete denudation of the shoreline as these reservoirs are at low stages. Because its primary purpose is flood control, Lake Kaweah (Kaweah River, Tulare County, CA) is maintained at a very low level or empty for most of the year, and generally only fills between May and June. Due to the limited capacity of the reservoir, large spills of floodwater often occur after large rain floods, causing damage downstream. Water is generally released as quickly as possible to maintain flood-storage space in the reservoir. During floods in 1997, the reservoir filled and emptied twice because of this operation regime.

Where these processes have been well studied, regulated rivers can dramatically alter ecosystems above and below dams—unfortunately, Ecologists had to study these systems after the dams were built, so when the plans to build 12 dams in Sarawak was proposed in the 1980s, the literature that inform policymakers about the real costs based on the ecological outcomes was quite limited.

Because of its socio-economic visibility, the impact of dams on fish and fisheries were some of the first studies published. In fact, the results of experiments to lift fish above high dams had already been published. By the 1950s, researchers had noted how velocities based on water releases can influence community assemblages. Species diversity can also depend on the amount of flows, where in some examples high species diversity was associated with high flow conditions.

In general, the ecological implications have been evaluated in two ways: the change in the riparian vegetation downstream of the dams and the impact on the watershed integrity as whole.

## FLOODING AND GEOMORPHOLOGY

One of the most common ecological impacts downstream of dams is the lack of flooding. Although touted as a positive for most projects, i.e., lower flood risks for human communities, the impact on floodplain ecology is devastating. Floodwaters do not only deposit sediments and carry nutrients in the forests, but also fish get access to a dramatic increase in habitat area and food items. In some cases, the fish allowed access to the floodplain can grow dramatically faster, which increases their survival rates and fecundity, e.g.in the Consumnes River (Chapter 12).

## GEOMORPHOLOGY AND DAM CAPACITY

With the construction of a dam, sediments fall out of the water column and are stored behind the dam. Each year the reservoir behind the dam loses storage volume with deposition of sediment (Fig. 9.12). Thus the lifespan of any given dam depends on the amount of sediment being deposited from upstream. For the Bakun Dam, the lifespan of the reservoir is approximately 50 years. River water released from the dam are low in suspended sediments and might even be sediment-starved, thus eroding the channel. This might change channel bank formations and river migration, and even downcut the channel (incisement). Thus dams redistribute sediments—blocking them upstream and eroding them downstream.

In the bottom of a reservoir, these trapped sediments are composed of mineral and organic matter. The organic matter is subject to decomposition. Decomposing organic matter consumes oxygen, and oxygen concentrations in the benthos and water column can become zero if the decomposition rates are low relative to the water mixing with the atmosphere. When the sediments have no oxygen or are anaerobic, bacteria in the water or sediments transform nitrate to $N_2$ or $N_2O$, organic carbon to

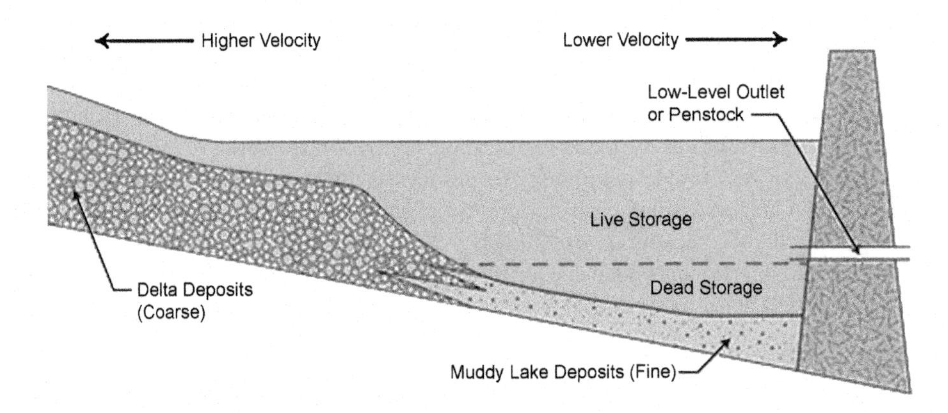


**FIGURE 9.12**

Sedimentation patterns in reservoirs where coarse materials are deposited as river waters enter the reservoirs (adapted from Morris, G.L. and J. Fan).

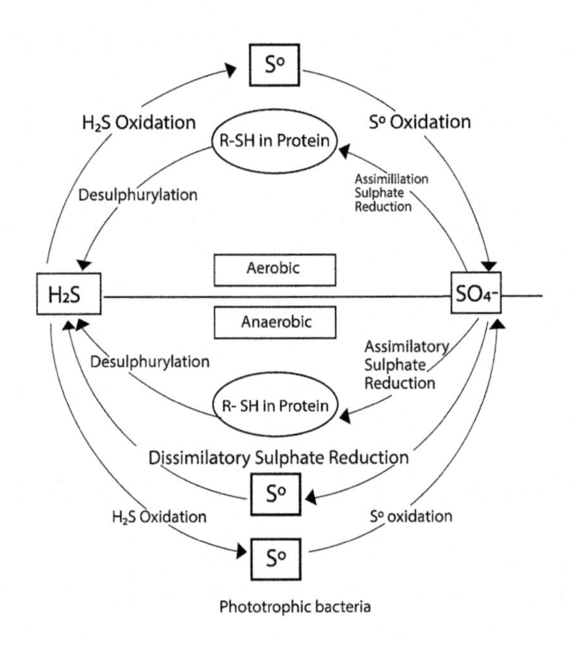


**FIGURE 9.13**

The sulfur cycle depends on the composition of the bacteria and the redox of the environment.

methane, and sulfur compounds to hydrogen sulfide. Methane and nitrous oxide are greenhouse gases and to claim that hydroelectric, especially in the tropics, is a "trivial" emitter of greenhouse gases is patently false. Hydrogen sulfide is toxic to many aquatic organisms (Fig. 9.13).

## DISCONTINUITY IN HYDROLOGICAL PROCESSES

The quality and quantity of water that enters a reservoir differ from the water leaving a reservoir. Downstream water is often cooler, where released water is often taken near the bottom, where cooler water can be found. The oxygen concentrations might be low. Depending on the nutrient status of the reservoir, algae blooms may occur and, in some cases, these blooms may produce toxic compounds, which might also affect aquatic taxa in and below the reservoir.

The San Francisco Bay watershed represents 40% of California's drainage area. As the two major rivers, the San Joaquin and the Sacramento Rivers are major sources of freshwater draining into the bay. From the 1940s to the 1960s, nineteen of the twenty major tributaries draining the western slopes of the Sierra Nevada Mountains (SNM) were dammed. These reservoirs were created to control floods, store water for a wide range of communities, and generate electricity.

The Consumnes River is the only river not impeded by reservoirs, whereas the Mokelumne River, a river just south of the Consumnes has multiple reservoirs. Whereas the Consumnes experiences peak streamflows as a result of precipitation events, the Mokelumne River sustains a relatively homogeneous flow after precipitation events. These disparities influence sedimentation and nutrient concentrations spatially and temporally in the streams.

The reservoirs on the Mokelume, as well as those on other rivers in the Sierra Nevada Mountains, capture sediment and nutrients exported from western slopes of the Sierra Nevada Mountains. Downstream of the dams, the unnaturally stable streamflows increase siltation within the gravel beds that lead to habitat degradation and reduction in biodiversity. With higher TSS, warmer waters, and elevated nitrate concentrations, the downstream waters experience algal blooms downstream, which can then reduce spawning habitat value for migratory fish species.

## BLOCKED MIGRATION ROUTES AND NON-NATIVE SPECIES

Dams are barriers. Dams block the flow of water, but also block the movement of migrating animals. In some cases, fish require migration to reproduce or spawn, but with the presence of a dam, that upstream habitat is no longer available to fish in rivers are not adapted to lake conditions, thus many reservoirs find themselves subject to fish introductions, i.e., nonnative fishes that are adapted to lake conditions. The populations of the migrating fish declines, whereas nonnative fish populations increase. Unfortunately, the ecological impact of dams on migrating fish has not been evaluated in the Sarawak. However, based on examples in the Colorado, Yangtze, and Amazon Rivers, some riverine fish are negatively impacted.

Anadromous American Shad (*Alosa sapidissima*, Fig. 9.14) is native to the East Coast and has been declining due to dam installation without fish ladders, and large commercial shad fisheries. However, installing fish ladders has been a key strategy to improve their changes of recovery (Fig. 9.15). Shad was introduced to the Sacramento River in 1871, and populations ironically began to explode after dam installation and correlated with a decline in native salmonids populations.

## NEW HABITAT AND FISH INTRODUCTIONS

Fish introductions are ubiquitous. But changes have been made to improve the outlook for some natives. For example, Yosemite National Park stopped stocking fish in 1990. Some lakes have reverted to a fishless state naturally after the cessation of fish stocking due to poor spawning habitat. Many lake

**FIGURE 9.14**

American Shad (*Alosa sapidissima*).

**FIGURE 9.15**

Maryland's Conowingo Dam with a fish lift at the center of the image (Source: Wikipedia).

fish population declined naturally without stocking. But nine nonnative fish species exist in Yosemite, including bluegill, smallmouth bass, five trout species, and two trout hybrids. Whereas the removal of nonnative fish may be key to restore aquatic habitat, the task is daunting. The park began an experimental restoration project in 2007 to remove trout from selected lakes that are crucial for the survival of the Sierra Nevada yellow-legged frog. The success of the program remains an active area of monitoring and is documented with periodic public announcements.

In rivers, the nonnative fish removal is nearly impossible, but water management might be used to improve native fish success. For example, restoration efforts in the lower Sacramento River and San Francisco Estuary have focused on increasing native fish and other aquatic species populations by increasing phytoplankton biomass. Based on this research, restoration efforts includes the natural river flood cycle as an important restoration activity.

## GROUNDWATER AND SURFACE WATER CONNECTIONS

### GAINING AND LOSING STREAMS REVISITED

The hyporheic zone is a region beneath and lateral to a stream bed, where there is mixing of shallow groundwater and surface water. The flow dynamics and behavior in this zone (termed hyporheic flow) is recognized to be important for surface water/groundwater interactions, as well as fish spawning, among other processes.

Deep sediments provide aquifers (water bearing geologic formations) for human development, but they represent a long-term dynamic of surface waters leaving the channel to go into the formations called losing reaches, and then at times to come to the surface as springs or channels called gaining reaches. In some cases, streams might be both gaining and losing in different sections, or even change depending on the time of the year. It is ironic that with an increased pavement, less water is capable of infiltrating into the groundwater during storms and more water runs off, which means less water stored in groundwater for drinking and more potential for flooding.

## MANAGED AQUIFER RECHARGE

Managed aquifer recharge is the purposeful recharge of water to aquifers for subsequent recovery or environmental benefit. Aquifers, permeable geological strata that contain water, are replenished naturally by way of rain soaking through soil and rock to the aquifer below, or by infiltration from streams. The human activities which enhance aquifer recharge can be put into three categories as follows:

***Unintentional*** such as through clearing deep-rooted vegetation, by deep seepage under irrigation areas, and by leaks from waterpipes and sewers;

***Unmanaged*** including stormwater drainage wells and sumps, and septic tank leach fields, usually for disposal of unwanted water without thought of reuse;

***Managed*** through mechanisms such as injection wells, and infiltration basins and galleries for rainwater, stormwater, reclaimed water, mains water and water from other aquifers that is subsequently recovered for all types of uses.

Note that there are opportunities to convert from unmanaged recharge to managed recharge with the aim of protecting the environment and using the recovered water.

Enhancing natural rates of groundwater recharge via MAR provides an important potential source of water for urban and rural Australia. This paper addresses all forms of MAR, but the emphasis is on urban applications. MAR can be used to store water from various sources, such as stormwater, reclaimed water, mains water, desalinated seawater, rainwater, or even groundwater from other aquifers. With appropriate pre-treatment before recharge and sometimes post-treatment on recovery of the water, MAR may be used for drinking water supplies, industrial water, irrigation, toilet flushing, and sustaining ecosystems.

Common reasons for using MAR include the following:

- securing and enhancing water supplies,
- improving groundwater quality,
- preventing salt water from intruding into coastal aquifers,
- reducing evaporation of stored water, or
- maintaining environmental flows and Groundwater-Dependent Ecosystems, which improve local amenity, land value, and biodiversity.

Consequential benefits may also include

- improving coastal water quality by reducing urban discharges,

- mitigating floods and flood damage, or
- facilitating urban landscape improvements that increase land value.

MAR can play a role in increasing storage capacity to help city water supplies cope with the runoff variability in Australian catchments exacerbated by climate change. It can also assist in harvesting abundant water in urban areas that is currently unused.

## ENVIRONMENTAL FLOWS

### HUMAN DEVELOPMENT AND WATER SUPPLIES

Rain falls from the sky, but comes diffusely distributed across the globe. With some predictability, some seasons in many regions produce more rainfall although without the predictability of sun and moon. And yet, as depressions collect water, these waters flow to the ocean or terminal lakes, or are appropriated by the more powerful from the less powerful.

### MINIMUM FLOWS AND OTHER ENVIRONMENTAL ALLOCATIONS

Riverine Ecosystems are among the most impacted ecosystems worldwide. Human activities, such as the construction of dikes, dams, barrages and weirs, the straightening and deepening of river channels, the conversion of floodplains to agricultural land, water abstraction, water transfer and pollution, etc., have heavily modified most large lowland rivers. As a result, a large number of fish species are threatened or endangered, and the fish productivity of most Riverine Ecosystems has declined. This is all the more troubling since, large lowland rivers support a significant proportion of the world's fish diversity and their fisheries provide a major source of food, employment, and income to society.

The flow regimes of most large lowland rivers have been heavily manipulated to serve the needs of society. To balance the interests of different stakeholder groups, many countries have implemented water resources management plans.

Water policy decisions are typically made with little or no consideration of fish conservation and fisheries, despite a high public perception of fish. The reason is not only the relative lower socioeconomic importance of fisheries compared to ecosystem services, such as flood protection and navigation.

The optimal flow regime for navigation, agriculture, and flood protection (in terms of magnitude, timing, and duration of flow events) is comparably easy to determine, the consequences for fish diversity and fisheries are much more difficult to quantify.

Environmental flow refers to the water considered sufficient for protecting the structure and function of an ecosystem and its dependent species. It means enough water is to be released in the downstream of the river system after utilizing the water for the development projects to ensure downstream environmental, social, and economic benefits.

Realizing its importance, several countries have made ensuring environmental flows mandatory. For example, The Mekong River Agreement, 1995; South Africa's National Water Act, 1998, and the Swiss Water Protection Act, 108. These legislations attempt to ensure required minimum flow in the river system to sustain ecosystem services. A wide range of environmental flow methodologies have been developed to determine flow thresholds for various objectives, such as the preservation of natural conditions, the maintenance or restoration of ecological integrity and cultural and recreational values.

Most of these methods were developed primarily to protect endangered fish species and to maintain fisheries resources in human-modified rivers.

## CLIMATE CHANGE AND WATER SUPPLIES

### FLOODING AND CLIMATE

Will a warming climate affect river floods? River-flood risks are expected to rise as climate change intensifies the global hydrological cycle and more people live in floodplains. But the real question that needs to be answered is how will flood timing and magnitude change for each region and because of what mechanism. For example, based on climate and flooding records, climate change has increased earlier spring snowmelt floods in northeastern Europe, earlier winter floods in western Europe caused by earlier soil moisture maxima, and later winter floods around the North Sea and parts of the Mediterranean coast as a result of delayed winter storms.

### COLORADO RIVER AND OVER-APPROPRIATION

The Colorado River Basin covers about 246,000 square miles, including parts of the seven "basin States" of Arizona, California, Colorado, Nevada, New Mexico, Utah, and Wyoming and also flows into Mexico. The river supplies water to more than 30 million people, irrigates nearly 4 million acres of cropland in the United States and Mexico, and supplies hydropower plants that generate more than 10 billion kilowatt-hours annually.

But the river's capacity to support the native species richness and their habitats is severely constrained.

Ironically, the compact that was designed to allocate the water was done under high water flow conditions and under the current climate regime, the allocations cannot be met (Fig. 9.16).

### SNOW MELT AND STORAGE CAPACITY

Historically, California has buffered its water supplies and flood risks both within—and beyond—the Delta's catchment by developing many reservoirs, large and small, high and low. Most of these reservoirs carry water from wet winter seasons—when water demands are low and flood risks are high—to dry, warm seasons (and years) when demands are high and little precipitation falls. Many reservoirs are also used to catch and delay (or spread in time) flood flows that otherwise might cause damage to communities and floodplains (Fig. 9.17).

All currently available climate models predict a near-surface warming trend under the influence of rising levels of greenhouse gases in the atmosphere. In addition to the direct effects on climate—for example, on the frequency of heatwaves—this increase in surface temperatures has important consequences for the hydrological cycle, particularly in regions where water supply is currently dominated by melting snow or ice. In a warmer world, less winter precipitation falls as snow and the melting of winter snow occurs earlier in spring. Even without any changes in precipitation intensity, both of these effects lead to a shift in peak river runoff to winter and early spring, away from summer and autumn when demand is highest. Where storage capacities are not sufficient, much of the winter runoff will immediately be lost to the oceans. With more than one-sixth of the Earth's population relying on glaciers

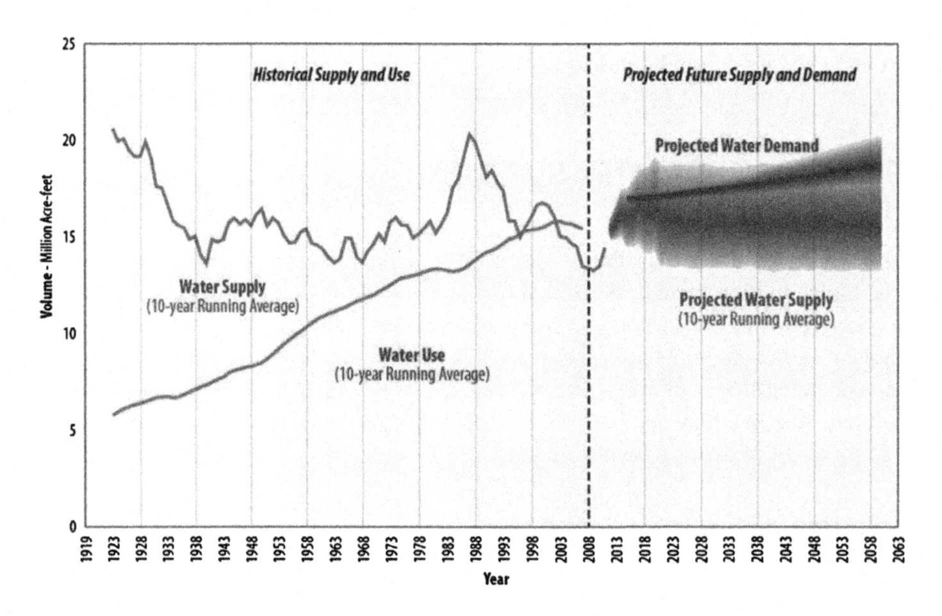


**FIGURE 9.16**

Projections for water supply for the Colorado River (Source: US Bureau of Reclamation).

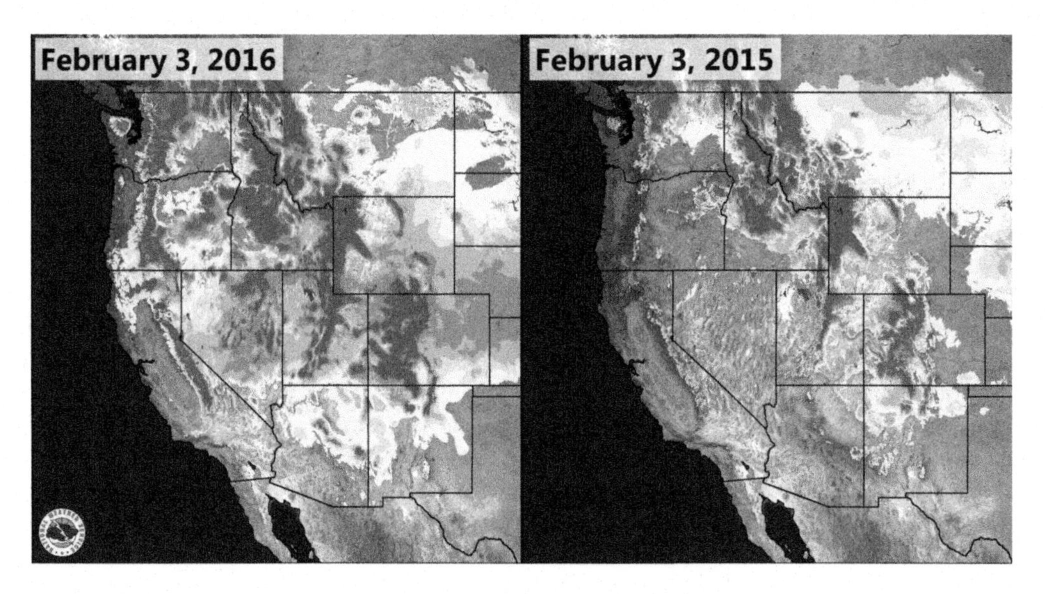


**FIGURE 9.17**

Snow pack estimates in the Sierra Nevada Mountains (February 2015 and 2016) (Source: NOAA).

**Table 9.1 Selected irrigation districts.**

| Founded | State | District | Notes |
|---|---|---|---|
| 1909 | California | South San Joaquin Irrigation District | Southern San Joaquin County |
| 1910 | Washington | Quincy-Columbia Basin Irrigation District | Delivers irrigation water to farmland in the Columbia Basin |
| 1911 | California | Imperial Irrigation District | Water is transported via All-American Canal, a 130 km long aqueduct, located in southeastern California. It conveys water from the Colorado River into the Imperial Valley and to nine cities |
| 1914 | Ohio | Miami Conservancy District | Great Miami River and its tributaries |
| 1918 | Oregon | Central Oregon Irrigation District | Provides irrigation water for Central Oregon |
| 1918 | Nevada | Truckee–Carson Irrigation District | Supports agriculture in Lyon County and Churchill County |
| 1921 | California | Nevada Irrigation District | Nevada County and portions of Placer and Yuba Counties |
| 1933 | Ohio | Muskingum Watershed Conservancy District | Muskingum River Watershed |
| 1949 | New Mexico | Carlsbad Irrigation District | Declared a National Historic Landmark in 1964 |
| 1925 | New Mexico | Middle Rio Grande Conservancy District | Rio Grande in the Albuquerque Basin section |
| 1951 | Utah | Jordan Valley Water Conservancy District (Salt Lake County Water Conservancy District) | Mainly operates in Salt Lake Country |

and seasonal snow packs for their water supply, the consequences of these hydrological changes for future water availability—predicted with high confidence and already diagnosed in some regions—are likely to be severe.

## IRRIGATION: FOOD DEMAND AND WATER DIVERSION

### WATER USE EFFICIENCIES

Historically, irrigation was the basis for economies and societies across the globe, from Asia to the Southwestern United States. Irrigation is the foundation for social development in arid climates, but it is certainly not limited to these regions.

To reduce the uncertainty and increase the total volume of available water has been a key driver to "develop" water supplies. For example, in the Sierra Nevada Mountains, groups of farmers banded together to create irrigation districts (Table 9.1). By investing money and obtaining public funding, these irrigation districts were able to build reservoirs, canals, and control structures. For many of these districts, they hold primary water rights, thus the conflicts with new land uses (and their water demands). The recognition of water to support ecosystem services has become a complex area of law, with all the associated battles between state versus federal authorities, government regulations versus

**FIGURE 9.18**

Stamens extending from a rice flowers.

farming profitability, etc. Instead of getting bogged down in these issues, we will focus on the Science of Irrigation and impacts on ecosystems as a whole—first in terms of supply and then in terms of water quality in Chapter 10.

Irrigation is the application of controlled amounts of water to plants at needed intervals. In contrast, agriculture that relies only on direct rainfall is referred to as rain-fed or dryland farming. One particularly interesting example is rice production in wetlands.

Irrigation systems are also used for cooling livestock, dust suppression, disposal of sewage, and in mining. Irrigation is often studied together with drainage, which is the removal of surface and subsurface water from a given area.

## RICE AND WETLANDS

Rice (*Oryza sativa*) is one of the most important crops in the world. As a C3 grass (Poaceae) (Fig. 9.18), the productivity of rice is more sensitive to temperature and water availability compared to C4 plants, which have the more efficient photosynthetic pathway. As a result, *Oryza sativa* have two major subspecies: the sticky, short-grained *japonica* or *sinica* variety, and the nonsticky, long-grained *indica* variety. Japonica varieties are usually cultivated in dry fields, in temperate East Asia, upland areas of Southeast Asia, and high elevations in South Asia, whereas indica varieties are mainly lowland rices, grown mostly submerged, throughout tropical Asia.

By the year 2020, the demand for rice is likely to increase by about 50%. Based on this projection, several concerns arise with respect to the sustainability of the aquatic variety:

- Wetland rice fields and irrigation schemes favor the propagation of aquatic invertebrates (mosquitoes and snails), vectors of human diseases (Malaria, Schistosomiasis, and Japanese Encephalitis).
- Lack of training and equipment for the proper use of pesticides has been associated with new rice technologies, resulting in significant off-field environmental impacts through their effects on non-target ricefield fauna: accumulation in the food chain, runoff from the fields, transportation to the water table, and detrimental effects on farmers' health.

**FIGURE 9.19**

Sacramento and Yolo Bypass.

- Ricefields are a major source methane emissions. Crop intensification will increase emission if mitigation techniques adoptable by farmers are not developed.

The linkages between water supply, energy, and even recreation has even become more dominant than the linkages between water and the green revolution, recall the Aswan Dam (see Chapter 2, section "Green Revolution: Power and Irrigation"). However, some models might be used as alternatives. For example, Yolo Bypass along the Sacramento River might be considered a useful model.

### YOLO AND WETLAND CONSERVATION

The Yolo Bypass is one of two flood bypasses in California's Sacramento Valley located in Yolo and Solano Counties. Through a system of weirs, the bypass diverts floodwaters from the Sacramento River away from the state's capital city of Sacramento and other nearby riverside communities (Fig. 9.19).

Sacramento experienced several severe floods prior to construction of the bypass, called the Yolo Bypass. The bypass forms a valuable wetland habitat when flooded during the winter and spring rainy season. In the summer, some areas of the bypass are used for agriculture.

Crops in the bypass include rice, safflower, processing tomatoes, corn, sunflower, and irrigated pasture. And because of its tolerance for cold weather, about 50% of the rice grown is wild rice. Most of the farming is mainly done in late spring and summer when there is little chance of flooding. After the crops are harvested, crop residue is left on the surface and creates foraging area, and food opportunities, e.g.ring-necked pheasants.

## FISHERIES

### INLAND FISHERIES CRISIS

In contrast to marine fisheries, inland water fisheries receive considerably less attention in discussions of the global fisheries crisis, although they provide important sources of protein, jobs, and income, especially in poor rural communities of developing countries. For example, Lake Chilwa (Malawi) supports a large subsistence fishery (18 million USD), whereas Lake Naivasha supports an export industry (620 million USD). One of the most important lakes is Tonle Sap in Cambodia, which feed about 14 million people and represents 15% of the Cambodian gross domestic product.

Tonle Sap is supported by high flows in the Mekong River, where the river actually backs up and flows into Tonle Sap. Thus, overfishing along the Mekong River threatens not only large species of fish and overall catch in the Mekong, but also in the Tonle Sap. For example, fishers report that catches of river catfish have dropped by 90% in some fishing lots of the Tonle Sap from about 100 metric tons 20 years ago to just 5 metric tons, or even 1 metric ton, today.

## FISHERIES AND SUSTAINABLE YIELDS

In a 1883 inaugural address to the International Fisheries Exhibition in London Thomas Huxley asserted that overfishing or "permanent exhaustion" was scientifically impossible, and stated that probably "all the great sea fisheries are inexhaustible." Unfortunately, marine fisheries were already showing the signs of collapse.

Although we have seen how population biology might be used to create concepts of sustainable yield, the concept has been poorly implemented and in many cases, does not provide useful management strategies.

Sustainable fishing means leaving enough fish in the inland waters, respecting habitats, and ensuring people who depend on fishing can maintain their livelihoods. MSY doesn't consider the complexities of these criteria.

The notion of sustainable development is sometimes regarded as an unattainable, even illogical notion because development inevitably depletes and degrades the environment.

However, when human dimensions of fishery management include cultural, social, and economic values; individual and social behavior; demographics; legal and organizational frameworks of management; communication, education, and citizen participation, and decision-making processes, then sustainable fisheries may be possible. However, few have invested in the complex social-natural science projects that include these dimensions.

## FISHERIES IN THE GREAT LAKES

Prior to environmental protections, the ecology of the Great Lakes had received decades of pollutant loading. Over the years, the Great Lakes have been characterized as "dead" or in crisis. Whereas these descriptors make headlines, the ecological state of the lake is complicated. For example, even at its worst, the Great Lakes support approximately 140 native species, including important fisheries.

The re-establishment of native deepwater fishes research theme is intended to coordinate research related to lake trout (Salvelinus namaycush), ciscoes (Coregonus spp.), sculpin (Cottus spp.) and Myoxocephalus spp. re-establishment in the Great Lakes. Impediments to re-establishment are still largely unknown, fundamental questions exist with respect to the biological, ecological, and genetic diversity of extant cisco populations, and relatively little is known about the ecological role of sculpins in deepwater food webs (Fig. 9.20).

**FIGURE 9.20**

Asian Carp in the Great Lakes (Source: Detroit Free Press).

## FISHERIES AND DAMS

Proponents of large dams place considerable emphasis on the potential which a dam's reservoir offers for the setting up—or, indeed, the expansion—of fishing industries. Even those who are critical of many aspects of large dams see such fisheries as providing major benefits. Professor Ackermann, for instance, regards the boost given by dams to fishing as "one of the more gratifying aspects of man-made lakes". Among other things, he sees such fisheries as slowing down the migration of young people to the cities, and considers that the new fishing opportunities may even lure back those who have already left an area.

Undoubtedly, when a large reservoir is filled, there is likely to be a dramatic rise in the population of those fish species which are favored by the new lacustrine conditions—although, those fish which are adapted to a riverine environment will tend to disappear. All in all, however, the actual number of fish is likely to increase quite substantially as advantage is taken of the vastly expanded aquatic environment. So, too, the release of large quantities of nutrients from the rotting vegetation and soils, which have been submerged by the reservoir—together with the increased populations of those microorganisms favored by the new conditions—will encourage the expansion of fish populations.

## ANADROMOUS FISHERIES IN CONTEXT

Among the most productive salmon runs occurs in the Eel and its tributaries with runs of both coho and Chinook salmon (*Oncorhynchus tshawytscha*), accounting for an estimate in 1980 at 103,000 and 42,000 fish, respectively. All the native fish populations in the Eel have declined including four species of salmon, two species of trout, a sturgeon, stickleback, and smelt. Probably much of the decline is due to habitat degradation from the erosion that has occurred as a result of timber harvest, and a lesser amount due to the introduction of exotic species, especially the squawfish (Fig. 9.21).

**FIGURE 9.21**

Chinook Salmon, *Oncorhynchus tshawytscha* (Source: Evermann and Goldsborough, 1907).

**FIGURE 9.22**

The vermilion darter is federally listed as endangered, and it only lives in Alabama (Source: Outdoor Alabama).

## WATERSHED MANAGEMENT

### EXTREME FLOOD EVENTS

In east Arizona the Salt River drains into the Colorado River. In 1952 there was a major flood with a discharge rate of 2900 $m^3/s$. However, on the basis of evaluating fine-grained deposits at the channel margins, Earth scientists believe the source was from floods well above the 1952 event that may have been as much as 4600 $m^3/s$. These flood events might have return intervals of 1000–2000 years. This begs the question, what should be the planning horizon when developments occur in the flood plain.

### ALABAMA RIVER AND EXTINCTIONS

Alabama ranks fourth in species diversity in the United States, after Hawaii, Florida, and California. With regard to the number of species per acre, Alabama ranks second. Furthermore, the state of Alabama has the highest diversity in the nation for freshwater mussels, freshwater turtles, freshwater snails, and crayfish. Approximately 180 mussel species, or 60 percent of the nation's mussels, occur in Alabama, and several species are endemic. Of all snails in the United States, 43 percent are found in Alabama and roughly 102 species are endemic. With more than 132,000 miles of perennial and intermittent rivers and streams, and more than 500,000 acres of standing water, it is not surprising that Alabama has so many water-dependent species (Fig. 9.22).

Alabama has approximately 117 endangered or threatened species listed and ranks third in the highest number of threatened and endangered species, behind Hawaii and California, but species are constantly being removed from or added to the list. Estimates suggest that nearly 100 species have become extinct in Alabama since colonial times. About half of the threatened and endangered species in Alabama are mussels. Ray-finned fishes form the next largest class of threatened and endangered animals. More than 20 percent of all species listed in Alabama were added to the list in 2001. This sizable growth reflects an increase in scientific knowledge about species distributions and habitat requirements. Of the 105 animal species, 21 are no longer found in Alabama. These species are listed because they occurred historically in Alabama or because potential habitat in Alabama borders their current range.

Most of the state's listed species are aquatic or dependent on water for feeding or reproduction. There are 14 river basins in Alabama. The Mobile River System, which includes the Tombigbee, Mobile, Alabama, Cahaba, Black Warrior, Coosa, and Tallapoosa sub-basins, supports a great diversity of species, several of which are endemic. Dams, canals, mining, dredging, and direct and indirect pollution threaten this system and the species depending on it. During the past few decades, at least 12 species of mussels and 42 species of aquatic snails are presumed to have gone extinct. Currently, 37 aquatic animal species endemic to the Mobile River Basin are now protected under the ESA, including 2 turtles, 11 fish, 14 mussels, and 7 snails. A six-year recovery plan, established through a joint effort of the USFWS and the Mobile River Basin Coalition has been developed for the entire Mobile River Basin, focusing on 22 animal species, including 4 fish, 11 mussels, and 7 snails. This recovery plan focuses on protecting the habitat's integrity and quality by working with the community on issues such as pollutants, research, education, technology, and restoration.

## TOILET-TO-TAP

Secondary effluent obtained during wastewater treatment is subjected to additional treatment prior to its reuse as "reclaimed water." This chapter gives an overview of the treatment technologies used to produce reclaimed water followed by descriptions of reclaimed water applications in the United States. Such applications include environmental, urban, agricultural, and industrial reuse. In addition, reclaimed water can be utilized for potable reuse following recharge, or even direct potable reuse following advanced treatment. Recycled water regulations and water quality standards are also reviewed, and finally microbial water quality aspects of recycled water are discussed, including changes in water quality within distribution system pipes, as a function of residence time.

## DEVELOPMENT AND WATER SUPPLY

To negotiate the pressure for continued economic development, a sustainable water supply system will require political ways to address conflict of human water needs and the needs of the environment (Fig. 9.23).

In a region of Brazil, water withdrawals from the São Francisco River generates electricity, supplies drinking water, and maintain fisheries. However the final stretch of the river flow through the semiarid region of the country and is prone to drought during El Niño events. Because water is being diverted to other watersheds in northeast Brazil, the region's socioenvironmental systems have become vulnerable to a potential collapse. How will the conflicting demands across multiple hydrologic basins be resolved?

**FIGURE 9.23**

A woman scoops water in a dry riverbed near Kataboi village in remote Turkana in northern Kenya (UK Department for International Development/CC BY 2.0).

According to several forecasts, many regions will experience lower than predicted streamflows and retention levels. Policymakers should look globally for solutions. Regions in Australia face many of the same issues that planners in other Mediterranean climates encounter, such as high agricultural demand, groundwater overdraft, and the obligation to sustain viable environment while meeting the needs of a growing human population.

Australia utilizes water markets to force efficient use. The government set up water markets, ensuring environmental health through set allocations by watershed for environmental sustainability. Theoretically, these markets promote efficient water use through fluctuating water price based on demand for entitlements to fixed water allocations. A case study focusing on water markets and their environmental impacts suggests that water markets will limit the effectiveness of legislation directed toward restoration of natural flows due to insufficient weightings of environmental allocations.

Without a doubt, anthropogenic impacts on lotic systems impair the water flows and their ecology. It is a major challenge to find environmentally conscious and sustainable approaches to water management.

## NEXT STEPS

### CHAPTER STUDY QUESTIONS

1. Describe 5 lessons we can draw from past hydromodification projects that can be used to inform proposed instream modifications?
2. Describe how water management goals can be consistent with the Flood Pulse Concept.
3. Describe how ground-surface water interactions influence water supplies.
4. Summarize how Aquatic Ecosystems are modified by regulated rivers.
5. What are environmental flows and how can they be used to improve ecological outcomes?
6. Summarize watershed management concepts and describe how they can be used to improve water supply management goals.

## LITERATURE RESEARCH ACTIVITIES

1. Select several lakes where the sulfur cycle has been evaluated. Using the readings and based on Fig. 9.18, create a figure that represents how the sulphur cycle works in these lakes.
2. Using MSY as a starting place, evaluate the use of fisheries modeling in three different lakes. When was the modeling begun, has it been successful, and what improvements should be made?

CHAPTER

# WATER QUALITY AND CATCHMENTS

# 10

*Water is the most critical resource issue of our lifetime and our children's lifetime. The health of our waters is the principal measure of how we live on the land.*

**Luna Leopold**

## CONTENTS



Ecology and Management of Inland Waters. https://doi.org/10.1016/B978-0-12-814266-0.00024-6

Approximately 3 hours north of San Francisco, Clear Lake along with its surrounding wetlands and uplands supports several important bird species, including the western grebe (*Aechmophorus occidentalis*) (Fig. 10.1), Clark's grebe (*Aechmophorus clarkii*), double-crested cormorant (*Phalacrocorax auritus*), great-blue heron (*Ardeaherodias*), osprey (*Pandion haliaetus*), and bald eagle (*Haliaeetus leucocephalus*). The native Clear Lake gnat (*Chaoborus astictopus*) larvae historically provide food for fish species, but was considered a nuisance due to large swarms and exterminated using larvicides, including the chlorinated hydrocarbon DDD.

Clear Lake is well-known among Entomologists for the Clear Lake gnat (*Chaoborus astictopus*) and historical control efforts. Although the gnat resembles a tiny mosquito, the midge is a nonbiting Dipteran. Clear Lake gnat hatches start anytime from March through June, depending on weather. The larvae, often referred to as "phantom midge", are transparent and very difficult to see in the water column.

Before pesticide use began in the 1940s, the adult gnat was so abundant that large piles of dead gnats appeared beneath streetlights each summer. Some reported that gnat swarms could become so dense that windshields and headlights could be covered with gnats making it impossible to see, and pedestrians needed to cover their faces to avoid inhaling the gnats.

In 1949, as part of an effort to boost tourism and improve the local economy, DDD (dichlorodiphenyldichloroethane) was applied to the lake in heavy doses to eradicate the gnats that were driving summer tourists away from the lake. The treatment succeeded in controlling the gnats that year, and for the following year, however, in 1953, the gnat population rebounded, prompting another large application in 1954. The final application of DDD to Clear Lake was made in 1957. However, the western grebe population was decimated and numerous dead carcases were found. High concentrations of DDD

**FIGURE 10.1**

Western grebe (*Aechmophorus occidentalis*) is the largest North American grebe (Source: Wikipedia). It is black and white, with a long, slender, swan-like neck, and red eyes. Western Grebes nest in colonies on lakes that are mixed with marsh vegetation and open water. Western Grebe nests are made of plant debris and sodden materials, and the nest building begins roughly around late April through June. The construction is done by both sexes and is continued on throughout laying and incubation. This species of waterbirds is widespread in western North America, so there is no specific place of abundance.

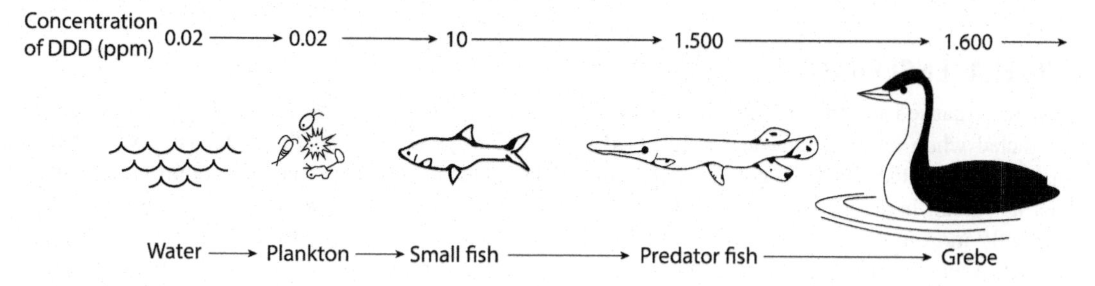


**FIGURE 10.2**

DDD concentrations in various trophic levels in Clear Lake. Clear Lake provided early evidence of the dramatic impact of biomagnification.

in their tissues was the cause—an example made famous by Rachel Carson's book, "Silent Spring," prompting the eventual banning of DDD/DDT applications in the USA.

The bioaccumulation of such persistent chemicals led to deleterious effects on the ecosystem, including mortalities of western grebes (Fig. 10.2).

Water quality is a value based-measure, relative to our expectations of and goals for our Aquatic Ecosystems. We prize high quality water, but what we mean by that can vary by culture and political jurisdiction. For example, California recognizes 23 different beneficial uses, which in some cases, might be in conflict with each other for any given water body. For example, water quality to meet agricultural uses, drinking water, or cold-water fishes are going to require different standards and protections.

Water is a resource, but can also be contaminated, and it is a potential hazard. Whether from geologic sources, biological diseases, or anthropocentric pollution, various attempts are made to treat and reduce our risk and exposure to poor water quality. To appreciate our efforts to manage water quality, the reader should be able to

1. Summarize the ecological processes that promote water-borne diseases;
2. Describe how geochemistry and nutrients influence water quality;
3. Describe the fate, transport, and toxicity of pollutants in aquatic systems and their impacts;
4. Summarize efforts to assess and improve the water quality.

## PUBLIC HEALTH AND CONTAMINATED WATERS

### SUPPLIES OF CLEAN WATER

The UN World Water Development Report (2003) from the World Water Assessment Program indicates that, in the next 20 years, the quantity of water available to everyone is predicted to decrease by 30%. Currently, 40% of the world's inhabitants have insufficient fresh water for minimal hygiene. More than 2.2 million people died in 2000 from diseases related to the consumption of contaminated water or drought. Children are often the most vulnerable, where over 5 thousand die every day from 15 easily preventable water-related diseases; often this means lack of sanitation infrastructure.

### CHOLERA, ENCEPHALITIS, AND MALARIA

Cholera is caused by an infection of the intestines by the bacterium *Vibrio cholerae*. It is usually contracted when untreated sewage is released into waterways and affects the water supply for food washing or shellfish beds. In general, Cholera has become rare in industrialized countries, but that does not mean high incident regions, such as the Indian Subcontinent and sub-Saharan Africa, are inherently more prone to Cholera. On the contrary, recent analyses suggest that Cholera was spread from SE Asia via trade and colonialism. Thus incidence rates are also the result of globalization. (See Fig. 10.3.)

In contrast, several forms of encephalitis, inflammation of the brain, can be caused by parasites in birds that are then transferred to humans via a vector. For example, the *Culex tarsalis* is also known as the Western Encephalitis Mosquito. The range of this mosquito includes much of the United States, especially west of the Mississippi River and is the primary vector of Western Equine Encephalitis, St. Louis Encephalitis, and more recently, West Nile Virus. Although usually preferring to seek birds as a blood meal source, they also readily bite other animals (e.g., horses), transferring diseases from bird reservoirs to mammals, including humans (Fig. 10.4).

Efforts to control the disease often focus on the vector, i.e., the mosquito and their aquatic larvae. But the source from reservoirs to infected species can be a bit complicated. For example, some evidence suggests that the virus may depend on *Culex tarsalis* to maintain viral load within the riparian zone. However, *Culex quinquefasciatus* may transfer the virus from the Riparian Zone into human residential communities. Thus understanding the relationship to the vector relative to other species populations may be a key component to protect public health.

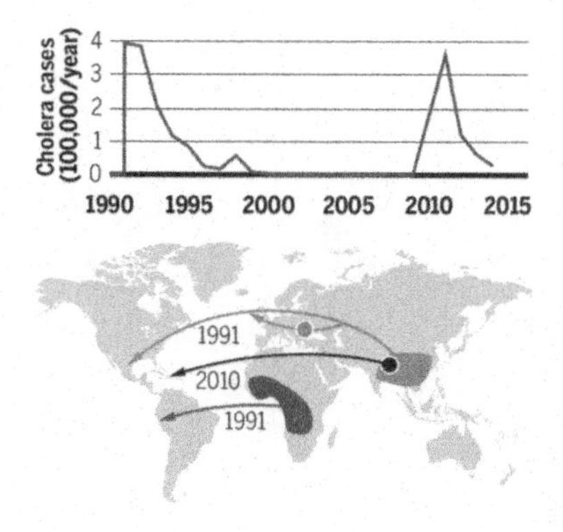


**FIGURE 10.3**

The two major cholera epidemics that occurred in the Americas in the past 50 years were both caused by imported strains. So were 11 African epidemics. (Source: Kupferschmidt, 2017).

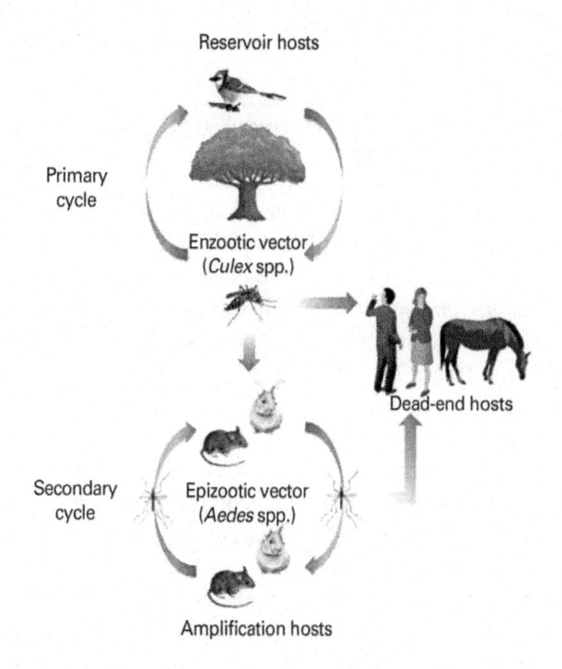


**FIGURE 10.4**

Western Encephalitis Mosquito vector path. (Source: Go et al., 2014).

The vast majority of malaria cases and deaths occur in Africa, but malaria is widespread in tropical and subtropical regions, including North America. In fact, in the 1910s, California had the highest rate of malaria in the United States until most of Central Valley Wetlands were destroyed.

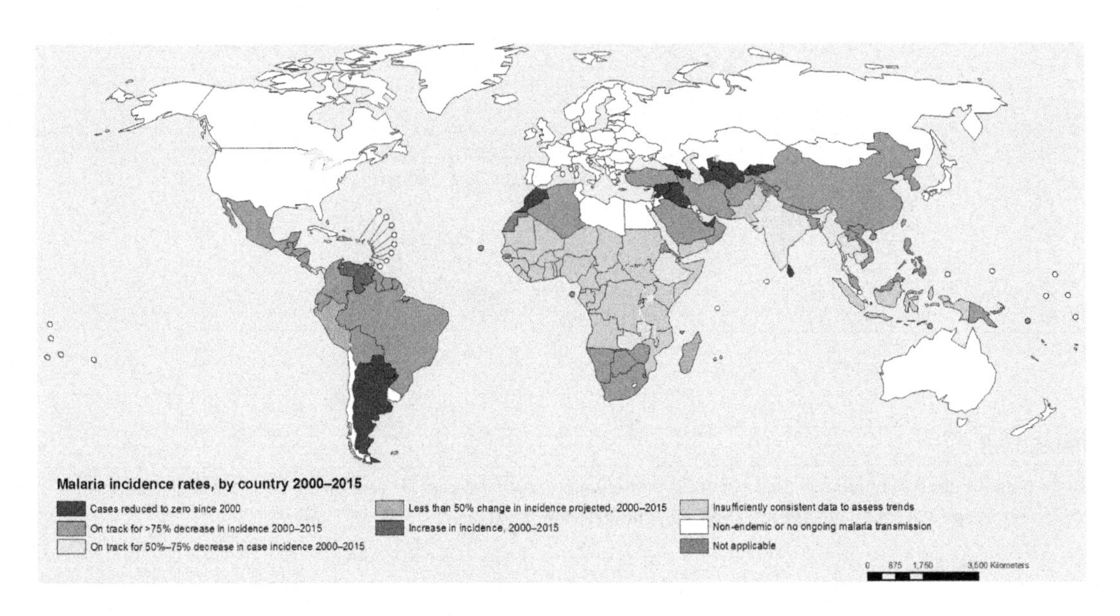


**FIGURE 10.5**

Map of malaria incidence rates, by country 2000–2015 (Source: World Health Organization et al., 2015).

In 2015, almost 300 million cases of malaria occurred world-wide and resulted in over 700,000 deaths (Fig. 10.5). Although it is often associated with poverty and has a negative effect on economic development, the spread of the disease can be traced to colonization.

Malaria is caused by a *Plasmodium* spp. and member of the Alveolata (SAR: Stramenopiles, Alveolata, and Rhizaria). *Plasmodium* requires two different hosts to complete its life cycle: *Anopheles* mosquitoes and mammals (Fig. 10.6). The phylogeny of these malarial parasites suggests that the *Plasmodium* of mammalian hosts forms a well-defined clade strongly associated with the specialization to the *Anopheles* mosquito vector.

The genus *Anopheles* is in the Diptera Order and has about 460 recognized species; whereas over 100 can transmit human malaria, 30–40 commonly transmit the *Plasmodium* parasites. *Anopheles gambiae* is one of the best known, because of its predominant role in the transmission of the most dangerous malaria parasite, *Plasmodium falciparumspecies* to humans.

After the Second World War, mosquito control became increasingly reliant on pesticides, e.g., DDD and DDT. However, with mosquito resistance and biomagnification, their use in the United States was prohibited in 1972. As an alternative, the Western Mosquito Fish (*Gambusia affinis*) was introduced to control aquatic larvae populations (Fig. 10.7) and remains an important strategy, although the ecological impacts have been not without costs and criticism.

Alternatively, some regions relied on an indirect method to reduce malaria risks: reduce the prevalence of breeding habitat. Draining wetlands, ponds, and even puddles is another strategic solution. Although the draconian destruction of wetlands to reduce mosquito breeding grounds remains active in some areas, a more appropriate and effective method has been to identify and remove places where water collects in developed areas, such as roadside potholes, discarded tires, and buckets.

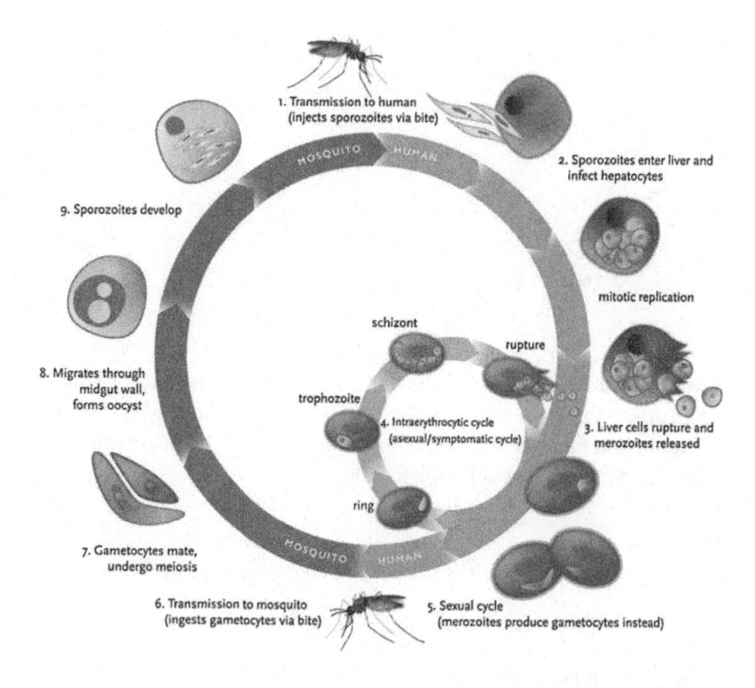


**FIGURE 10.6**

Plasmodium lifecycle (Source: Center For Disease Dynamics, Economics & Policy).

**FIGURE 10.7**

Western Mosquito Fish (*Gambusia affinis*) (Source: Los Angeles County West Vector & Vector-Borne Disease Control District).

Finally, many regions rely on therapeutic treatments of malaria. However, increasing rates of drug and multidrug resistance suggest the efficacy of these treatments is limited. As an alternative, Scientists

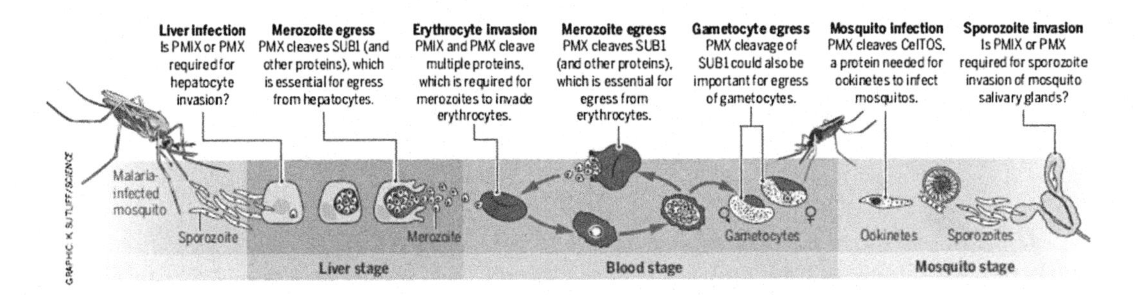


**FIGURE 10.8**

Plasmpsin role in the Plasmodium lifecycle (Source: Boddey, 2017). One approach to combat Malaria could rely on disabling some of the more than 100 proteases *Plasmodium* uses throughout its lifecycle. For example, if a drug can stop the expression plasmepsin X in merozoites, then merozoite egress would be halted from hepatocyte (liver) and erythrocyte (red blood) cells, and infection would be impossible.

have evaluated drugs to knockout specific *Plasmodium* proteins and prevent its ability to complete its lifecycle (Fig. 10.8).

## HUMAN WASTE AND TREATMENT

Throughout history, waste has been generated by humans. In areas with low population density waste production was probably negligible. But as population densities increase, the risk of being exposed to disease also increases. Thus the success of advanced societies depends on how biodegradable waste is addressed.

Burial and burning have been and are still common ways get rid of waste. For example, the Maya of Central America (300–600 CE) had a monthly ritual, where villagers gathered together to burn their rubbish in large dumps. However, in other cases, the lack of adequate disposal methods have created a record of surfacewater contamination issues (Fig. 10.9).

The connection between wastewater and drinking water was recognized and forced people to consider more advanced approaches than dumping untreated waste into waterways. Early technologies relied on septic tanks, a watertight chamber made of concrete and clay pipes, as part of a sewage treatment system.

Originating in France, the septic tank was invented by John Mouras around 1860 for his home. When the sewage overflowed from the tank, it would be released into a cesspool. After nearly ten years, Mouras decided to open the tank to see how his prototype was holding up. To his surprise, it was virtually completely free of solids. Based on a modified design with the assistance of Abbè Moigno, he obtained a patent in 1881 (Fig. 10.10).

The settling and anaerobic processes in septic tanks reduce solids and organics, but the treatment is only moderate. The treated liquid effluent is commonly disposed in a septic drain field, which provides further treatment. Septic systems continue to be used in rural areas, especially in Maine and Vermont, that are not connected to a sewerage system. However, groundwater can become contaminated in areas with septic tanks.

**FIGURE 10.9**

Nor Loch c 1690 is one of the first documented waste problems in Europe. The Nor Loch was initially a marsh, and part of the natural defense of the Edinburgh Old Town. In 1460 King James III ordered a hollow to be flooded to strengthen the castle's defenses. The loch was formed by creating an earthen dam to block the progress of a stream that ran along the north side of the castle. As the Old Town became ever more crowded during the Middle Ages, the Nor Loch became polluted by sewage, household waste, and general detritus thrown down the hillside (Source: Part of an engraving by John Slezer).

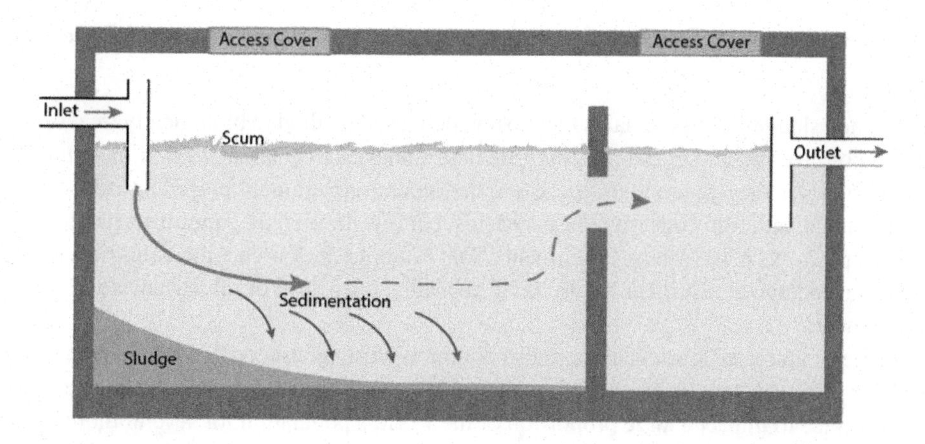


**FIGURE 10.10**

Early French "la fosse Mouras" septic tank design (Source: Wikipedia).

In 1914 Edward Ardern and William Lockett presented their classic paper, "Experiments on the Oxidation of Sewage Without the Aid of Filters," where they described how an activated sludge process can accelerate oxidation of sewage. Using aeration and a biological floc composed of bacteria and

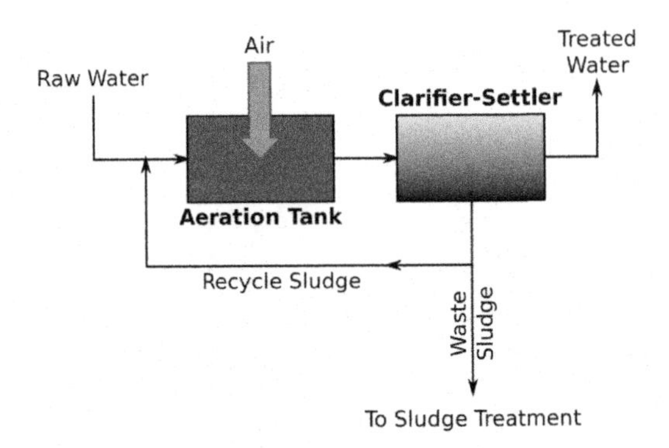


**FIGURE 10.11**

Activated sludge as part of the sewage treatment designs (Source: Wiki Commons).

protozoa, the activated sludge process oxidizes carbonaceous and nitrogenous biological matter and ammonia. In general, the activated sludge process (Fig. 10.11) includes the following:

- Aeration tank, where air (or oxygen) is injected in the mixed liquid;
- Settling tank (usually referred to as "final clarifier" or "secondary settling tank") to allow the biological flocs (the sludge blanket) to settle, thus separating the biological sludge from the clear treated water; and
- Treatment of nitrogenous matter or phosphate involves additional steps, where water is left in anoxic condition.

The activated sludge process remains the most widely used method to biologically treat sewage, but requires a major investment. On average, high-income countries treat about 70% of the municipal and industrial wastewater they generate. In the USA, the federal government played an important role in subsidizing municipalities to construct these systems, but few developing countries have the economic or political capacity to make robust investments. For example, in low-income countries, only 8% of the wastewater undergoes treatment of any kind. Globally, over 80% of all wastewater is discharged without treatment.

The release of untreated wastewater remains common practice, especially in developing countries, due to lacking infrastructure, technical and institutional capacity, and financing. Thus water-borne diseases will continue to impact a large proportion of the world's population for several more decades.

## PHARMA, INDUSTRY, AND FARMS: EMERGING CONTAMINANTS

Emerging Contaminates, or better, Contaminants of Emerging Concern, fall into three categories: 1) new compounds or molecules that were not previously known, or that just recently appeared in the scientific literature; 2) contaminants of emerging interest, which were known to exist but for which the environmental contamination issues were not fully realized or understood, and 3) legacy contaminants, where new information has forced us to revise our understanding of the risks related to these legacy contaminants. Pb may be the first compound to be a containment of emerging concern.

Contemporary Contaminants of Emerging Concern are contaminants introduced into water sources by way of agricultural runoff, personal care pollutants, and industrial waste. Pharmaceuticals, personal care products, endocrine disruptors, and antibiotics are prime examples of emerging contaminants. Even caffeine has significant effects in aquatic organisms. Up to 90% of oral drugs pass through the human body and end up in the water supply. Contaminants of Emerging Concern usually enter the water cycle after being discharged as waste that later enter surface or ground waters by effluent discharge, by seepage and infiltration into the water table, eventually entering the public water supply system.

Hundreds of unregulated emergent contaminants have been observed mainly in the $ng \cdot L^{-1}$ to $mg \cdot L^{-1}$ range in surfacewaters throughout the industrialized countries. For example, over 200 different pharmaceuticals alone have been reported in river waters globally, with high concentrations of some antibiotics. A South Korean study found 14 pharmaceuticals, 6 hormones, 2 antibiotics, 3 personal care products, and 1 flame retardant in surfacewaters and wastewater treatment plant effluents.

Industrial sources of water contamination has a long history. For example, Vat dyes are a class of dyes that are made in a bucket or vat. Almost any dye, including fiber-reactive dyes, direct dyes, and acid dyes, can be used in a vat dye. Cotton, wool, leather, and other fibers can be all dyed with vat dyes. The original vat dye, indigo, was replaced by analine compounds and led to contaminated surfacewaters in the Rhine, Switzerland, Cuyahoga River, Cincinnati, and Toms River, New Jersey. The detailed story is well described by Dan Fagin in the book **Toms River**.

Some emergent contaminants disrupt endocrine activity (Fig. 10.12), and others are toxic. The USEPA has recognized a few as carcinogens. Besides an increasing catalog of compounds found in surfacewaters, many of these compounds pose a risk to human health and the environment. However, a better understanding of the sources and how to control the compounds will require decades of research. Unfortunately, conventional drinking water treatment methods are ineffective at removing these types of chemicals, so new removal technologies have been proposed. For example, granular activated carbon is an expensive but effective technology to remove many of these compounds.

In addition to water contaminants is Concentrated Animal Feedlots (CAFOs) discharge into surfacewaters. To prevent rampant disease outbreaks, many of these systems rely on antibiotics and growth hormones to promote rapid growth. The concentrations of these compounds has been increasing dramatically in surfacewaters, where these agricultural systems are common, e.g., hog farms in the United States Midwest, poultry in the United States Southeast. Microbes in selected Chinese lakes have become resistant to antibiotics, potentially rendering specific antibiotic useless to prevent disease severally compromised, and some fear this could cause future epidemics.

## ECOLOGY OF CLEAN DRINKING WATER

### ECOSYSTEM SERVICES AND WATER QUALITY

Once in water, pathogens and other contaminants that are harmful to humans can be difficult to remove; but natural purification processes can often keep them from even reaching drinking water sources. For example, the intestinal parasite *Giardia lamblia* is difficult to remove from drinking water sources using

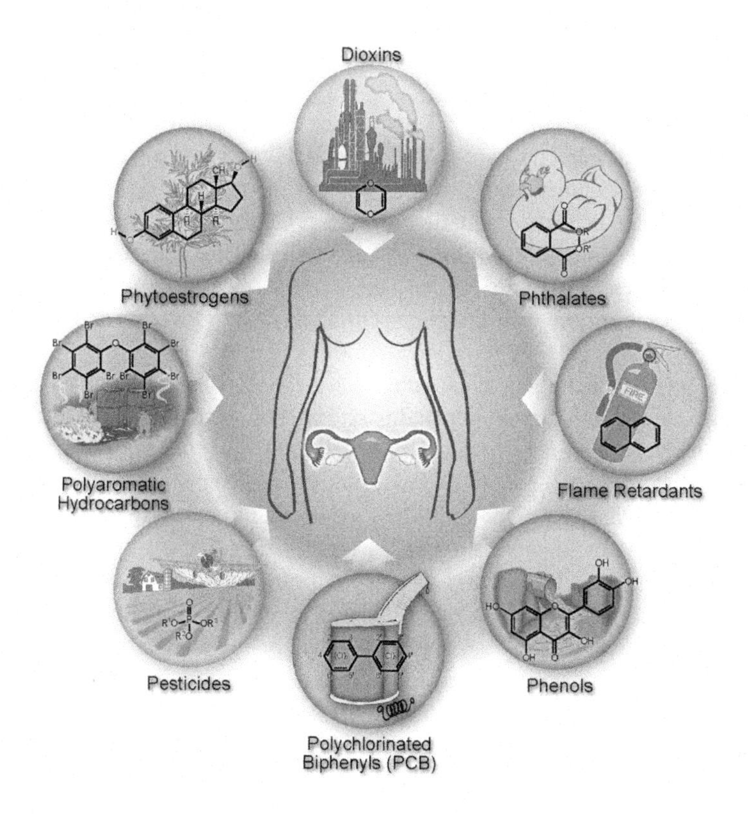


**FIGURE 10.12**

Endocrine disruptors affect the human body from a wide range of sources. In addition, these endocrine disrupting chemicals have diverse impacts on female reproduction (Source: Grindler et al., 2015).

conventional treatment systems, but concentrations decline after flowing through forested watersheds. Could we rely on various ecosystems to improve our water quality?

Water purification can be considered an ecosystem service. Pollutants such as metals, viruses, oils, excess nutrients, and sediment are "processed" and "filtered out" as water moves through wetland areas, forests, and Riparian Zones. For example, wetlands can remove 20 to 60% of metals in the water; trap and retain 80 to 90% of sediment from runoff, and eliminate 70 to 90% of entering nitrogen. Although these processes may provide clean drinking water and other uses, including habitat, the internal processes within these systems need to be understood.

In these systems, water purification is a function the system's capacity to adsorb or transform compounds via their contact with soil particles and organic matter, and subject to biological processes. Riparian Forests can intercept and absorb sediments (i.e., drop out of the water column) by providing contact with leaf litter and areas where surface runoff velocity is reduced and water can infiltrate into the ground. Since a high proportion of P is sediment bound, this process also reduces P loading by ~50% into surfacewaters. In this example, the forest acts as a filter to trap particles. It is important to acknowledge that the P may still accumulate, thus the P is redistributed in the landscape and reduces its

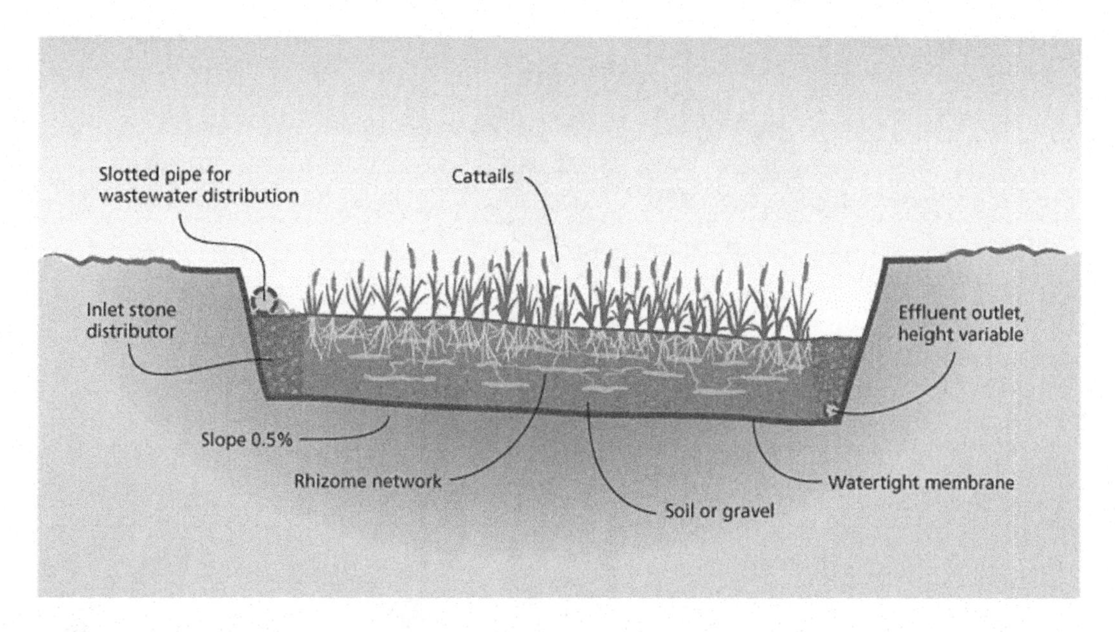


**FIGURE 10.13**

Cross-section of constructed wetland designed for subsurface flow and to process agricultural discharges. (Source: Copyright © 2011 Regents of the University of California. Used by permission.)

ecosystem impacts. In contrast to a filtering mechanism, dissolved reactive nitrogen forms ($NH_4^+$ and $NO_3^-$) can be transformed by microbes, where in some cases up to 90% of the inorganic nitrogen is removed via denitrification. However, even in this case, some proportion of microbial processes might generate NO, an ozone-depleting compound, and $N_2O$, a greenhouse gas.

Constructed wetlands mimic some of the filtration of natural systems. If designed and constructed properly, they can dramatically improve water quality (Fig. 10.13). However, they do not replace natural wetlands and rarely provide other wetland services, such as flood control and habitat.

## DEVELOPING ECOSYSTEM SERVICES

Although the United States spends more than $2 billion annually for clean water initiatives, it is easier and less costly to prevent pollution than to clean contaminated water. For example, in the late 1990s, the New York State initiated a project to provide drinking water to NYC by restoring the Catskills Watershed for $1 billion instead spending $8 billion on a water treatment facility (Fig. 10.14). Catskills residents have now become stewards of the NYC Water Supply that can result in complex political dynamics.

## THE BUILT ENVIRONMENT AND DEGRADED ECOSYSTEM SERVICES

Human activities that compact soil, contaminate the water, or alter the composition of organisms degrade the capacity of ecosystems to improve water quality. In an effort to reduce flooding, the built

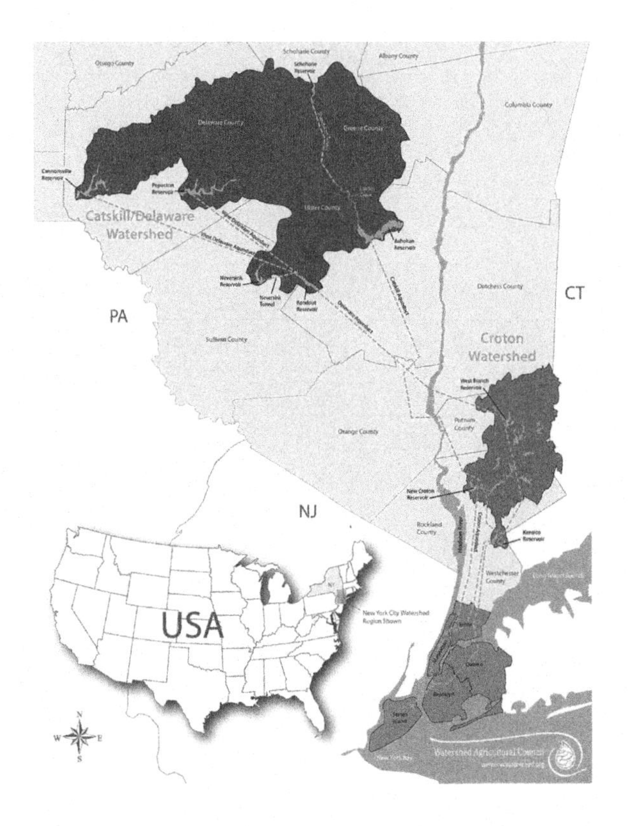


**FIGURE 10.14**

Map of NYC Water Supply (Source: NY Department of Environmental Protection).

environment had unintentionally reduced the effectiveness of these ecosystem services in a wide range of ways:

***Paved surfaces*** Paved parking lots, roads, and highways allow contaminated water to directly enter lakes, streams, rivers, and coastal waters. Urban development and sprawl reduced the area of natural wetlands and Forest Ecosystems. To combat this problem and comply with the Clean Water Act, many cities in the United States have developed bioswales, low areas that can capture and trap runoff from paved surfaces (Fig. 10.15).

***Channelization*** When water is channelized, the connection to the flood plain is severed. Channelized water might be diverted for agricultural, industrial, or navigational use, thus waterflow through the system might reduce the potential for water quality improvements. In the last 100 years, the world's freshwater waterways altered for navigation purposes have increased more than 50 times. Reversing this trend is expensive, but some regions have developed plans to restore riverine processes by increasing the channel width to create low-flow and high-flow channels. During

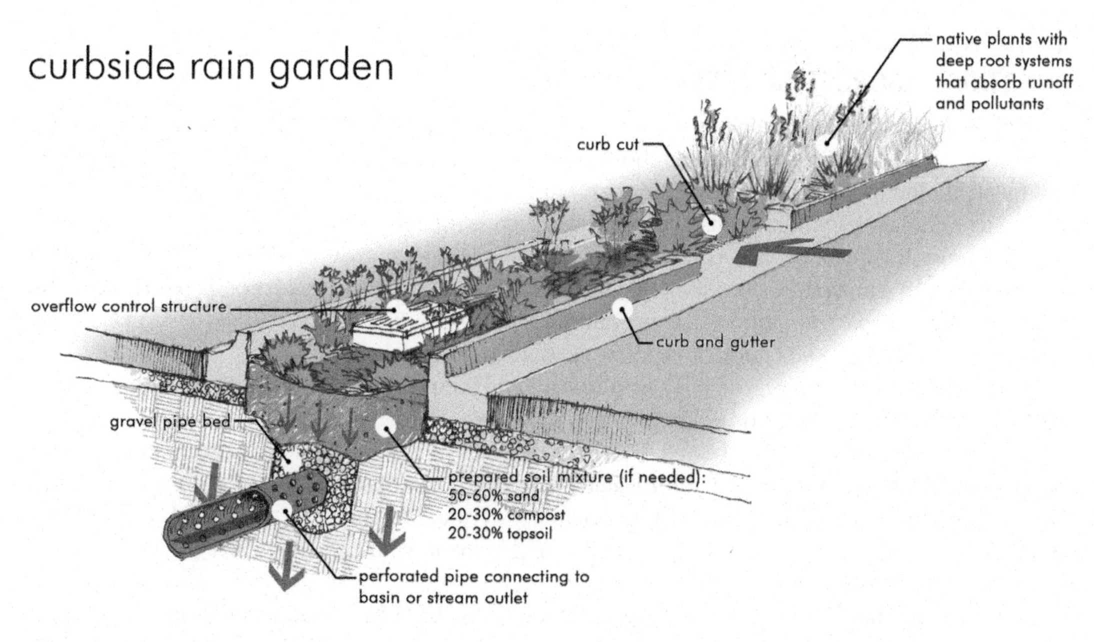


**FIGURE 10.15**

Bioswale design (Source: Earth Wind Water Civil Engineering).

high flow, water velocities are reduced and the lost connection between the channel and flood plain is partially mitigated.

***Loss of Riparian Areas*** The removal of Riparian Forests for streamside agriculture or impervious surfaces, such as roads and parking lots reduces rivers' and streams' natural capacity to process pollutants. When water (and pollutants) interact with the landscape adequately, water quality is often improved. Thus various plans in urban environments have been proposed that include riparian restoration to improve stream ecosystem functions.

***Species composition shifts*** The introduction of exotic species of plants and animals can eliminate native species and alter an aquatic system's capacity to cycle nutrients. Florida alone spends more than $7 million per year to combat invasive aquatic weeds that interfere with navigation and reduce water quality.

***Nitrogen and phosphorous pollution*** Although nitrogen and phosphorous can be mitigated with wetlands and Riparian Zones, concentrations can exceed the assimilatory capacity of the system and alter composition of the community. In the case of N, these can be a source of greenhouse gases and reduce overall air quality.

These impacts fall into the characteristics of urban stream syndromes, which will be further discussed in Chapter 12.

# THERMAL POLLUTION

## INDUSTRIAL SOURCES OF THERMAL POLLUTION

Thermal pollution is the degradation of water quality as a result of changes to the ambient water temperature. Warm water increases respiration rates, reduces the amount of $O_2$ water can hold, promote harmful algal blooms, and shifts species composition. A common cause of thermal pollution is the use of water as a coolant by power plants and industrial manufacturers.

When water used as a coolant is returned to the natural environment at a higher temperature, the sudden change in temperature decreases oxygen supply and affects ecosystem composition. Fish and other organisms adapted to particular temperature range can be killed by an abrupt change in water temperature (either a rapid increase or decrease) known as "thermal shock."

Thermoelectric generation, which is one of the several industries that use water, accounts for more than 50% of the use of surface or groundwater as a coolant. The industrial use of water, especially in rapidly developing countries, is expected to grow further and aggravate an already precarious situation concerning availability of and demand for water. Accelerated growth in the power generation industry alone will account for a major share of this demand. A typical thermal power plant of ~2000 MWe capacity can require 65 $m^3/s$ of cooling water; the requirement would be about 50% more in the case of a nuclear power plant.

By increasing the water temperature, aquatic organisms have increased respiration rates (which requires oxygen), whereas the oxygen saturation concentration declines. Using river water to cool power plants has become an increasingly contested issue in the permitting process for power plants in the United States.

## MITIGATING WARM WATER DISCHARGES

In November 2006, the Washington Department of Ecology adopted temperature criteria in the surfacewater quality standards. The temperature criteria limit the allowable temperature increase of the receiving water due to human-caused impacts, including point-source discharges.

Based on regulations such as these, a wide range of mitigation methods have been employed to reduce the impact of thermal pollution. Methods to reduce thermal impacts pretreatment and treatment mitigation measures and catchment scale activities. For example, pretreatment mitigation might reduce solar radiation on storage ponds, creating wetlands, or constructing cooling towers (Fig. 10.16). At the watershed scale, riparian restoration or reclamation, and reuse where water temperature is not critical.

Mitigation approaches have pros and cons, but from a surfacewater perspective, pretreatment and treatment measures can have substantial effects on the habitat quality.

## SHADING AND SALMONID BEHAVIOR

In smaller streams, water temperatures can be affected by landuse patterns. Though the amount of energy to heat water is much higher than air, inland water temperatures can be subject to large temperature changes based on seasonal and diurnal scales. These water temperature changes can have direct impacts on water quality and habitat. For example, the loss of riparian shading allows solar radiation to warm the stream waters, which can reduce dissolved $O_2$. Reduce $O_2$ availability can severely reduce

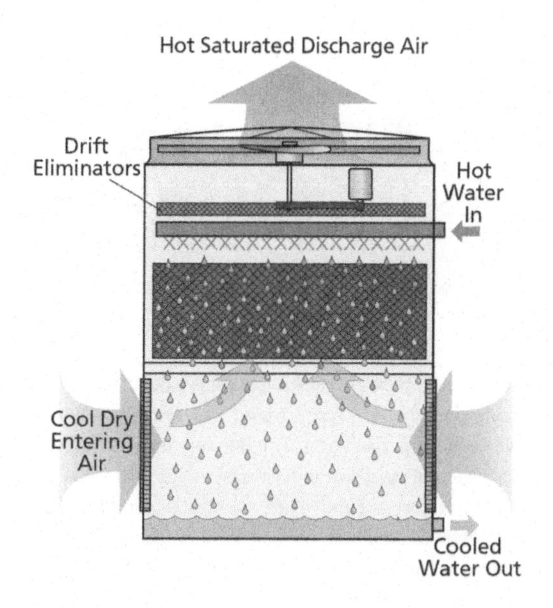


**FIGURE 10.16**

Evapco (Westminster, MD) uses a patented technology to reduce that amount of water evaporated, while removing the heat from the remaining water. Warm moist air is drawn to the top of the cooling tower by the fan and discharged to the atmosphere. The cooled water drains to the basin at the bottom of the tower and is returned to the heat source to be used again.

the population size of macroinvertebrates and reduce fish fitness (body fat, egg size and number, etc.). In addition, the thermal tolerance for many aquatic organisms is narrow.

The distribution, health, and survival of Pacific Northwest salmonids depend on water temperatures. Certain activities in the life of a salmonid are triggered by water temperature, such as cooler river water temperatures in the fall, which signal the time for upstream migration. To some degree, fish can tolerate the seasonal swings in temperature and the more dramatic variations in climatic conditions that push temperatures outside the optimal range. When ambient water temperatures become too warm, salmon respiration rates increase and their weights decline. However, in some cases, fish can maintain cooler temperatures than the ambient temperature, thus mitigate the negative physiological effects of warmer waters. For example, spring Chinook salmon in the Yakima River, Washington can maintain body temperatures ~2.5°C below the ambient stream temperature. How can an ectotherm keep their temperatures below the ambient water temperatures? These fish use behavior to "regulate" their body temperatures by seeking out colder waters in a stream, capitalizing on the spatial variation within a stream reach, e.g., areas of hyporheic exchange.

Nevertheless, rising temperatures in many Northwest streams are correlated with the decline of populations in the same areas. Numerous studies since 1985 have documented declines in Oregon and Washington salmonid populations, where temperature was identified as a contributing factor. Without their behavioral capacity to mitigate water temperatures, these fish might be more vulnerable than they already are.

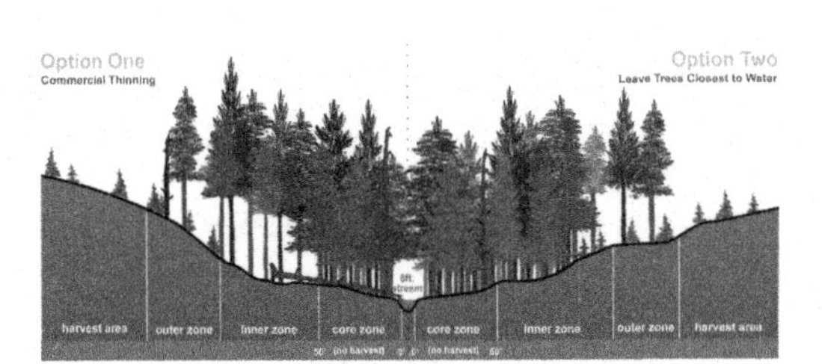


**FIGURE 10.17**

Buffers in managed forests (Source: Washington Forest Protection Association). Forests are complex and dynamic environments. Variations in topography, soil type, stream size, and other conditions affect the number of trees foresters are required to leave in the Riparian Area. The law requires leaving the Riparian Area in a condition today that will grow to replicate natural stands of older forest at age 140 years. A certain number of trees and canopy cover need to be left within the Riparian Zone to achieve this Desired Future Condition (DFC). Both of the proposed buffer options depicted below will attain the DFC and either can be used depending upon site specific conditions. Both have a 50-foot core "no harvest zone". Option 1 calls for thinning to encourage growing large trees faster. Option 2 leaves more trees closer to the stream.

Urban runoff-stormwater discharged to surfacewaters from roads and parking lots-can be a source of elevated water temperatures, but in many areas the removal of forest cover has a more direct impact. In many states, where forestry is practiced, strict rules maintain shading for streams (Fig. 10.17). For example, stream margins are protected from tree cutting to minimize the loss of shade.

## SEDIMENT AND TURBIDITY

### WATERSHED SOURCES OF SEDIMENT AND STREAM CHANNEL PROCESSES

Sediment transport is a natural process and many have argued that the point of rivers is to move sediment downstream. However, with land use changes, e.g., deforestation and construction; agricultural practices; and development activities, accelerated erosion rates is ubiquitous. Sediment in the water column reduces transparency and can be deposited downstream and exacerbate flooding. Three principal sources of sediment are the following:

1. Soil erosion of upland areas from overland flow, including farmed areas (Fig. 10.18);
2. Head-cut or knickpoint migration in degrading low-order systems; and
3. Remobilization of stored sediment through channel processes acting on floodplains or other storage sites, including channel migration, bank widening, and avulsion (Fig. 10.19).

In general, low-order streams and fluvial erosion of storage areas are poorly understood and may represent significant, if not major, sources of sediment in many streams. Similarly, these two processes may strongly affect nitrogen and phosphorus dynamics and delivery to receiving waters; erosion of storage areas exerts an equally strong influence on retention times.

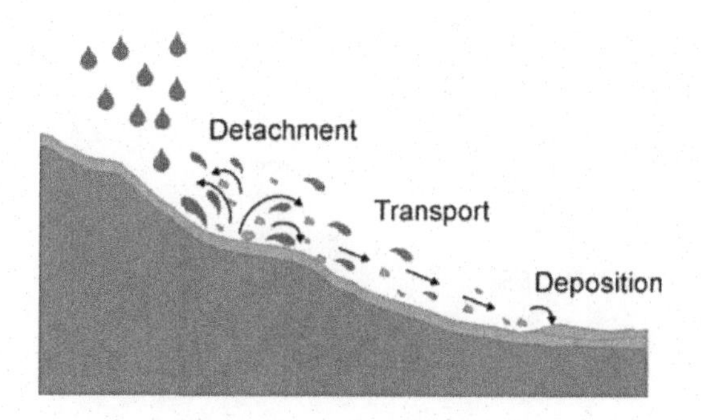


**FIGURE 10.18**

Upland erosion processes.

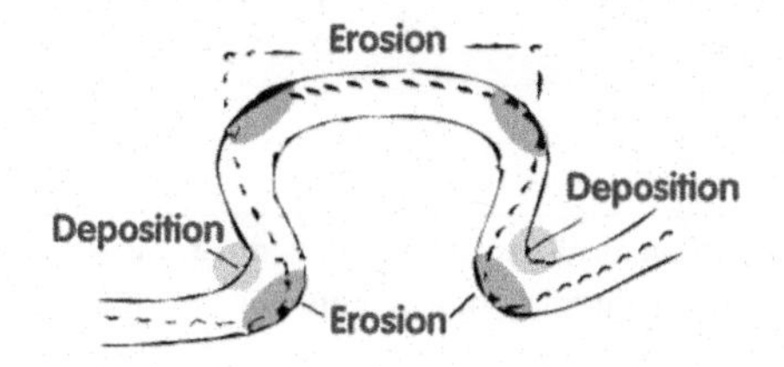


**FIGURE 10.19**

Streams have a complex response to sediment load, topographic gradient, and water velocity. When accelerated erosion puts more sediment in a stream, these dynamics can easily be disrupted.

## CARP AND TURBIDITY

Although turbidity is often associated with water inflow sources (nutrient loading and algae growth or suspended sediments), other biotic factors can also play a role. For example, carp can be a source of turbidity. The common carp (*Cyprinus carpio*) is native to Europe and Asia, and has been introduced to almost every part of the world. They can eat a herbivorous diet of water plants, but prefer to scavenge the bottom for insects, crustaceans (including zooplankton), crawfish, and benthic worms. By feeding in lake bottoms they tend to disrupt the muds by scrounging for food. Common carp destroy vegetation, increase water turbidity, and promote eutrophication of ponds and lakes. In addition, the effects of common carp can even reduce *largemouth bass* catch rates because of turbidity enhancement by high biomass of common carp.

## ACID RAIN: LONG-RANGE TRANSPORT AND LONG-TERM DATA

### EARLY OBSERVATIONS AND ASSOCIATIONS

Acidic precipitation was documented as early as 17th Century with observations linked to corrosion of marble statues. But it was not until the 1950s when researchers began documenting region-wide precipitation of acid rain. In 1963 Ecologists initiated the Hubbard Brook Ecosystem Study to assess

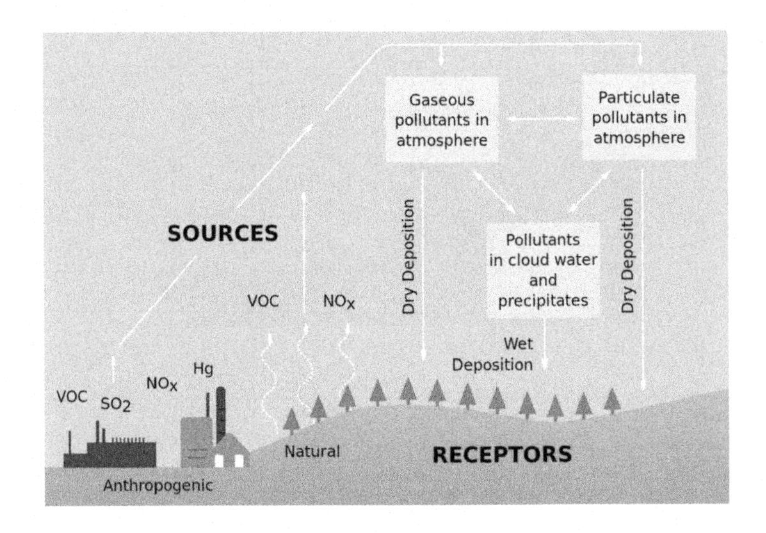


**FIGURE 10.20**

Processes involved in acid deposition (Source: Wikipedia). Although $CO_2$ lowers the pH of rainwater, anthropogenic sources of $SO_2$ and $NO_x$ are the main cause of acid rain.

mass balance in water and chemical budgets using gauged watersheds. From the study's inception, rain and snow inputs to Hubbard Brook were unusually acid, and researchers worked to evaluate the ecological impacts, e.g., forest health. Nevertheless, the extent of that surface geology vulnerability to acid precipitation was not appreciated at first. For example, pH changes in Alpine Lakes in California began in the mid 19th century, coinciding with increased settlement and energy use during the California gold rush, whereas paleolimnological research suggests that Sierra Nevada Ecosystems began to be consistently affected by acid deposition by the 1920s.

## MECHANISMS OF ACIDIFICATION

As we discussed in Chapter 6, rain water is slightly acidic naturally. The pH can be further reduced by volcanic eruptions. In fact, acidic deposits have been detected in glacial ice thousands of years old in remote parts of the globe.

Modern anthropogenic acid deposition increased dramatically in Europe and eastern North America after World War II, as the combustion of fossil fuels dramatically increased. To avoid local air pollution problems, emissions are released via high smoke stakes, which allow gases to be carried hundreds of kilometers in the atmosphere before they are converted to acids and deposited (Fig. 10.20).

The principal causes of acid rain are nitrogen and sulfur compounds from the combustion of fossil fuels:

***Nitric acid production*** When burning fossil fuels, $N_2$ can react with the fuels in the presence of $O_2$, producing several types of nitrogen oxides. These oxides, which are usually referred to as NOX,

include a family of seven nitrogenous oxide compounds, i.e., $N_2O$, $NO$, $N_2O_2$, $N_2O_3$, $NO_2$, $N_2O_4$, and $N_2O_5$.
When any of these oxides dissolve in water and decompose, they form nitric acid ($HNO_3$) or nitrous acid ($HNO_2$).

***Sulfuric acid production*** There are a large number of aqueous reactions that oxidize sulfur from Sulfur (IV) to Sulfur (VI), leading to the formation of sulfuric acid. The most important oxidation reactions are with ozone, hydrogen peroxide and oxygen; reactions with oxygen are catalyzed by iron and manganese in the cloud droplets.
Sulfur (IV) dioxide dissolves in water and then, like carbon dioxide, hydrolyzes in a series of equilibrium reactions:

$$SO_2(g) + H_2O \leftrightharpoons SO_2{\cdot}H_2O \tag{10.1}$$

$$SO_2{\cdot}H_2O \leftrightharpoons H^+ + HSO_3^- \tag{10.2}$$

$$HSO_3^- \leftrightharpoons H^+ + SO_3^{2-} \tag{10.3}$$

In the gas phase, sulfur dioxide is oxidized to S(VI) by reaction with the hydroxyl radical via an intermolecular reaction:

$$SO_2(g) + OH^{\cdot -} \leftrightharpoons HOSO_2^{\cdot}(g) \tag{10.4}$$

which is followed by

$$HOSO_2^{\cdot}(g) + O_2 \leftrightharpoons HO_2^{\cdot} + SO_3(g){\cdot} \tag{10.5}$$

In the presence of water, sulfur trioxide ($SO_3$) is converted rapidly to sulfuric acid:

$$SO_3(g) + H_2O(l) \leftrightharpoons H_2SO_4(l){\cdot} \tag{10.6}$$

When clouds are present, the loss rate of $SO_2$ is faster than can be explained by gas phase chemistry alone. This is due to reactions in the liquid water droplets. Just as nitrogen oxides are referred to as NOX, SOX encompasses the family of sulfur oxides.

## WET AND DRY DEPOSITION

After 18 years of continuous measurement to fit a linear regression (Fig. 10.21), researchers at the Hubbard Brook Ecosystem Study demonstrated that sulfate deposition at Hubbard Brook was strongly related to $SO_2$ emissions hundreds or thousands of kilometers distant. However, as researchers tried to conduct mass balances of this deposition, they made the distinction between wet and dry deposition.

Wet deposition of acids occurs when any form of precipitation (rain, snow, hail, etc.) removes acids from the atmosphere and delivers it to the Earth's surface. This can result from the deposition of acids produced in the raindrops (see aqueous phase chemistry above) or by the precipitation removing the acids either in clouds or below clouds. Wet removal includes both gases and aerosols.

Acid deposition also occurs via dry deposition in the absence of precipitation. This can be responsible for as much as 20 to 60% of total acid deposition. This occurs when particles and gases stick to the ground, plants, or other surfaces.

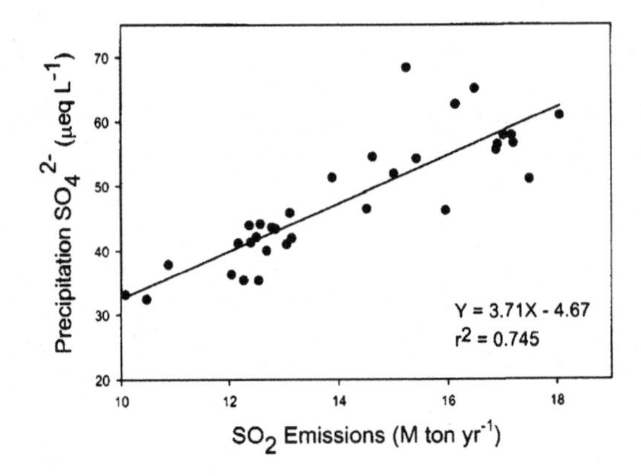


**FIGURE 10.21**

Stream $SO_4^{2-}$ as a function of $SO_2$ emissions (Source: Driscoll et al., 2001).

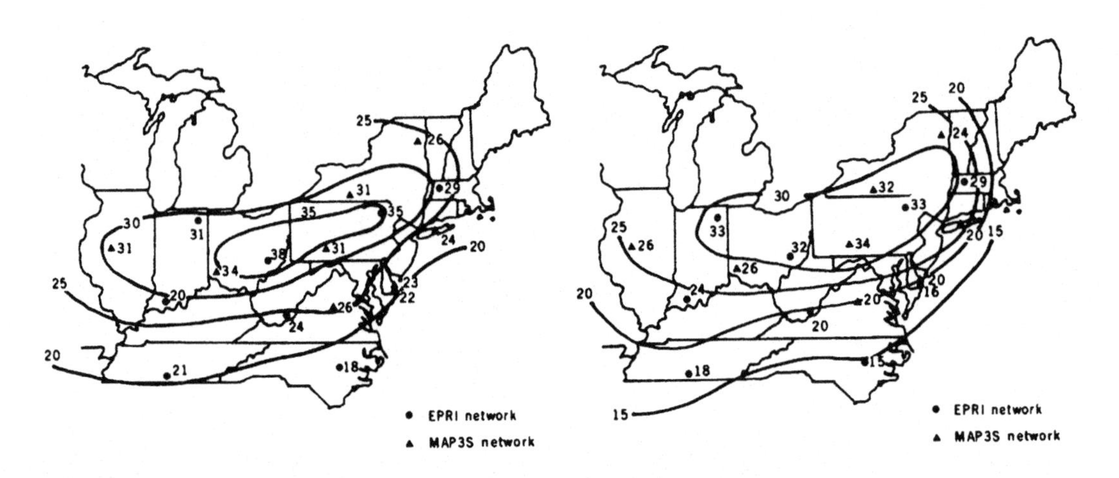


**FIGURE 10.22**

Average $SO_4^{2-}$ (left panel) and $NO_3^-$ (right panel) (in micromoles per liter) for the period August 1978 through June 1979. Average concentrations for individual sites are plotted adjacent to the site locations (Source: Pack, 1980).

## CONTINENT LEVEL DEPOSITION

Once large scale monitoring began, the continental scale of low pH deposition could be clearly documented, especially northern Europe and northeast of North America (Fig. 10.22). In spite of continental wide scale, ecological impacts were required before regulatory action was implemented, which were done at the watershed level at Hubbard Brook.

**FIGURE 10.23**

Forest damage from acid rain (Source: Lovecz, Wikimedia.org. Public domain).

## SOIL CHEMISTRY AND SOIL FERTILITY

The Hubbard Brook Ecosystem Study documented some of the most compelling ecosystem impacts. Because the soils, forests, and streams were part of a long-term study, researchers could document ecosystem-wide impacts of acid rain. For example, not only was stream pH values declining, but various cations were leached at higher rates than soil regeneration rates. Thus soil fertility and forest health were being affected by acid rain.

Cations, such as calcium and magnesium can be leached from soils, effectively replacing the cations with protons on soil clay binding sites:

$$2\,H^{+}(aq) + Mg^{2+}(clay) \rightleftharpoons 2H^{+}(clay) + Mg^{2+}(aq) \tag{10.7}$$

$$2\,H^{+}(aq) + Ca^{2+}(clay) \rightleftharpoons 2H^{+}(clay) + Ca^{2+}(aq) \tag{10.8}$$

As of result of acid rain, the Hubbard Brook studies documented calcium and magnesium leaching, a situation researchers hypothesized reduced forest growth (Fig. 10.23). To test this hypothesis, wollastonite (a calcium silicate mineral) was added in 1999 to an entire watershed in an amount roughly equivalent to the amount estimated to have leached in the previous 50 years. Results suggest positive survival and growth responses in sugar maple (*Acer saccharum*).

Acid rain can also have direct effects on trees. For example, when calcium is leached from the needles of red spruce, the trees become less cold-tolerant and exhibit winter injury and even death. Plant growth reduction has been well documented and has led to a decline in forest health.

Besides the loss of Ca and Mg, $Al^{3+}$ was lost from soils. But instead of being a soil fertility issue, $Al^{3+}$ can be toxic to aquatic organisms.

## AQUATIC EFFECTS OF ACID RAIN

Because of the preponderance of coal plants in eastern United States, impacts of acid deposition was easily documented in lakes and creeks. The declined habitat quality could be directly attributed to acid rain.

Both the lower pH and higher aluminum concentrations in surfacewater that occur as a result of acid rain can cause damage to fish and other aquatic animals. At pH lower than 5, most fish eggs will not hatch and lower pHs can kill adult fish. As lakes and rivers become more acidic biodiversity declines.

Acid rain has eliminated groups of aquatic insects some fish species, including brook trout (*Salvelinus fontinalis*) in some lakes, streams, and creeks in geographically sensitive areas, such as the Adirondack Mountains of the United States. Alternatively, changes in species composition has also been documented. In a small granitic catchment in France ephemeroptera abundance declined in streams that had a pH of 4.9, whereas plecoptera larvae were dominant. These data suggest that stoneflies may be better adapted to an acidic condition.

According to the Environmental Protection Agency, acid rain caused acidity in 75% of surveyed acidic lakes and about 50 percent of surveyed acidic streams. In addition, high altitude forests are especially vulnerable as they are often surrounded by clouds and fog, which are more acidic than rain. Lakes at these elevations may also be even more poorly buffered and sensitive to pH changes.

## REGULATORY FRAMEWORKS AND CURRENT SITUATION

International cooperation to address air pollution and acid deposition began with the 1972 United Nations Conference on the Human Environment. In 1979 the Geneva Convention on Long-range Transboundary Air Pollution created the framework for reducing air pollution and acid deposition in Europe. The convention produced the first legally binding international agreement to reduce air pollution on a broad regional basis. This first agreement has been extended by several protocols since its original inception. After some several years of debate and President Reagan's strategic use of uncertainty to stall, the United States began addressing its emissions with agreements with Canada and regulatory mechanisms in the Clean Air Act.

The Clean Air Act (CAA) Amendments (1990) established a cap and trade program to address acid deposition nationwide by reducing $SO_2$ and NOX emissions from coal-fired power plants. In contrast to traditional regulatory methods that establish specific emissions limitations, a cap and trade program introduced a novel allowance trading system that harnessed market incentives to reduce pollution. The EPA has capped annual $SO_2$ emissions at 8.95 million tons, a level of 50% below the 1980 emissions of the power sector.

Reductions in emissions, primarily of $SO_2$ due to federal regulations, caused ~60% decline in acidity at Hubbard Brook. NOX and $SO_2$ emissions reductions have led to a significant drop in acid deposition, as well as sulfate ($SO_4^{-2}$) and nitrate ($NO_3^-$) deposition (Fig. 10.24). However, ammonia ($NH_3$) and ammonium deposition continue to increase in some parts of the United States, especially in areas with intensive agriculture and livestock production.

There remains a highly critical view of the acid rain regulations and on-going acid rain produced by some countries, such as China. The reader is encouraged to read more about interaction between Environmental Science and policy with respect to this issue. In addition, there are legacy effects of acid rain (depleted $Ca^{2+}$ soils and reduced stream and lake $Ca^{2+}$ to support the food web) that we are just starting to understand, thus remaining an important research and management area.

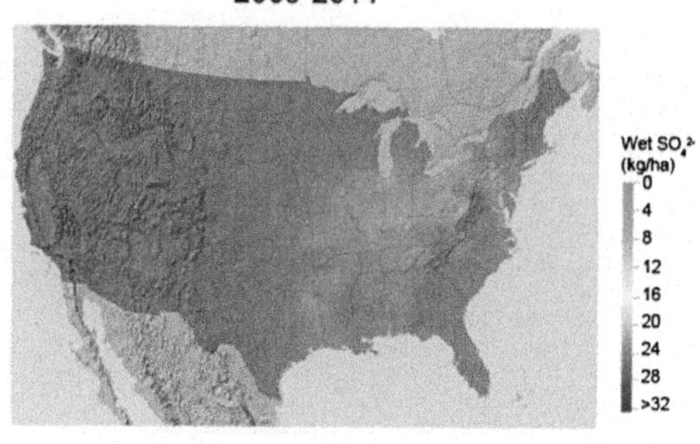


**FIGURE 10.24**

Mean wet sulfur deposition comparing 1989–1990 and 2009–2011 (Source: Burns et al., 2011).

## TROPHIC STATUS REVISITED: CULTURAL EUTROPHICATION

### CULTURAL EUTROPHICATION

In the 1970s, the OECD (Organization for Economic Cooperation and Development) provided one of the first definitions of cultural eutrophication as "an enrichment of water by nutrient salts that causes structural changes to the ecosystem such as: increased production of algae and aquatic plants, depletion of fish species, general deterioration of water quality and other effects that reduce and preclude use."

Although natural and slow eutrophication process occurs in many water bodies, cultural eutrophication is the result of a continuous increase in nutrient loading, mainly nitrogen and phosphorus, until a threshold is reached—dramatic changes in the lake or stream.

With a significant increase of algae biomass as a result of greater nutrient availability, algae grows in an uncontrolled manner or "blooms". As biomass for heterotrophs, the algae fuels oxygen consumption in the water body, which can then go anoxic. In stratified lakes or deep rivers, microorganisms under anoxic conditions can produce or release toxic compounds, such as $NH_3$ or $H_2S$. In addition, the hypoxia reduces plant and animal biodiversity and is associated with fish kills. Hypoxic conditions generally occurs when the rate of degradation of the algae by microorganisms is greater than that of oxygen regeneration. These conditions are exasperated during summer conditions when water temper-

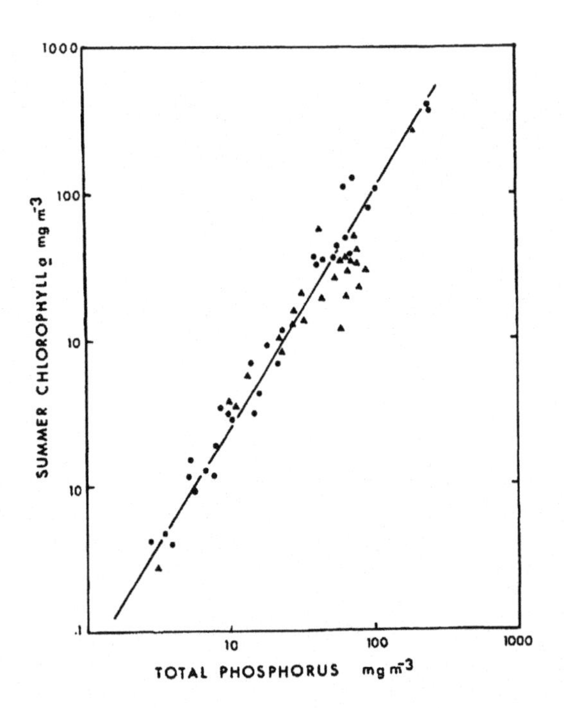


**FIGURE 10.25**

Relationship between phosphorous concentration and chlorophyll in Canadian lakes (Dillon and Rigler, 1974).

atures are warm, water capacity to hold oxygen is low, and lakes are highly stratified. Without mixing, a hypolimnion might remain anoxic all year around.

Although lake productivity varies, the excessive input of nutrients has become a nearly universal problem in Aquatic Systems with a wide range of impacts. According to a report by the European Union, eutrophication affects 54% of Asian lakes, 53% of those in Europe, 48% of those in North America, 41% of those in South America, and 28% of those in Africa. In addition, streams also suffer from eutrophication.

## LAKE PROCESSES IN CONTEXT

Although by the mid-1950s, some evidence suggested that temperature and precipitation could predict lake productivity, most Ecologists believed that nutrients drove primary production. The phenomenon was first documented in Japan, but soon a similar positive relationship between phosphorus and phytoplankton biomass for lake ecosystems was identified throughout the Northern Hemisphere in Canada (Fig. 10.25). In fact, the relationship describing phosphorus and phytoplankton biomass for Japanese and Canadian lakes was nearly identical.

The data from Japan and North America strongly supported the hypothesis that nutrients, particularly phosphorus, control phytoplankton biomass in lake ecosystems. In fact, there is a strong positive correlation among chlorophyll concentrations, photosynthetic rates, and total phosphorus concentration. Nevertheless, because correlation is not causation, the relationship needed to be tested using a predictive causative mechanism.

## NUTRIENT LOADING AND LAKE EUTROPHICATION

Based on the hypothesis that total phosphorous limits algae growth, W. Thomas Edmondson predicted that eutrophic conditions in Lake Washington were the result of nutrient loading from sewage discharge.

As the city of Seattle grew in population, so did the volume of sewage being discharged into the lake. As a result, the water quality of the lake began to change, e.g., loss of clarity and blooms of the Cyanobacteria *Oscillatoria rubescens*. By monitoring the lake's water quality, Edmondson predicted that the Cyanobacteria blooms would decline with the diversion of the waste-water.

By the 1963, raw sewage was being diverted from the lake, and Edmondson's prediction could be tested. In the following years, chlorophyll *a* concentrations began to decline and subsequent decline in total phosphorous, dissolved phosphorous, and the Cyanobacteria dominated by *Oscillatoria rubescens* (Fig. 10.26).

## TESTING THE P AS A CONTROL OF PHYTOPLANKTON GROWTH

Just as Edmondson was documenting the changes in Lake Washington with the reduction of nutrient loading, an experiment that tested nutrient additions was beginning in Canada. In the case, Limnologists designed whole lake experiments based on nutrient additions.

For example, a lake called Lake 226 was split in two by a vinyl curtain. Each subbasin of Lake 226 was fertilized from 1973 to 1980 and monitored until 1983. In one basin, a mixture of sucrose and nitrate was added and in the other basin sucrose, nitrate, and phosphate. Both sides of Lake 226 responded to nutrient additions and the phytoplankton biomass surpassed that in reference lakes also being monitored. Phytoplankton biomass remained elevated in Lake 226 until the experimenters stopped adding fertilizer at the end of 1980 (Fig. 10.27). Then, from 1981 to 1983 the phytoplankton biomass in Lake 226 declined significantly, just as they had in Lake Washington.

In conclusion, the experimental data support the correlative relationship between phosphorus concentration and rate of primary production and the cause and effect that nutrient, especially phosphorous, availability controls rates of primary production in freshwater ecosystems.

## PHOSPHORUS DYNAMICS: INTERNAL LOADING

Phosphorus availability in the water column depends on a number factors that include surfacewater input, water mixing, and sediment fluxes. We have discussed water mixing in Chapter 6, section "Lake Mixing", but now we need to dig a bit deeper into the sediments of the lake bottoms. The bottom of lakes is composed of detritus, i.e. organic matter, that has sank and settled on the lake bottom and at various stages of decomposition. Joining the material at the lake bottom is a host of detrivores, such as oligochaetes, nematodes, protozoan, but mostly bacteria. These organisms consume oxygen and when measured in the sediment, oxygen concentrations is often unmeasurable within a cm or two from the water-sediment interface. In fact, depending the organic contribution, oxygen demand from sediments can remove oxygen from the hypolimnion.

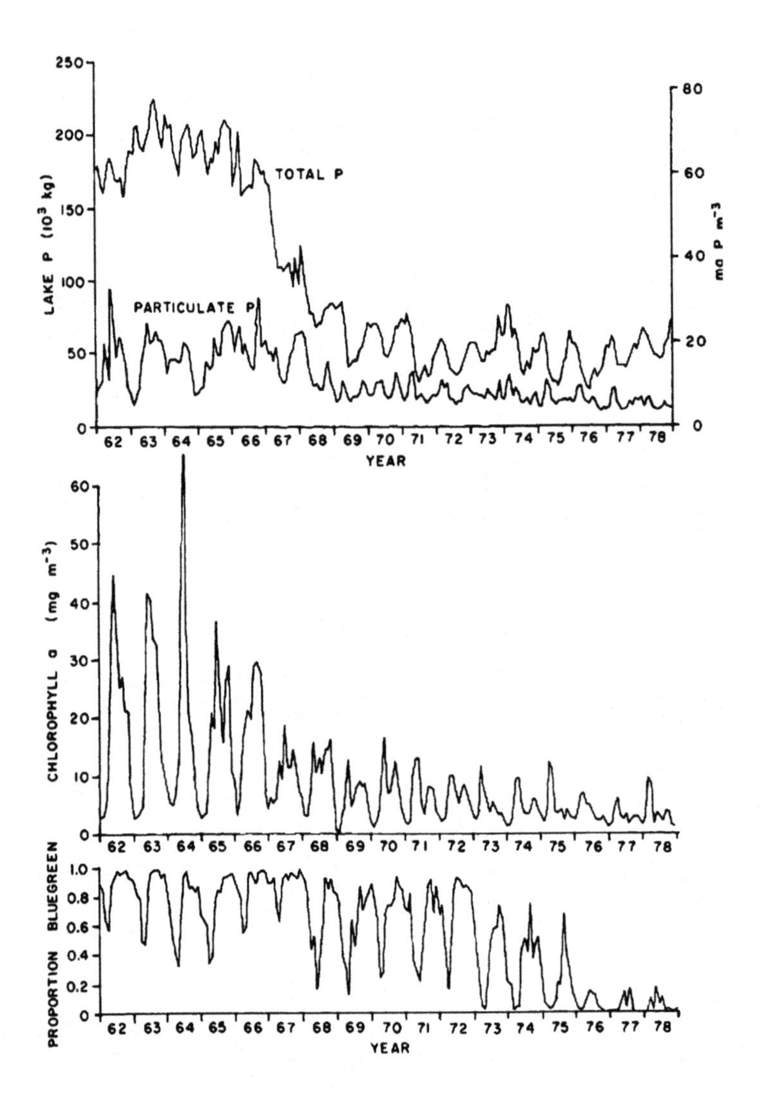


**FIGURE 10.26**

Phosphorous, chlorophyll, and Cyanobacteria (referred to as the proportion of "Bluegreen") concentrations in Lake Washington (Edmondson and Lehman, 1981).

Orthophosphate or $PO_4^{2-}$ is the inorganic form of phosphorus taken up by algae, but inorganic phosphorous can also occur as $PO_4^{3-}$. Since common analytical methods measure both, we usually refer to these as Soluble reactive phosphorus, or SRP. SRP also reacts and precipitates with metals, especially the oxidized state of iron or iron III ($Fe^{3+}$). When the sediments are oxidized, iron III precipitate with phosphate forming iron-containing minerals. However, when the sediments are anoxic, bacteria use iron as an electron acceptor, convert it to Fe II ($Fe^{2+}$), which is soluble, and phosphate is

**FIGURE 10.27**

A whole lake experiment shows the effect of nutrient additions on average phytoplankton biomass (Findlay and Kasian, 1987).

released and free to diffuse out of the sediments into the water column. The cycle of P precipitation under aerobic conditions and mobilization under anaerobic conditions is called internal loading—in contrast to watershed sources of P, such as runoff, streams, or other discharges.

## NITROGEN AND NITROGEN FIXATION

On a superficial level, most freshwaters may be considered P-limited, but data suggest that we consider a more nuanced approach. In fact, it is more useful to consider how nutrients shift species composition and productivity in uneven ways. For example, with added P, some lakes experience blooms of nitrogen-fixing Cyanobacteria. When N-fixing bacteria are favored over non-N-fixing bacteria, then it implies that N is limiting, once the bacteria are released from the P limitation.

## P AND PRIMARY PRODUCTIVITY REVISITED: SCATTER AROUND THE LINEAR RELATIONSHIP

Although the relationship between P and primary productivity is a powerful predictive relationship, Environmental Scientists must also pay attention to the variation associated with these relationships. For example, Fig. 8.14 suggests a strong relationship. However, the axes are log-transformed, which masks orders of magnitude of variation. When residual variation remains unexplained by the predictor, we should consider other factors that might influence the response. In other words, besides P, are there other explanations for eutrophic conditions in surfacewaters?

***Phosphate physiology*** The uptake and use efficiency of P varies with algal species and even within genotypes. As nutrient concentrations change, different efficiency genotypes might be selected and respond to nutrient inputs. Therefore nutrient responses in a surfacewater may depend on range of the efficiencies, where some genotypes will be favored over others, a phenomenon that might explain variation in eutrophic conditions.

***Bouyancy*** The vertical distribution algae can depend on the carbohydrate ballast dynamics and the colony size. When thermal stratification occurs, algal colonies can exhibit diurnal vertical migration. But winds and lake mixing can also affect these distribution factors that might promote or reduce eutrophic conditions.

***Grazing/Herbivory dynamics*** As described earlier, Lake Washington's blooms were the result of bottom-up controls. However, this conclusion is an over simplification. What if the phytoplankton could not be consumed by the zooplankton filter feeders? For example, some Cyanobacteria produce a variety of toxins, which can disrupt specific eukaryotic cellular functions and reduce zooplankton health. Or as was documented in Lake Washington, *Oscillatoria rubescens*, a Cyanobacteria species produces mucilage (i.e., slime) that gums up filter feeder appendages, making them less susceptible to grazing. In other cases, by growing in a filamentous habitat, filter feeder cannot easily consume these algal species. In any case, in contrast to more palatable algae species, these examples demonstrate that blooms might partially depend on phytoplankton characteristics.

***Cascading effects*** One of those factors is the intensity of predation on the zooplankton that feed on phytoplankton. As discussed in Chapter 8, Trophic Cascades can control primary productivity and may augment eutrophic conditions. However, short of introducing new predators, which may have a wide range of ecological impacts, the role of top-down controls are difficult to demonstrate.

***Climate change*** has a wide range of direct and indirect impacts on eutrophication. However, one direct impact of warmer climate is that Cyanobacteria seem to be better suited for warmer waters, thus their growth and impact on surfacewaters may continue to expand with a warming planet.

Overall, these explanations suggest a combination of top-down and bottom-up controls. Although top-down and bottom-up dichotomy provides a reasonable model of lake phytoplankton biomass predictions, a better question might be "under what circumstances are bottom-up controls active" and "under what circumstances are top-down controls active"? Anwering such questions will help us better understand and facilitate the mitigation of cultural eutrophication.

## NUTRIENT SOURCES THAT DRIVE EUTROPHICATION

***Agricultural practices*** and the use of fertilizers contribute to N and P loading. When these nutrients exceed the capacity of a farm to assimilate them, they can be exported into surfacewaters.

***Discharge of wastewater into water bodies*** Wastewater directly discharged into surfacewaters release nutrients and promotes algae growth. Waste-water treatment facility construction in the industrialized world coincided with rapid and sustained economic growth. However, without these economic conditions, the costs to construct and manage treatment facilities is prohibitive or unappealing politically, in spite of the direct human health implications and maintaining functioning surfacewaters.

***Reduction of assimilation capacity*** After years of nutrient inputs and/or the loss of wetlands associated with streams and lakes, these systems may reach a tipping point. Without the capacity to assimilate or to 'self-purify,' these systems maybe more vulnerable to nutrient loading. Thus developing effective predictive models may be useful to limit eutrophication.

## IMPACTS AND MANAGEMENT OF EUTROPHICATION

Eutrophication conditions have numerous undesirable impacts on both the surfacewater itself and the ecosystem services that local communities rely on. For example, some of the most common impacts of eutrophication include the following:

1. Increase particulate substances (phytoplankton, zooplankton, bacteria, fungi, and debris) that reduce the clarity of the water and light penetration, reducing photosynthesis and further exasperating low water column $O_2$ concentration;
2. Production of inorganic chemicals, such as $NH_3$, $NO_2^-$, and $H_2S$, which can form harmful substances, such as nitrosamines, suspected of mutagenicity in drinking water treatment plants;
3. Creation of organic substances or presence of particular algae that give the water disagreeable odors or tastes. In addition, these complex chemical compounds can accelerate corrosion and limit the flow rate in treatment facilities;
4. Reduction or collapse of a fishery;
5. Production of toxins by algae that can affect human and nonhuman health;
6. Decline of tourism or other economic impacts, e.g., due loss of fisheries;
7. Reduction of oxygen concentration, especially in the deeper layers that can lead to fish kills.

These effects generally impact local communities, but the policies and regulations needed to mange such situations reside with jurisdictions, typically with mixed agendas; thus creating the political will to address these impacts is often a challenge.

## CARBON, NITROGEN, AND PHOSPHOROUS: WASTEWATER TRENDS

Sewage emissions of nitrogen (N) and phosphorus (P) and organic carbon constitute an important source of nutrients in freshwater and coastal marine ecosystems at local, regional, and even global scales.

Since the 1960s, sewage N and P discharges to surfacewater are a function of increasing population and economic growth; changes in diet, urbanization; regional climate and geologic history; and construction of sewerage and treatment systems. P discharges from sewage depend on the similar process. Since P loading has major effects on lake water quality, controlling P sources has been an important regulatory agenda. For example, the use of P-based detergents in laundry washing machines and dishwashing machines has a consequential role in P loading. In the mid 1970s, P-free detergents based on zeolites were introduced and progressively replaced sodium tripolyphosphate ($Na_5P_3O_{10}$) detergents. In some parts of the world, P-free detergents based on zeolites make up 80 to 100% of all laundry detergent. However, the use of automatic dishwashers has increased, and they rely on P-based detergents and have again become an important source of P-loading to surfacewaters. Thus in spite of some progress in waste treatment and source control, relatively subtle consumer choices can have important effects on lake water quality.

## CONTROLLING OR MANAGING PHOSPHORUS

Traditional eutrophication reduction strategies range from addressing the symptoms to addressing the cause. For example, to treat anoxic conditions, physical mixing of the water or aeration is used to increase water column oxygen concentrations. But if the water body is large or hypolimnion is deep, this is far from practical. Copper is relatively toxic to algae and has been used in lakes and reservoirs for decades to prevent the growth of algae in drinking water supplies. In some cases, nutrient inputs reductions have been attempted, but this requires a high proportion of cooperation within the catchment area. In general, these strategies are usually ineffective, expensive, and impractical for surfacewaters.

Nevertheless, using a suite of activities, the most effective strategy to maintain healthy waters is to prevent eutrophication:

- Improving wastewater treatment plants performance;
- Restoring or expanding ecosystems that remove or trap nitrogen and phosphorus in run-off water;
- Limiting the amount of phosphorus in detergents;
- Improving agricultural practices by providing appropriate incentives for farmers to reduce soil and nutrient losses;
- Promoting animal husbandry practices to limit or treat animal wastewater.

However, in many cases, the surfacewaters have been too compromised and require more costly solutions, such as the following:

- Removing and treating hypolimnetic water rich in nutrients;
- Remove top 10–20 cm of sediment with high P and subjecting the same to regular redox fluctuations;
- Aerating hypolimnion to prevent reducing conditions and sediment release of P and toxic substances;
- Chemical precipitation of phosphors by adding iron, aluminum salts, or calcium carbonate to reduce free P in the water column.

Since about 50% of the lakes in the United States show signs of eutrophication, the task will require local, regional, state, and federal responses to maintain and possibly restore the water quality of these lakes.

## HARMFUL ALGAL BLOOMS: HABS

### HARMFUL ALGAL BLOOMS

Cyanobacteria become Harmful Algal Blooms (CHABs) when toxin-producing strains dominate surfacewaters. Although their impact is often associated eutrophication, they pose a particularly dangerous outcome because of their toxicity. Thus Cyanobacteria have become the most studied groups of phytoplankton.

Cyanotoxins can cause illness and death when toxin-containing water or animal tissue is ingested or through incidental contact. There are four categories of toxin effects, which are usually associated with specific toxins:

***Hepatotoxins (microcystins, cylindrospermopsin)*** liver damage can result from ingestion; binding to protein phosphates inhibits normal cellular function;
***Nerotoxins (anatoxins and axitoxins)*** results in an adverse effect on the structure or function of the central and/or peripheral nervous system;
***Dermatoxins (lynbyatoxins and aplysiotoxins)*** is a toxic chemical that damages skin, mucous membranes, or both, often leading to tissue necrosis;
***Skin irritants (all cyanotoxins)*** give rise to rashes associated with exposure or contact with cyanobacteria.

The taxa that can produce toxins is extensive and the chemical diversity of harmful chemicals is almost inconceivable, where *Anabaena circinalis* can produce Anatoxin-a, ~90 variants of mycrocystin, and saxitoxin, whereas *A. flos-aqua* only produces Anatoxin-a, Homoanatoxin-A, and Anatoxin-a(S). A study evaluating microcystin-LR in four California reservoirs identified different Cyanobacterium strains: 11 contained high concentrations of microcystin-LR.

Historically, three Cyanobacteria groups had been identified as problems: *Anabaena* spp., *Aphanizomenon* spp., *Microcystus* spp. and known euphemistically as 'annie, fanny, and mike'. Added to the most common sources of CHAB, are *Plantothrix* spp. and Oscillatoria, perhaps, adding 'thrixie' and 'ozzie' to our list of culprits. In spite of the play on words, the deadly impact of these compounds on humans and water-dependent organisms is real.

Pinto Lake, a small tectonic lake along the Central Coast of California, is known for regular blooms of Cyanobacteria and high concentrations of microcystins. Microcystins are peptide chains that can bioaccumulate with toxic effects across the food web (Fig. 10.28). In this case, Pinto Lake has been implicated as the source of microcystins in the Monterey Bay and death of over 20 threatened southern sea otters. The lake experiences CHABs in the summer and late fall, but outflow to the ocean is minimal because of the Mediterranean-type climate. It has been hypothesized that with the first rains, high concentrations of microcystins enter the coastal zone and the marine food web and sea otter prey (Fig. 10.29).

Unfortunately, using microscopy to identify the cyanobacteria, predict their population dynamics, and assess their potential to produce toxins remains an impractical approach to address the risks of CHABs. However, molecular biology tools have been increasingly applied to Cyanobacteria, where the genes producing the toxins can be quantified from environmental samples. More generally, using molecular biology tools, environmental scientists may be able to determine the potential production of toxins by analyzing the toxin-producing genes (or gene clusters) extracted from cells in the water column, or in the benthos. (See Fig. 10.30.)

## PESTICIDES AND POISONS

### STREAM RECOVERY FROM TOXIC SPILLS: DUNSMUIR 1991

In 1991 a Southern Pacific freight train derailed and a damaged tank car leaked 12,000 gallons of Metam Sodium, a potent herbicide and pesticide, into the Sacramento River. Dissolved in water, Metam Sodium forms several highly toxic compounds.

Volatile compounds created a toxic cloud that affected the surrounding communities, notably the small town of Dunsmuir, with reports of headaches, shortness of breath, chest pains, rashes, dizziness, and vomiting.

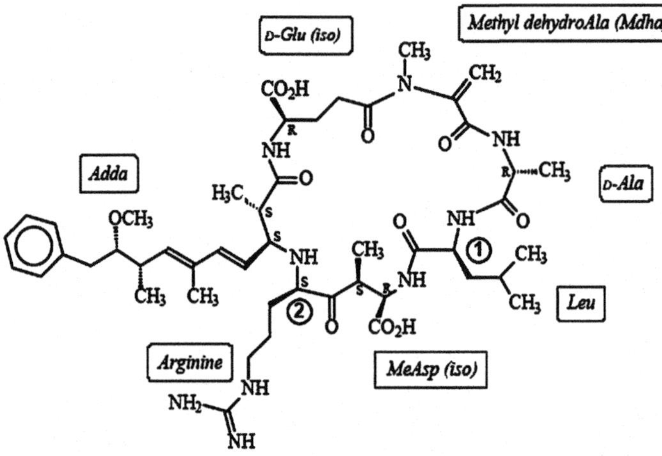

**FIGURE 10.28**

The generic structure of a microcystin (Source: Harada et al., 1996). Variations occur primarily at positions 1 and 2. For example, Microcystin-LR contains the amino acids leucine (L) and arginine (R) at positions 1 and 2 respectively. Microcystin-RR has arginine at both positions. Nodularins are similar, with the five amino acids Adda-$\gamma$Glu-Mdhb-$\beta$MeAsp-Arg making up the core ring system.

The spill killed most of the riparian and aquatic life along a 41 mile stretch of river to Lake Shasta. Riparian trees and streamside vegetation were defoliated. Hundreds of thousands of willows, alders, and cottonwoods eventually died. Many more were severely injured. Without the Riparian Habitat, insects, bats, otters, and mink starved or emigrated to find food sources. It has been estimated that over a million fish, and tens of thousands of amphibians and crayfish were killed. The aquatic invertebrates, such as aquatic insects and mollusks, were wiped out.

However, the river began the recovery process almost immediately. After the leak of Metam Sodium ceased and breakdown chemical concentrations declined, various taxa began reemerge. But the recovery was uneven. For example, aquatic taxa that relied on allochthonous sources of carbon, such as leaf litter from the riparian vegetation required several more years to recover. As in other Riverine Systems, in spite of the initial damage, streams can usually recover when a stressor is removed.

## PESTICIDES, INVERTEBRATE TOXICITY, AND TMDLS

The Salinas River drains one of the most productive agricultural regions of the United States. Producing over 70% of the leafy greens and generating millions of dollars in receipts, the value of these crops relies on pest control. Unfortunately, the lower Salinas River Watershed is characterized with widespread pyrethroid pesticide pollution and sediment toxicity to aquatic invertebrates (Fig. 10.31).

The concentration and toxicity exceed the water quality standards established by the Regional Water Quality Control Board of the United States. Thus the lower River is classified as impaired and listed on the Federal 2010 Clean Water Act section 303(d) List of Impaired Waters.

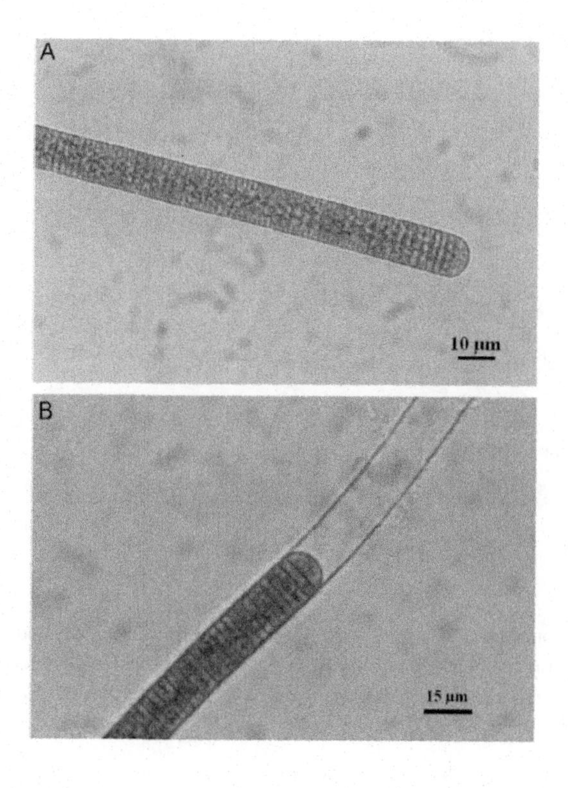


**FIGURE 10.29**

Photomicrographs of two strains of microcystin producing in the Oscillatoriaceae: LM603a (A) and DVL1003c (B) (Izaguirre et al., 2007).

To address this, the RWQCB adopted a plan to reduce the amount of pesticide discharged into the river in what is known as a Total Maximum Daily Load (TMDL). The TMDL is then implemented by the municipalities, county, and farmers through a suite of regulatory mechanisms. For example, for owners and operators of irrigated agricultural lands, a conditional waiver (usually referred to as an Agricultural Order) in the lower Salinas River Watershed was designed to reduce toxic discharges.

Implementation of the conditional waiver includes the use of best practices, such as grass buffers, bioswales, constructed wetlands, reduced and improved application of pyrethroids. Although some of these practices have evaluated scientifically, many have not. More importantly, few of these practices are reliable and often difficult to establish or maintain in the context of working farms. Thus engaging land managers to develop and test the effectiveness of various practices remains a high priority for Environmental Scientists to protect surface- and groundwater quality.

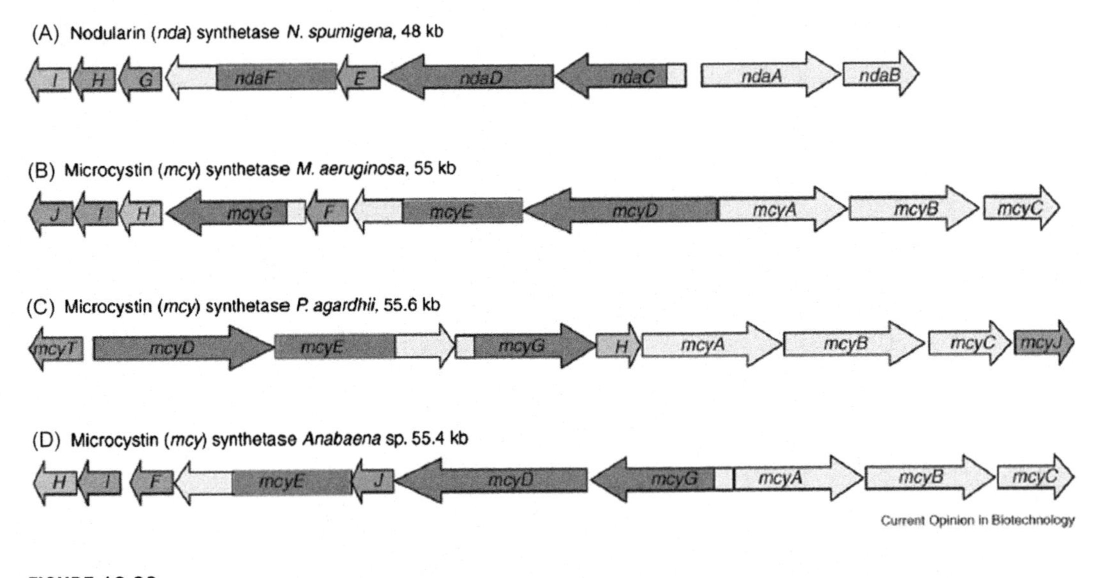


**FIGURE 10.30**

Hepatotoxin gene clusters from various cyanobacteria. Structures of the microcystin and nodularin gene clusters of (A) N. spumigena, (B) M. aeruginosa, (C) P. agardhii, and (D) Anabaena sp. 90, showing genes encoding polyketide synthases (red), non-ribosomal peptide synthetases (yellow), tailoring enzymes (green), and ABC transporters (blue). Diagram not drawn to scale (Source: Pearson and Neilan, 2008).

## RADIONUCLIDES

Established in 1943 as part of the Manhattan Project in Hanford, the Hanford Project soon operated nine nuclear reactors and five large plutonium processing complexes, which produced plutonium for most of the more than 60,000 weapons built for the Unites States nuclear arsenal. With a sense of urgency, safety procedures and waste disposal practices were inadequate, and Hanford's operations released significant amounts of radioactive materials into the air and the Columbia River. Hanford is currently the most contaminated nuclear site in the United States and is the focus of the nation's largest environmental cleanup.

From 1944 to 1971, water from the Columbia River was used to cool the reactors (Fig. 10.32) and then returned it to the river. Because neutrons passed through the water, dissolved solids and particulates in the cooling water became radioactive. In addition, radioactive material was released through corrosion. Substantial amounts of radiation was released from the site, but the radioactive levels of the effluent may not have exceeded permissible levels; however, because of the secrecy of the operation, the voracity of these claims may never be established.

In January 1971 the system of once-through cooling waters was terminated and a monitoring program was implemented to document the decline of radionuclides in Columbia River biota.

The amounts in seston and periphyton decreased rapidly and were measurable only until the spring of 1973. The highest concentration of $^{65}Zn$ were found in the biota, but varied by species. $^{65}Zn$ in caddisfly larvae was not measurable by February 1973, but concentrations in McNary Chironomids fluctuated between unmeasurable levels to 24 pCi/g dry weight. As Chironomids ingest sediment,

**Pyrethrum-derived**

Pyrethrin

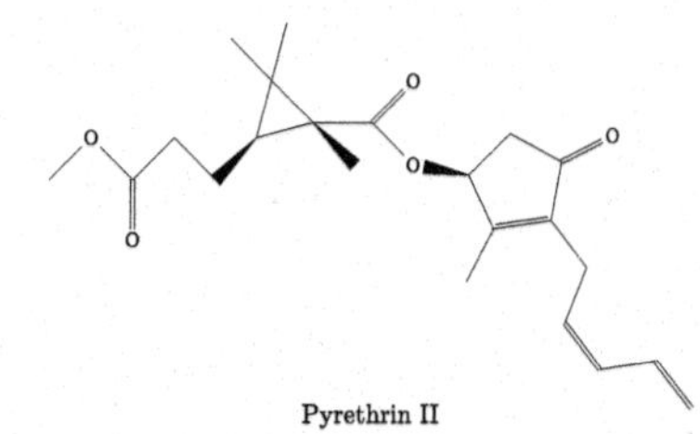


Pyrethrin II

**1st generation synthetic pyrethroids**

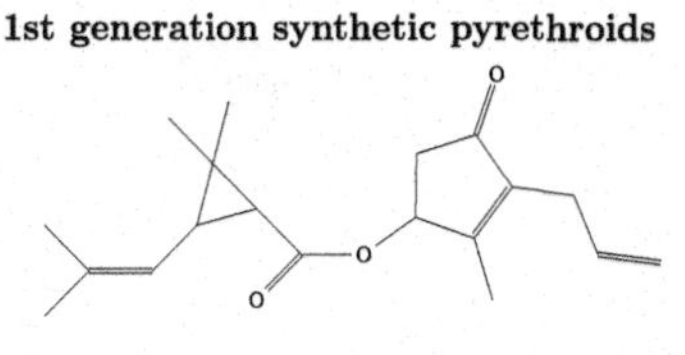


Allethrin

Resmethrin

**2nd generation synthetic pyrethroids**

Type I

Type II

**FIGURE 10.31**

Structure of pyrethrin and pyrethroids. Initially isolated in 1924, two pyrethrin compounds as the active agent pyrethrum's insecticidal action. In the first generation, the synthetic pyrethroids were synthesized, which included allethrin, tetramethrin, and resmethrin (discovered at Rothamsted Experimental Station). In terms of insecticidal potency, the introduction of a cyano (CN) residue enhanced insecticidal activity roughly 3–6 fold compared to noncyano pyrethroids.

stomach contents of contaminated sediments explained concentrations rather than incorporation into larval tissues. Whereas many radionuclides declined in various taxa in the river after the cooling waters were changed, concentrations of $^{60}$Co did not.

$^{60}$Co has a half-life of 5.27 years and emits $\beta^-$ and $\gamma$ particles (Fig. 10.33). $\beta^-$ particles barely penetrate beyond the surface of organism, but can alter molecular structure and cause cancer or death. If the struck molecule is DNA, $\beta^-$ particles can cause spontaneous mutation. $\gamma$ particles are less energetic, but can penetrate through the entire body and cause damage at the cellular level, causing diffused damage throughout the body.

In the Colombia River, levels of $^{60}$Co in fish showed some decreases, but obvious trends were not present. $^{60}$Co remained elevated in the seston, periphyton, and invertebrates, which posed a new set of questions about $^{60}$Co. Some of the $^{60}$Co was seeped into the river from a leaking disposal trench

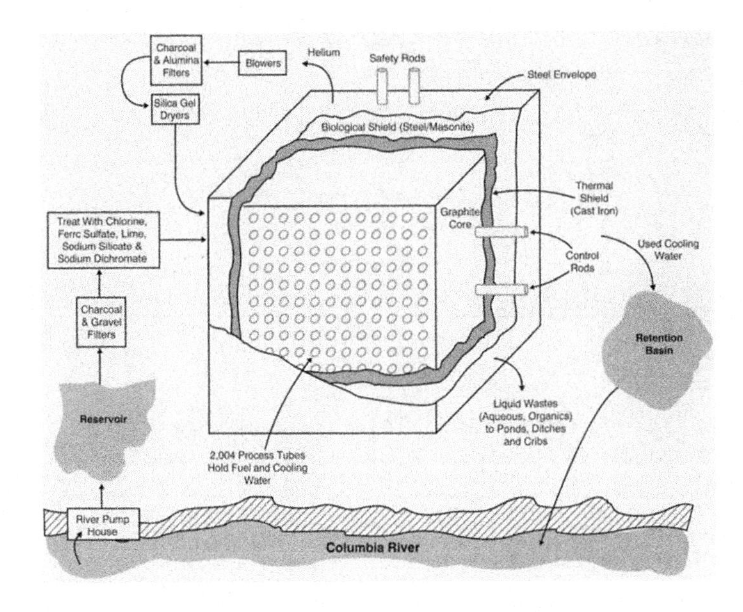


**FIGURE 10.32**

Hanford cooling system (Source: US Department of Energy).

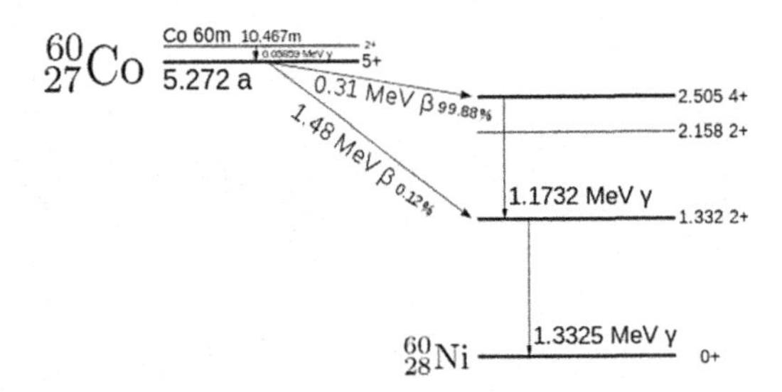


**FIGURE 10.33**

Decay sequence of $^{60}$Co (Source: Wikipedia).

near an operating reactor. These results demonstrate the power of monitoring programs that can inform management activities well outside for what they were originally designed.

It is also with great irony that the advances in Aquatic Ecology used radionuclides, in some cases with planned releases and in other cases, fortuitously, tracing fallout. For example, radioactive fallout from above-ground nuclear testing lofted $^{137}$Cs into the atmosphere. $^{137}$Cs returned to the surface with rainfall and adhere to soil and sediment particles. As these particles are eroded and deposited into basins, the $^{137}$Cs remains attached. The radioactivity of sediment cores can then be used to quantify sediment deposition since the peak of above-ground testing in 1968.

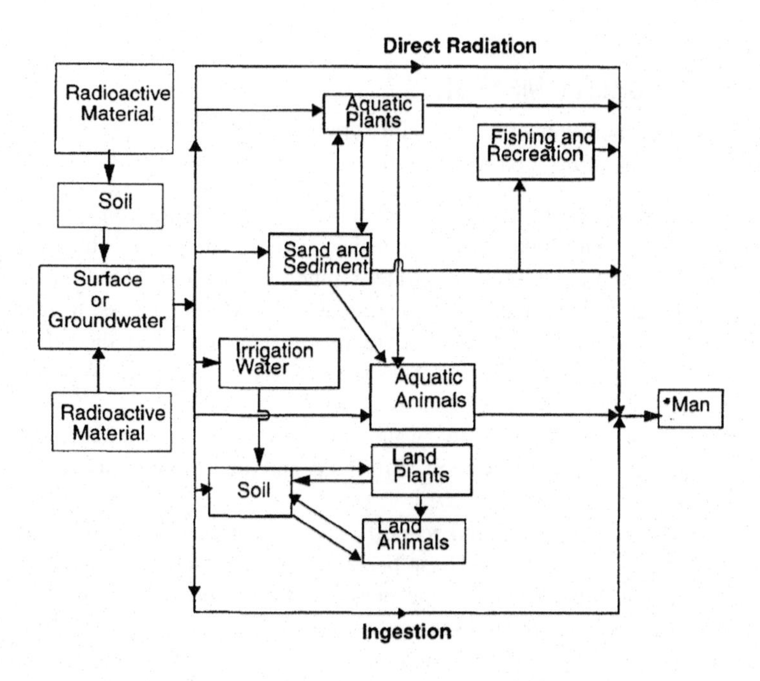


**FIGURE 10.34**

Fate of radionuclides in the Savanah River ecosystem. (Source: Carlton, 1999).

As a response to the public outcry and contamination of the Marshall Islands, their inhabitants, and a Japanese fishing boat with radionuclides, the Atomic Energy Commission sponsored "Radioecology" research at National Laboratories and approximately fifty universities.

Some of the earliest work in Radioecology consisted of studying the effects of radiation by exposing plants and animals to controlled radiation doses. At the time this work was done in the 1950s and 1960s, little was known about the effects of ionizing radiation, or what the ecological consequences might be of a nuclear reactor accident, or nuclear war.

An enthusiastic supporter of the program, H.T. Odum, argued that radionuclide could be used to gain a holistic view of an ecosystem. Thus research should not be directed at individual organisms, but whole ecosystems. In 1957 such an experiment was initiated at Savannah River, Georgia.

One of the most successful approaches was to inject $^{32}P$ into plant stems and trace the fate of the radioactivity. As a result, Odum and his colleagues demonstrated how trophic levels were organized. For example, they were able to document the predatory activities of certain spiders of their prey (leafhoppers and beetles), and prey herbivory of the injected radioactive plant.

These methods were applied to test Lindeman's Trophic Models (Fig. 10.34), using independent tracers that could be combined with biomass and productivity measures. Overall, most cite their methods as simple and crude, but their work became a foundation on which to improve.

# WATER QUALITY PARAMETERS

## IN-STREAM WATER QUALITY MEASURES

Some of the most common measures of water quality are listed below, with an explanation of why they are important to the health or utility of a water body.

***Temperature*** is a physical property of water that has a profound effect on organisms that live or reproduce in the water, particularly coldwater fish, such as salmon and steelhead and some amphibians (frogs and salamanders). When water temperature becomes too high, salmon and trout suffer a variety of ill-effects, ranging from decreased spawning success, to increased susceptibility to disease and toxins, to death. Water temperature also reduces the solubility of oxygen, on which aquatic life depends, and increases the toxicity of ammonia. Water temperature may enhance sensitivity to other toxic substances as well.

***Turbidity*** is a measure of the clarity of the water. Fine sediment suspended in the water column usually causes turbidity in Lotic Systems. But POM and algae commonly cause clarity changes in lakes.

***Bacteria (Fecal Coliform, E. coli, and Enterococci)*** are measured to determine the relative public health risk with contact with water. These bacteria originate from the wastes of warm-blooded animals; thus their presence indicates the possibility of pathogens reaching a body of water. Sources include inadequately treated sewage, improperly managed animal waste from livestock, pets in urban areas, aquatic birds and mammals, or failing septic systems. Unfortunately, the tests are not very selective and with a delay in getting the results; thus, the effectiveness of these simple tests indicators could use a dramatic improvement.

***Dissolved oxygen*** in water is necessary for aquatic animal life, just as oxygen in air is necessary for human life. The concentration of Dissolved Oxygen (DO) is a single, easy-to-measure characteristic of water that correlates with the health of aquatic life in a water body. Low DO is often related to an excess of nutrients and eutrophic conditions. Other forms of organic matter, such as sewage or food waste, can also lead to low DO. For this reason, Biological Oxygen Demand in water (i.e., how much oxygen it would take to decompose all the organic matter present) is a common measure.

***pH*** indicates the balance between hydrogen ions, which are acidic, and hydroxide ions, which are basic. A balance of the two is at a pH of 7. Most aquatic organisms prefer a pH of 6.5 to 9. Like temperature, water pH is a fundamental controlling property that affects other chemical constituents (e.g., $NH_3$ and metal solubility), as well as important biological processes such as the level of permeability of fish gills and amphibian skins (e.g., how well gases can flow through the gills/skin, allowing the organism to breathe).

***Nutrients*** in water include various forms of nitrogen and phosphorus. In general, it is useful to measure all of the inorganic forms ($NH_4^+$, $NO_3^-$, $NO_2^-$, SRP), but also the total, and organic fractions. Excess nutrients are associated with eutrophication and CHABs, and is important to measure both the concentration and the load (e.g., kg/day in a stream) to assess downstream impacts.
$NH_3$ is toxic to aquatic life. Since $NH_3$ will vary with the pH and temperature of the water, Environmental Scientists measure the total $NH_3$ and then use equations using temperature and pH to estimate water column $NH_3$ at the time of sampling.

***Sediment*** In flowing waters, a dynamic balance exists between the supply of sediment from natural erosion and the energy of the moving water that carries and redistributes the sediment load.

Furthermore, this balance varies from place to place within the stream channel. Sediment balance determines the very character of many streams and their suitability for various forms of aquatic life.

***Toxic substances*** A toxic substance is any substance, material, or disease-causing agent, that upon exposure, ingestion, inhalation, or assimilation into an organism will cause death, disease, malignancy, genetic mutation, or other abnormalities in affected organisms or their offspring. Unfortunately, it is impossible to evaluate all known toxic substances. Thus analytic methods are based on watershed land use and typical pollutants associated with them.

***Metals*** such as arsenic, cadmium, copper, mercury, and lead pose a threat to aquatic life, domestic water supplies, livestock, and human health. Eating fish contaminated with certain metals, arsenic and mercury in particular, can cause the metals to accumulate in human tissue, posing a significant health threat. Potentially dangerous levels of metals are identified mainly through chemical analysis of water, but sediment and fish tissue analyses are also used.

Most of the mercury in surfacewaters remains inorganic, but in certain environments (low pH, low dissolved oxygen, and high dissolved organic matter), as are found at the bottoms of lakes, marshes, and wetlands, some of it is converted to a much more toxic organic form—methylmercury. Methylmercury accumulates in muscle tissue of fish that most people prefer to eat, and thus is of particular interest from a human health standpoint.

***Organic compounds*** include household and agricultural pesticides, many solvents, and other household and industrial chemicals. Polychlorinated biphenyls (PCBs), for example, are industrial chemicals that are toxic and carcinogenic. Although banned in the United States in 1977, PCBs persist in the environment, and they accumulate in fish and human tissues when consumed. Some organic compounds are highly bioaccumulative and may be measured in sediment or tissue, as well as water.

***Emerging pollutants of concern*** Pharmaceuticals and Personal Care Products (PPCPs) are emerging pollutants of concern. PPCPs can affect reproductive and developmental processes in fish and wildlife and the insect life they depend on for food. They include a wide variety of synthetic chemicals, such as antibacterial compounds in soaps, hormones, fragrances, cleaning agents, and both over-the-counter and prescription medications. Steroids, nonprescription drugs, sunscreen, and insect repellant are commonly found in surfacewaters.

## PHYSICAL HABITAT

Physical Habitat Assessment Methods aim to identify, survey and assess physical habitats and/or the overall functioning and conditions of rivers and streams. They are mainly applied at a local/reach scale, consider all the spatial components of a river corridor (channel, Riparian Area and floodplain), and assess the hydromorphological state at present time.

Unfortunately, most of these methods are not suitable to understand physical processes and causes of river alterations, because of a series of reasons, including the scale of investigation (too small), the survey resolution (too accurate), the temporal scale (not taken into account), the variability of river systems (not covered).

## BIOASSESSMENT

### LINKING WATER QUALITY TO BENEFICIAL USES

In 1997 staff working for the Regional Water Quality Control Board for the Central Coast Region of California had reasons to believe that agriculture runoff was the source of nitrogen in surfacewaters, which exceeded the drinking water standard (10 mg/L nitrate-N). However, citing the mixed land uses in the region that included highly urbanized areas and sewage treatment facilities, many farms were unconvinced. To resolve the question, farmers requested a local university to evaluate the role of land use and water quality in several watersheds.

The results were unsurprising and similar to other parts of the country—nitrogen concentrations did increase in areas of urbanization, but the highest concentration increases were associated with agricultural land uses. What was surprising however, was that specific areas had concentrations between 40–50 mg/L Nitrate-N, and in some areas were as high as 80–110 mg/L nitrate-N. With a drinking water standard of 10 mg/L nitrate-N, the regulators were concerned. However, because the waters measured were not drinking water source, the application of the drinking water standard was a bit problematic.

Following a national trend, the Regional Board developed alternative numeric criteria based on other beneficial uses, e.g., Aquatic Habitat. In contrast to lakes, the ecological impacts on eutrophication are a bit subtle, so many jurisdictions rely on several water quality indicators to evaluate if the water meets the beneficial uses levels.

Biological Assessment (bioassessment) is an evaluation of the condition of a waterbody based on the organisms living within it. It involves surveying the types and numbers of organisms present in the water and comparing the results to established swamplogobenchmarks of biological health. Scientists and managers around the world use this approach to directly and quantitatively measure the ecological health of a waterbody and to monitor the cumulative impacts of environmental stressors on surfacewaters.

### LINKING ECOLOGICAL TOLERANCES TO MANAGEMENT THRESHOLDS

An ecological threshold defines a point when an ecosystem quality might experience an abrupt change. For species living in these systems, the ecosystem change might be outside the ecological tolerances; thus unable to maintain healthy populations. As species assemblages change, the analysis of thresholds may demonstrate nonlinear dynamics, and by multiple factor, controls that operate at diverse spatial and temporal scales.

Nevertheless, linking the ecological tolerances with thresholds in aquatic environments, at ecosystem, landscape and regional scales may be useful to maintain the value of the ecosystem to a diverse set of aquatic species.

### BMI

Benthic macroinvertebrates (BMIs) and benthic algae are the primary biota used for bioassessments in California. BMIs are a diverse group of small but visible animals that live at the bottom of rivers and streams. They are comprised mostly of aquatic insects, but also include crustaceans, mollusks, and worms. BMI assemblages are found in most waterbodies and are reliable indicators of biological health, because they are relatively stationary and respond predictably to a variety of environmental stressors. Benthic algae are also sensitive to environmental stressors and provide environmental condition information that is often complementary to that derived from BMI assemblages. Because of their short lifespans and rapid reproduction rate, algae can respond quickly to changing water conditions. They are also more directly responsive to nutrients (such as nitrogen and phosphorus) and are therefore suited for monitoring nutrient runoff, one of the major environmental stressors in California.

### PERIPHYTON

Benthic algae (periphyton or phytobenthos) are primary producers and an important foundation of many stream and lake food webs. These organisms also stabilize substrata and serve as habitat for many other organisms. Because benthic algal assemblages are attached to substrate, their characteristics are affected by physical, chemical, and biological disturbances that occur in the stream reach during the time in which the assemblage developed.

Diatoms, in particular, are useful ecological indicators because they are found in abundance in most lotic ecosystems. Diatoms and many other algae can be identified to species by experienced phycologists. The great numbers of species provide multiple, sensitive indicators of environmental change and the specific conditions of their habitat. Diatom species are differentially adapted to a wide range of ecological conditions.

Periphyton protocols may be used by themselves, but they are most effective when used with one or more of the other assemblages and protocols. They should be used with habitat and Benthic Macroinvertebrate Assessments, particularly because of the close relation between periphyton and these elements of Stream Ecosystems.

### DIVERSITY MEASURES

Although diversity indices have a long-history in Ecology, their use to evaluate surfacewater conditions is relatively recent. Score-based biotic indices are widely used to evaluate the water quality of surfacewaters. Using these indices requires detailed knowledge of the taxonomy, distribution and tolerance to pollution in the region. Using richness and Shannon–Weaver Indices can provide valuable information, but they are snap-shot measures. Droughts, severe sedimentation, floods, etc., can alter aquatic composition for a period of time, and also there are seasonal differences in their life cycles. Thus it is important to develop a reliable, reproducible method that allows comparable results.

## NEXT STEPS

### CHAPTER STUDY QUESTIONS

1. Summarize the ecological processes that promote 3 different water-borne diseases.
2. Describe how Geochemistry influences water quality.
3. Describe the fate, transport, and toxicity of pollutants in Aquatic Systems and their impacts.
4. Summarize efforts to assess and improve the water quality.

## LITERATURE RESEARCH ACTIVITIES

1. Blooms of Cyanobacteria occur in many parts of the world on a regular basis. For example, Microcystus blooms occur every summer since 1999 in the northern reach of the San Francisco Bay Estuary. High microcystin concentrations have occasionally been measured during these blooms. *M. aeruginosa* has the potential to negatively effect the bay-delta.
   Create a list of possible chemicals to screen for 5 lakes that have different species compositions in their blooms.
2. Select 5 regional water bodies and review sources of literature on their water quality. Summarize the water quality issues in a series of info-graphics.

CHAPTER

# BIOGEOCHEMISTRY AND GLOBAL CHANGE

# 11

*Many scientists are now warning that we are moving closer to several "tipping points" that could – within as little as 10 years – make it impossible for us to avoid irretrievable damage to the planet's habitability for human civilization.*

**Al Gore, NYU Law School speech (2006)**

## CONTENTS




Ecology and Management of Inland Waters. https://doi.org/10.1016/B978-0-12-814266-0.00025-8

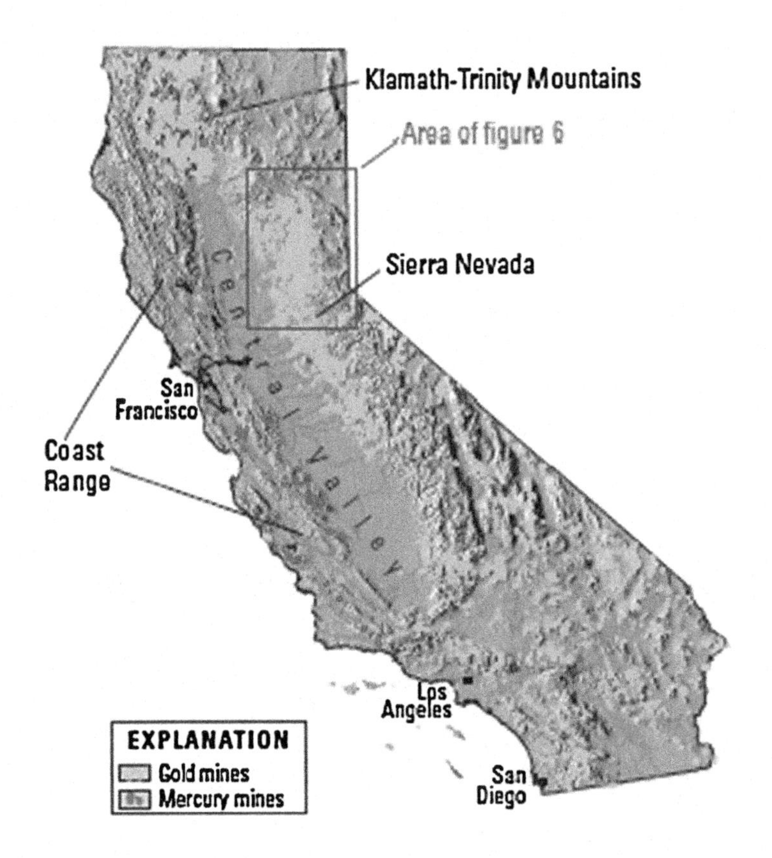


**FIGURE 11.1**

Mercury and gold mines in California (Source: USGS).

Clear Lake in California has been the symbol of some of the worst environment outcomes of any lake, including applications of DDD, contaminated runoff, species introductions, and toxic algal blooms. Rated as one of the best bass fishing lakes in the United States, the prize is the nonnative largemouth bass (*Micropterus salmoides*). A prize that might have potentially toxic levels of mercury and should not be consumed, because even small amounts may cause serious health problems for animals, including humans, especially for in utero development and early in life. Mercury can have toxic effects on the nervous, digestive and immune systems, and on lungs, kidneys, skin and eyes. Thus an understanding of the biogeochemistry of mercury is key to protect human health and ecosystem functioning, and potentially reduce the risks of fishing in Clear Lake and beyond.

Understanding Hg in waters is not a simple measure of concentration in the water column, but must be put in context of biogeochemistry associated with the geology (Fig. 11.1), atmosphere, water, sediment, and biota in aquatic systems – re-enforcing a clear theme throughout the text – we rely on a complex set of disciplines to understand aquatic systems and biogeochemistry is one way of combining disciplines.

This chapter explores the physical, chemical, biological, and geological processes and reactions that govern the composition of and changes to Aquatic Systems. Depending on the redox, pH, and chemical concentrations, chemicals might be transformed that ultimately influence the patterns and processes in Aquatic Systems.

After reading this chapter, the reader should be able to

1. Describe the patterns and processes that influence chemicals in Aquatic Systems; and
2. Summarize how biogeochemical cycles influence the use of Aquatic Systems.

# THE REDOX ENVIRONMENT AND MICROBIAL COMMUNITY

## EVOLUTION OF AUTOTROPHY AND ANAEROBIC FIXATION

Autotrophic Carbon Fixation is essential for sustaining life on Earth. Though $CO_2$-fixation pathways began evolving while the atmosphere was $O_2$ free, understanding the diversity of these pathways still have implications for Environmental Scientists.

For example, through an oceanic expedition, a novel thermophilic bacterium (*Thermosulfidibacter takaii* (strain ABI70S6(T)) was isolated from a deep-sea hydrothermal field in the Southern Okinawa Trough. This bacterium, an obligate anaerobic chemolithotroph, couples $CO_3^-$ fixation with S reduction and hydrogen oxidation,

$$18H^+ + S + 6CO_3^- \longrightarrow 12H_2O + S^{2-} + C_6H_{12}O_6 \tag{11.1}$$

and may provide a link to the evolutionary history of $CO_2$ fixation prior to the great oxygen catastrophe (Fig. 11.34).

But we do not need to go to deep sea vents to find fixation pathways that differ from eukaryotes. Inland waters display a wide diversity of microbial processes and coupled reactions that fix or produce $CO_2$. These alternative pathways are not biogeochemical curiosities, but drive global nutrient cycles for carbon, oxygen, nitrogen, and sulfur. Furthermore, some biogeochemical processes release or transform toxic compounds, e.g., mercury and sulfide, and contribute to or mitigate greenhouse gas emissions.

Microbes inhabit every nonsterile surface on Earth, including rocks; soils and sediments, surface and ground waters, plants and animals. In each habitat, a mix of microorganisms form a microbial community assembled by immigration, emigration, and persistence. But because the scale and their biogeochemical role, Environmental Scientists often use the redox environment to explain which biogeochemical processes.

## REDOX REACTIONS

Redox reactions include all chemical reactions, in which atoms have their oxidation state changed. In general, redox reactions involve the transfer of electrons between chemical species. The chemical species from which an electron is stripped is oxidized, whereas the chemical form to which an electron is added is reduced. In other words,

***Oxidation*** is the loss of electrons or an increase in oxidation state by a molecule, atom, or ion;
***Reduction*** is the gain of electrons or a decrease in oxidation state by a molecule, atom, or ion.

As an example, when $N_2$ is fixed, it is converted to $NH_3$. The $N_2$ molecule has two atoms with a redox state of 0, and each atom gains three electrons to have a redox state of −3. Thus the reaction reduces N and need 6 electrons to accomplish this. But this is only half of the reaction—the electrons must be donated from another oxidizing process. The N-fixation is a coupled reaction that includes a reduction and an oxidation reaction. When sugars are oxidized via cellular respiration, the product is $CO_2$, where the carbon has been oxidized and $O_2$ is reduced to $H_2O$.

Although oxidation reactions are commonly associated with the formation of oxides derived from oxygen molecules, oxygen is not necessarily involved in these reactions. Other chemicals can serve as a source or acceptor of electrons. For example, in anaerobic environments, prokaryotes may rely on alternative compounds to serve the same function.

## REDOX AND BIOLOGY

The biochemical basis of all life is based on redox reactions, where the foundation begins with making organic carbon molecules that provide the structure of cells, the formation of macromolecules (DNA, proteins), and making carbon-based sugars that can store energy.

Oxygenic photosynthesis can be summarized by the following equation:

$$6\,CO_2 + 6\,H_2O \xrightarrow{h\nu} C_6H_{12}O_6 + 6\,O_2 \tag{11.2}$$

where $h\nu$ is the energy from light. To build our case about redox and carbon-fixation pathways in deep sea vents and wetlands mud without the use of light, we need to disaggregate the equation to the processes we think occur in the cells:

$$2\,H_2O \longrightarrow 4\,e^- + 4\,H^+ + O_2 \tag{11.3}$$

This reaction requires energy, approximately 29.9 kcals. Solar radiation provides the energy and freed electrons to reduce $CO_2$ to sugars, but not in a direct way.

For the Archaeplastida (Chapter 3), the Calvin Cycle is used to fix and reduce $CO_2$ and requires about 370 Kcals to produce glucose (Fig. 11.2). Though electrons are produced by light, the Calvin Cycle relies on intermediaries to transfer the electrons (NADPH) and the energy to drive chemical reactions (ATP) .

NADPH is oxidized and it reduces 1,3-biphosphateglycerate to glyceraldehyde-phosphate:

$$NADPH \longrightarrow NADP^+ + H^+ + 2\,e^- \tag{11.4}$$

With n supply of ATP and NADPH, the Calvin Cycle is independent of the water-splitting process that relies on the light, thus is referred to as the "dark reaction."

To maintain the Calvin Cycle process, NADPH must be regenerated. Thus the light reaction uses the electrons freed by splitting water to reduce the $NADP^+$ (Fig. 11.3):

$$NADP^+ + H^+ + 2\,e^- \longrightarrow NADPH. \tag{11.5}$$

Although it is relatively easy to appreciate the impact of oxygen production on the biosphere, many questions remain concerning how oxygenic photosynthesis evolved. One theory suggests that it was a

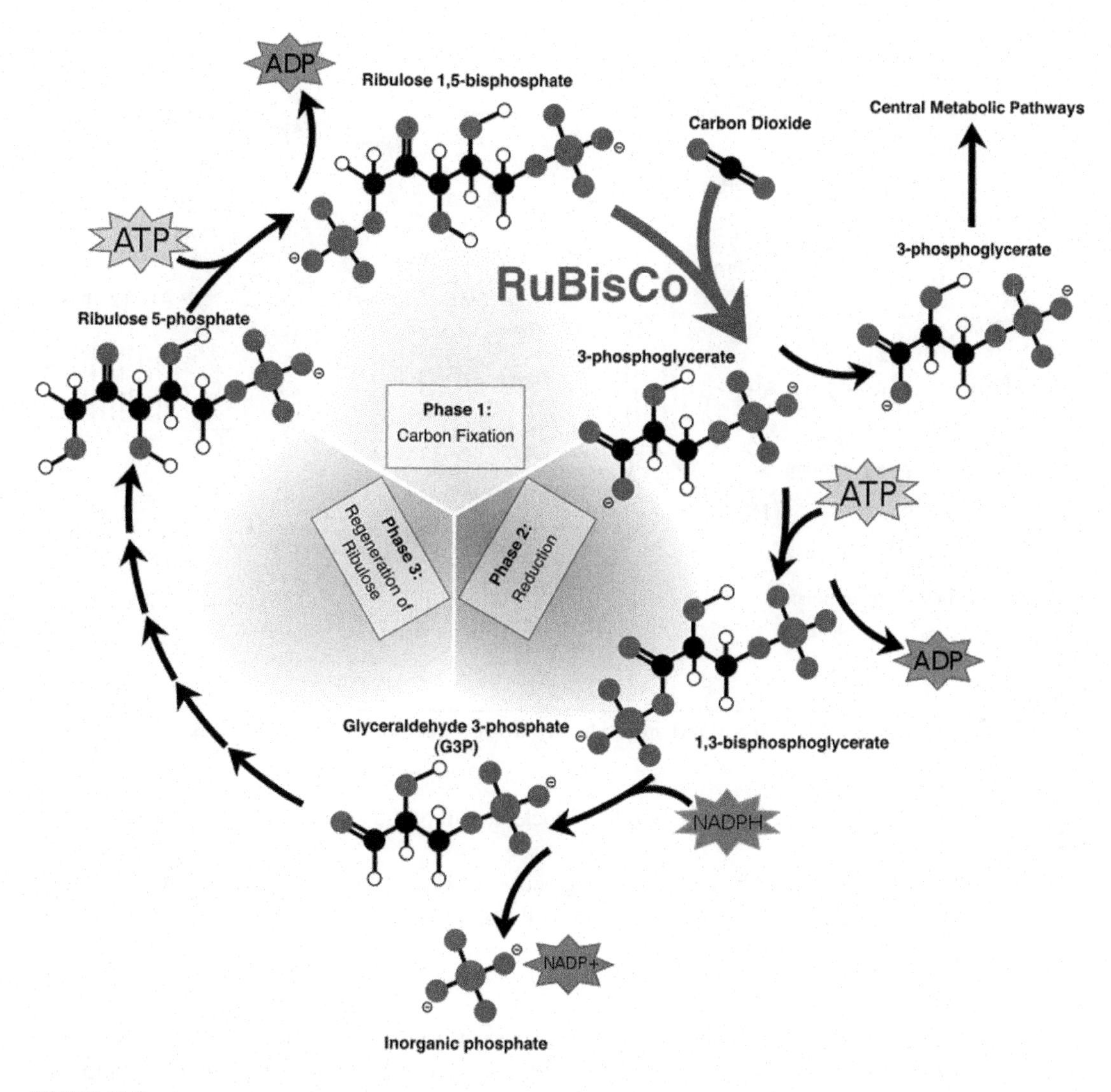


**FIGURE 11.2**

Overview of the Calvin Cycle pathway (Source: Wiki Commons).

fortuitous adaptation of primitive pigments that had been used for infrared thermotaxis. By using the sun's energy, chemolithotrophic bacteria became less dependent on the heat of hydrothermal vents in the early Archean Eon—possibly 3,800 Mya. Thus this is the first leap toward inland water habitats.

Glucose and related compounds act as an energy molecule. In multicellular organisms, these molecules can be transported to other cells that need energy, e.g., plant roots, and then be oxidized to drive other chemical reactions or in more complex polymers, e.g., starch and lipids. But glucose

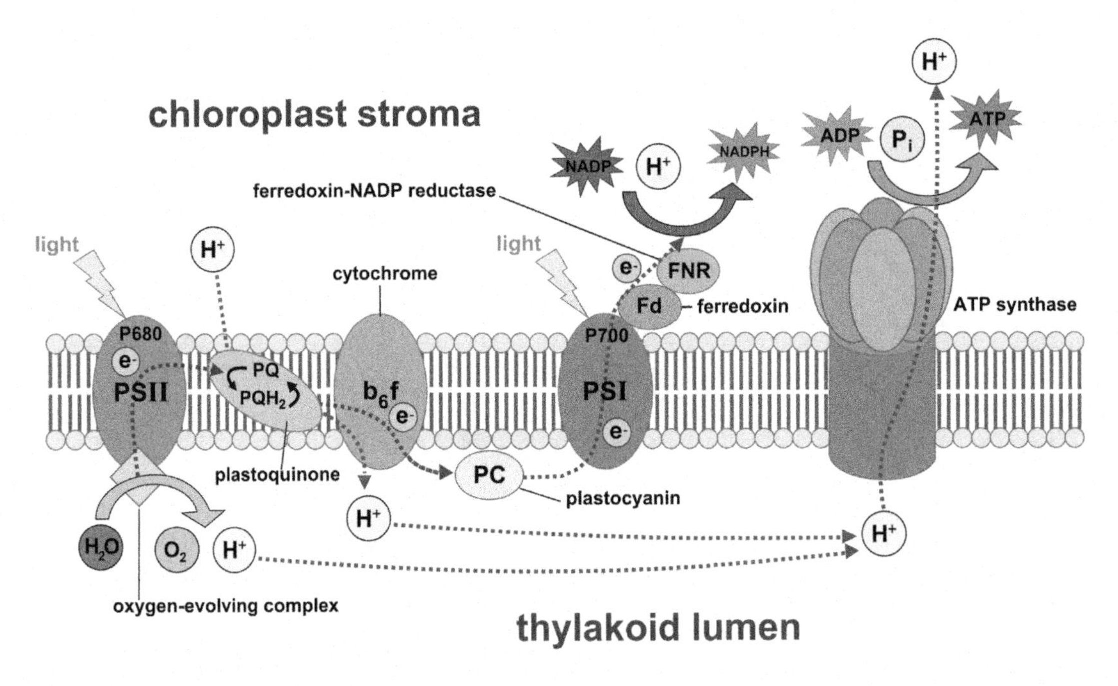


**FIGURE 11.3**

Photosystem I and II, where ATP and NADPH are produced via an $e^-$-transport chain (Source: Wikicommons).

and fixed carbon, in general, become the building blocks for all of the cellular structures, proteins, and DNA.

Cellular respiration is one pathway to oxidize glucose ($C_6H_{12}O_6$) to $CO_2$, coupled with the reduction of oxygen to water. The summary equation for cell respiration is

$$C_6H_{12}O_6 + 6O_2 \longrightarrow 6CO_2 + 6H_2O + \text{energy}. \tag{11.6}$$

Although the reaction is highly simplified, it is important to appreciate the redox process, where sugar is oxidized and $O_2$ is reduced, while 686 kcal/mole of energy is released. Although photosynthesis and cellular respiration can be viewed a complementary, cellular respiration is not the reverse of the photosynthesis. Respiration relies on different enzymes than those used in photosynthesis, and in eukaryotes the entire process is conducted inside a specialized organelle, the mitochondria.

There are four general stages in cellular respiration: glycolysis, Krebs Cycle, the electron transport chain, and ATP production. Glycolysis is a redox reaction, where glucose is oxidized to pyruvate and $NAD^+$ is reduced (Fig. 11.4). But the real energy-driver in respiration is due to the Krebs Cycle (Fig. 11.5), where three NADH, one $FADH_2$, and one GTP are produced—an equivalent to about 30–38 ATPs.

Now on our final note on this reaction, concerns how $O_2$ is consumed and ATP is produced. Similar to how $O_2$ production is facilitated through membrane-bound enzymes in the light reaction, $O_2$ consumption is associated with a membrane-bound enzymes driven by NADH. As NADH is oxidized, the

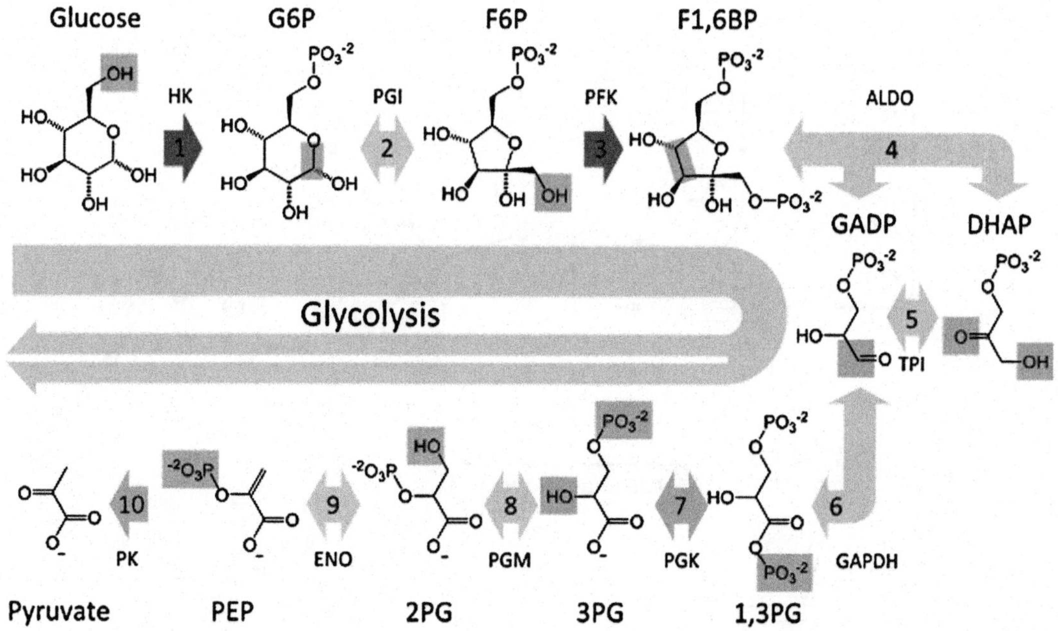

**FIGURE 11.4**

Glycolysis is a catabolic reaction that breaks bonds in the sugar molecule to release energy (Source: Wiki Commons). The metabolic pathway of glycolysis converts glucose to pyruvate by via a series of intermediate metabolites. Each chemical modification (red box) is performed by a different enzyme. Steps 1 and 3 consume ATP (blue), and steps 7 and 10 produce ATP (yellow). Since steps 6–10 occur twice per glucose molecule, this leads to a net production of ATP.

electrons used to reduce $O_2$ and the protons are transported across the membrane. The proton gradient is used to produce ATP (Fig. 11.6).

In contrast to the light and dark reactions, which can operate relatively independently of each other, Krebs Cycle (TCA) depends on $O_2$ for two key reactions. Should oxygen become limiting in the mitochondria, $NAD^+$ becomes limited and the succinate $\longrightarrow$ fumarate reaction will stop, grinding the cycle to a halt. For plants not adapted to flooding, this can mean death. For wetland plants, alternative metabolic pathways continue producing ATP under low oxygen conditions.

The origins of this reaction occurred before there was free $O_2$ in the atmosphere, a more reduced environment. In fact, the reaction may have been developed for very different purposes, e.g., as an antioxidant when free $O_2$ was toxic, and only later were enzymes available for a complete cycle. In any case, as we describe biogeochemical processes of inland waters, where $O_2$ is limited, we should also appreciate the flexible nature of these respiratory processes when $O_2$ availability is limited, similar to a preoxidized atmosphere.

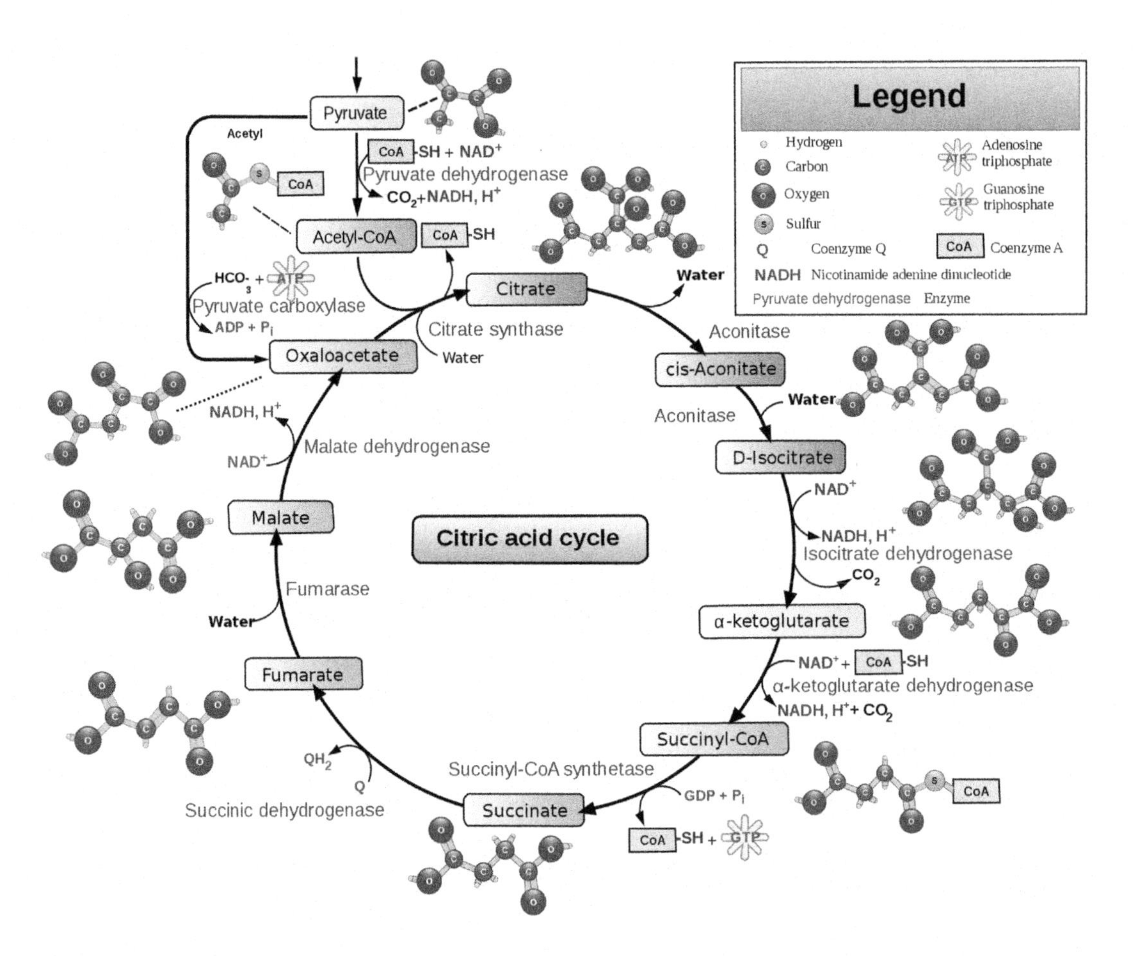


**FIGURE 11.5**

Tricarboxylic Acid Cycle (also known as the Citric Acid Cycle) and some preceding steps (Source: Wiki Commons).

# REDOX GRADIENTS

## THE REDOX SEQUENCE

The energy required or gained from redox reactions depend on numerous conditions. But if we assume standard pressure and temperatures, a comparison of the energy released or required can be used to explain the energetic value for various biochemical processes, and even help us understand how microbes function in selected redox environments.

As articulated in electrical chemistry, redox potential (Oxidation/Reduction Potential, ORP) measures tendency of a chemical species to acquire electrons from or lose electrons to an electrode. Redox potential is usually measured in millivolts (mV). Each element/molecule has a unique redox potential.

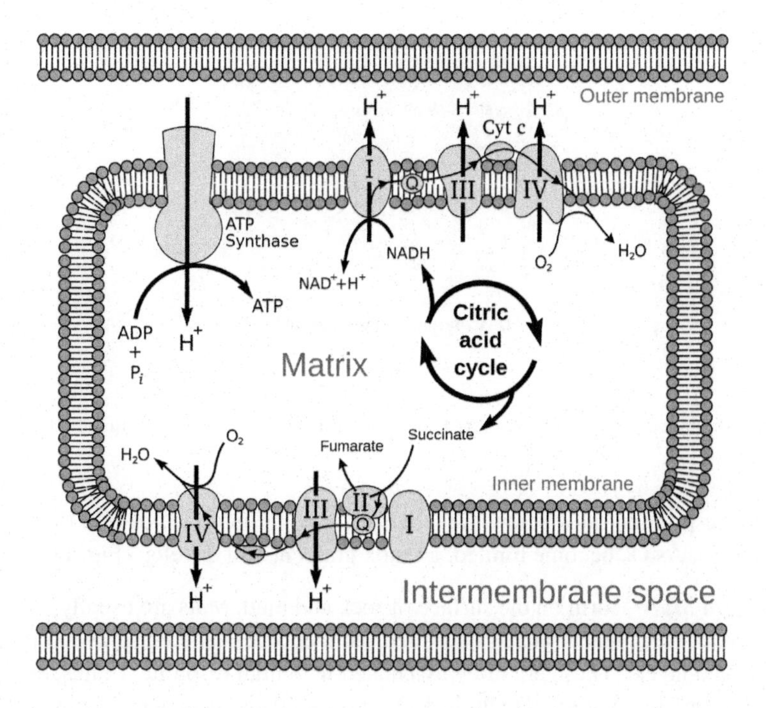


**FIGURE 11.6**

The electron transport chain in the mitochondrion is the site of oxidative phosphorylation in eukaryotes. The NADH and succinate generated in the Citric Acid Cycle are oxidized, providing energy to power ATP synthase.

**Table 11.1 Redox half cell reactions (Source: Schlesinger, 1997).**

| Electron donating half-reactions | Eh (mV) | Notes |
|---|---|---|
| $O_2(g) + 4H^+ + 4e^- \longleftrightarrow H_2O$ | 812 | $p_{O_2} = 0.2$ bar |
| $2NO_3^- + 6H^+ + 6e^- \longrightarrow N_2(g) + 3H_2O$ | 747 | $[NO_3^-] = 10^{-3}$ M; $p_{O_2} = 0.2$ bar |
| $MnO_2 + 4H^+ + 2e^- \longrightarrow Mn^{+2} + 2H_2O$ | 526 | $[Mn^{+2}] = 10^{-4.74}$ M; $p_{N_2} = 0.8$ bar |
| $NO_3^- + 2H^+ + 6e^- \longrightarrow NO_2^- + H_2O$ | 431 | $[NO_3^-] = [NO_2^-]$ |
| $Fe(OH)_3 + 3H^+ + e^- \longrightarrow Fe^{+2} + 3H_2O$ | −47 | $[Fe^{+2}] = 10^{-4.74}$ M |
| $SO_4^{-2} + 10H^+ + 8e^- \longrightarrow H_2S + 4H_2O$ | −221 | |
| $CO_2 + 8H^+ + 8e^- \longrightarrow CH_4 + 2H_2O$ | −244 | |

A positive redox potential describes an affinity for electrons and tendency to be reduced, whereas a negative redox potential describes an adversion to elections and a tendency to be oxidized (Table 11.1).

## MAPPING SPATIAL GRADIENTS

Oxygen is well mixed in the atmosphere, but in water the mixing rate, as measure by its diffusion rate, is about 10000x slower. In addition, below the sediment-water interface diffusion rates are even slower.

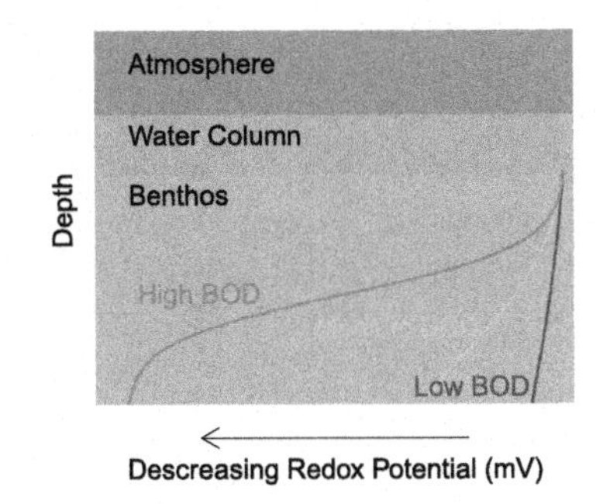


**FIGURE 11.7**

Hypothetical redox gradient changes with depth as a function of Biological Oxygen Demand (BOD).

Meanwhile, if the respiration rates below the atmosphere-water interface exceed the rate of diffusion, $O_2$ will be consumed. As $O_2$ become limited, a redox gradient will develop (Fig. 11.7).

***Microbial and Algal Mats*** form on the surface of rock and mud. Mats are usually composed of alternating layers of Cyanobacteria and deposited sediments deposited or growing in place, creating dark-laminated layers. These are often associated with high respiratory rates, especially at night and have a strong redox gradient within 1–2 mm from the atmosphere-mat interface (Fig. 11.8).

***Lentic Anaerobic Sediments and Lotic Hyperheic Zones*** Lentic sediments usually have high organic matter, thus a high demand for $O_2$, which can create a strong gradient of ORP that gets more negative with depth. In the same, way Lotic System Sediments might have a ORP gradient with depth. However, as waters flow through the sediments and are subject to respiratory oxygen demand, a redox might develop along two axes, with depth and along the interstial flow lines in a stream.

***Wetland Soils*** Many wetlands experience seasonal flooding. During such times, the soils experiences a dramatic change in redox potential in a relatively short period of time. As they are saturated, soils can become anaerobic within hours, days, or weeks, depending on the biological activity in the wetland. The $O_2$ demand can create a strong redox gradient until soils dry.

***Stratified Surface Waters*** Just as stratified lakes include a strong temperature gradient with depth, redox potentials can also vary dramatically with depth. The epilimnion is usually well-mixed and in contact with the atmosphere is oxygenated. However, at the thermocline, very little water mixes between the epilimnion and hypolimnion, thus $O_2$ can become quite limited in the hypolimnion, especially if there is a strong demand in the sediments or water column itself. Under these conditions the water and habitat quality of the hypolimnion and epilimnion can very dramatically (Fig. 11.9).

## LINKING CHEMOLITHOTROPHS AND BIOGEOCHEMICAL CYCLES

As $O_2$ gradients are established and the redox gradient becomes more negative, a spatial pattern of microbial communities and biogeochemical processes reenforce these patterns (Fig. 11.10).

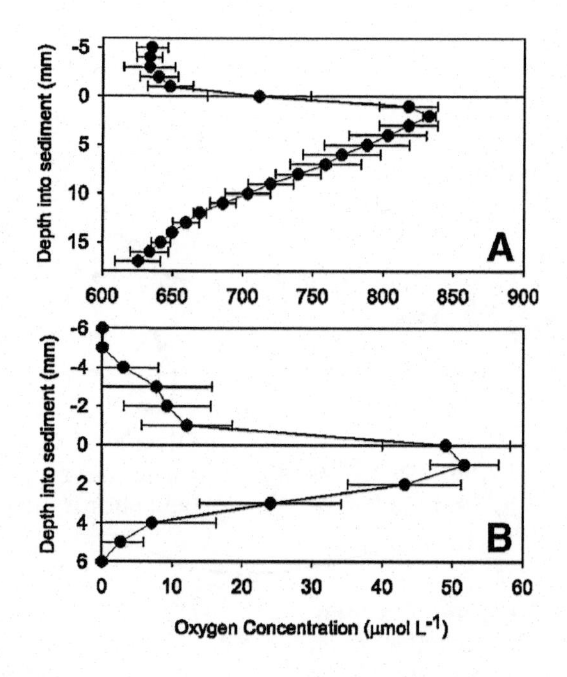


**FIGURE 11.8**

Antarctic Microbial Mats: A modern analog for Archean Lacustrine Oxygen Oases (Source: Sumner et al., 2015).

Whereas phototrophs get energy from light, chemotrophs derive energy by coupling oxidation and reduction equations based on the available oxidants and reactants. Chemotrophs can be either autotrophic or heterotrophic. In the case of heterotrophs, using alternatives electron acceptors is the key (e.g., alternatives to $O_2$). For autotrophs, capitalizing on available oxidants to fix $CO_2$ is an important strategy (e.g., FeII $\longrightarrow$ FeIII). Chemotrophs have chemical, ecological, and management implications to every inland water.

***Nitrogen Cycle*** Nitrogen can be transformed into a variety of compounds and plays a critical role in the biosphere (Fig. 11.11). Nitrogen has 9 redox states: $N^{-3}$, $N^{-2}$, $N^{1-}$, $N^{0}$, $N^{+1}$, $N^{+2}$, $N^{3+}$, $N^{4+}$, and $N^{5+}$, determined by the orbital structure of the element.

The majority of Earth's atmosphere (78%) is nitrogen. However, atmospheric $N_2$ (nonreactive N) is not directly available for biological use, leading to a scarcity of nitrogen in many types of ecosystems. Thus the conversion to reactive nitrogen is a key process in the biosphere. Biologically mediated nitrogen biogeochemical processes include fixation, ammonification, nitrification, anammox, denitrification, and dissimilarity nitrate reduction to ammonium.

The Nitrogen Cycle is of particular interest to Ecologists because nitrogen availability can affect the rate of key ecosystem processes, including primary production and decomposition. Specifically, inorganic nitrogen can be taken up by microbes and autotrophs to build nitrogen-containing compounds, such as amino acids, proteins, peptides, etc. Once these microbes and autotrophs are dead or eaten, the nitrogen can be released back into the environment to be taken up again. But this description misses how microbial processes drive the nitrogen cycle at microscopic, stream

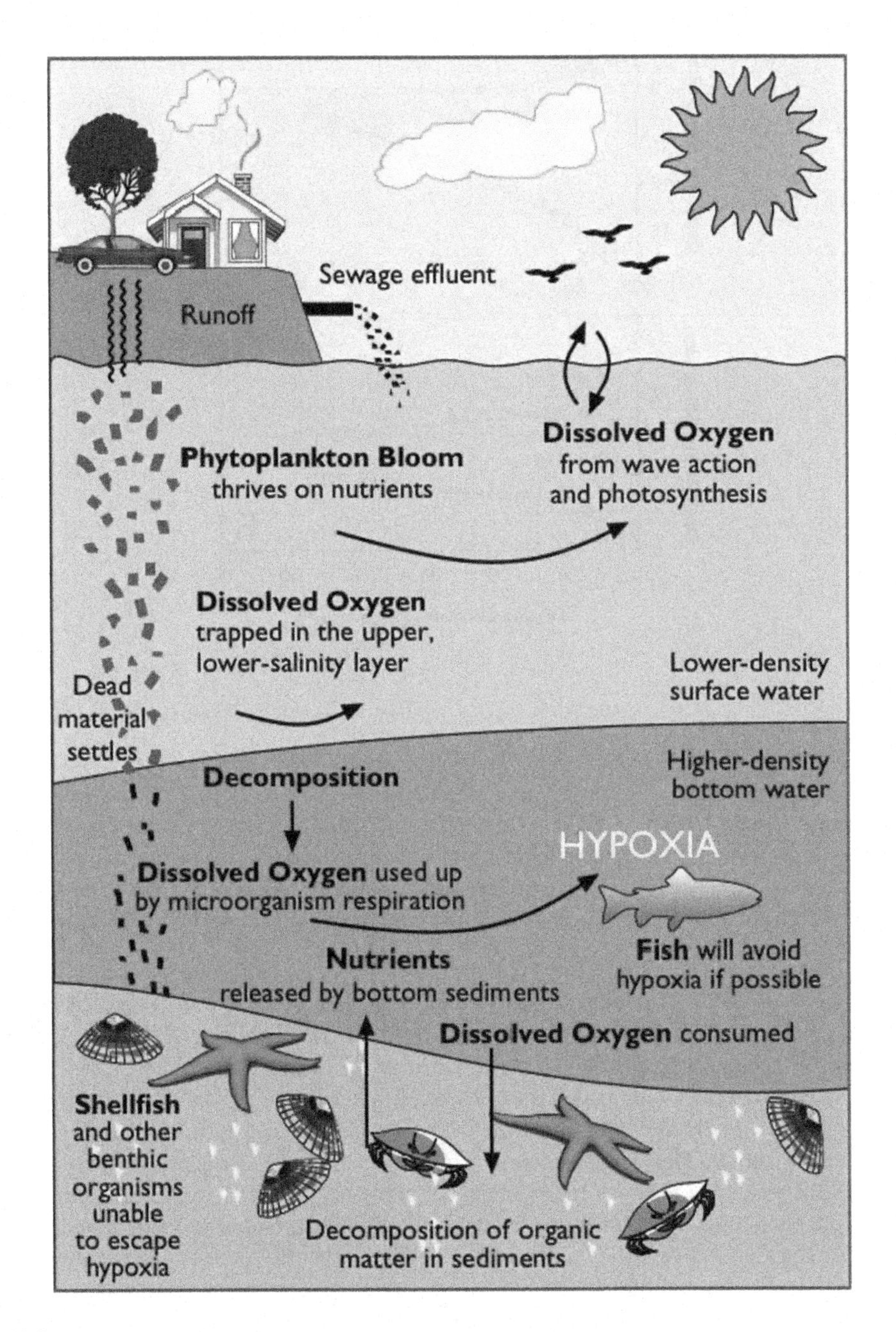


**FIGURE 11.9**

Diagram of a stratified lake an hypoxic hypolimnion (Source: US EPA).

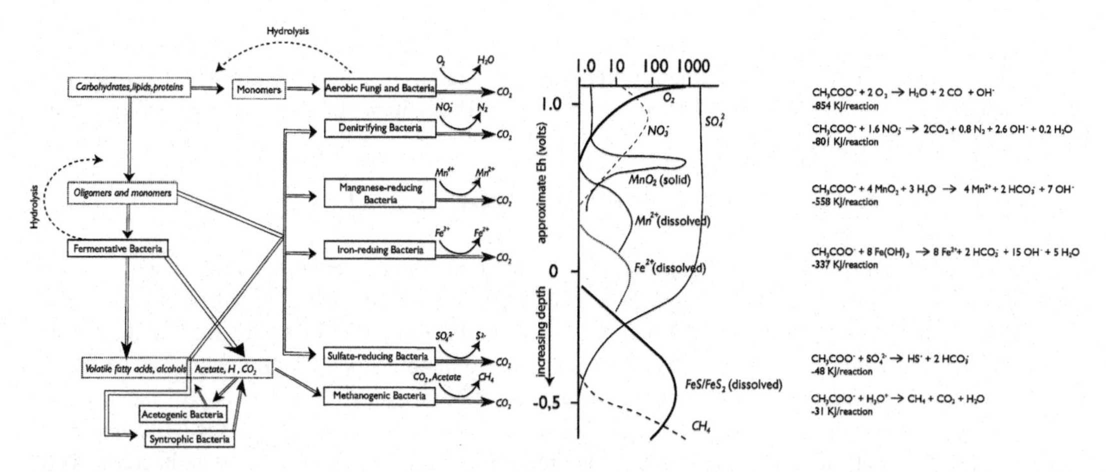


## FIGURE 11.10

Simplified sequence of microbial redox processes in a stratified sediment (Source: Rincón-Tomás et al., 2016, redrawn with modifications). For better comparison, the Gibbs free energy change at standard conditions in kJ/reaction for the complete oxidation of acetate is shown. The approximate redox potentials at increasing sediment depths and concentrations of involved redox partners are indicated. The succession of electron acceptors from oxic (high redox potential) down to anoxic (low redox potential) conditions—oxygen, nitrate, $Mn^{4+}$, $Fe_3^+$, sulphate, protons—lead to lower energy yields.

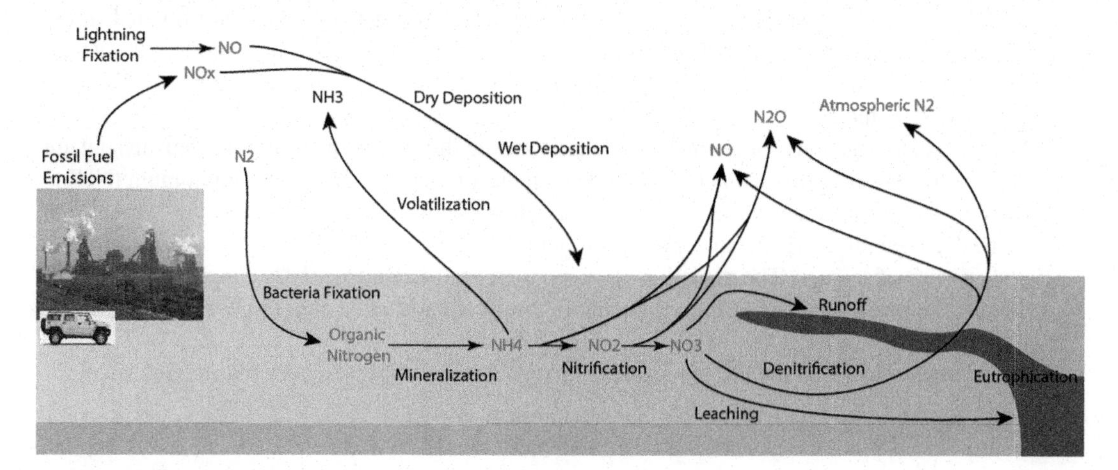


## FIGURE 11.11

Diagram of the Nitrogen Cycle (Source: Wikipedia).

reach, watershed, and global scales. Furthermore, human activities, such as fossil fuel combustion; production and use of synthetic nitrogen fertilizers, and nitrogen-rich wastewater discharges have dramatically altered the global Nitrogen Cycle by releasing reactive nitrogen into the atmosphere. In fact, humans create, use, and discharge more N than the natural nitrogen cycle.

Reactive nitrogen are oxides or reduced forms of N. To create reactive nitrogen, biological fixation reduces $N_2$ to $N^{-3}$. Producing $NH_3$ and $NH_4^+$ and $NH_2$-R, allows nitrogen to become assimilated into intercellular biochemical pathways. About 50% of the nitrogen produced on the planet is by biological fixation:

$$N_2(g) + 8H^+ + 8e^- + 16ATP \rightleftharpoons 2NH_3 + H_2 + 16ADP + 16Pi, \tag{11.7}$$

whereas some is produced by lightning, most of the remaining is industrially produced.

To break the triple bond in $N_2$ requires a lot of energy. Although every organism needs N, relatively few organisms can fix N. Rhizobia and cyanobacteria are known for their ability to fix N and usually do this in an $O_2$-free or limited environment. Since $O_2$ competes for the active site in nitrogenase, the enzyme that catalyzes the N≡N cleavage, the efficiency of the reaction declines in the presence of $O_2$. Nitrogen fixation rates are highest when there is a physical barrier or chemical gradient to limit $O_2$ concentrations. For example, some Cyanobacteria form heterocyst, cells that specialize in N-fixation. These cells provide an $O_2$-free environment for N-fixation (Fig. 11.12)

The bacteria and fungi that can reduce $N_2$ are quite diverse, common players include Rhizobia, Azospirillum, Azotobacter, Beijerinckia, Clostridium, and Cyanobacteria. In some cases, these are free-living, in others they form dependent, epigenetically controlled symbionts. *Azolla* may not be an obligate symbiont, but N-fixing capability of *Azolla* means that it is widely used as a fertilizer, especially in parts of southeast Asia (Fig. 11.13). Indeed, the fern has been used to bolster productivity in China for over a thousand years. When rice paddies are flooded in the spring, paddy waters are inoculated with Azolla, which quickly multiplies to cover the water, suppressing weeds. Once the plant dies, the decomposition process releases N, and the N can be assimilated by the rice.

Once fixed, nitrogenous compounds enter the food web through a variety of mechanisms, but are also excreted or otherwise mineralized. One main decomposition product is ammonium ($NH_4^+$). In addition, surfacewater sources of $NH_4^+$ include agricultural runoff, nitrogenous waste from human sewage, discharge from confined animal feedlots, or even atmospheric deposition from a wide variety of sources. The oxidation of ammonium produces nitrite ($NO_2^-$) and nitrate ($NO_3^-$), which is referred to as nitrification. In general, nitrification is a two-step process and carried out different groups of bacteria.

*Nitrosomonas* and *Comammox* spp, motile bacteria with a flagellum, carry out the first step,

$$NH_4^+ + \frac{3}{2}O_2 \longrightarrow NO_2^- + 2H^+ + H_2O \tag{11.8}$$

which only releases 65.7 kcal/mol of energy. As chemoautotrophs, these bacteria oxidize the nitrogen to produce energy using power-generating membranes (Fig. 11.14). The energy drives $CO_2$ fixation using the Calvin Cycle in bacteria and other pathways in archaea.

The second step is carried out by several groups of bacteria, e.g., *Nitrobacter*, *Nitrospira*, and *Comammox*,

$$NO_2^- + \frac{1}{2}O_2 \longrightarrow 2NO_3^- + \text{energy} \tag{11.9}$$

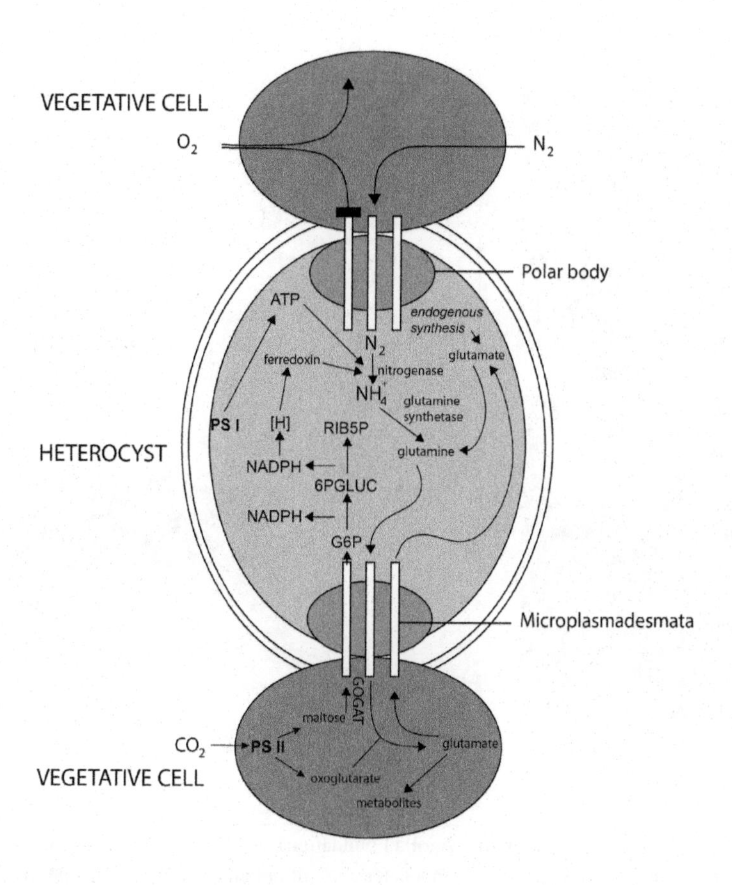


**FIGURE 11.12**

Schematic of Cyanobacteria heterocyst (Redrawn from Heimann and Cirés, 2015).

but it only generates 17.5 kcal/mol. Thus this reaction produces 25% less energy than the first step, meaning that the bacteria grow slower with the same amount of substrate available. But also means, when $NO_2^-$ is present, these bacteria are quick to use it. Therefore we often see low $NO_2^-$−N concentrations in surface waters.[1] However, in waters that have high rates of nitrogen loading, $NO_2^-$ can be surprisingly high, >3–20 ppm $NO_2$−N.

As a general rule, most surfacewaters without anthropogenic sources of N have $NH_4^+$−N and $NO_3^-$−N and are characterized by less than 1 ppm and $NO_2^-$−N is an order of magnitude lower. Because $NH_4^+$ is positively charged, it tends to be held by the negative charges in soils and less subject to leaching. Whereas the negative charge on $NO_3^-$ limits soil adsorption, its capacity to travel with water and leach is increased. Thus nitrate contamination of surface- and groundwater

[1] Note: when measuring inorganic nitrogen compounds in surface waters, the concentrations are reported in terms of N, because N is often a limiting element and allows and maintain a consistent measure of the N as it is converted from $NH_4^+$ to $NO_2^-$ to $NO_3^-$.

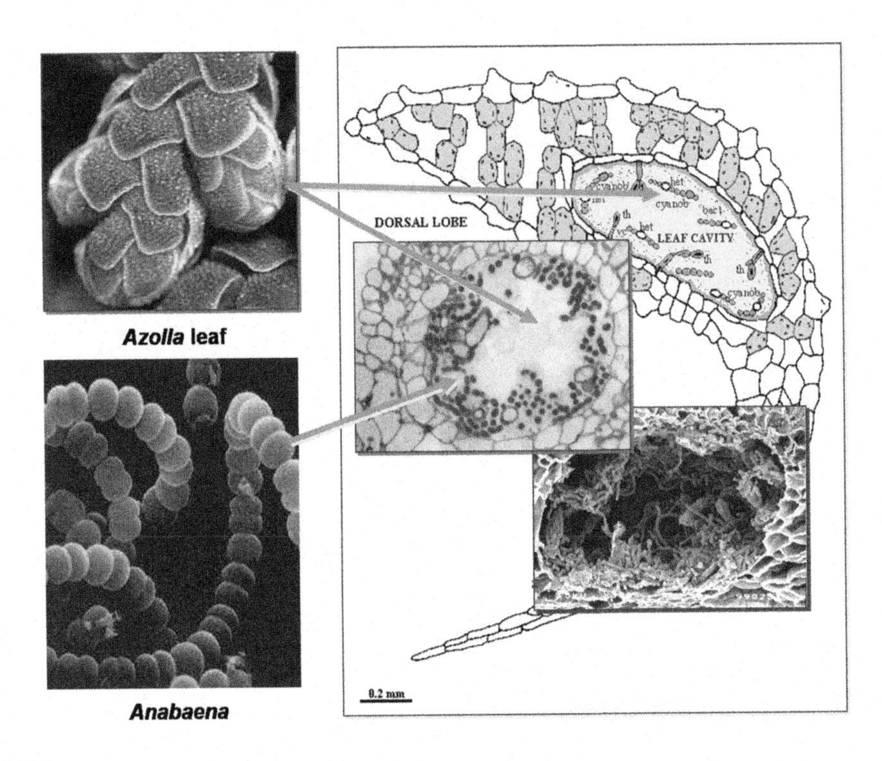


**FIGURE 11.13**

Image of Azola and Cyanobacteria (Source: The Azolla Foundation). As symbionts, the Azollo can take up nitrogen fixed by the Cyanobacterium. Meanwhile, the Cyanobacteria might obtain carbohydrates from the Azolla and some protection from oxygen.

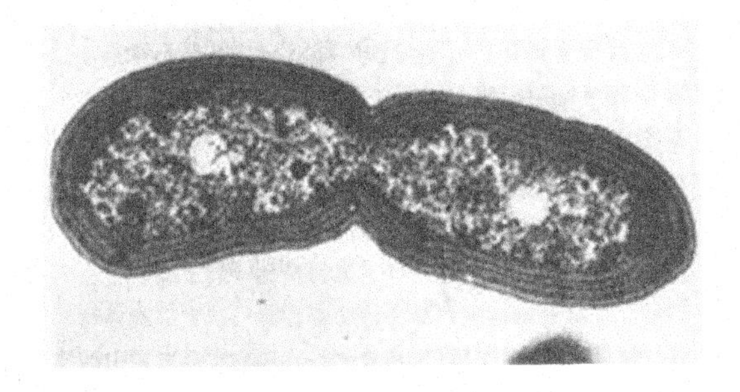

**FIGURE 11.14**

Nitrosomonas has a characteristic lamellar membrane structure, with specialized membranes that oxidize ammonia and pump $NO_3^-$ and NO away from the cytoplasm (Source: Yuichi Suwa).

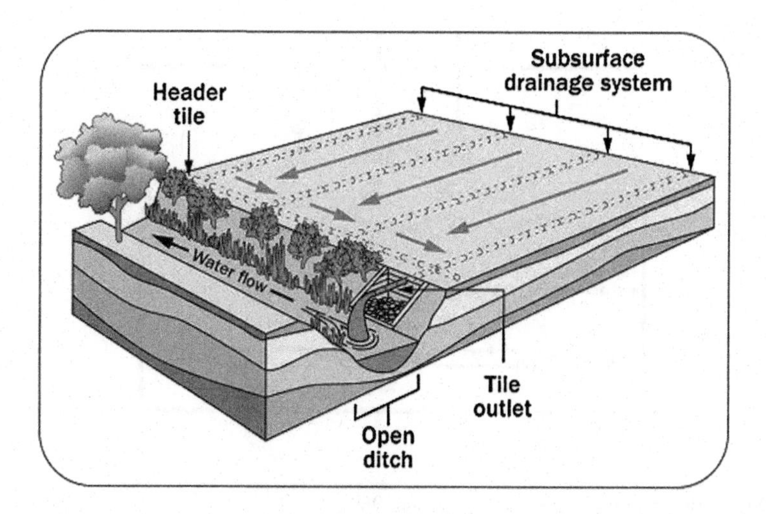


**FIGURE 11.15**

Subsurface drainage network are common in soils with poor drainage, often glaciated or floodplain soils with fine textured particles below the soil surface (Source: Ontario Ministry of Agriculture, Food, and Rural Affairs).

is quite common. The high nutrient loading areas ($NH_4^+$–N can be 1–5 ppm and $NO_3^-$–N, often 2–50 ppm), can exceed 100 ppm in farmed areas that have poor drainage and subsurface drainage systems (Fig. 11.15).

Unfortunately, these reactions do not tell us a complete story about the conversion from $NH_4^+$ to $NO_3^-$. Nitrification is not 100% efficient and some intermediate molecules escape the reaction (Fig. 11.16). Importantly, for Environmental Science, these compounds—nitric oxide (NO) and nitrous oxide ($N_2O$)—have air quality impacts (Fig. 11.16). In the troposphere, NO can promotes smog formation and in the stratosphere it reacts with and destroys ozone, reducing the capacity of the Ozone Layer to filter UV radiation. Nitrous oxide is a potent greenhouse gas, in fact it absorbs infrared radiation 278x better than $CO_2$.

One key message at this point is that these reactions occur in the presence of oxygen—in fact, they require oxygen. Furthermore, as these chemoautotrophic bacteria oxidize nitrogen as an energy source, they fix carbon and make sugars and microbial biomass.

In contrast, heterotrophic bacteria will oxidize organic carbon sources bacteria, but many Aquatic Systems experience limited oxygen concentrations. In these cases, a diverse group of bacteria can use alternative electron acceptors. Recall from above that the regeneration of NADH and succinate-fumarate reaction depend on the electron transport chain, where oxygen is used as the electron acceptor. But some bacteria can use $NO_3^-$ as an electron acceptor, we call these denitrifyers.

One of the more common reactions of biological denitrification can be summarized as

$$NO_3^- + 6H^+ \longrightarrow \frac{1}{2}N_2 + 3H_2O. \tag{11.10}$$

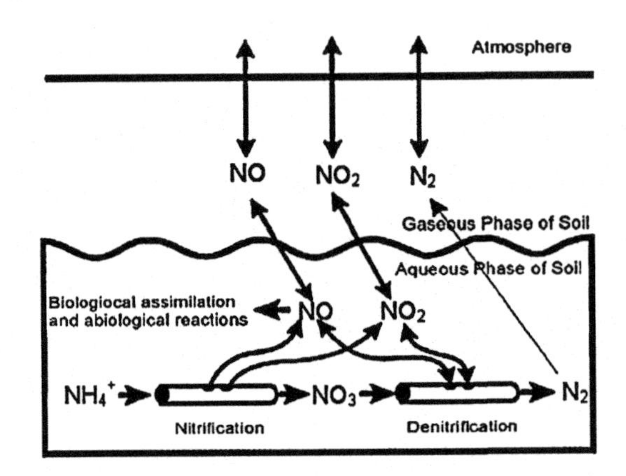


**FIGURE 11.16**

Hole in the Pipe Model (Source: Firestone and Davidson, 1989).

However, the energy obtained from the nitrate reduction is lower than oxygen reduction. Thus this less preferred electron acceptor, while allowing microbe to continue to metablize do so at a reduced rate. Therefore in completing the Nitrogen Cycle, biological denitrification releases $N_2$ gas, the dominant atmospheric gas.

Anammox and Dissimilatory Nitrate Reduction to Ammonium (DNRA) are two other important biogeochemical pathways in the nitrogen cycle (Fig. 11.17). Although the reactions increase the complexity of the N-Cycle, their role in inland waters is pretty obscure because methods to study their relative importance is still developing.

Although other biogeochemical processes occur in the Nitrogen Cycle, biogeochemical pathways discussed above form the basics of the cycle with an additional caveat. Just like the "leaky" oxidation of ammonium to nitrate, denitrification also releases intermediate compounds—in fact, the same two, NO and $N_2O$ (Fig. 11.16).

***Fe and Mn Cycles*** Both the oxidation and the reduction of iron and manganese are mediated by microbes, but abiotic transformations are also important and may compete with the biological processes.

4.3% of the continental crusts are composed of iron. With two redox states, FeII and FeIII, iron is used as a source or as an acceptor of electrons. Although the abundance of manganese is 50-fold lower in the Earth's crust relative to iron, Mn is the second most abundant redox-active metal occurring as $Mn_2^+$, $Mn_3^+$, and $Mn_4^+$. As $O_2$ and $NO_3^-$ are consumed as electron acceptors, Fe and Mn become important for specific redox reactions. Both of iron and manganese compounds play an important role in the microecology of aquatic systems, and can even be diagnostic in wetland soils. In addition, in the reduced form, they are soluble, but in the oxidized form they precipitate. This mobility explains how these elements become concentrated in specific redox conditions.

Dissimilatory iron- and manganese-reducing microorganisms catalyze the reduction of Fe (III) to Fe (II), and Mn (IV) or Mn (III) to Mn(II). Under anaerobic conditions, the Electron Transport Chain reduces oxidized manganese or iron and formed a variety of oxides. For example, the

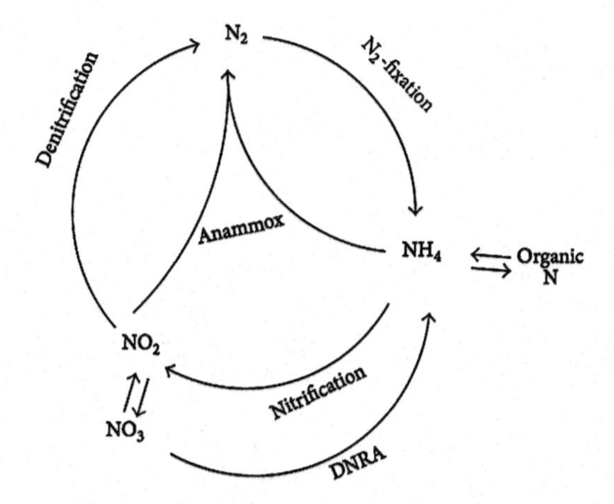


**FIGURE 11.17**

The Nitrogen Cycle also has two other important pathways (Source: Wiki Commons), but these are much harder to measure, thus their role in the environment has only been recently elucidated. Anammox combines $NO_2$ and $NH_3$ to form $N_2$ and maybe is responsible for 30–50% of the $N_2$ produced in the oceans. DNRA (Dissimilatory Nitrate Reduction to Ammonium) generally occurs under reducing conditions, where high organic carbon concentrations exist.

reduction of $MnO_2$ to $Mn_2O_3$ to $Mn_3O_4$ requires a redox state change for Mn:

$$Mn^{4+} + e^- \longrightarrow Mn^{3+} + e^- \longrightarrow Mn^{2+} \tag{11.11}$$

Because these redox reactions are sensitive to pH (Fig. 11.18), a combination of abiotic and biotic forces drive these redox reactions, creating a complex set of reaction in surface waters, wetland soils, and aquatic sediments. In addition, as free ions, $Mn^{2+}$ and $Fe^{2+}$, both manganese and iron exhibit phytotoxicity. Thus under anaerobic conditions, plant fitness is reduced for sensitive species.

Nevertheless, these reactions produce indicators of wetland soils that experience seasonal flooding. Fe and Mg can be reduced, translocated, and concentrated into a nodule form. The metals form precipitated ferromanganiferous nodules in the soil (Fig. 11.19). Nodules are used as a diagnostic tool to delineate the legal boundaries of wetlands.

***Sulfur Cycle*** Sulfur naturally occurs as a mineral, inorganic ions, and ion complexes and incorporated into organic molecules. Sulfur has oxidation states ranging from +6 in sulfate, $SO_4^{2-}$, to −2 in sulfides. Therefore elemental sulfur can donate or accept electrons, depending on the redox environment and is transformed, assimilated, and deassimilated by a complex array of processes and organisms (Fig. 11.20).

Minerals such as pyrite ($FeS_2$) comprise a major pool of sulfur on earth. But throughout the Earth's history, the amount of mobile sulfur has been slowly increasing as it has been released from the mantel via volcanic activity, and from the crust via weathering. The oceans are the most important sink for sulfur in the form of $SO_4^{2-}$, where it is the major oxidizing agent in sediments.

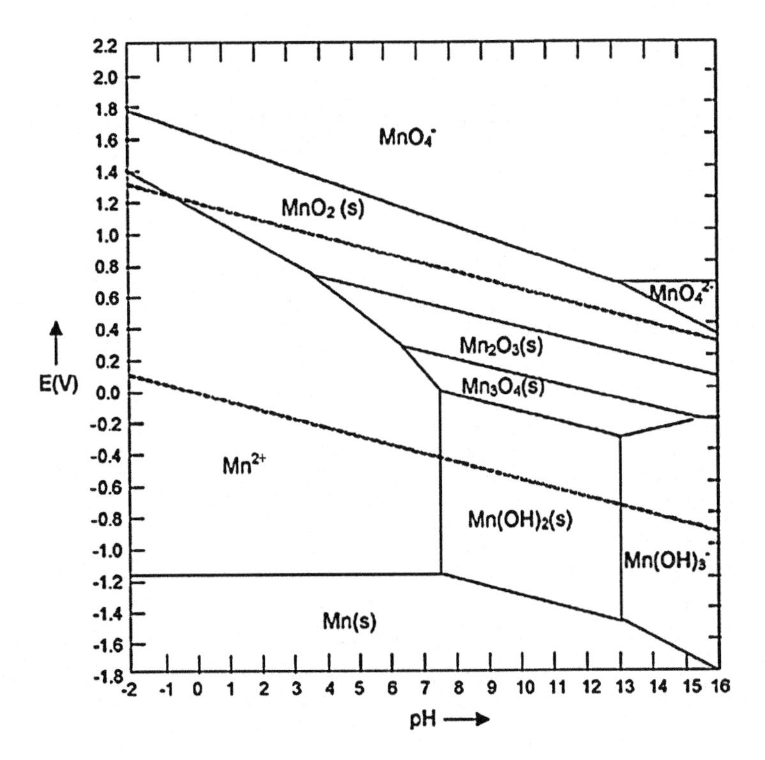


**FIGURE 11.18**

Pourbaix diagrams plot electrochemical stability for different redox states of an element as a function of pH (Source: http://www.wou.edu/las/physci/ch462/redox.htm). As noted above, these diagrams are essentially phase diagrams that map the conditions of potential and pH (most typically in aqueous solutions), where different redox species are stable.

In recent times, coal and fossil fuels combustion have added a substantial amount of $SO_2$ to the atmosphere, which is an important cause of acid rain, as described in Chapter 10.
Desulfurization, the process in which organic molecules containing sulfur can be desulfurized, produces hydrogen sulfide gas ($H_2S$, oxidation state $= -2$). An analogous process for organic nitrogen compounds is deamination.
In addition, the Sulfur Cycle includes a diverse set of photosynthetic bacteria. For example, some chemolithotrophs use hydrogen sulfide as an electron donor. These bacteria, such as heliobacteria, acid bacteria, photosynthetic green and purple sulfur bacteria, and filamentous anoxygenic phototrophs, each have highly conserved mechanism to free electrons and fix $CO_2$ (Fig. 11.21). By comparing these pathways, we have a window to understand how photosynthesis evolved in a preoxygen world and learn about how these cycles work today.
For example, purple sulfur bacteria are often found in stratified water environments, including hot springs, stagnant water bodies, as well as microbial mats in Intertidal Zones. Since purple sulfur bacteria do not use water as their reducing agent, they do not produce $O_2$, thus are non-

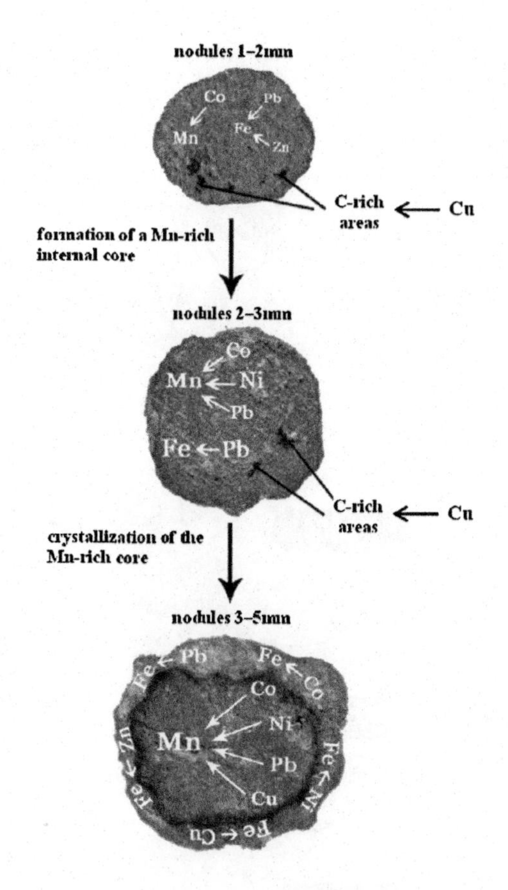


**FIGURE 11.19**

Schematic diagram showing association between trace elements and Fe and Mn during nodule formation and growth (Source: Timofeeva et al., 2014).

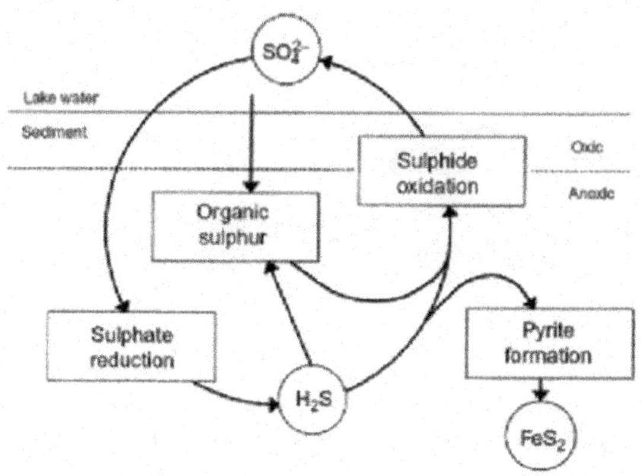


**FIGURE 11.20**

Sulfur cycle includes oxic and anoxic processes.

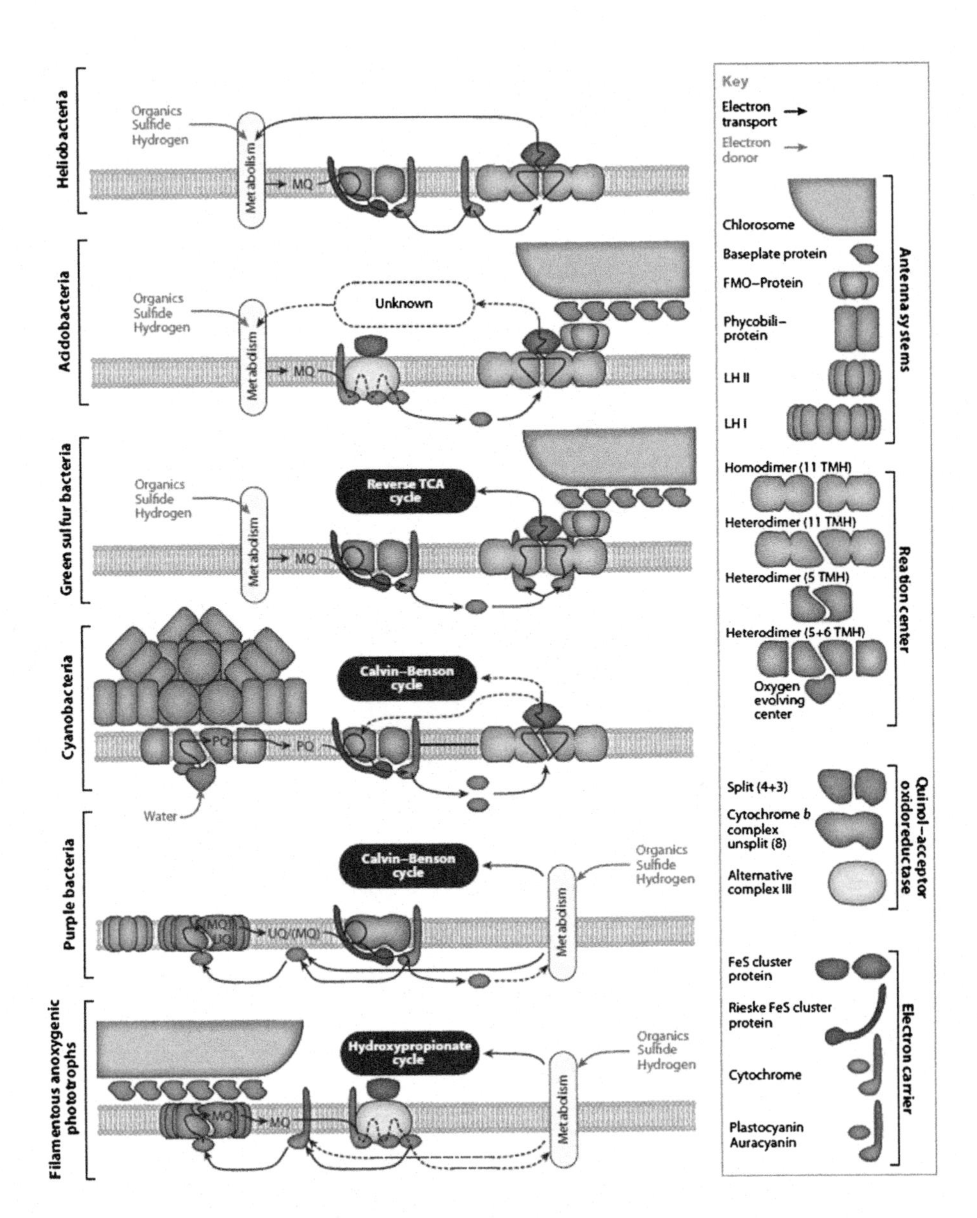


**FIGURE 11.21**

Photosynthetic machinery and electron transport of photosynthetic bacteria, including a description of photosynthetic complexes (Source: Hohmann-Marriott and Blankenship, 2011). Abbreviations: LH, light harvesting; MQ, menaquinone; PQ, plastoquinone; TCA, Tricarboxylic Acid; TMH, transmembrane helix(ces); UQ, ubiquinone.

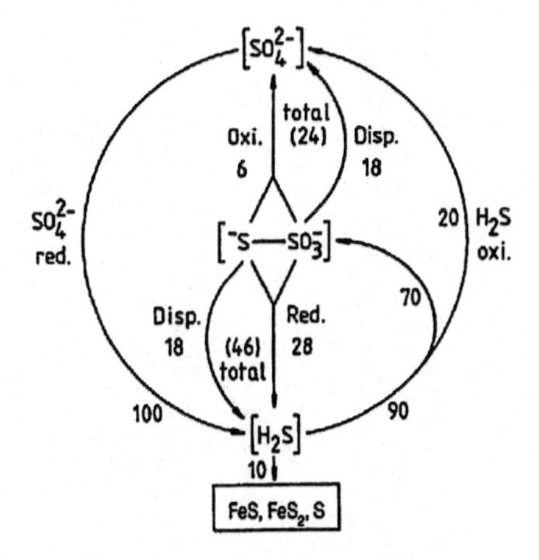


**FIGURE 11.22**

Summary of Sediment Sulfur Cycle (Source: Jørgensen, 1990). Process rates are expressed as percent of the sulfate reduction rate (Red = reduction; Oxi = oxidation, and Disp. = disproportionation).

oxygenic. Instead, they use sulfide or thiosulfate as the electron donor in their photosynthetic pathway. As sulfide oxidize, granules of elemental sulfur are produced, which can be further oxidized to form sulfuric acid.

The oxidation in elemental sulfur by sulfur oxidizer bacteria produces $SO_4^{2-}$:

$$HS^- + 2O_2 \longrightarrow SO_4^{2-} + H^+ \tag{11.12}$$

In sediments or surface waters, this occurs in the presence of $O_2$, but $SO_4^{2-}$ can diffuse into the sediments and become the electron acceptor for anaerobic bacteria. Dissimilative Sulfate Reduction is the process by which sulfate reducers generate hydrogen sulfide from sulfate under anaerobic conditions. These processes can have dramatic environmental impacts on the pH of waters and have been linked to the release of toxic metals from mine tailings.

The sulfur is a further complicated by sulfur compounds that can "shuttle" or transfer electrons from one redox environment to another. For example, thiosulfate can be an intermediary between an oxidized and reduced environment (Fig. 11.22). Although originally associated with marine sediments, the thiosulfate shunt probably occurs in a wide range of wetlands with significant sulfur oxidation and reduction processes.

Added to the thiosulfate shunt, the sulfur cycle also interacts with iron redox reactions that include both biotic and abiotic processes that produce FeS compounds, Fe oxides, and Fe oxides-phosphorous complexes (Fig. 11.23). Oxidized iron can precipitate with P and remove P from the water column, but under anaerobic condition, Fe is reduced and P is released. Thus hypoxia exacerbates internal P loading in lakes.

***Carbon and Methane*** Methane is, after water vapor and carbon dioxide, the third most important greenhouse gas in the atmosphere. Its concentration in the atmosphere has more than doubled since preindustrial times. Human energy production and use, landfills and waste, cattle raising, rice production, and biomass burning are primary causes of the increase. However, 40% of

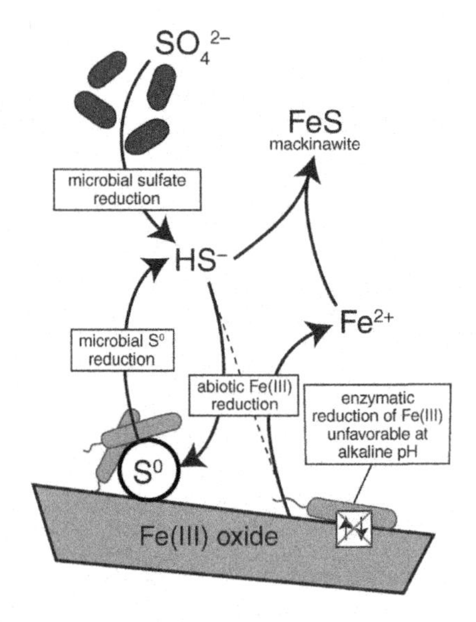


**FIGURE 11.23**

Diagram illustrating $S_0$ mediated Fe(III) reduction under alkaline conditions (Source: Flynn et al., 2014).

current global methane sources are natural. Most natural emissions come from anaerobic decomposition of organic carbon in wetlands, with smaller contributions from the ocean, termites, animals, wildfires, and geological sources.

Under highly reduced conditions, microbes reduce $CO_2$ to $CH_4$ gas. The process is called methanogenesis. However, as methane diffuses through the redox gradient, methanotrophs can use $CH_4$ as an electron source and oxidize it. Although methane production can be high from wetlands, without methanotrophs $CH_4$ production would be much higher.

In addition, some sulfate-reducing bacteria play a role in the anaerobic oxidation of methane:

$$H_4 + SO_4^{2-} \longrightarrow HCO_3^- + HS^- + H_2O. \tag{11.13}$$

Therefore linking sulfur biogeochemistry to $CH_4$ production and oxidation and greenhouse gas fluxes.

# WETLAND ECOLOGY AND REDOX

## FLOODING, REDOX, AND PLANT PHYSIOLOGY

Plants have diverse responses to flooded conditions. Some plants are intolerant and even short-term flood with lead to death, while others can tolerate extensive flooding, and some depend on aquatic conditions with a range of physiological, anatomical, and morphological adaptions. Since saturated conditions in soils impede gas exchange, soil waterlogging leads to hypoxia and progressively to anoxia and high $CO_2$ in the root zone, which is often accompanied by increased mobilization of 'phytotoxins' in reduced soils, such as $Fe_2^+$, $Mg_2^+$, and $S^{2-}$. These phytotoxins affect root metabolism, nutrient acquisition and thus growth (and survival) of plants in these conditions.

Most wetland-adapted species have developed a variety of mechanisms, including morphological (emission of adventitious roots), anatomical (aerenchyma formation in roots) and physiological changes, which may be rapid or slow in response to flooding.

Plants respond to several indicators of flooding that include reduced oxygen levels, accumulation of nitric oxide and ethylene, and low energy status. Although not a direct measure of flooding, these indicators result from flooding. For example, as we have noted, glycolysis in eukaryotes requires $O_2$ for $NAD^+$ and succinate reduction. Without $O_2$ ATP production via glycolysis is severely constrained. The reduced energy status initiates downstream induction signals that up- or down-regulate protein production, modifies cell metabolism, changes developmental processes that result in anatomical and morphological changes. If these changes are sufficient for the flooding regime and able to cope with an anaerobic rhizosphere then plant will survive flooding.

## PLANT PHYSIOLOGY AND ANAEROBIC CONDITIONS

In fact, most plants are irreversibly damaged with water logging—where the internal cells become acidified, aquipors change their confirmation, and water can no longer enter the plant—and in a strange ironic twist the plants starts to wilt as a sign of water stress, and the plant may not recover.

Under hypoxic conditions, alternative mechanisms to obtain energy from glucose, such as ethanolic fermentation (Fig. 11.24), is usually a short-term response. However, the increase in ethanol concentrations become toxic.

Whereas physiological adjustments are more transient for species that can withstand short-term flooding, a change in long-term metabolism may only be observed in species that are highly tolerant to flooding (Fig. 11.25).

## PLANT MORPHOLOGY ADAPTIONS

Many wetland-adapted species have developed a variety of morphological adaptions or even responses to flooding. For example, under flooded conditions, lowland rice stems will elongate by increasing internode distances. Or some plants will initiate the development of adventitious roots from shoot that might access more aerobic conditions. Some morphological adaptation can be quite subtle. For example, the submerged leaf surface have hairs or waxy cuticles that create a gas film or air gap between the leaves and the water. Plants that with a gas film have enhanced $CO_2$ and $O_2$ exchange through the leaf stomata compared to plants without a gas film, increasing the effective diffusion rates of these gases in water.

## PLANT ANATOMICAL ADAPTIONS

Anatomical adaptions include tissue or intracellular changes that usually improve $O_2$ exchange or reduce oxygen loss for specific portions of the root.

***Aerenchyma*** Oxygen needs to be transported to the roots in order to avoid root anoxia and ethanol toxicity and death. Under partial submergence the least part of the shoots are above water and the

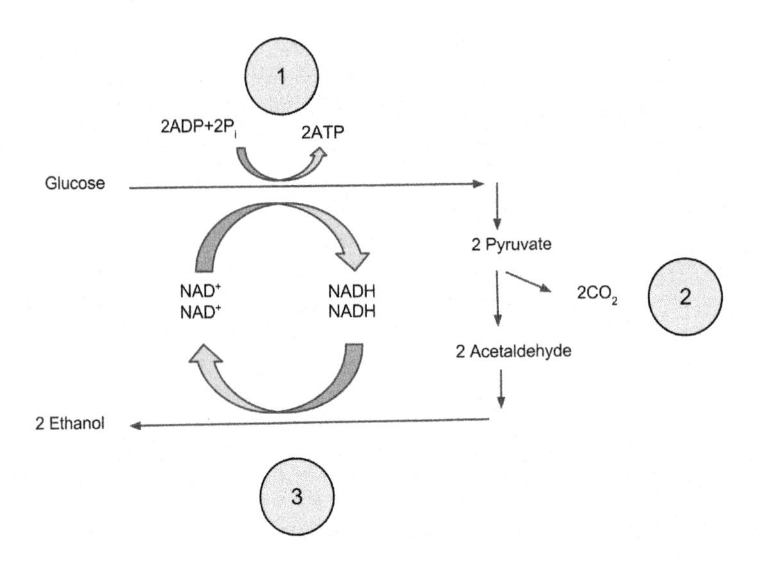


**FIGURE 11.24**

Ethanol fermentation. One glucose molecule breaks down into two pyruvates via glycolysis. The energy from these exothermic reactions is used to bind inorganic phosphates to ADP and convert $NAD^+$ to NADH. The two pyruvates are then broken down into two acetaldehyde and give off two $CO_2$ as a waste product. The two acetaldehydes are then reduced to two ethanols, and NADH is oxidized back into $NAD^+$.

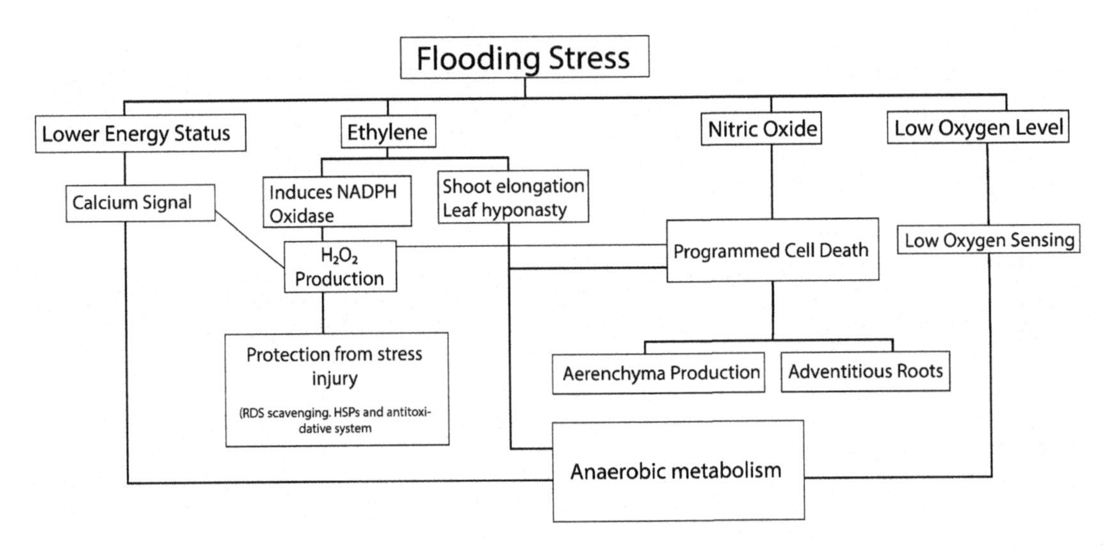


**FIGURE 11.25**

Hypoxia signaling in plants (redrawn by Luyi Huang from Lekshmy et al., 2015). With the rapid decline of $O_2$, the energy status of the plant is reduced. In addition, ethylene and NO accumulates that imitate a cascade of signals that influence developmental and regulatory adaptions to tolerate flooding.

**FIGURE 11.26**

Arechemy formed by lysigeny, a process that involves the collapse and death of cells in the cortex zone (Source: https://en.wikipedia.org/wiki/Aerenchyma#/media/File:Aerenchyma2).

capture of atmospheric oxygen by leaves is possible, but plant root do not have this luxury. With the accumulation of ethylene, downstream transduction signals initiate programmed cell death in the shoot interiors. With cell death, empty spaces develop in the stem called aerenchyma, thus facilitating $O_2$ movement to the roots (Fig. 11.26).

In plant species having (or developing) aerenchyma, the magnitude of oxygen reaching the root apex depends on the effectiveness of longitudinal transport. Extensive aerenchyma in rice provide a low-resistance internal pathway for $O_2$ transportation towards underground organs, but also allow $CH_4$ emissions that bypass methanotroph oxidation, thus increasing the GHG contribution in rice production.

***Suberine and ROL*** In all roots, oxygen is required for respiration to provide sufficient energy for growth, but 30–40% of the $O_2$ supplied via the root aerenchyma of wetland plants is lost to the soil, a process defined as Radial Oxygen Loss (ROL). Root Apex Oxygenation is crucial for continuing with root elongation and soil exploration under flooding conditions.

When tissue respiratory demands are satisfied along the root, such effectiveness is mostly dependent on the loss of oxygen towards the rhizosphere. The loss of oxygen from the root depends on the presence of barriers impeding its leakage towards the soil (Fig. 11.27).

## RHIOZPHERE TOXICITY AND ROL

The rhizosphere, i.e., the soil or sediment directly influenced by plant roots and play an important role in plant metabolism and growth. As soils are flooded and become anaerobic, the redox environment can produce mobile, toxic chemicals, such as $Mn^{2+}$, $Fe^{2+}$, and $S^{2-}$. But roots often leak a substantial amount of $O_2$, thus mitigating the anaerobic environment that might influence the redox state of

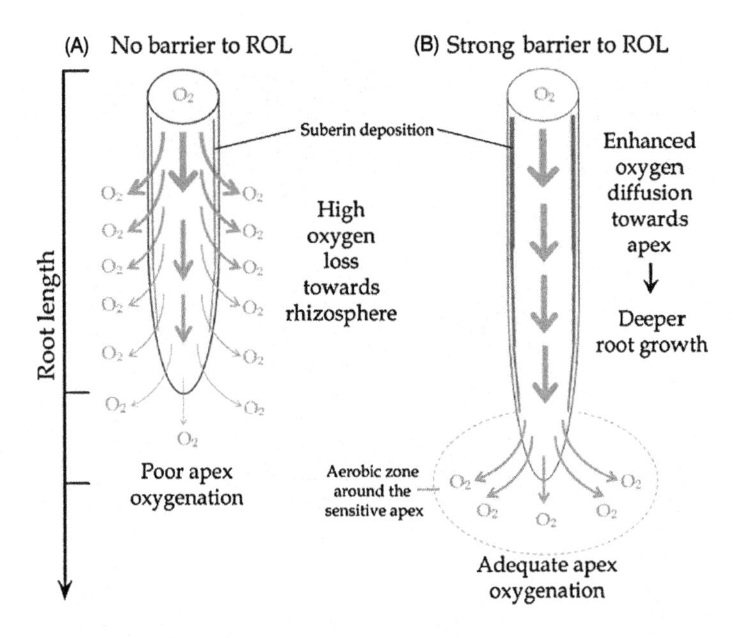


**FIGURE 11.27**

Scheme showing two different patterns of Radial Oxygen Loss (ROL) from roots (Source: Colmer and Voesenek, 2009, Striker, 2012). Root without barrier to ROL in the outer cortex (A) loses oxygen along the root, creating $O_2$ deficiency at the apex and short roots in anoxic soils. (B) Root having a strong barrier to ROL: $O_2$ is transported to the apex allowing deeper root growth in flooded soils. The loss of oxygen is circumscribed to the apex, which generates an aerobic zone that diminishes entry of potentially toxic compounds ($Fe^{2+}$, $Mn^{2+}$, $S_2^-$) in highly reduced soils. The thickness of the grey color arrows indicates the amount of oxygen available.

metals. By creating an oxidized zone around the roots, toxic compounds can be oxidized and rendered nontoxic.

For example, Fe plaque can form on root surfaces in wetland plant roots. Fe plaque is formed where the oxygen leaking from roots reacts with reduced soluble $Fe^{2+}$ to form a reddish precipitate or plaque coating on root surfaces.

During the process of ROL and rhizosphere oxidation, substantial quantities of Fe are transferred to the plaque, leading to a well-defined zone of ferric hydroxide accumulation.

## PLANT PHENOLOGY: SEASONAL AND DIURNAL/TIDAL RESPONSES

The hydrology of wetland varies spatially and temporally—and these define the spatial distribution of the vegetation. Seasonal wetlands have plant zonation gradients that might reflect the depth of water, time, or longevity of flooding. For example, the seasonal flooding or tidal flooding (twice daily; Fig. 11.28) will select for different zonation patterns that depend on the amount of time of innundation and the redox response of the rhizoshere.

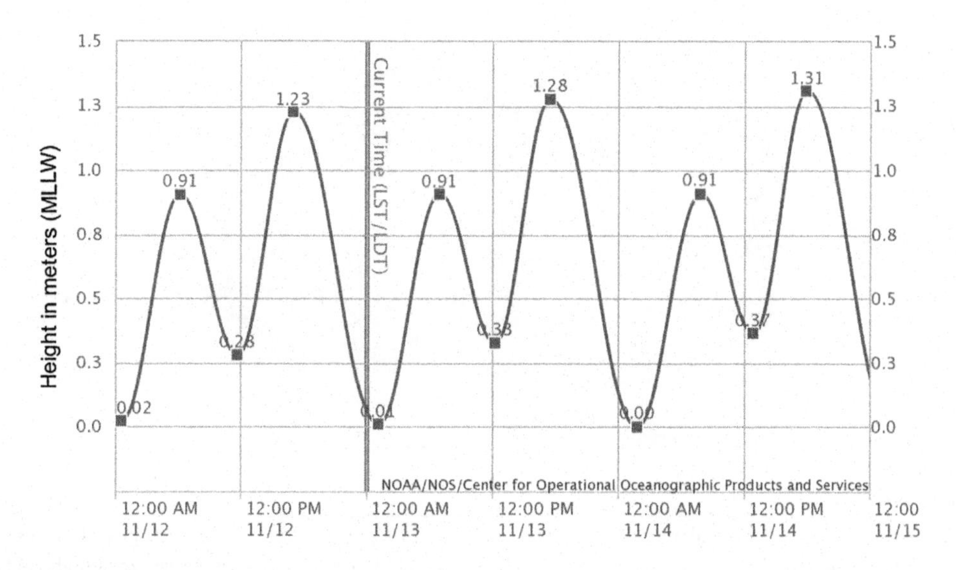


**FIGURE 11.28**

Tidal cycle, Port of Stockton, San Joaquin River (Source: NOAA).

To avoid potential toxicity of flooded soils, plants might grow early and set seed before anaerobic conditions can have long-term impacts. Alternatively, plants might get established during drawdown and where soil oxygen concentrations are increasing with time as root development occurs. Based on individual tolerances, zonation patterns can become established and reenforced by flood events.

For example, the vegetation of the Pantanal display a complex set of zonation patterns depending on the hydroperiod, flooding depth, and interannual variation. These patterns are the result of balance between flood and drought tolerance for each individual plant.

# CARBON CYCLING IN WATER

## CARBON AND FERMENTATION

Fermentation is another process in which cells gain energy from organic compounds. In this metabolic pathway, energy is derived from the carbon compounds without the use of oxygen and often called anaerobic respiration. This process can occur throughout the redox gradient, but is generally associated with hypoxic soils.

Beside $CO_2$, fermentation produce either ethanol ($C_2H_5OH$) or Lactic Acid ($CH_3CH(OH)COOH$). This respiration pathway is important in waterlogged ecosystems, such as in peat, bogs, and wetlands.

## BIOGEOCHEMISTRY OF DIC

Survival of aquatic macrophytes depends on the ability to obtain carbon from a carbon-limited environment. Similar to $O_2$, inorganic carbon can become limited to phytoplankton, periphyton, and

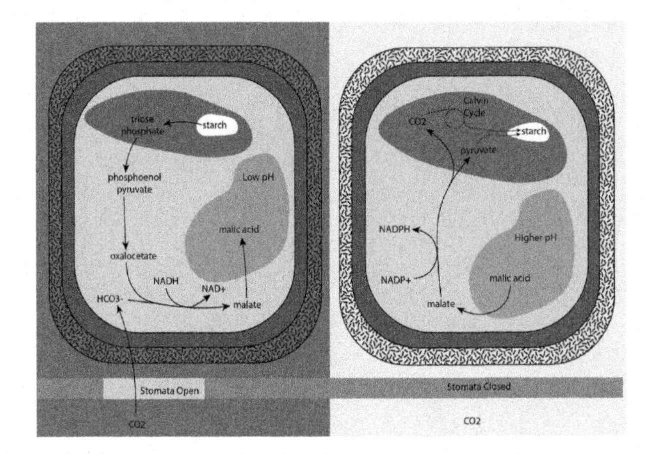


**FIGURE 11.29**

Anatomy and physiology of CAM plants (Source: Arizona Board of Regents / ASU Ask A Biologist). In contrast to CAM plants that live in the desert, to conserve water, submerged CAM plants seem to photothesize at night and has high DIC demand during the day.

submerged plants. Limitations are at their highest in alkaline conditions, or when potential fixation rates exceed regeneration rates of $CO_2$.

The availability of inorganic carbon can have a dramatic impact on the growth and development of autotrophs. In some cases, plants have developed specific $CO_2$-concentrating pathways to avoid inorganic carbon limitations. For example, there are three types of photosynthesis pathways in terrestrial plants that confer adaptive advantages. For instance, in warmer climates, the C4 pathway is often associated with greater water-use efficiency and reduction in photorespiration compared to the C3 pathway (Fig. 11.29). CAM-type photosynthesis capitalizes on the independence of the light and dark reaction by using a modified pathway at night to fix $CO_2$ and minimize water losses. But how these pathways function in an Aquatic System is only partially understood.

Stable Carbon Isotope Ratio ($\delta^{13}C$) can distinguish between the biochemical processes that are responsible for the transformation of carbon dioxide/bicarbonate to sugars. However, compared to their terrestrial cousins, the $\delta^{13}C$ patterns in aquatic plants is more variable (Table 11.2). Whereas terrestrial plants have a relatively narrow range, ecosystem processes can differentiate based on photosynthetic pathway and the subsequent discrimination of carbon isotopes. Whereas all three photosynthetic pathways can be found in different aquatic macrophytes, the relationship between the type of photosynthesis and $\delta^{13}C$ is poorly constrained. $\delta^{13}C$ in Aquatic Systems depends on more parameters than in terrestrial systems. For example, the sources inorganic carbon (groundwater, atmosphere, respiration) can vary, carbon assimilation rates fractionate stable carbon isotopes, and hydrology parameters (velocity, turbulence, diffusion rates) influence $\delta^{13}C$ availability.

Thus, the same species from the same site with the same sources of carbon may also have a different $\delta^{13}C$ due to microhabitat conditions that surround the aquatic plant leaves. Because of the unique environment between and within Aquatic Systems, the study of biochemical pathways of plants, espe-

**Table 11.2 Stable Isotope Fractionation (Source: Richet et al., 1977).**

| Pathway | $\delta^{13}C$ ‰ |
|---|---|
| C3 | −20 to −37 |
| C4 | −12 to −16 |
| CAM | −10 to −20 |
| Phytoplankton | −18 to −25 |
| Aquatic Plants | −4 to −50 |

cially as with anthropogenic impacts on water quality is an important area of research, so we can better understand how aquatic photoautotrophs tolerate stress associated DIC biogeochemistry.

## BIOGEOCHEMISTRY AND ANTHROPOCENE

### RIPARIAN VEGETATION AND GROUNDWATER

Early in the 1970s, environmental scientists observed lower $NO_3^-$ concentrations in streams with riparian forests. In evaluating this pattern, groundwater monitoring demonstrated a decline $NO_3^-$ as groundwater flowed from the forest edge to the stream, but the reduction or attenuation was uneven and depended on a combination of hydrology, geomorphology, and ecological parameters. Partially based on the biogeochemical processes in riparian zones, management activities have been recommended to maintain their ecosystem functions (Fig. 11.30).

Nitrogen attenuation of shallow groundwater flowing through forested Riparian Buffer Zones can be quite substantial. Under optimal conditions, removal rates can range between 50% and 90% with buffer widths between 5–30 m. Although plant uptake may partially remove N, denitrification is the dominant removal mechanism—converting nitrate to $N_2$. Highest removal rates are usually associated groundwater flow above shallow, impermeable soil layers that maximize water residence time and contact with plant roots and organic-rich soils to fuel denitrifiers. In fact, the tree roots may play a key role—not by direct uptake, but as a source of organic carbon. Where groundwater bypasses the root zone and surface soil layers with limited organic carbon, N attenuation is minimal.

Riparian (and grass) stream buffers adjacent can protect water quality. Besides nitrate attenuation, buffers have other ecosytem functions, such as sediment trapping, phosphorus filtering, and mitigation of toxics. On the other hand, these buffers may be a substantial source of $N_2O$, contributing to global warming. To promote these practices various economic and regulatory incentives promote the protection or restoration of Riparian Vegetation. Meanwhile, Environmental Scientists need to monitor these systems for their effectiveness and evaluate potential tradeoffs.

### HYPERHEIC EXCHANGE AND NITROGEN BIOGEOCHEMISTRY

The amount of nitrogen exported via a stream is always less then the total inputs into the watershed. Some of the missing nitrogen is exported in biomass, lost from gas emissions (e.g., $N_2$, $NH_3$, $N_2O$, and NO), or stored in groundwater. Besides $NO_3^-$ attenuation in buffer zones, observations that nitrate con-

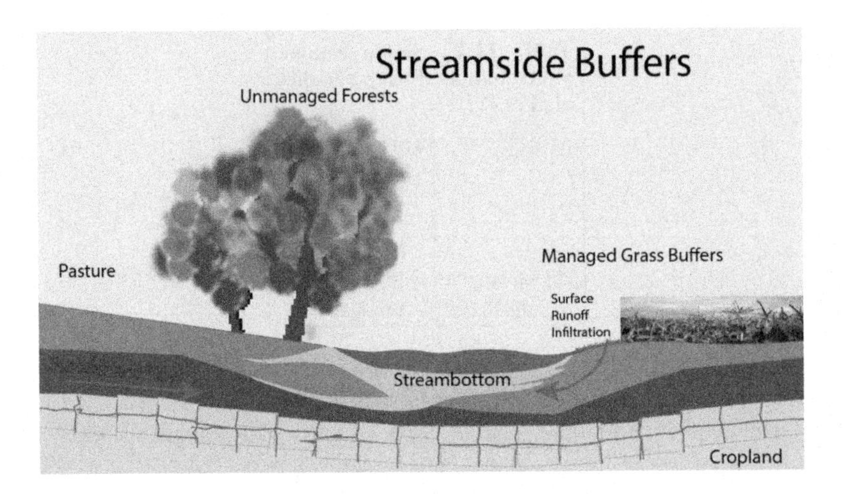


**FIGURE 11.30**

Riparian or Forested Buffer Zones (Source: Stroud Water Research Center). Zone 1 – from stream edge to 15 feet, native riparian forested trees and shrubs, no harvesting zone. Zone 2 – from edge of Zone 1 out another 20 feet to 35 feet or more, fruit and nut trees and shrubs, non-mechanical harvest allowed. Zone 3 – from edge of Zone 2 out another 50 to 100+ feet, woody florals and forbs, including biomass crops. Mechanical harvest allowed.

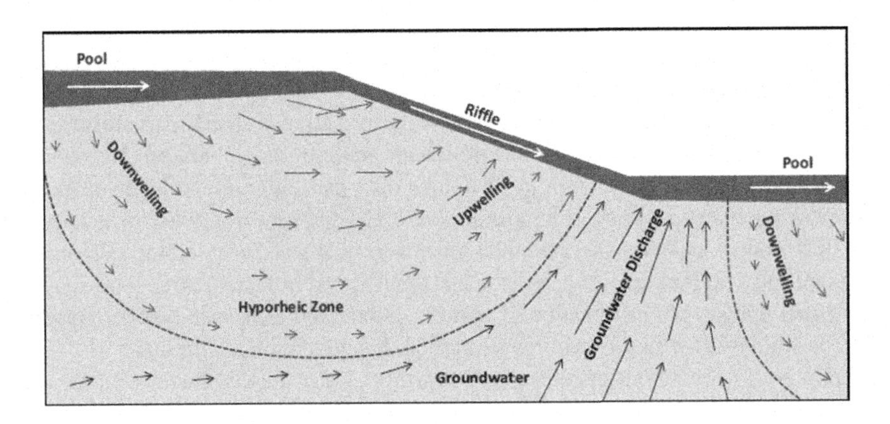


**FIGURE 11.31**

Diagram of hyporheic exchange through a pool-riffle sequence (Source: American Geophysical Union).

centrations decline from upstream to downstream locations suggest that channel attenuation processes also occur.

Like many Riparian Zones, there is a zone in the benthos of water exchange (Fig. 11.31). With DOC, microbial films, and limited O that promote denitrification, the Hyporheic Zone can remove 50% of the stream $NO_3^-$.

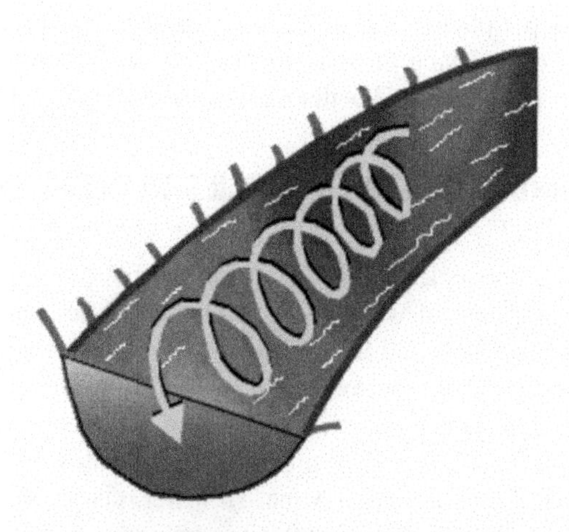


**FIGURE 11.32**

Nurrient spiraling. (Image courtesy of: Hebert, P.D.N, ed. Canada's Aquatic Environments [Internet]. CyberNatural Software, University of Guelph. Revised 2002.)

Similar to Riparian Forests, removal of nitrate depends on effective contact with groundwater exchange, rich in organic matter, and the redox environment. In addition, processes that attenuate N are going to be different for P. To better understand how streams assimilate and release nutrients, researchers have developed and parameterized a conceptual model to evaluate how the nutrient removal rates vary.

As dissolved forms of nutrients travel downstream they are transformed or assimilated into the biomass. As the biomass is consumed, the nutrients are released or further assimilated into the food web. In fact, these nutrients are usually passed through several links in a stream food web, where microbial uptake, grazing, excretion, decomposition, mobilization, and reassimilation may occur many times. In the context of advective flow, this process can be conceptualized as a spiral-like model (Fig. 11.32).

By measuring nutrient uptake rates and traveled distance of nutrient molecules, we can quantify nutrient spiraling ($S$),

$$S = Sw + Sb, \tag{11.14}$$

where ($S$) is the average distance a nutrient molecule travels downstream during one cycle. The cycle begins with the availability of the nutrient in the water column in inorganic form, and includes the distance traveled in the water ($S_w$) until its uptake ($U$) and assimilation by an organism, where the nutrient becomes part of an organic molecule. Additional distance traveled as part of the biota ($S_b$) completes the distance as the nutrient molecule are remineralized and released.

Nutrients are likely to be initially incorporated into autotrophs or bacteria associated with the streambed and then be consumed by benthic meofauna and invertebrate before eventually being released. The distance traveled in the biota ($S_b$) can be subdivided in various ways, for example CPOM, benthic and suspended FPOM, periphyton, and an array of consumer and predator compartments (benthic macroinvertebrates, fish, etc.). Using these compartments, we can estimate how different biota affect nutrient transport in Lotic Systems and better understand the impacts of nutrient loading.

## MICROBES, MERCURY, AND INTRODUCED SPECIES IN CLEAR LAKE: TROPHIC DYNAMICS

Clear Lake faces multiple anthropogenic stresses, including introduced species, mercury contamination, noxious Cyanobacteria, and loss of wetlands. Dramatic land-use changes in the vicinity affect lake ecology by altering runoff patterns and increasing erosion. In many parts of the world, mercury mining, past and present, pose significant environmental and human health effects. The legacy of mines in California includes tailings that leach mercury, contaminated sediments, and biomagnification.

Beginning in the 1860s, mining surface deposits of elemental sulfur in the Clear Lake watershed was the primary activity for the Sulphur Bank Mine. However, deeper deposits from the site were contaminated with cinnabar (HgS), and in 1872 the site was converted to mercury mining and was renamed to the Sulphur Bank Mercury Mine. At that time the Sulphur Bank Mercury Mine accounted for 9% of the nation's mercury production and was used to enhance the recovery of gold from mines in the Sierra Nevada Mountains. Long after the mine had closed, elevated concentrations of mercury were found in fish in 1970s, and the Sulphur Bank Mercury Mine was placed onto the National Priority List as an Environmental Protection Agency Superfund Site in 1990.

Inorganic mercury concentrations in lakebed sediments vary spatially. Fore example, concentration exceed 400 $mg{\cdot}kg^{-1}$ close to the mine and decline exponentially with distance from the mine.

Whereas atmospheric deposition is a major source of mercury in Aquatic Systems, mercury leached from mine spoiling to mercury laden sediments are the deposition mechanisms into Clear Lake. Once in surfacewater, mercury enters a complex cycle. The various forms of mercury can be converted from one to the next; most important is the conversion to methylmercury ($CH_3Hg^+$), a very toxic form. Mercury can be deposited into the sediments by particle settling and then later released by diffusion or resuspension. It can enter the food chain, or it can be released back to the atmosphere by volatilization (Fig. 11.33).

Numerous water quality parameters influence mercury concentrations, availability, and toxicity. For example, DOC and pH have a strong effect on the fate of mercury in Aquatic Systems. In some cases, some fish have higher concentrations of Hg with lower pH and/or higher DOC. Thus an increase in acidity and DOC levels may enhance Hg mobility in the environment, increasing its likelihood to enter the food chain.

In 2015 Lake Titicaca (Bolivia-Peru) water column experienced a spike of methyl-mercury (MeHg) and $H_2S$ concentrations associated with an algal bloom. At the high lake elevation, UV-A and UV-B radiation was much higher than at sea level. But the bloom reduced water clarity and UV-A and UB-B penetration. By reducing UV-A and UV-B radiation, photodemethylation of MeHg was reduced, leading to dramatic spike in the MeHg concentrations. Since MeHg bioaccumulates, the spikes of MeHg posed a serious health risk for those that rely on the fishery for food.

In addition, food web dynamics may also play an important role in Hg biogeochemistry. Threadfin shad (*Dorosoma petenense*) is a pelagic feeder and was first seen in Clear Lake in 1986. As its populations dramatically increased, other pelagic fish, such as largemouth bass (*Micropterus salmoides*)

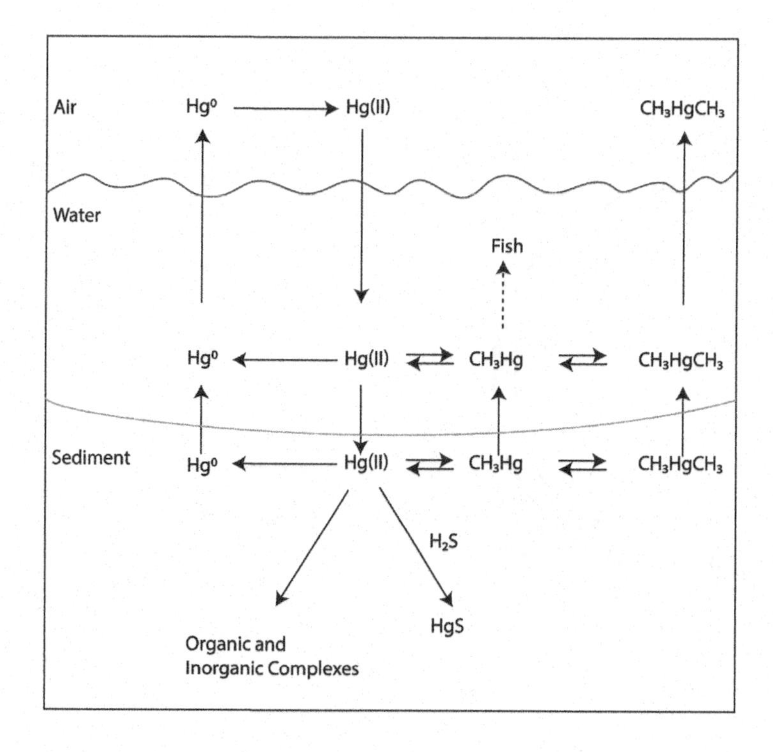


**FIGURE 11.33**

Diagram of Mercury Cycle (Redrawn by Luyi Huang).

were displaced and forced to forage the benthos, where MeHg concentrations are high, thus becoming increasingly contaminated with MeHg.

In a parallel example, feathers of grebes and ospreys were collected between 1967 and 2006 and analyzed for total mercury, TotHg. TotHg concentrations in grebe feathers declined from 23 $mg \cdot kg^{-1}$ in 1969 to 1 $mg \cdot kg^{-1}$ in 2003, but increased to 7 $mg \cdot kg^{-1}$ in 2004–2006. Osprey feathers showed a similar pattern of decline from 1992 to 1998, rebounding by 2003. Although the decline was probably due to EPA remediation activities, the later increase seems to coincide with the Threadfin Shad population spike in the early 2000s.

Clear Lake and Lake Titicaca are both ancient lakes. Clear Lake is probably 480,000 years old and Lake Titicaca is nearly 3 millions years old. Because of sediment deposition and human land use, Clear Lake is eutrophic, whereas Lake Titicaca is mesotrophic with increasing signs of eutrophic conditions near human settlements. Both lakes are the result of tectonic activity. Clear Lake was formed by volcanic activity in an exorheic basin and ~400 meters above sea level, whereas Titicaca with a height of ~3800 meters above sea level was created in an endorheic basin and high plateau associated with the orogeny of the Andes. Thus the fate of Hg in the Andes will largely remain within the lake ecosystem, whereas Clear Lake's Hg has been exported to the Sacramento River and marshlands of the Bay-Delta Ecosystems.

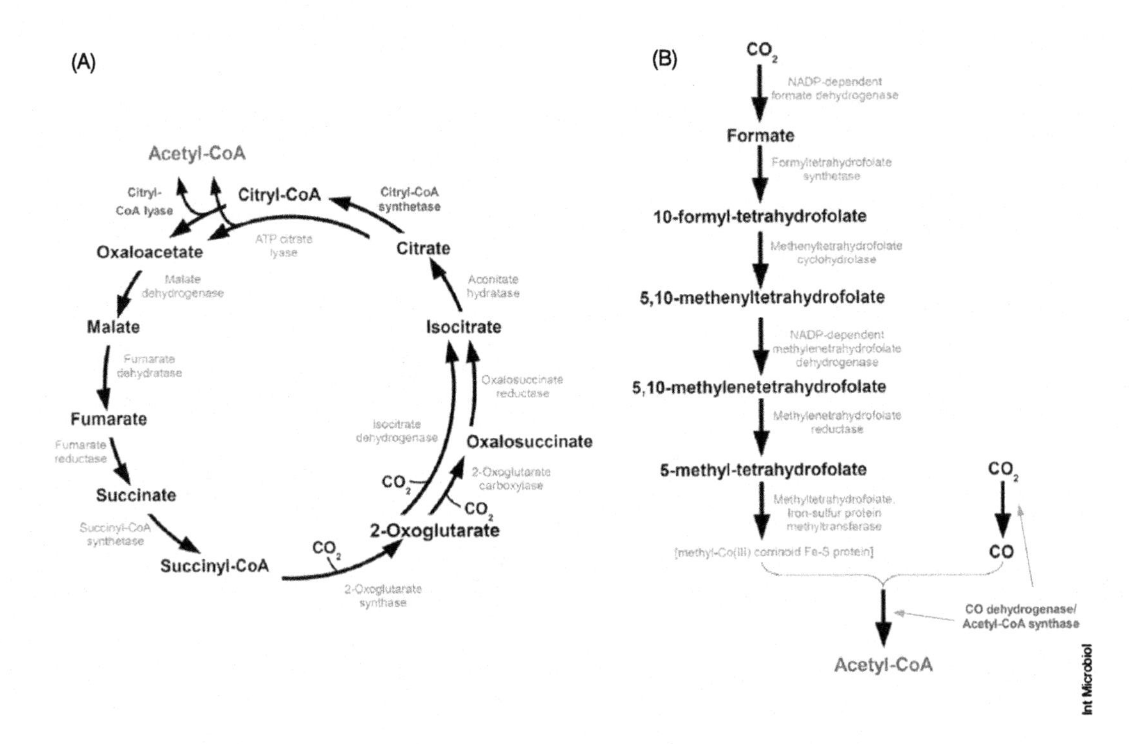


**FIGURE 11.34**

The reductive Tricarboxylic Acid (rTCA) Cycle and Wood–Ljungdahl pathway are probably ancient carbon fixation pathways to ensure the biosynthesis of the five universal precursors of anabolism: acetyl-CoA (coenzyme A), pyruvate, phosphoenolpyruvate, oxaloacetate, and 2-oxoglutarate (1–5).
However, the adenosine 5'-triphosphate (ATP)-dependent citrate cleavage enzymes, ATP citrate lyase (ACL) or citryl-CoA synthetase (CCS)/citryl-CoA lyase (CCL), seem to have emerged at a later stage and from a different pathway (Nunoura et al., 2018) (Modified from Mailloux et al., 2007).

The future of these lakes and health of the people who depend on them will require ongoing research to understand Hg biogeochemistry and develop strategies to mitigate or minimize the health risks.

# EXTREMOPHILES AND BIOGEOCHEMISTRY

## ANCIENT ATMOSPHERE AND ANAEROBIC BACTERIA

The first autotrophs probably evolved using sulfur and protons to reduce $CO_2$. Without free $O_2$, these archaea used similar biochemical pathways found in eukaryote, but used them in very different ways. For example, the reductive Tricarboxylic Acid Cycle (rTCA) and Wood–Ljungdahl Pathway are probably ancient carbon fixation pathways, but have been since repurposed by some microbes. To appreciate how redox conditions can drive biochemistry evolution also informs us about how microbes respond

**FIGURE 11.35**

Emerald Pool Hotspring, Yellowstone Park (Source: https://pixels.com/featured/emerald-pool-hot-springs-yellowstone-gary-whitton.html).

and survive in extreme conditions, some that are natural and some that are anthropogenically made. For example, similar biogeochemical processes occur in hot springs and acid mine drainage.

## HOT SPRINGS: YELLOW STONE NATIONAL PARK

Reduced species of sulfur (e.g., sulfide, elemental sulfur, and thiosulfate) are prevalent in geothermal habitats, and numerous hyperthermophiles have been shown to utilize various sulfur species as either electron donors or acceptors.

Yellowstone National Park contains a significant number of geothermal waters (Fig. 11.35) with high (e.g., $> 10\ \mu M$) concentrations of dissolved sulfide. The interaction between sulfidic waters and atmospheric oxygen produces of a variety of oxidized sulfur species and sulfur-arsenic complexes. The result is the multitude of possible abiotic and biotic reactions with sulfur species. The role of micro-organisms in high-temperature sulfidic sediments remains an active area of research as we learn more about the diversity of biogeochemical processes on the planet.

## ACID MINE DRAINAGE: PH AND REDOX

Iron Mountain Mine (California) was a massive sulfide ore deposit mined for iron, silver, gold, copper, zinc, and pyrite from the 1860s to 1963. Twenty years later, the abandoned Iron Mountain Mine was listed as a Superfund Site, because it generates some of the most toxic water in the country (Fig. 11.36).

FIGURE 11.36

Water drainage from Iron Mountain (Source: USGS).

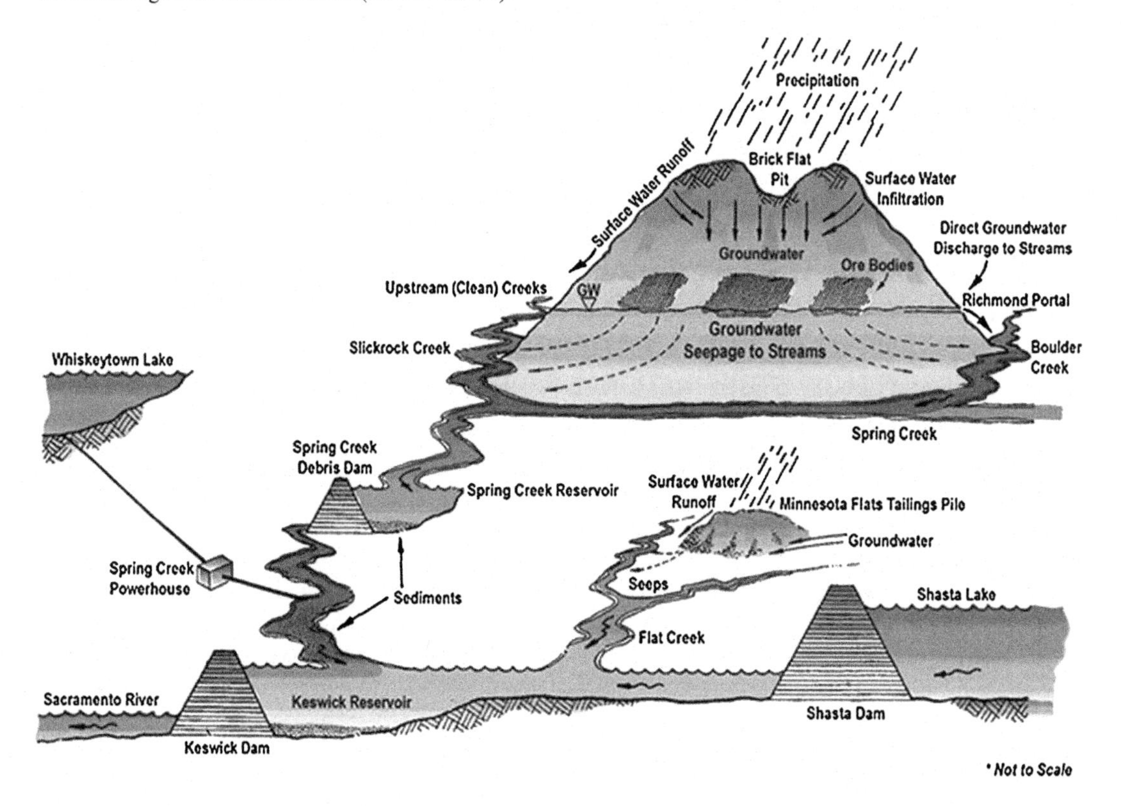


FIGURE 11.37

Diagram of Acid Mine Drainage from Iron Mountain. (Source: courtesy of USGS.)

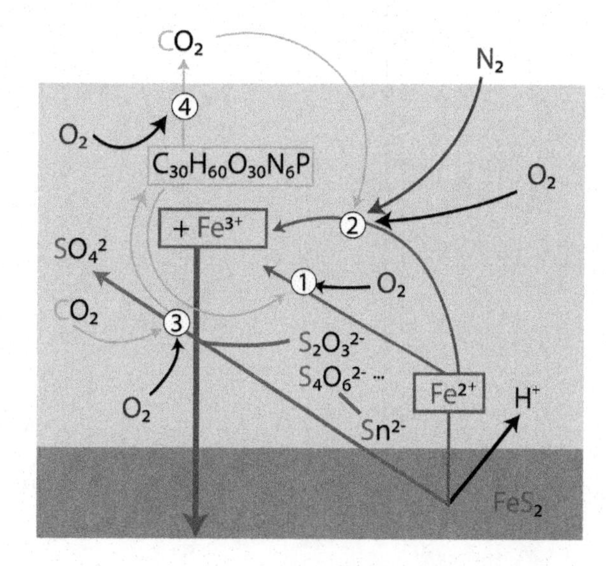


**FIGURE 11.38**

Potential iron, sulfur, and carbon cycling based on known metabolic capabilities (1, 2, 3, and 4) associated with AMD members (redrawn from Baker and Banfield, 2003). Crystalline pyrite ($Fe_2S$) is in yellow at the bottom and green is representing AMD solution. Elemental sulfur is shown at the pyrite-water interface as a possible inhibitor of surface dissolution. The overall oxidation of pyrite is shown at the bottom, with $Fe_3^+$ indicated as the primary oxidant. Intermediate sulfur compounds are indicated as follows: $S_2O_{32}^-$ being thiosulfate and $S_4O_{62}^-$ is tetrathionate. $C_{30}H_{60}O_{30}N_6P$ indicates organic carbon compounds. (Color online.)

When exposed to water sulfide ores (pyrite), oxidize to produce very acid water. The drainage water may be the most acidic water on the planet. Samples in 1990 and 1991 had a pH of 3.6.

The mine drains into Spring Creek Reservoir and Keswick Reservoir of the Sacramento River and an important source of drinking water. Before being treated, mine drainage caused significant fish kills in the Sacramento River since 1899. High heavy metals and low pH have eliminated most of aquatic life in several creeks, including Spring Creek. All mine effluent is now treated with lime (CaO). Although controlled-treated and diluted releases are scheduled, accidental releases also occur and discharge heavy metals into the Sacramento River (Fig. 11.37).

Sulfides such as pyrite ($FeS_2$) that are exposed to air and water through mining activities undergo oxidative dissolution to generate sulfuric acid. But the reactions occur much faster when catalyzed by specific microbes.

The oxidation of pyrite can be separated into several steps, where some produce acid and others do not. A general equation for this process is

$$2FeS_2 + 7O_2 + 2H_2O \longrightarrow 2Fe^{2+} + 4SO_4^{2-} + 4H^+ \tag{11.15}$$

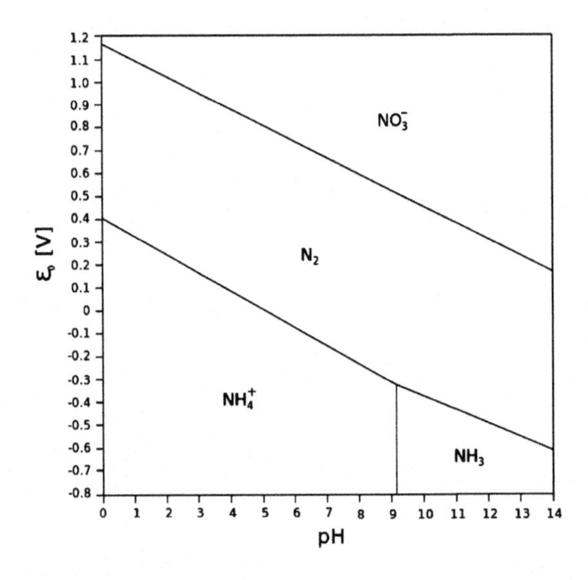


FIGURE 11.39

Nitrogen Pourbaix diagram.

But the bacteria *Acidothiobacillus* is often associated with this portion of the reaction. With the production of sulfate, the Fe II can remain soluble. The Fe II can be oxidized to Fe III:

$$4\,Fe^{2+} + O_2 + 4\,H^+ \longrightarrow 4\,Fe^{3+} + 2\,H_2O \quad (11.16)$$

Either of these reactions can occur spontaneously or can be catalyzed by microorganisms that derive energy from the oxidation reaction. However, the microbial process accelerates the process dramatically.

Finally, the Fe III becomes an exceptional oxidizer, where microbes oxidize additional pyrite:

$$FeS_2 + 14\,Fe^{3+} + 8\,H_2O \longrightarrow 15\,Fe^{2+} + 2\,SO_4^{2-} + 16\,H^+ \quad (11.17)$$

The net effect of these reactions is to release $H^+$, which lowers the pH and maintains the solubility of the ferric ion.

Of course, as basis of the process it is more complicated. Microbes live in a complex community or microbial mat. A microbial biofilm consortium allows for a wide range of reducing and oxidizing reactions within vary small spatial scales. These microbial communities at Iron Mountain derive their energy from iron and sulfur oxidation, but rely on the air ($O_2$, $CO_2$, $N_2$) and AMD (P) solutions for other elements. For example, most of the prokaryote that can oxidize sulfur also can fix $CO_2$ (Fig. 11.38).

Over the years, we have been learning more about the microbial community and how they cope with these extreme conditions. For example, the archaea of Iron Mountain lack cell walls, suggesting that membrane composition and construction are key factors for acid tolerance. Extremophiles are often

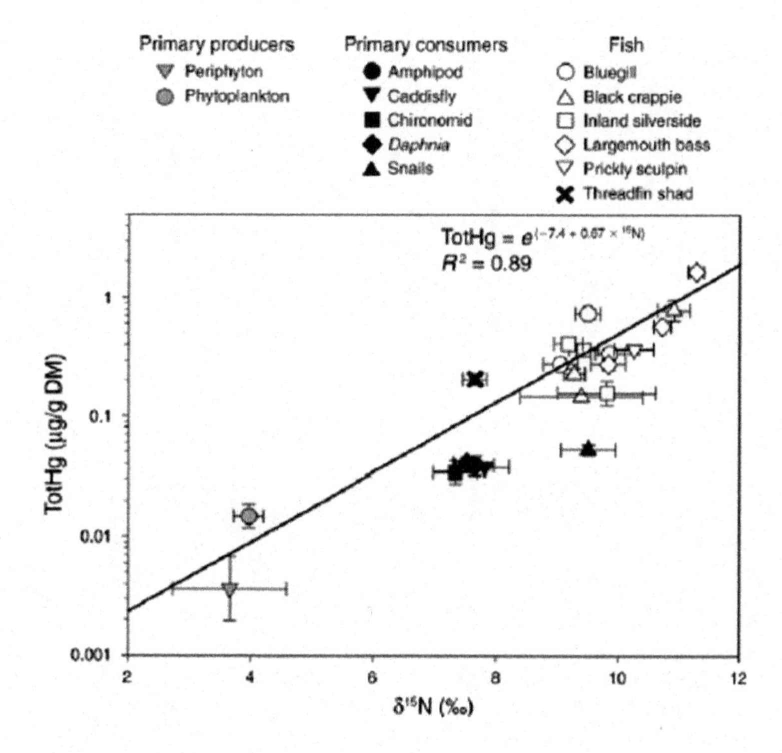


**FIGURE 11.40**

Total mercury (TotHg) concentration (DM, dry mass) vs. $\delta^{15}N$ in the Clear Lake food web (Source: Eagles-Smith et al., 2008). Each symbol represents geometric (for TotHg) or arithmetic (for $\delta^{15}N$) means. Error bars represent ±SE.

confined to areas of small geographic extent, such as hot springs and hydrothermal vents, however, they can have significant global impact in iron and sulfur cycling. Furthermore, where locally active, they can have dramatic impacts on inland waters as seen in Iron Mountain.

## NEXT STEPS

### CHAPTER STUDY QUESTIONS

1. Describe the patterns and processes of redox reactions in wetlands and their influence on wetland plant distributions.
2. Summarize how mercury biogeochemistry can affect Aquatic Systems.

### PROBLEM SETS

1. Explain the simplified Pourbaix diagram for nitrogen (Fig. 11.39). Discuss the reactions that are implied by the lines, and explain why they have the slopes they do.

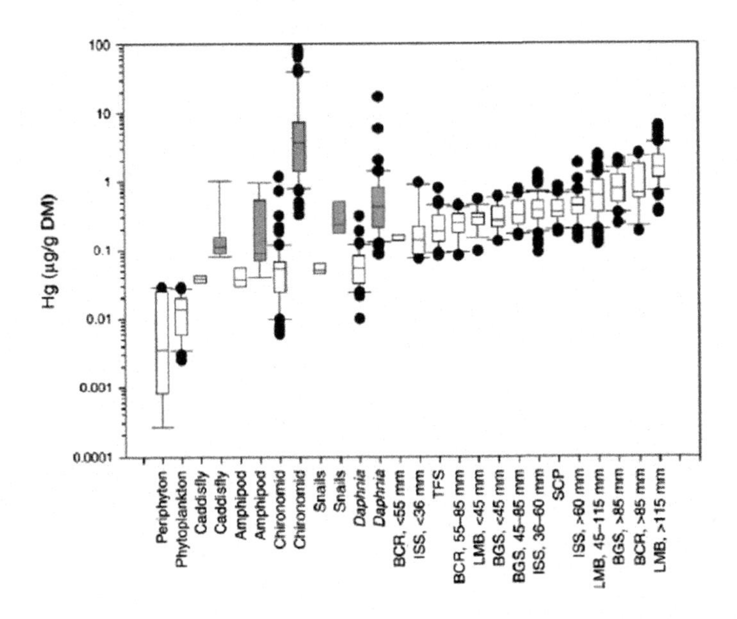


**FIGURE 11.41**

Boxplots of mercury concentrations in the Clear Lake food web (Source: Eagles-Smith et al., 2008). Upper and lower box boundaries represent the 25th and 75th percentiles, and the center box line represents median values. Whiskers represent 10th and 90th percentiles. Open boxes for periphyton, phytoplankton, caddisfly, amphipod, chironomid, snails, and Daphnia are methylmercury (MeHg) concentrations, and gray boxes are total mercury (TotHg) concentrations. All fish values are TotHg concentrations (DM, dry mass). Species abbreviations are: BGS, bluegill; BCR, black crappie; ISS, inland silverside; LMB, largemouth bass; SCP, prickly sculpin; TFS, threadfin shad.

2. Create a diagram that describes the relationship between Fe and S along an aerobic-anaerobic transition.
3. Based on what you know about Clear Lake, describe the processes that might explain the following figures (Fig. 11.40 and Fig. 11.41).

CHAPTER

# CONSERVATION AND RESTORATION

# 12

*Shall we surrender to our surroundings or shall we make our peace with nature and begin to make reparations for the damage we have done to our air, to our land and to our water?*
**Richard Nixon (1913–1994), 37th U.S. President, State of the Union Message, 22 Jan 1970**

*Many estuaries produce more harvestable human food per acre than the best midwestern farmland.*
**Stanley Cain, testimony, U.S. House of Representatives, Merchant Marine and Fisheries Subcommittee, March 1967**

## CONTENTS



Ecology and Management of Inland Waters. https://doi.org/10.1016/B978-0-12-814266-0.00026-X

The "urban stream syndrome" describes the observed ecological degradation of urban streams. Symptoms of the urban stream syndrome include a flashier hydrograph, elevated concentrations of nutrients, altered channel morphology, and reduced biotic richness, with increased dominance of pollution- tolerant species and invasives.

But these are only a few of the stressors. Shopping carts, tires, and plastics are ubiquitous in urban streams. Often storm drains or storm drains combined with sanitary outflows discharge greases and oils; human, pet, and pest feces; sediment; heavy metals; industrial chemicals; trash; various endocrine disruptors, and gardening chemicals, such as fertilizers or pesticides.

The mechanisms driving the syndrome are complex and interactive, but most impacts can be ascribed to poorly developed waste management options for urban dwellers and hydraulically efficient drainage systems that carry storm water and waste to streams (Fig. 12.1).

The Los Angeles River is perhaps one of the most notorious urban streams: a concrete river. With development along the margins, the channel is lined with concrete and is cut off from the floodplain; similarly, fences restrict access to the public as these streams have become public hazards. In an attempt meter flow through these channels, upstream dams store rainfall, modifying the hydrograph, disrupting sediment flow, altering geomorphology, preventing animal migration.

Nevertheless, even causal observers note that the Los Angeles Channel is filled with sediment ranging from sands to cobbles and boulders. The sediment supply has been sufficient to create some channel complexity and allow willows to become established. Filamentous algae grows on the channel bottom, even attaching itself to the concrete. In addition, at the right time of the year, aquatic insect adults emerge and swarm along the edges of the river to mate. Thus this river is not without ecological capacity and even limited activities could improve the ecological value of such an altered river.

As an ecosystem that integrates natural processes into a built environment, restoring the structure and function of urban streams requires a broad approach that aligns ecological goals with diverse cultural values, economic constraints, and political capacity.

Restoration Ecology has become a well-developed field that provides numerous tools to improve the ecological structure and function of inland waters. With threats to species diversity, habitat quality, and ecosystem services, it is critical to appreciate the lessons learned from past restoration activities and refine tools for future efforts. This chapter showcases a number of restoration efforts and how their success might be evaluated.

**FIGURE 12.1**

Image of an urban stream, Second River, Orange, New Jersey (Source: Jim Henderson).

After reading this chapter, you should be able to

1. Describe the ecological consequences of invasive species to inland waters;
2. Summarize strategies to manage and control invasive species;
3. Describe the challenges and successes of efforts to restore wetlands, streams, and lakes in California;
4. Compare and contrast various definitions of restoration success, and
5. Evaluate various monitoring methods to gauge the success of restoration.

# BIODIVERSITY TO ECOSYSTEM FUNCTIONS

## THREATENED AND RARE TAXA

Whereas Freshwater Systems only cover ~2.3% of the Earth's surface, about 6% of the described species are found in freshwaters. Many regions have high amounts of species diversity, but some areas are exceptional and taxa-specific. For example, 50% of the freshwater vertebrate diversity in the Amazon Basin are composed of amphibians. However, global measures of freshwater populations and their diversity has been in steep decline. For example, between 1970 and 2014, freshwater populations have declined by 83%. The proportion of freshwater animals threatened is between 23% and 37%, for every bioregion, excluding Oceania. The greatest threats are in the Indo-Malaya and Netropic Regions, but the situation is critical on a global scale (Fig. 12.2).

The southeastern USA has both the highest number of endemic freshwater fish, crayfish, and mussel species in the United States. In addition, the region has experienced the highest growth rates in the

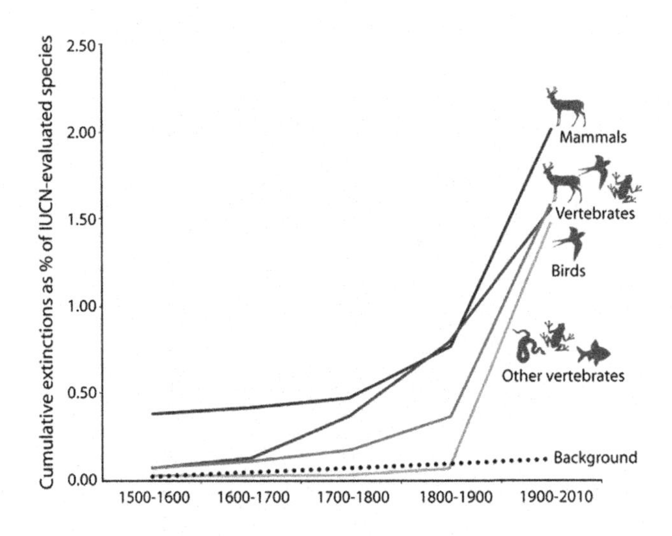


**FIGURE 12.2**

Cumulative vertebrate species recorded as extinct or extinct in the wild by the IUCN (2012). Only the conservative estimate is displayed (Source: Ceballos et al., 2015).

country over the last 60 years. Thus special efforts are needed in the region to protect the diversity in spite of the growth of human development.

After the passage of the Endangered Species Act (ESA), endangered species became a symbol of legal conflicts between conservation goals and economic development. For example, even though the snail darter (*Percina tanasi*) has a limited range in East Tennessee freshwater rivers, its protection case was argued in the Supreme Court. Discovered in 1973, the snail darter was listed as endangered under ESA and the United States Fish and Wildlife Service identified the completion of the Tellico Dam as a 'taking,' because it would block the darter's migratory route. A lawsuit ensued and was one of first major tests of ESA. In 1975 the Supreme Court ordered to halt construction of the dam, thus enforcing the ESA and protecting the snail darter. Later, Congress passed a rider to exempt the Tellico Dam from ESA requirements, and the dam was completed by 1979. Whereas the ESA remains a powerful tool for conservation, this and other legal battles exposed the weaknesses of the law to navigate conflicts between single-species conservation and development projects.

## ESA AND THE COLORADO PIKEMINNOW

The Colorado Pikeminnow was the Colorado River's top predator in the early 1900s and had been known to take anglers' bait in the form of mice, birds, and even small rabbits. With a lifespan of 40 years, the pikeminnow can grow to nearly 1.8 meters in length and a weight of 45 kg (Fig. 12.3). However, since the 1990s, few adult Colorado pikeminnow are over 1 meter in length.

Colorado Pikeminnow was once abundant in the Colorado River and most of its major tributaries in Colorado, Wyoming, Utah, New Mexico, Arizona, Nevada, and California. The pikeminnow is adapted to warm rivers and a hydrologic cycle characterized by peak flows from snowmelt runoff and low, relatively stable base flows. This fish is also known for long-distance spawning migrations of more than

**FIGURE 12.3**

Image of Colorado Pikeminnow (Source: Joe Ferreira). The Colorado pikeminnow is one of the largest minnows in the world.

300 km in late spring and early summer. But recent surveys have detected only two extant populations in the river basin—one in the upper Colorado River System and one in the Green River System.

The pikeminnow had been listed as endangered before the ESA was passed, but given full protection after 1973. With several major dams on the river, the capacity to migrate to spawn has been severely impeded. In addition, the water released from the bottom of reservoirs is cold and far from the pikeminnow's optimum.

The recovery options for the species are quite limited. Colorado River dams will not be removed to protect the pikeminnow. Thus recovery efforts focus on dam operations that might create more natural flow patterns, improve fish passage up- and downstream, and restrict stocking of nonnative fish to reduce negative interactions with the pikeminnow. Thus far, the progress to recover the pikeminnow has been ambiguous.

## EXPANDING THE FOCAL SPECIES CONCEPT

One early criticism of the ESA has been the single species focus, symbolized by derision of private land owners who question the economic value of a single species relative to economic returns on development activities. We now know endangered species indicate habitat degradation, which may cause other species to be at risk. Thus single species concerns should be thought of as indicators and used to protect a range of taxa and habitats. Furthermore, by comparing the benefits of protecting habitats and associated ecosystem, services may dramatically outweigh conservation costs of a single species focus.

Over 50% of the large freshwater fauna (i.e., megafauna) are listed as threatened (Critically Endangered, Endangered, or Vulnerable). However, these taxa may provide particularly robust justifications for conservation beyond their own recovery. To explore this concept, the following examples demonstrate how their status might be used for conservation goals:

**Chinese Paddlefish *(Psephurus gladius)*** is the world's longest freshwater fish, but has not been observed since 2003.

**Baiji or Yangtze River Dolphin *(Lipotes vexillifer)*** which could represent the first human-caused extinction of a Cetacean species (Fig. 12.4).

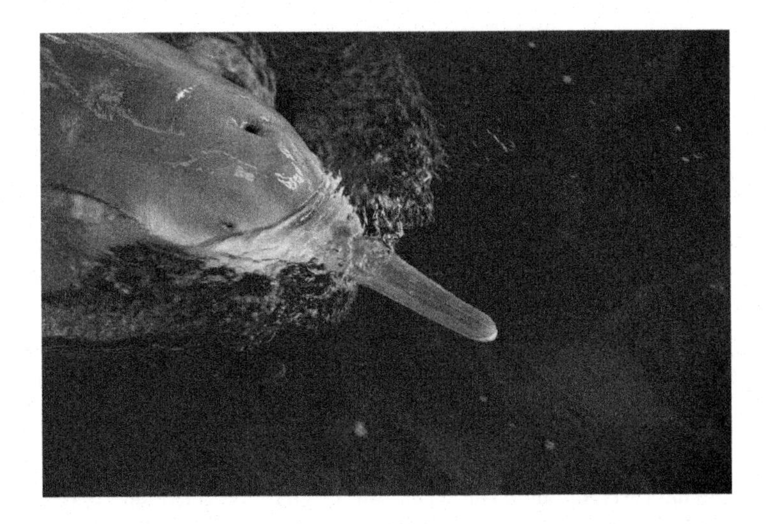

**FIGURE 12.4**

Image of Yangtze River Dolphin (Source: National Geographic).

**Yangtze Sturgeon *(Acipenser dabryanus)*** has experienced drastic declines due to overfishing, including the overharvesting of juveniles. But other stressors include river fragmentation (from dams such as the Gezhouba Dam and Three Gorges Dam), habitat degradation from landuse changes, and an increase in pollution from wastewater and runoff.

**Adriatic Sturgeon *(Acipenser naccarii)*** is native to the Adriatic Sea and large rivers that drain into the Adriatic, but is bred in captivity. However, the populations have dwindled and has been extirpated from many former river habitats. The sturgeon may be functionally extinct in the wild, as no recent spawning has occurred.

Instead of focusing on these animals in isolation, Conservation Biologists argue that we can frame these taxa within a certain context to justify their protection beyond their individual existence value.

***Flagship species*** are charismatic species that act as ambassadors to promote broadscale conservation, to raise conservation funding, and to attract public attention. Examples of these species include river dolphins, Chinese Giant Salamander (*Andrias davidianus*), hippopotamus, sturgeons, and paddlefishes. Mammals are exceptionally important flagship species because they share many behaviors that human recognize and value. However, flagship species have also been used in theme parks and zoos, which have been strongly criticized because they rely on captivity.

***Keystone species*** play critical and unique ecological roles and have disproportionate importance relative to their abundance. Because of their visible role in the ecosystems, their protection has demonstrable ecosystem importance. Keystone species include ecosystem engineers such as beavers, crocodilians, and hippopotamus (e.g., pigmy hippopotamus (*Hexaprotodon Liberiensis*)), but also species that exhibit top-down, trophic cascade roles, such as the wolf.

***Umbrella species*** have large habitat area requirements, for which conservation action potentially benefits other co-occurring species. For example, river dolphins, sturgeons, and paddlefishes have migratory patterns that can serve to protect many different habitats and be used to restrict dam construction.

However, some ecosystems do not have megafauna and conservationists link smaller fauna or flora to justify ecosystem protection. For example, the delta smelt (*Hypomesus transpacificus*) is critically endangered. Endemic to the upper Sacramento-San Joaquin Delta of California, it mainly inhabits the freshwater-saltwater mixing zone of the estuary, except during its spawning season, when it migrates upstream to freshwater following early spring, high discharge flow events (around March to May). As an indicator species for the overall health of the Sacramento-San Joaquin Delta Ecosystem, this species should be seen as an umbrella species instead of being used to pit farmer livelihoods against ESA and symbol of government overreach.

### REMOVING A STRESSOR: MUSSEL RESTORATION

In reality, protecting endangered species takes a suite of strategies. Protecting megafauna might provide long-term, future protection for members of whole ecosystems, but many species experience ongoing stressors that cannot wait for the political will to establish robust regional conservation outcomes. For example, when specific stressors affect species, Conservation Scientists identify specific management changes to address the stressor.

For example, surveys of the Duck River in Tennessee expected to find the endangered Cumberland Monkeyface Mussel (*Theliderma intermedia*). However, downstream of the Normandy Dam, no mussels were found. The problem seemed to be caused by the chronically low oxygen levels released from the dam. Once the problem was identified, the TVA began aerating the released water in the early 1990s. By the 2000s, the population was in the tens of thousands, extending for 50 river kilometers. This success story highlights how addressing a stressor can have a dramatic impact on an endangered or threatened species in spite of the infrastructure constraints, i.e., dams.

## TALES OF SPECIES INTRODUCTIONS AND GENETIC EROSION

### AN UNDER APPRECIATED COMMONS

Invasions into streams and lakes by nonnatives are ubiquitous. The number of introduced species to Aquatic Ecosystems is mind-boggling and likely exceeds 2000 species. Some introductions are purposeful, others are accidental. However, the impacts of species introductions is highly varied and depends on the life-history of interacting species, ecological tolerances, and trophic relations. Whereas some have had long-term and dramatic impacts throughout the world, e.g., rainbow trout, others might have localized community effects. For example, the *Mysis diluviana* (opossum shrimp, Fig. 12.5) was introduced to Lake Tahoe in the mid 1960s to "improve" food resources for another introduced species: *Oncorhynchus nerka*, Kokanee Salmon. The opossum shrimp fed on *Daphnia* and *Bosmina* and drastically reduced their population sizes ever since.

As function of globalization of trade and transport the rate of introductions will increase. However, the ecological and economic effects of invasive plant and animal species are often observed and handled at the local scale, rather than at the international level. Ecological and economic damages wrought by invasive species are on par with those caused by other global-scale problems, such as climate change,

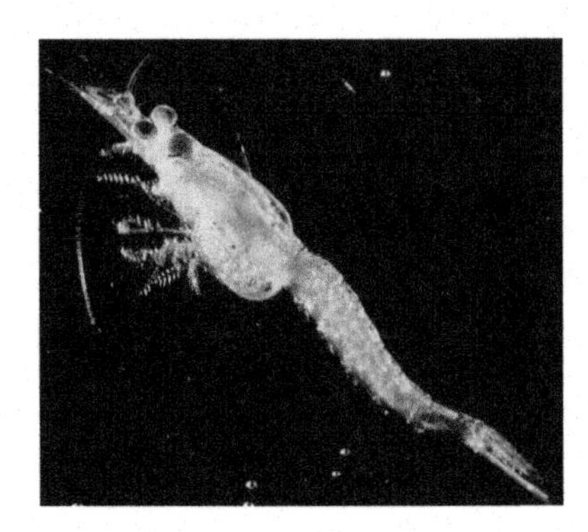

**FIGURE 12.5**

Image of opposum shrimp (*Mysis diluviana*) is a mysid Crustacean (opossum shrimp) found in freshwater lakes of northern North America (Source: Wiki commons).

but only a few regulations govern the movement of exotic species and the agencies that implement these regulations do not have the resources to respond to the threats.

## AQUARIUM ALGAE AND SANTA ANA SUCKER

The Santa Ana Sucker (*Catostomus santaanae*) is a threatened species found only in the lower portion of the Santa Ana watershed and the upper portions of the Los Angeles and San Gabriel Rivers. Its native habitat has been severely fragmented. One population lives in a short, perennial segment along the lower Santa Ana River. Since the flow is maintained by treated wastewater, the water temperature is warm and nearly constant (think warm showers, dishwaster, and laundry waste).

Unfortunately, these water temperatures are ideal for the tropical filamentous Red Algae (*Compsopogon coeruleus*, Fig. 12.6). Probably introduced by an aquarium owner who was no longer interested in maintaining an aquarium. Now this Rhodophyte displaces native diatoms species, the preferred food of the Santa Ana sucker, adding a new dietary stressor on threatened the fish.

After enough precipitation, the Santa Ana River will flow and overwhelm the warm, treated wastewater. The flow scours the channel and remove a significant amount of the *Compsopogon coeruleus* biomass. This process may give managers clues how the invasive species can be managed to protect the sucker population.

Aquaria dumping is the subject of urban myths, e.g., alligators in the NY sewage system. Although reproducing populations of alligators in sewage system is unlikely, other bodies of water constantly receive unwanted pet additions. For example, when a lake in Golden Gate Park was drained to be restored, Biologists identified over 100 species of fish. Instead of proper disposal, aquaria owners conflate humanitarian values with dumping potential pests into surface waters.

## WATER HYACINTH: CHOKING SLOW MOVING WATERS

Invasive aquatic plants can have dramatic effects in streams, lakes, and wetlands, such as altering food webs, reducing Dissolved Oxygen Concentrations, and disrupting navigation. Water hyacinth

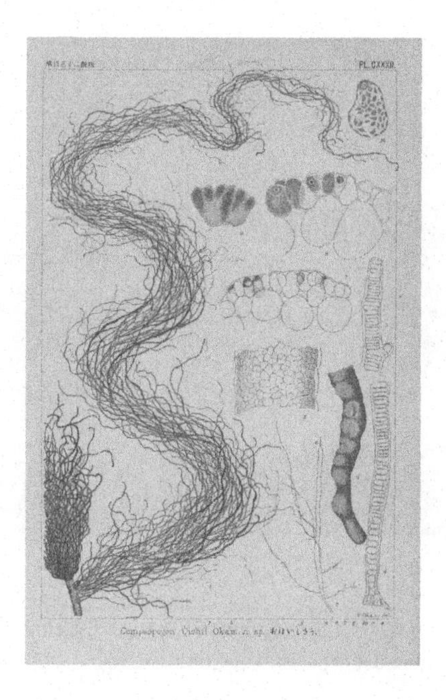

**FIGURE 12.6**

Image of invasive red algal, *Compsopogon coeruleus* (Source: Kintaro Okamura, 1867–1935).

(*Eichhornia crassipes*) is one of the more famous aquatic invaders. As a native of Amazon Basin, water hyacinth was introduced at the World's Fair in New Orleans in 1884 and was soon shutting down shipping in Louisiana, and choking Florida's waterways. At one point, the United States Congress nearly passed a bill to import hippopotamus to control the weed. The bill failed. But over 100 years later and unable to find successful alternatives, California resource managers began using Hippos (*Hippopotamus amphibius*) as a biocontrol agent to clear the channels (Fig. 12.7).

## RIPARIAN TREES AND ALTERED GEOMORPHOLOGY

Tamarisk, (*Tamarix* spp.) commonly known as salt cedar, is a prolific nonnative tree or shrub that displaces native vegetation and animals; alters soil salinity; lowers the water table; and increases fire frequency. Salt cedar is an aggressive competitor, often developing monoculture stands (Fig. 12.8).

Tamarisk, introduced to the United States in the 19th century as an erosion control agent, spread through the west and caused major changes to natural environments. Tamarisk reached the Grand Canyon area during the late 1920s and early 1930s, and by 1963 had become a dominant Riparian Zone species along the Colorado River. Tamarix initially colonized bare in-stream sand deposits (i.e., islands and bars), but soon displaced other vegetation and altered the river geomorphology.

Several government agencies and nonprofit organizations have been working to control Tamarisk. For example, Tamarisk Leaf Beetles have been imported and used as a successful biocontrol agent, but eradication is probably impossible. Yet, the river's vegetation now includes some natives in spite of the Tamarisk's competitive advantage.

**FIGURE 12.7**

A “bloat” of hippopotamuses from Botswana was imported to eat water hyacinth (*Eichhornia crassipes*) that might clog intake pipes for the California Aqueduct (Source: US Coast Guard).

**FIGURE 12.8**

Tamarix growing along the Colorado River at the mouth of the tributary Paria River (center). The sediment-laden water of the Paria River enters and mixes with the clear Colorado River water released from Glen Canyon Dam (May 17, 1993) (Credit: David J. Topping, USGS. Public domain).

## SHOTHOLE BORER: HABITAT STRUCTURE AND FUNCTION LOSSES

At the time of this writing, an undescribed beetle has become a new insect pest in Southern California (Fig. 12.9). Related to *Euwallaecea fornicatus* from Vietnam, the polyphagous shothole borer drills small holes into the trunks of trees and creates a gallery inside to lay its eggs. The beetle also inoculates the galleries with a pathogenic fungus (*Fusarium euwallacea*) for food for their larvae. The Fusarium

**FIGURE 12.9**

Female polyphagous shot hole borer (Source: Photo by Gevork Arakelian, LA County Agricultural Commissioner).

**FIGURE 12.10**

A Least Bell's Vireo brings food to her young nestled away in a well hidden nest (Source: USFWS).

infects the wood of the tree and kills susceptible species. As of fall 2014, over 200 species of trees in Southern California have been attacked. Thirty-three tree species have been confirmed as reproductive hosts, including the native Riparian species Coast Live Oak (*Quercus agrifolia*), California Sycamore (*Platanus racemosa*), Fremont Cottonwood (*Populus fremontii*), and Red Willow (*Salix laevigata*), and White Alder (*Alnus rhambifolia*).

One of the most disturbing sites to witness is the near complete mortality of the willows in Tijuana Estuary. Willows, an important Riparian shrub/tree, are key habitat for the endangered Least Bells Vireo (*Vireo belli*, Fig. 12.10). Least Bells Vireo populations plummeted with stream channelization

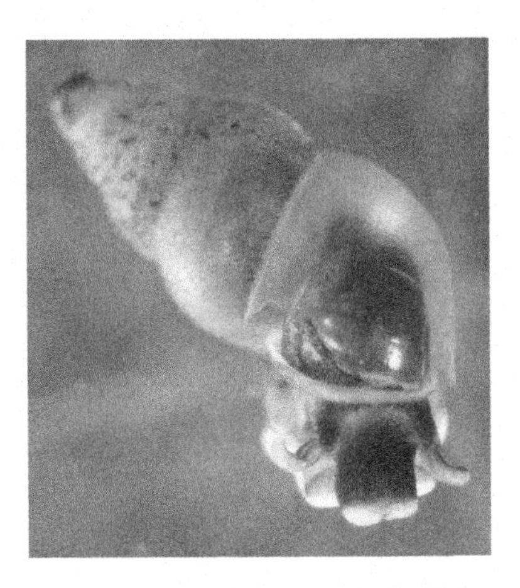

**FIGURE 12.11**

Image of the New Zealand Mud snail (Source: Dan Gustafson).

projects over the last 150 years because of the loss of Riparian Habitat. With the current rapid decline of the willows, the Riparian-dependent vireo faces a new stressor.

## NEW ZEALAND MUDSNAIL: FOOD WEB MODIFICATIONS

The Freshwater New Zealand Mudsnail (*Potamopyrgus antipodarum*) has been introduced to dozens of countries (Fig. 12.11), where populations can reach phenomenal densities.

This macroinvertebrate often outcompetes and displaces native species to reach snail densities of over 750,000 individuals per $m^2$. They can consume up to half of the food resources in a stream and have been linked to reduced populations of aquatic insects, including mayflies and caddisflies prey for trout and salmon. In their native habitat, the snails pose no problem because of a trematode parasite that sterilizes many snails, keeping the populations at moderate levels. However, in the absence of these parasites, they have become an invasive pest in many inland waters (Fig. 12.12).

## ANGLER'S DELIGHT

Although inland water fishing might not even produce 1% of the revenues of marine fisheries, many more people are participating in the activity. Substantial fisheries exist, e.g., the Great Lakes, Mississippi, Tonle Sap, and African Rift Zone lakes.

In 2006, the Unites States Fish and Wildlife Service estimated that of the 30.0 million US anglers, 85% fished in freshwaters, took 337 million fishing trips and spent $26.3 billion. What are the environmental implications for this fishing effort?

One significant aspect of freshwater fisheries is that many waters are stocked. In fact, fish are transported and introduced to surface waters all around the world to "develop" or "improve" fishing. Even places, such as Yosemite, where the first recorded fish stocking occurred in 1877 before the national park was established has not been immune. Two years later, 20,000 trout were planted in valley streams.

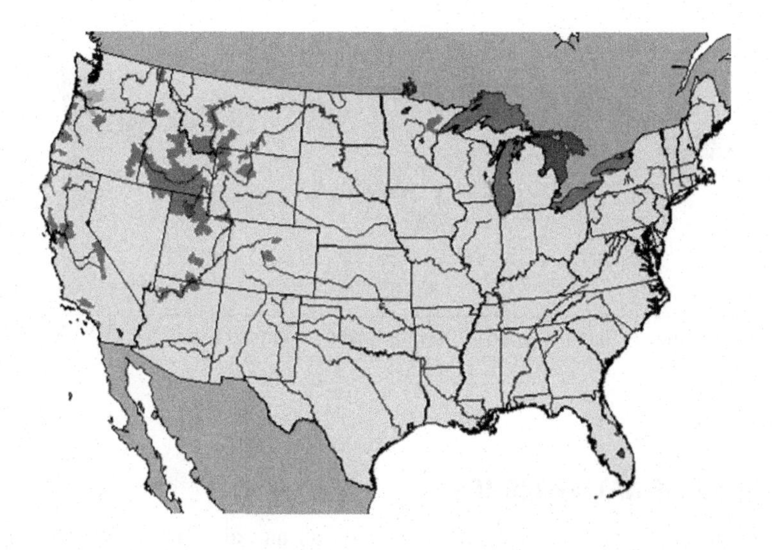

**FIGURE 12.12**

Map of the New Zealand Mud Snail distribution (Source: USFWS, 2009).

Historically, all of Yosemite lakes were fishless due to natural barriers, e.g., water falls. In fact, many lakes in the western United States were fishless because of natural barriers. To 'remedy' this situation, fish were carried in jugs, coffee cans, and by mule to remote steams and lake, until aerial stocking began in 1952. 60% of the naturally fishless lakes in the western lakes now contain nonnative trout. Many of these lakes are in national parks or wilderness areas. In an almost pathological behavior, reservoirs, lakes, and streams are seeded continuously with nonnative, game fish in the hopes to maintain an exciting and active fishery, while catching the associated tourist dollars.

Because many high Alpine Lakes have low species diversity and simple food web linkages, the effects of introduced trout in Alpine Lakes affected the whole communities. Rainbow Trout (*Oncorhynchus mykiss*), the most widely introduced fish in the world, eat native fauna, triggering large ecological changes that include the elimination of amphibian and reptile populations, changes in zooplankton and benthic macroinvertebrate structure, and influences nutrient cycling. The decline of Mountain Yellow-legged Frog (*Rana muscosa*) and Pacific Treefrog (*Pseudacris regilla*) are associated with trout introductions.

In 1969 the National Park Service (NPS) acknowledged that their mission to conserve the natural resources was in conflict with fish stocking. After over 100 years of the practice, fish stocking was ended in 1991; however, established populations of fish continue to influence community structure in numerous locations.

Rainbow trout is now considered to be the most invasive invertebrate and continues to impact the native species throughout the world.

**FIGURE 12.13**

Image of *Salmo salar* or Atlantic Salmon is one of the most common farmed fish in North America and Europe (Source: Timothy Knepp).

## GENETIC EROSION OF NATIVE FISHES

Atlantic Salmon (*Salmo salar*) is an important farmed fish and also one of the best researched fish (Fig. 12.13). Native salmon populations are typically genetically distinct from each other and locally adapted. Farmed fish represent a limited set of source populations that have been through a domestication selection process for several generations. Consequently, farmed and wild salmon differ in a range of traits, including genetic polymorphisms, growth, morphology, life history, behavior, physiology, and gene.

Since the 1970s, tens of millions of farmed salmon have escaped into the wild, where they interbreed with native fish. For example, fish surveys demonstrated half of ~150 Norwegian populations had interbred with farmed fish with an introgression average of 6.4%, i.e., percentage of farmed genes in wild populations. As a consequence, wild population productivity had been reduced. Unless there is a dramatic reduction in the number of fertile escapees, the genetic erosion of wild fishes may be unrecoverable.

# WETLAND ECOSYSTEM RESTORATION

## WETLAND LOSSES

Over 30% of the wetlands on the planet have been lost. The greatest losses have occurred in Asia, but many believe the situation is most serious in Europe in terms of the number of critical habitat remaining.

Since the 18th century, wetland area in the United States decreased from nearly 890,000 $km^2$ in the lower 48 states to 436,000 $km^2$ in 2004. Since the 1950s, over 50% of this loss has come from wetlands being transitioned to agricultural lands. Other contributing factors to wetlands loss include, but are not limited to development and forestry. The United States Fish and Wildlife Service estimates that up to 43% of threatened and endangered species rely directly or indirectly on wetlands for their survival, thus wetland habitats constitute a critical habitat.

## TYPE CONVERSIONS AND STRESSORS IN WETLAND SYSTEMS

There is no comprehensive wetland protection law in the United States. However, Section 404 of the Clean Water Act specifies that permits are required to alter wetlands by filling with or removal of sediments (dredging). As a sequel to the Rivers and Harbors Appropriation Act of 1899, the law was designed to maintain shipping and navigable waters. Now under the jurisdiction of the Army Corp of Engineers, the law has become an important mechanism to protect wetlands—with some controversy about to what extent wetlands must be connected to navigable waters.

Outside of agricultural conversions, wetland losses occurred because of channelization and dike building and filling in wetlands with sediment and soil for development. In other cases, dredging is used to increase the depth of shipping channels, but reduces the habitat value of benthos and connected to a loss of the shallow water habitats. These types of conversions have accounted for much of the wetland loss until the permit process was enforced.

In spite of the permit process, wetlands were being lost at an alarming rate until 1988 when President GH Bush adopted a national policy of "no net loss". In other words, if a wetland is filled, there must be mitigation to create or restore similar wetlands of a similar area. The implementation details are more complex and not always successful, but this was one of the first policy statements highlighting the value of wetlands in the Unites States.

***Hyrdologic Modifications*** All wetland habitats depend on specific hydrological characteristics, but these can be easily compromised. For example, wetlands in an urban environment are often subject to flashy storm flows, received too much water, or are cut off, receiving too little water to support the habitat. Although naive as a complete restoration plan, restoring the hydrology of wetlands is the basic foundation for wetland restoration. Without the appropriate hydrology, wetland restoration cannot succeed.

***Fragmentation*** reduces the capacity for species to move within and among wetlands. Based on Island Biogeography Theory, populations have a range size and that can be fragmented to the extent that the habitat cannot support the population. If individuals can travel between "island habitats" then they might maintain viable populations. However, if too many of these habitats are lost or they are too far apart or the matrix is too hard to cross, then the population might not be viable (Fig. 12.14).

To address and remediate these impacts, Landscape Ecologists work closely with Restoration Ecologists, resource managers, and private landowners to prioritize areas to create corridors to move between habitats or restore key areas that might reduce the impacts of fragmentation. Although dispersal disruptions are regularly cited as a consequence of habitat fragmentation, fragmentation also creates abiotic and biotic edge effects, such as temperature gradients, entry points for invasive species, and domestic animal predation and harassment (e.g., domestic cats).

***Species Introductions*** Although it is an estuarine example, the case of the introduction of *Spartina alterniflora* provides a well-documented cautionary tale. The species was discovered in a restored wetland in the South Francisco Bay in 1990s. But because the East Coast Tidal Wetland Grass was well-integrated and even hybridizing with the native *Spartina foliosa,* making eradication challenging. Moreover, California Ridgway's Rail (*Rallus obsoletus obsoletus*), an endangered and secretive lives within both the native and invasive grasses (Fig. 12.15). Thus removing the damaging invasive grass species may conflict with protecting an endangered bird species that uses the grass as habitat. The most effective management and restoration approach should not

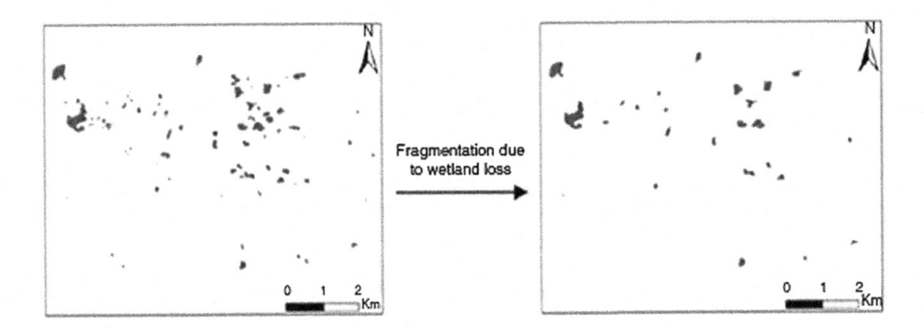


**FIGURE 12.14**

Diagram of demonstrating wetland habitat fragmentation, where habitat isolation of wetlands increases after wetland loss.

**FIGURE 12.15**

Image of Ridgway's Rail at High Tide (Source: Chris Cochems). In the past, thousands upon thousands of California clapper rails foraged, mated, and nested in the extensive marshes along San Francisco Bay.

eradicate the invasive grass as quickly as possible, but rely on the slow removal of the invasive, allowing the birds to readapt and colonize the native *Spartina*.

## VERNAL POOLS

In many regions around the world, vernal pools are some of the most vulnerable of wetlands. Vernal pools are seasonal wetlands formed in a depression that fill with rainfall and dry during periods of

**FIGURE 12.16**

Image of Female San Diego Fairy Shrimp (*Branchinecta sandiegonensis*) with eggs (Source: Joel Sartore). Endemic to Southern California and restricted to vernal pools in coastal Southern California and Northwestern Baja California, Mexico. Upon drawdown, eggs will become cyst and can remain dormant until wet conditions return.

drought. They are usually shallow (0.1 to 1 meter) and vary in size from 50 $m^2$ to 5000 $m^2$. They exist on low permeable soils or layers, such as claypans, hardpans, or bedrock. Because vernal pools can only exist where these special geologic conditions prevent the water from infiltrating, they are often found in clusters.

Vernal pools are important ecologically in California because they are one of the few types of ecosystems dominated by native species and have substantial amount of rare and endemic species. For example, numerous endemic Crustaceans and annual forbs occur in vernal pools and many are listed as endangered (Fig. 12.16). Unfortunately, ~90% vernal pools in California have been destroyed.

Vernal pool plants and animals are very sensitive to the duration and timing of ponding. For example, the San Diego Fairy Shrimp (Fig. 12.16) hatch from cyst, eggs that can withstand desiccation between pond fillings. However, if the duration of ponding is too short, the shrimp cannot complete its lifecycle, and there is a no replenishment of the cyst-bank. To avoid the risk of local extinction, a small percentage of the cyst bank hatches. This bet-hedging strategy allows the Fairy Shrimp to maintain a cyst-bank even after several bouts of filling and drying.

The conservation and restoration of vernal pools requires careful planning to ensure the duration of filling is adequate and that population retains genetic diversity. In some cases, to restore nonfunctioning or new vernal pools, soil inoculum is used to introduce the species via the cyst-bank and seed bank. To ensure these pools have the bet-hedging cyst and seed dormancy mechanisms, the soil inoculum needs to contain cyst and seeds of varying age and the genetic diversity to maintain the dormancy mechanisms.

Efforts to restore vernal pools are further complicated by the effects of grazing. In an effort to improve vernal pool habitat, cattle have been excluded from selected pools. In some instances, the exclosure reduced endemic and rare annual plant cover. In addition, cattle has been linked to "unburrying" Crustacean cysts and improving their recruitment. The example demonstrates that restoration work requires careful attention to avoid making assumptions about what might improve habitats without careful monitoring to gauge success. Currently, managers believe that carefully managed grazing can improve Vernal Pool Habitats, but monitoring should be used to evaluate potential for adverse effects.

## RESTORING LOTIC SYSTEMS

### ADDRESSING STRESSORS IN LOTIC SYSTEMS

The threats to streams vary dramatically, based on landscape position (headwaters, floodplain, etc.) and landuse patterns. Developing robust restoration activities is constrained by landscape development patterns that are, in most cases, impossible to address in any kind of systematic fashion. Nevertheless, many localized efforts can reduce the impacts on Lotic Systems.

***Water quality*** Nutrient loading and subsequent algae growth can reduce $O_2$ availability and lead to mortality or sublethal effects on fish and bethic invertebrates. The success to reduce nutrient loading relies on watershed source control, fate and transport pathways, and buffer zone integrity. Reducing watershed loading usually require a diverse set of targeted strategies and use of incentives.

***Minimum flows*** Some regulated rivers can be managed to maintain minimum flows that protects high-quality spawning habitat for salmonids. In many cases, we do not have the stream gauge records that link discharge to habitat quality in a spatially explicit way. However, numerous modeling attempts successfully predict minimum flows and the area of quality spawning habitat. With good monitoring, these models can be fine-tuned to maintain the balance between minimum flows and water storage.

***Geomorphology*** Regulated rivers often have sediment budget imbalances. In some cases, the stream might be sediment-starved, leading to bank erosion or downcutting. To combat this, gravel augmentation has been used to improve Benthic Habitat, but this practice has numerous drawbacks, including the financial costs, disruption of the Benthic Habitat, and the remedy might need to be regularly repeated.

***Connectivity*** Dams are the most important barrier for fish and invertebrate migration (Fig. 12.17 and Fig. 12.18). The use of fish ladders had been used since the 1910s. But their failure has been documented by 1925 because of their poor design, intermittent operation, and the lack of a strong, regular flow of "attraction water" that draw fish to the ladder. In general, only 3% of the fish pass through effective ladders. Furthermore, fish ladders are expensive. Replacing a fish ladder on the Russia River to improve its efficiency costed well over $3 million.

The lack of connectivity for migratory species will continue to vex the existence value of dams. With these and other associated costs, we are learning that the benefits-costs estimates for dam construction and maintanance are far less positive than we had thought.

***Riparian cover*** As streams have been modified and channelized, Riparian Cover has been lost. Most rivers' natural hydrology is affected by levees, dams, straightening, and riparian removal. The lack of cover increases stream temperatures, reduces allochthonous sources of organic matter, and reduces bird habitat. With an increased understanding of the value of Riparian Vegetation, various attempts to restore these habitats have been made. The Carmel River and Santa Lucia are examples of depleted groundwater areas with ensuing loss of Riparian Vegetation. With a substantial decline in the Riparian Corridor due groundwater pumping and development, Riparian Vegetation was on the verge of collapse in the 1970s. However, based on a better understanding of surface-ground water interactions, limitations on groundwater pumping and strategic restoration efforts have dramatically improved the Riparian Corridor along these rivers.

**FIGURE 12.17**

Image of Adult fish ladder at Ice Harbor Dam on the lower Snake River (Source: USACE Digital Visual Library, http://images.usace.army.mil/photolib.html).

## CHANNEL REACH RESTORATION

Like no other landuse, urbanization has had profound effect on streams. In contrast to most agricultural or rural impacts, urban areas may completely isolate many ecological processes from the stream channel (e.g., hyporheic zone), floodplain connection (e.g., concrete lined channels), and loss of Riparian Zones. Of course, some urban areas simple bury their creeks below concrete, it ends up looking like part of the storm drain plumbing of the city, and the ecological values are lost.

Daylighting describes projects that deliberately expose some or all of the flow of a previously covered stream or stormwater drainage. Daylighting reestablishes a waterway in its former channel where feasible, or in a new channel that weaves through the built environment.

Strawberry Creek in Berkeley, California was daylighted in 1984. Whereas other projects were completed in the 1970s, the Strawberry Creek project is widely considered an exemplar (Fig. 12.19).

In spite of the success of Strawberry Creek, restoring the stream channel geomorphology may be the most difficult part of urban stream restoration. And the success has been uneven. For example, restoration of Uvas Creek in South Santa Clara Valley was an attempt to increase steelhead trout habitat by redesigning the channel using natural stream design concepts.

An evaluation of the restoration has become part of a long and ongoing debate between Dave Rosgen, promoter of the most widely used river classification system, and Mathias Kondolf, Professor of Landscape Architecture at UC Berkeley. Kondolf criticizes the assumptions of Rosgen's classification system as applied to several restoration projects, highlighting the problems of Uvas Creek.

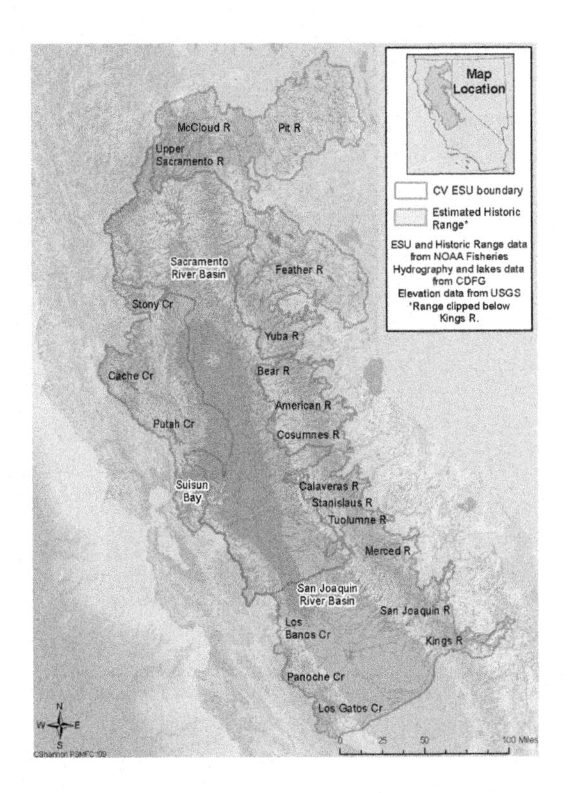


**FIGURE 12.18**

Major watersheds, where steelhead passages are severely constrained (Source: California Department of Fish and Wildlife). (Left) Historic distribution of steelhead in the Central Valley, with current distribution outlined in red (dark gray in print version). (Right) Dams block all fish access to areas upstream.

As designed by the United States Army Corp of Engineers, Uvas Creek was constructed to stabilized the stream channel, but within a couple of years a flood occurred and the stream abandoned the constructed channel, rendering the project a failure. The causes of this failure include the following:

1. lack of historic geomorphic analysis. Had a historic geomorphology analysis been conducted, planners would have known that the channel was a braided meandering stream,
2. channel stability is not a suitable goal since meandering rivers are inherently unstable, and
3. inappropriate assumptions concerning bankfull frequency and their implications on geomorphology in a semiarid region.

These criticisms do not necessarily negate the Rosgen methods, but how they were applied. In fact, Rosgen was a project reviewer and predicted the project would fail. The ACOE ignored this advice. These criticisms should, however, demonstrate that river restoration design and implementation should carefully evaluate historic information, carefully evaluate project assumptions, and enlist qualified expertise.

**FIGURE 12.19**

Daylighted section of Strawberry Creek, Berkeley, California (Source: Ferguson, 2016).

## AUGMENTED DAM DISCHARGES

In March 2014, the Morales Dam released water for 2 months to hydrate parts of the Colorado River in an attempt to restore downstream habitats in Mexico. This is one of the hundreds of examples, where managers of regulated rivers are experimenting with new flow regimes. In the West these are often associated with attempts to improve salmon and steelhead habitat and recruitment in what is sometimes called Instream Flow Needs (IFN).

IFN define the magnitude, frequency, timing, and duration of streamflows necessary to sustain organisms in a healthy ecosystem. In general, the goals of IFN will identify variable annual flow regimes that collectively recover natural ecological processes affecting fish habitat, Riparian Vegetation, stream channel morphology, and valley/floodplain morphology, rather than prescribe minimum baseflows only.

Releases from regulated rivers are typically based on power generation, irrigation water supplies, or to maintain flood-control capacity. IFN add additional constraints on how reservoir water releases are managed, but can have significant ecological value. In some cases, dams were not built with the flexibility to be meet IFN goals. For example, the high flows in the Colorado cannot be achieved because of dam design limitations. In contrast, the Itezhi-Tezhi Dam on the Kafue River in Zambia was constructed to hold more water than necessary for power generation, so specialized releases could maintain naturally flooded grasslands to support livestock herders. Research on these alternative release regimes will help us better evaluate to what extent these practices can mitigate ecosystems services lost by dam construction.

In the case of the Shasta River in northern California, spring releases are used to manage the water temperatures to support salmon. Whereas discharge is relatively predictable from mid-June through late-October, there is a great deal of interannual variability. Thus INF are designed to mimic annual spring pulse flows, to partially replace snowmelt runoff stored behind Dwinnell Dam, but also attempt to match characteristics of the water year totals.

## RESTORING LENTIC SYSTEMS

### REDUCING AND NEUTRALIZING ACIDIFIED SURFACE WATERS

Despite reductions in atmospheric $SO_4^{2-}$ deposition, acid rain has had a legacy effect. Many lakes continue to have low pHs, and there is some evidence that even more $SO_2$ reductions will be necessary to make further improvements.

In the meantime, many resource managers rely on limestone ($CaCO_3$) to increase the pH in lakes. This practice is more feasible in small lakes with high residence times. Whereas in large lakes or lakes with short residence times and low pH inputs, adding lime may be ineffective.

However, in spite of an increase in pH, the biological recovery in acidified lakes has been uneven. For example, when comparing lakes in Ontario, Canada, species richness has not recovered in lakes where the pH was increased above 6.5. These observations suggest that acidification will have longer-term impacts than anticipated and that restoring acidified lakes is more complex than adding lime.

### AERATION AND PHOSPHOROUS PRECIPITATION

When lakes are highly eutrophic and anoxic, aerating the water column has been a useful strategy for shallow, small lakes. Keeping the water column oxygenated will keep phosphorous precipitated with iron oxides and unavailable to phytoplankton. This strategy works for shrimp aquaculture, golf course hazards, and small urban ponds; it is not practical for most lakes.

Alternatively, aluminum sulfate ($Al_2(SO_4)_3 * 14\,H_2O$, called alum, can be added to remove phosphorous from the water column.

When aluminum sulfate is dissolved in alkaline water, aluminum hydroxide ($Al(OH)_3$) is formed, which is a solid precipitate:

$$\begin{aligned} Al_2(SO_4)_3{\cdot}14\,H_2O + 3\,Ca(HCO_3)_2 &\rightleftharpoons \text{intermediatesteps} \\ &\rightleftharpoons 2\,Al(OH)_3(s) + 6\,CO_2 + 3\,CaSO_4 + 18\,H_2O \end{aligned} \tag{12.1}$$

The solid precipitate forms a flocculent material, referred to as a floc, which has a high capacity to adsorb SRP. When enough is evenly added, the aluminum hydroxide blanket the benthos, effectively creating a barrier between the sediment from the water column and limits internal P loading (Fig. 12.20).

If the pH is too low or high, or there is too much bioturbation (e.g. by carp), alum will not be effective. Alternatively Fe (III) has been used to participate with phosphorous. Although this is a cheaper strategy, it relies on maintaining oxidized sediments to prevent the Fe from being reduced, releasing P into the water column. Eutrophic lakes rarely have oxidized sediments, thus this strategy has limited utility.

**FIGURE 12.20**

Image of a tanker barge preparing to treat Bass Lake (Plymouth, MN) with aluminum sulfate to control algae growth (Source: Daniel Ackerman).

In light of the watershed approach espoused above, there are many cases, where using the watershed approach exclusively may neither mitigate excess phosphorus in lakes nor even be feasible. Thus the use of alum may be the only practical way to accomplish meaningful and timely water quality improvements. Using alum as an element of a comprehensive watershed and lake management program will often be needed to achieve meaningful results in a timely and cost-effective manner.

Alum applications may be effective for 5 to 15 years. But when watershed phosphorus fails to reduce loading effectively, repeated applications may be necessary. In extreme cases, annual alum applications have been proposed.

Increased water clarity following alum applications will increase light availability. With more light, rooted macrophyte vegetation may increase and can provide additional benefits.

## CYANOBACTERIA, MACROPHYTES, AND LITTORAL HABITATS

Shallow lakes are the most abundant freshwater ecosystems on the planet, but many have become turbid and lost their submerged aquatic plant communities. With turbidity and reduced light penetration, PAR is insufficient for submerged aquatic plants to become established or maintain a positive carbon balance. Since the turbidity is caused by eutrophication, reducing the nutrient loading and a decline in water column algae growth and turbidity should lead to an increase in submerged plant cover.

But to make matters more complicated, macrophytes provide a means to improve water quality improvements. The presence of macrophytes can reduce nutrient concentrations (N and P), reduce phytoplankton growth (Chlorophyll *a*), and improve the trophic status of lakes. Historically, we thought there was a strong longitudinal gradient, where macrophytes were less effective at improving water quality in the tropics, but recent evidence does not support this view (Fig. 12.21).

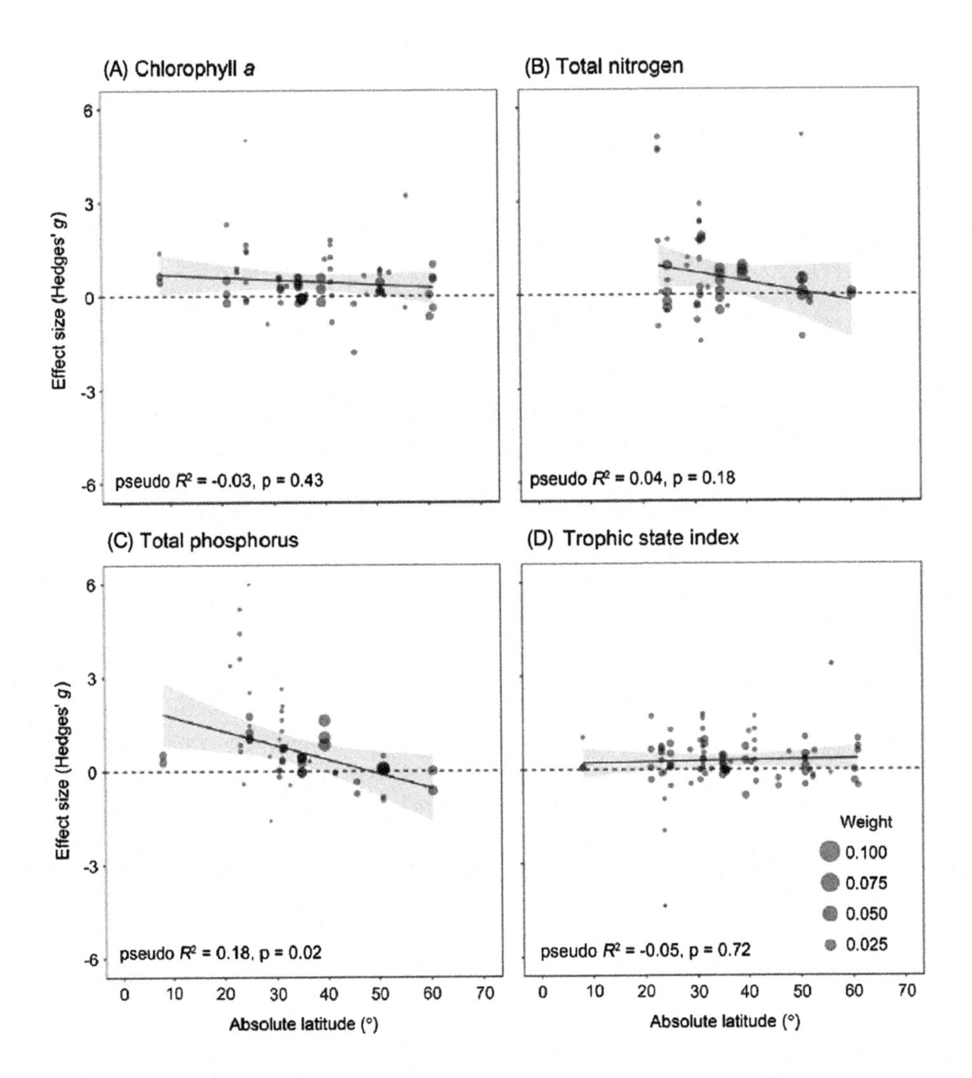


**FIGURE 12.21**

Effects of macrophytes on (A) Chlorophyll a Concentration, (B) Total Nitrogen Concentration, (C) Total Phosphorus Concentration, and (D) the Trophic State Index across latitudinal gradients (Source: Song et al., 2019). Each dot represents a pairwise comparison, and larger dots represent lower intra-study variance, which were used as a weighting factor in the analysis. The regression line represents the fitted values from the meta-regression, and the gray ribbon represents the 95% confidence interval.

Efforts to reduce eutrophic conditions and HABs and restore macrophyte vegetation generally rely on either external and/or internal restoration measures. External restoration measures involve the reduction of nutrient loading in the lakes. A reduction of direct discharge, i.e., sewage, can often yield rapid results when internal loading is not significant. Whereas watershed-based approaches to reduce

**FIGURE 12.22**

For many, Los Angeles is known for freeways and water imports. A car driving on a section of the aqueduct is fitting for this perception (Source: Getty Images).

N and P loading are more challenging, they are critical drivers. However, even with nutrient reductions, macrophyte community restoration success has uneven results. For example, external nutrient loading reduction can lead to an intermediate water clarity in the spring, which allows for macrophyte recolonization. But summers might still be subject to summer Cyanobacteria blooms and turbid conditions.

In contrast, internal restoration activities include biomanipulation or phosphorus precipitation in an attempt to control eutrophication blooms. Internal restoration measures, i.e., alum treatments or fish stock manipulations, often resulted in clear-water conditions both in spring and summer for a short period time, and then lakes often returned to turbid conditions.

These observations continue to drive research to evaluate mechanisms that might help to better address eutrophic conditions. But some preliminary conclusions suggest that macrophyte community restoration in turbid, eutrophic conditions may depend on remnant macrophyte stands, the specific restoration measure applied, and additional stochastic indirect influences on water clarity through food web processes.

Finally, we now believe that macrophyte recovery in shallow lakes may rely on a combination of external nutrient loading reductions and internal restoration measures.

## RESTORING ENDORHEIC LAKES: OWENS AND MONO LAKES

By the early 1900s the city of LA had grown to more than one hundred thousand inhabitants and was consuming more water than the Los Angeles River could provide. With predicted growth patterns, the city would need 220 million liters of water per day by 1925, and an alternate water source was necessary to maintain the desired lifestyle. To access more water, the Los Angeles Water and Power Company constructed aqueducts (Fig. 12.22) and canals for interbasin water transfers, first from Owens Valley, then the Colorado River, and finally from the Sacramento-San Joaquin Delta. Signaling the start of the long-running and bitter California Water Wars.

Starting in 1913, Los Angeles Aqueduct diverted water entering Owens Lake, which began the lakes' descent into becoming a dry lakebed (Fig. 12.23). And still needing more water, four creeks in the Mono Lake Basin were diverted into the aqueduct in 1941, drying up the streams below each diversion dam. The impact to the streams included the loss of the Riparian Vegetation and destruction

**FIGURE 12.23**

Satellite image of the Owen Lake Bed (Source: NASA Earth Observatory). When large portions of the bed are exposed, the Los Angleles Water and Power Company is required to "irrigate" the lakebed to reduce air quality issues associated with wind-born alkali dust on windy days.

of the fisheries. Without stabilized channels, floods tore through the desiccated floodplains plugging up side channels and turning creeks into wide, straight washes.

Without most of the inflow, Mono Lake experienced a 14 m drop in the lake level, a 50% reduction in lake volume, and salinity rose from 48 ppt to 100 ppt by 1982. The result was a fragmented and poorly functioning ecosystem. Islands where California Gulls nested became peninsulas accessible to predators. Wind-driven toxic alkali dust storms blew from exposed salt flats. The duck and geese population declined by 99%.

Salinity changes have obvious negative effects on many taxa. For example, Mono Lake Brine Shrimp (*Artemia monica*), a keystone species in the lake has specific salt tolerances. Increases in salinity have sublethal and lethal effects on the shrimp that include decreased growth rates, body size, hatching and reproductive potential along with increased female mortality. But the increase in salinity also have important indirect ecosystem effects. Benthic Nitrogen Fixation likely constitutes a substantial portion of Mono Lake's long-term nitrogen budget and the availability of nutrient supplies. Based on predicted salinity concentrations, if diversions were to continue, researchers found nitrogenase activity would be reduced by 50%. Thus salinity changes would have affected nutrient biogeochemistry and alter lake productivity, in addition to the direct impacts on species survival.

To avoid further impacts on the lake, the State Water Resources Control Board used these and other research results to set lake levels with an average permanent lake elevation at 1,949 m, with an expected salinity around 69 ppt. Although the lake will not be returned to the prediversion salinity levels, this decision was seen as a victory for those who had been working to save Mono Lake for decades.

Restoring the flow into a terminal lake is the obvious way to address declining water levels and increasing salinization. However, the diversion of water into terminal lakes continues to take its toll throughout the country and world—from the dramatic loss of area in the Aral Sea to the slow ongoing decline of Pyramid Lake (Nevada), Walker Lake (Nevada), and the Salton Sea (California). The causes are the same (i.e., upstream water diversions) and the ecological impacts are parallel—terminal lakes are a threatened habitat. Finding long-term solutions to maintain these lakes and ecological functions will require drastic changes in how water is used in arid climates.

## WATERSHED MANAGEMENT: LINKING TERRESTRIAL AND AQUATIC SYSTEMS

### RESTORING FLOOD PLAIN FUNCTIONALITY

In general, a fundamental part of restoring inland waters will rely on how water is managed at the watershed scale in the context of growing population, while planing for climate change impacts. For example, for Riverine Systems, floodplain connectivity needs to be evaluated at landscape scales to address each stressor and improve landuse practices.

Aquatic and Riparian Ecosystems benefit from hydrologic connectivity between rivers and their floodplains. In addition, the Riparian Zone can also facilitate linkages between the channel and floodplain.

Riparian species benefit from nutrients carried in by floodwaters, and aquatic species benefit by having access to floodplain for foraging, spawning, and taking refuge from high velocities. Floodplains can be particularly beneficial to juvenile anadromous salmonids, which use floodplains for foraging and refuge during their downstream migrations.

The Cosumnes River is the only major Central Valley river that retains relatively natural hydrologic conditions and capacity to flood into a portion of its floodplain. In winter 2004 and 2005, six enclosures were placed in each of the different habitats in the floodplain and two locations in the river. The ephemeral floodplain provided juvenile salmon habitat with a large amount of surface for forage, a variety of water velocities, warmer water temperatures, and areas of refugia. The difference in growth was dramatic (Fig. 12.24). Because the magnitude, duration, and timing of flows drive primary production in the Cosumnes Floodplain, fish will experience high intra- and interannual variation, which better mimics historic conditions.

Based on these observations, regional and state-wide programs have been developed to restore floodplain connectivity and habitat complexity to improve salmon fisheries. For example, one of the most significant and ambitious river restoration projects in the West is the restoration of the San Joaquin River, which had been characterized by 60-mile dry riverbed. The San Joaquin River—the second longest river in California—was once home to one of the largest populations of Spring-run Chinook Salmon, a species of fish that is classified as threatened under the Endangered Species Act. The San

**FIGURE 12.24**

Dramatic difference in fish sizes when fish are allowed to forage in the floodplain (Source: Jeff Opperman). Juveniles were placed in enclosures when wild salmon would naturally be rearing in the area. Found significant differences in growth rates between salmon reared in floodplain and river enclosures.

Joaquin River Restoration Program (SJRRP) was established as the result of a 2006 Federal Court Agreement that settled an 18-year battle between Environmentalists and federal water contractors over use of San Joaquin water below Friant Dam, part of the federal Central Valley Project.

The SJRRP, a comprehensive, long-term effort that was formed in 2007 with two major goals: 1) to achieve a naturally self-sustaining population of Chinook Salmon in the San Joaquin River from the Friant Dam to the confluence of the Merced River, and 2) to mitigate water supply impacts to water users as a result of the program's restoration flows. With augmented flows, removal of channel obstructions, and localized Riparian and floodplain restoration, the restoration project is a good example where regional partnerships, state and federal funding sources, and local restoration practitioners will have 'reinvented a river' within a few decades. In fact, Spring-run Chinook Salmon have successfully returned to the San Joaquin River for the first time in more than 65 years. Restoration work will also benefit other native fish species, including fall-run Chinook, Pacific lamprey, steelhead trout, and white sturgeon.

## EXPERIENCE, PLANS, AND FANTASIES OF DAM REMOVAL

Opposing dams has become symbols of the environmental movement in the United States. John Muir and the Sierra Club bitterly fought the O'Shaughnessy Dam and the plan to flood the Hetch Hetchy Valley (Fig. 12.25). Decades later, the Sierra Club successfully stopped dams from being built in Dinosaur National Monument and in Grand Canyon National Park. In a similar vein, calls to remove dams and restore rivers have been a consistent theme among many Environmentalists, NGOs, and popular environmental writers.

**FIGURE 12.25**

The O'Shaughnessy Dam was completed 1923 (Wiki commons) and the object of a major enviornmental fight in the early 20th Century.

In some cases dam removals have become a reality, but only for a small proportion of dams. There are over 75,000 dams in the United States and less than 500 of them have been removed. Most of the dams removed were small, less than 10 m high. And until 2012, no dam above 30 m had been removed until Elwha River was restored.

As dams reach the end of their useful lives, the pace and scale of dam removal will increase. However, with each dam, careful planning and monitor programs will allow Environmental Scientists to ensure that lessons learned can be used with each dam to maximize the environmental benefits and minimize negative impacts.

After two decades of planning, the Elwha Dam removal began in September 2011 and the Glines Dam was taken out a few years later. The Elwha River (Washington) now flows freely from its headwaters in the Olympic Mountains to the Strait of Juan de Fuca. However, before the removal began, Geomorphologists had to carefully evaluate the impact of releasing 13 million $m^3$ of sediments that had collected behind the dams.

Geomorphologists were cautiously optimistic that the river's capacity to transport the sediment would not damage fisheries or lead to flooding. Even within the first year, the river was behaving even better than expected, with new sandbars being recolonized by willows, alders, and other Riparian Vegetation. In addition the outlet of the river into the nearshore waters became a more productive habitat for crabs, fish, clams, and other species. In addition, salmon runs have been increasing each year.

However, it is important to note that although most commentaries and environmental organizations focus on the impacts of large dams, there is growing concerned that small hydropower dams may be as ecologically significant as large dams, because the shear number of them may be orders of magnitude higher. With few environmental assessment reviews before construction approval and post construction studies, we know very little about their impacts. The impact of small hydropower dams are likely to be highly localized, but could be more significant when compared to a large dam, especially if measured on a per megawatt. These concerns deserve a dramatic increase in research interests.

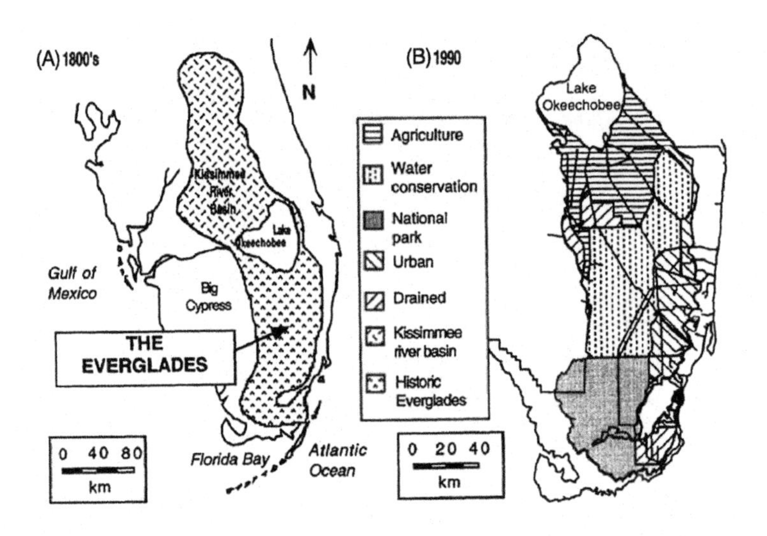


**FIGURE 12.26**

Historic Map of the Everglades (Source: Gunderson et al., 1995).

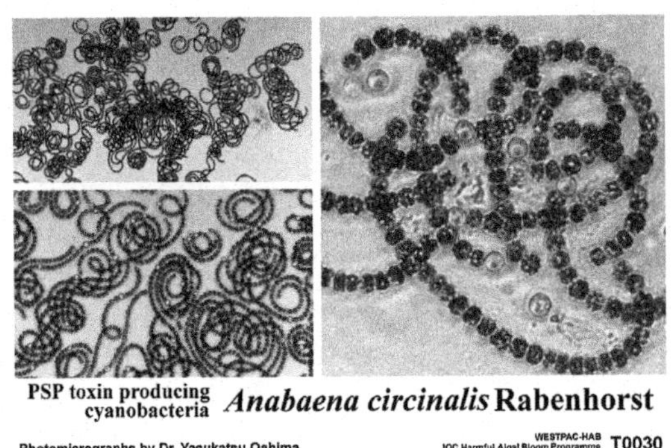


**FIGURE 12.27**

Image of Anabaena circinalis (Source: Yasukatsu Oahima).

## EVERGLADES: PHOSPHOROUS LOADING THE RIVER OF GRASS

The Everglades is 8000 km$^2$ wetland ecosystem that begins in Central Florida and empties into Florida Bay (Fig. 12.26). Behaving like a very slow moving river, extremely low levels of nutrients create a unique mosaic of sawgrass, tree islands, and open water. However, with excess P from agriculture, promotes the growth of vegetation, such as cattails, harmful algal blooms, and duckweed. In 1986 a widespread algal bloom that included *Anabaena circinalis* and covered one-fifth of Lake Okeechobee resulted from fertilizer runoff Fig. 12.27). Tested water had 500 ppb of phosphorus near farms, which is exceptionally high for the lake.

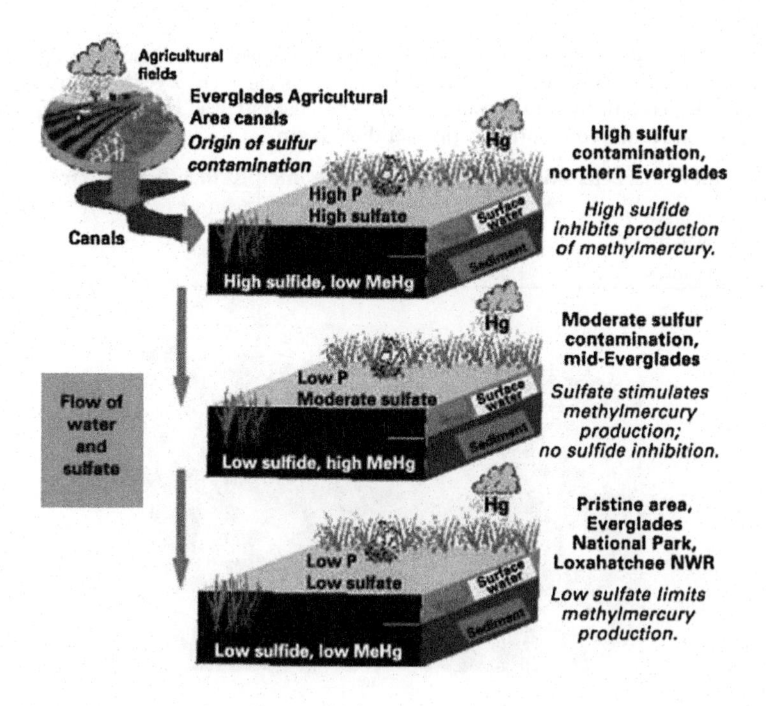


**FIGURE 12.28**

Role of phosphorus and sulfur in mercury biogeochemistry in the Everglades (Source: USGS).

The Cyanobacteria species, *Anabaena circinalis,* is particularly worrisome. Not only does this species produce toxins, it is also a nitrogen fixer. By providing the nutrients that increase *A. circinalis* populations, additional reactive N further disrupts the ecosystem.

Finally, through a complex series of biological and chemical processes, sulfur from agricultural runoff has increased the concentration of MeHg in specific portions of the Everglades (Fig. 12.28). Elevated concentrations of this neurotoxin have been detected in fish, birds, reptiles, and mammals, even in the endangered Florida Panther. Thus managing the Everglades require strict controls on agricultural runoff in a highly modified landscape.

## FOREST MANAGEMENT AND ORGANIC MATTER

Landuse can play a critical role in Stream Systems and affect a broad range of aquatic organisms. For example, forest provide allochthonous sources of organic carbon that is used in Lotic Food Webs and support fisheries. Therefore the impact of forest practices on streams has been an area of considerable research. As a result, forest management practices are highly regulated, especially around stream and Riparian Zones.

Nevertheless, the relationship between fishery productivity and forest management is far from intuitive. By comparing old growth and clear-cut tributaries in the Deschutes River Watershed (Oregon), researchers evaluated fish productivity relative to the amount of allochthonous and autochthonous

**Table 12.1 Organic matter changes in two tributaries in the Deschutes River watershed (Washington), where one tributary was clear-cut 7 years before the study (Source: Bilby and Bisson, 1992). The clearcut was conducted before regulations required a buffer to protect the stream.**

| Tributary Characteristic | Allochthonous Organic Matter ($g\,m^{-2}\,yr^{-1}$) | Autochthonous Organic Matter ($g\,m^{-2}\,yr^{-1}$) |
|---|---|---|
| Old-growth forest | 300 | 100 |
| Clear-cut forest | 60 | 175 |

**FIGURE 12.29**

A haul of fish being loaded from Tonle Sap in Kampong Khleang, Cambodia (Source: www.stanleyfoundation.org).

sources of organic matter. Although the combined allochthonous and autochthonous inputs were almost twofold greater in the old-growth site (Table 12.1), fish production was greater in the clear-cut site. Production of Coho Salmon (*Oncorhynchus kisutch*) and shorthead sculpin (*Cottus confusus*) during early summer was largely responsible for differences between sites. These observations have inspired further work to disentangle the mechanisms and provide additional information for forest practice rules.

## FISHERIES

The words nutrition and biodiversity are seldom linked, yet fish species diversity is a critical component in defining the nutritional profile of the many people (Fig. 12.29). Fish can be the principal source animal protein and fatty acids, and contribute vitamins and minerals. Under-resourced peoples, in particular, rely on a wide variety of species to meet their nutritional needs. In some cases, fishing communities might consume between 50 and 75 species of fish annually. Many of these species migrate between rivers, where they find refuge during the dry seasons, and floodplains, where they spawn and feed during the rainy season. Vulnerable communities suffer the most from nutritional losses caused by impeded fish migration. Thus fisheries makes a direct link between social justice and biodiversity.

Declining fish stocks and loss of biodiversity are rarely due to the activities of subsistence fishers. In the competition over freshwater sources, more powerful sectors, such as agriculture and energy limit the quality and quantity of water available for inland fisheries. These drivers have little to do with the populations that depend on these fish stocks. In fact, even where overfishing is well documented, the cause is often from overly capitalized fleets and advanced, highly efficient fishing methods. Thus dramatic improvements are needed to assess who use fisheries and population structure to better protect species diversity.

As a start, stock assessments is critical information to improve fishery management. Many freshwater fisheries are managed without stock assessments, even where the capacity for data collection and analysis exists. Management of such fisheries is likely to lose out on quantitative information that can support decision-making on issues ranging from the sustainability of fisheries exploitation, to the development of effective stocking and harvesting strategies for enhanced fisheries.

The lake sturgeon (*Acipenser fulvescens*) fishery in Lake Erie collapsed by the 1920s when < 1% of the previous peak catch was removed. Despite closures of the fishery, lake sturgeon remain rare in Lake Erie, indicating that other factors may be limiting their recovery. In some cases, reversing the collapse of a fishery is not simply a matter of removing the fishing effort, but requires additional restoration activities, e.g., habitat restoration.

# ADAPTIVE MANAGEMENT

## PROMISES AND PITFALLS OF ADAPTIVE MANAGEMENT

With nearly every project, Restoration Ecologists learn something new about ecosystem responses to the restoration process. These lessons are often based on unanticipated outcomes. As such, most restoration plans are augmented, adjusted, revised, or scrapped based on how the projects progress. Thus Restoration Ecologists practice Adaptive Management.

At its best, Adaptive Management is a structured, iterative process to make management decisions in spite of uncertainty. And when effective monitoring is in place, observations about how the ecosystem responds to restoration activities reduces the uncertainty. And hopefully, future management decisions will improve. The process is used by private and public resource managers and often is codified to ensure a consistent process if followed (Fig. 12.30).

Adaptive Management, however, is no panacea. First, Adaptive Management assumes that solutions are technical in nature. Second, rarely are all the pieces in place to get the most out of Adaptive Management. In the latter case, some see Adaptive Management as "management on the cheap" with a cynical view that project success was impossible by design because resources were inadequate from the start.

In addition, we now appreciate the inherent complexity of ecosystems and the inability to foresee all consequences of restoration activities across different spatial, temporal, and administrative scales. Thus we might see Restoration Ecology or Ecosystem Management, in general, as a "wicked problem" that has no clear-cut solution, e.g., the removal of the invasive *Spartina alterniflora*. Thus restoration workers acknowledge that no technical solutions are clearcut, and the complexity of the systems makes the outcomes rather uncertain. The challenge in using the Adaptive Management approach lies in finding the correct balance between gaining knowledge to improve management in the future and achieving the best short-term outcome based on current knowledge. Using an incremental and adaptive management, Restoration Ecologists have avoided the traps of falsely assuming a tame solution and inaction from overwhelming complexity.

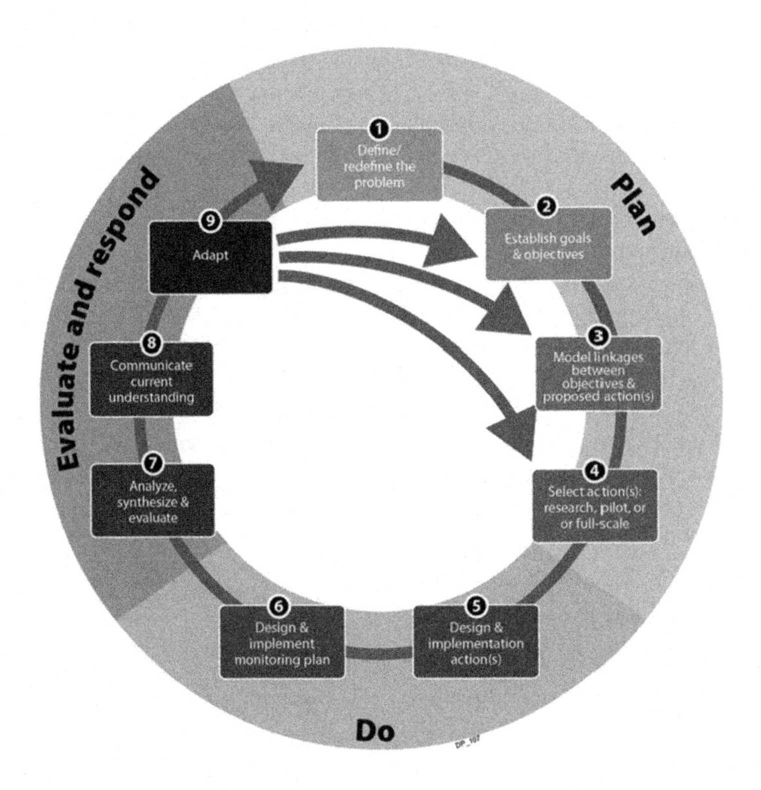


**FIGURE 12.30**

Schematic of Adaptive Management (Source: California Fish and Wildlife).

## EVALUATING RESTORATION SUCCESS

A wide range of habitats have been lost early in California's settlement history as a result of logging, mining, and even grazing. Even as the process continues in the form of development pressure, our understanding of pre-settlement or pre-European settlement is quite obscure. Even changes from one to another generation will shift the baseline of what is perceived as a pristine, natural, or functioning ecosystem. Thus the challenge of developing a sophisticated analysis of the site's history and a strong understanding how the ecosystem will respond to restoration explains why Restoration Ecologists are careful to define the goals for any restoration project.

Most restoration projects are assessed measures that can be categorized by three major ecosystem attributes: 1) diversity; 2) abiotic and biotic structure, and 3) ecological processes. For example, a framework proposed by Peter Moyle (1998) can be used to evaluate and categorize fish restoration into three tiers:

***Tier 1. Individual health*** Individual fish should have a healthy body conformation; should be relatively free of diseases, parasites, and lesions; should have reasonable growth rates for the region, and should respond in an appropriate manner to stimuli (e.g., predator avoidance).

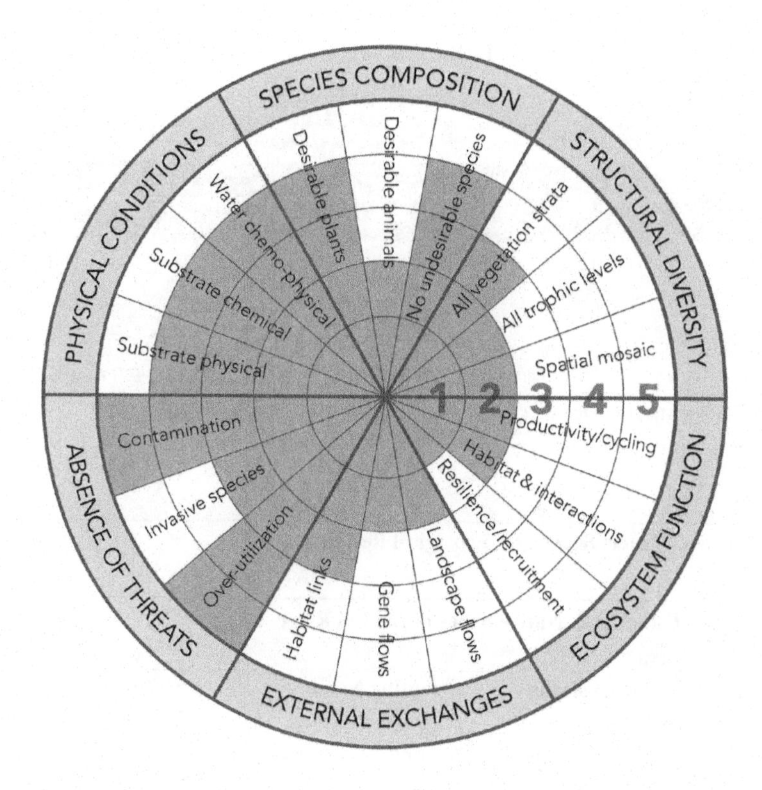


**FIGURE 12.31**

Diagram of Restoration Standards (Source: Society for Ecological Restoration).

***Tier 2. Population viability*** A fish population should include multiple age-classes, exhibit a viable population size, and be composed of healthy individuals. Since defining a viable population size is a challenge, two proxies can be used an alternative 1) extensive habitat available for all life-history stages, and 2) habitats should have a broad distribution for all life-history stages.

***Tier 3. Community*** A fish community in good health is (1) dominated by coevolved species, (2) predictably structured as indicated by limited niche overlap among the species, and by multiple trophic levels, (3) resilient in recovering from extreme events, (4) persistent in species membership through time, and (5) replicated geographically. In short, a dynamic fish assemblage that will predictably occupy a defined range of environmental conditions.

Although these criteria are designed for fish, they can be applied to most taxa, but is based on a single species focus.

As a more ecosystem-based approach, the Society of Restoration Ecologists promote a model that relies on the following categories: physical conditions, species composition, structural diversity, ecosystem function, eternal exchanges, and absence of threats (Fig. 12.31). Within each category, semiquantitative indices can be used to gauge restoration success. Frameworks such as these allow restoration projects to be robustly evaluated, where selected criteria might be biases. In addition, this

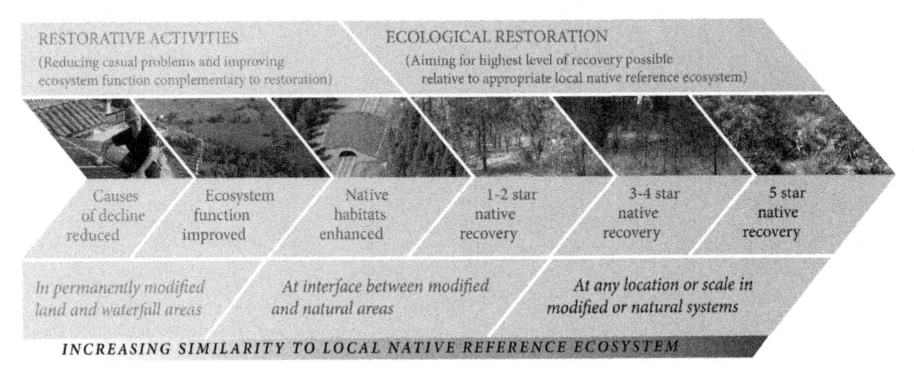


**FIGURE 12.32**

The restorative continuum (Source: Society for Ecological Restoration).

**Table 12.2 Native fishes of the Eel River.**

| Taxa | | |
|---|---|---|
| **Common Name** | **Scientific Name** | **Status** |
| Pacific lamprey | *Lampetra tridentata* | Declining |
| Pacific brook lamprey | *Lampetra richardsoni* | Unknown |
| Green sturgeon | *Acipenser medirostris* | Extinct |
| Longfin smelt | *Spirinchus thaleichthys* | Extinct |
| Coho salmon | *Oncorhynchus kisutch* | Declining |
| Chinook salmon | *Oncorhynchus tshawytscha* | Declining |
| Pink salmon | *Oncorhynchus gorbuscha* | Extinct |
| Chum salmon | *Oncorhynchus keta* | Extinct |
| Rainbow trout | *Oncorhynchus mykiss* | Declining |
| Cutthroat trout | *Oncorhynchus clarki* | Declining |
| Sacramento sucker | *Catostomus occidentalis* | Stable |
| Prickly sculpin | *Cottus asper* | Stable |
| Coastrange sculpin | *Cottus aleuticus* | Stable |
| Threespine stickleback | *Gasterosteus aculeatus* | Stable |

model can be modified to included other aspects of restoration, such as the single species focus of Moyle's model.

Another approach to evaluate restoration projects appreciated the temporal component of restoration, where the process might take years or decades (Fig. 12.32). Evaluating restoration progress parallels ideas of community development and recovery after a disturbance. By explicitly defining restoration goals over time, this model can be used to drive Adaptive Management activities to improve particular weaknesses in the restoration outcomes.

# NEXT STEPS

## CHAPTER STUDY QUESTIONS

1. Describe the ecological consequences of invasive species to inland waters.
2. Summarize strategies to manage and control invasive species.
3. Describe the challenges and successes of efforts to restore wetlands, streams, and lakes.
4. Compare and contrast various definitions of restoration success.
5. Describe and justify what might be included to monitor restoration success.

## ADVANCED APPLICATIONS

The Eel River drainage has been relatively protected from human diversion, with the exception of the Pillsbury Reservoir and the associated Cape Horn Dam built in 1921 in the upper reach of the mainstem. However, the status and trends of the fish reflect long-term decline (Table 12.2).

Evaluate the river based on the literature and develop a restoration plan to improve the biodiversity of fish in the river. Be sure to set out measurable goals.

CHAPTER

# CROSSING THE DIVIDES: THE MISMATCH BETWEEN POLITICAL AND WATERSHED BOUNDARIES

# 13

*Politicians are the same all over. They promise to build a bridge even where there is no river.*
**Nikita Khrushchev, Russian Soviet politician**

*We let a river shower its banks with a spirit that invades the people living there, and we protect that river, knowing that without its blessings the people have no source of soul.*
**Thomas Moore**

*Access to safe water is a fundamental human need and, therefore, a basic human right. Contaminated water jeopardizes both the physical and social health of all people. It is an affront to human dignity.*
**Kofi Annan, former Secretary-General (2001)**

## CONTENTS



Ecology and Management of Inland Waters. https://doi.org/10.1016/B978-0-12-814266-0.00027-1

According to the World Health Organization, each human requires at least 20 liters of fresh water per day for basic hygiene. But the actual amount of usage varies as a function of regional availability, access and development of water supply infrastructure. Most people in developed countries have treated systems and infrastructure to deliver high quality water to every home and sanitary systems to remove and treat waste. Meanwhile, vast numbers of people in Latin America, parts of Asia, South East Asia, Africa, and the Middle East either do not have sufficient water resources or the infrastructure to obtain reliable, high quality water and waste treatment. In many cases, these circumstances lead to conflict and negative health outcomes in terms of disease, malnutrition, or death. But these human health outcomes are also linked to landuse patterns that impact watershed integrity, hydrologic modifications, nonnative species introductions, biodiversity losses, and severely compromised ecosystems services. These linkages vary with settlement patterns, water resources, and regional political economy. Furthermore, future projections suggest that the situation will likely get worse (Fig. 13.1), unless dramatic changes occur in governance.

This chapter summarizes the regulatory and policy context for the allocation, use, treatment, oversight, and protection of inland waters. In a world where the water supply "pie" is shrinking, the institutional capacity must increase to meet increasing water demand and a changing climate. On the other hand, various partnerships have emerged to make significant progress to improve inland waters for the benefits of humans and nonhumans. Across the jurisdictions, several water-rights regimes, along with environmental legislation imposed by various state and federal agencies, create a dizzyingly complex political context; this primer will help readers to make sense of public policy and water. After reading this, you should be able to

**FIGURE 13.1**

Current and projected (2025) number of people who experience water stress.

1. Describe political conflicts and legal contexts for water supply;
2. Explain how endangered species protections influence the ecology of inland waters;
3. Describe the policy tools used to resolve conflicts and protect water supplies, water quality, and Aquatic Ecosystems.

This chapter presents examples of recent water-use policy changes in various countries (e.g., Australia and South Africa), demonstrating a range of opportunities for Californians to improve on our own poorly developed sustainable water-use policies.

## WATER RIGHTS AND RIGHTS TO WATER

### FRESHWATER AS A FUNDAMENTAL REQUIREMENT

In 2010 the UN General Assembly formally recognized the Human Right to Water and Sanitation. Although the HRWS has been recognized in international law through human rights treaties, declarations and other standards, the historical development of human rights is both a "fixed" concept with the development of nation-state legal frameworks (e.g., Amendments to the US Constitution), but also "fluid" as rights are extended and reinterpreted in light of changing conditions.

The tradition in the West was to guarantee rights for participation in governance via democracy. Based on participation, citizens could self-govern to improve human satisfaction or "the pursuit of happiness." Although humans need clean water and air, food and shelter, healthcare and education to participate in society, these needs have rarely framed as rights within developed countries. Thus in spite of the UN's recognition, conferring the Human Right to Water and Sanitation onto the anarchy of sovereign states is awkward. Nevertheless, some countries have developed 'rights-based' approaches for environmental resources, while more often environmental resources are protected in more limited sense, such as the Endangered Species Act or Clean Water Act in the United States.

### RIGHTS TO WATER

Access to water and water rights has a complex history. Depending on the historic settlement patterns and assumptions concerning the merits of water as a public good or a privately owned resource, a complex fabric of legal structures guide how water use is negotiated (Fig. 13.2). These legal structures vary dramatically across national and subnational jurisdictions.

In the United States, this 'right' to water is framed with a context of residence, where case law established some precedence. For example, in the case Pilchen v. City of Auburn, New York, shutting off Pilchen's water service was considered unconstitutional. But not because Pilchen should be guaranteed the right to water, but because the responsible party, the landowner in this case, failed to pay the waterbill. After Diane Pilchen, a single mother, was forced to move out because the city deemed the house inhospitable without water service, the court found that Pilchen could not be held accountable as the third party. So, in this case, the right to water was mediated with the contractual relationship between the water purveyor and property owner. Had the case involved a tenant that was contractually obligated to pay the water bill, no constitutional argument could have been claimed. A tenant that could not afford in his name would probably have no right to water from a constitutional basis.

The development of water infrastructure often creates a privatized commodity and reenforces inequalities (Fig. 13.3). Developing strategies to avoid these outcomes, while protecting water resources

**FIGURE 13.2**

Schematic of the components to provide the right to water (Source: Waterpreneurs). Declaring rights is easy, but actualizing them requires a robust set of components to ensure socially just and sustainable outcomes.

**FIGURE 13.3**

Cartoon contrasting rich and poor nations and how private sectors reinforce inequalities. Privatized water sources have become an important political issue (Source: Paresh Nath, Cagle Cartoons, 2010).

for non-human uses requires a high level of community participation, political will, and effective rule of law.

## WATER RIGHTS

The water code in California, for example, is a complex amalgam of common resource governance traditions: Roman, Spanish, English, and Indigenous. For example, Riparian Rights, an English water law, gained legal recognition after California was granted statehood. Under the law, owners of land that physically touches a water source have a right to use water from that source that has not been

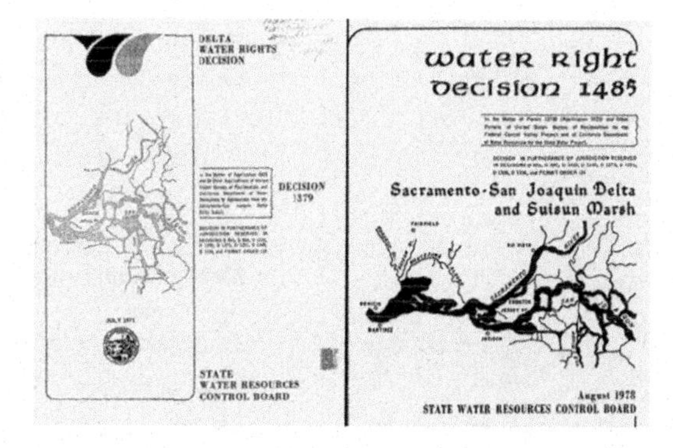


**FIGURE 13.4**

The State Water Resources Control Board made decisions to limit Delta exports based on environmental concerns that were later rendered ineffective by court battles and political interference, e.g., the 1971 (left) decision affirming the environment and threatening exports was stayed by a judge. In 1978 (right) the court sided with the Bureau of Reclamation that the state had no jurisdiction over their federal operation of diverting water for agriculture in spite of the damage to the delta.

deemed appropriated by another party. In contrast, appropriative rights allows water to be diverted and transported away from the sources. These two systems had led to barrage of legal battles through the last 19th and early 20th centuries until 1914 when the United States government tried to establish a more rational policy. After 1914 water rights were given by permit issued by the State Water Board. However, the permit system excluded groundwater, Riparian, and pre-1914 appropriative water right holders, creating a complex of layering of legal frameworks. Furthermore, the Water Board has the authority to consider the state's reasonable use and public trust doctrines for all water rights holders. Even without getting into the specifics, it should be clear that the legal system that governs is highly textured across the landscape based on settlement patterns, available water resources, water right claims, and the history of how capital was used to develop water, i.e., turn it into a resource.

One aspect of water policy in the state and most western states is that it is partially based on seniority. The first appropriator has legal claims and entitled to uses over more recent claims or junior claims. Importantly, environmental uses of water often have junior rights (Fig. 13.4).

In spite of the changes after 1914, water rights and permits have dramatically over-allocated the amount of available water in the state (Fig. 13.5). When precipitation totals are relatively high, this over-allocation does not have a major impact. However, in drought conditions the state has developed a system that is 'ripe' for legal and possibly extra-legal conflict.

## ESTABLISHING ENVIRONMENTAL FLOWS

An environmental flow describes the quantity, timing, and quality of water-flows required to sustain Aquatic Ecosystems and the people that depend on these ecosystems. However, flows for environmental purposes are junior to long-established water rights. In addition, the regulatory framework to protect streams has been based on stressors, e.g., streambed habitat for spawning, Dissolved Oxygen

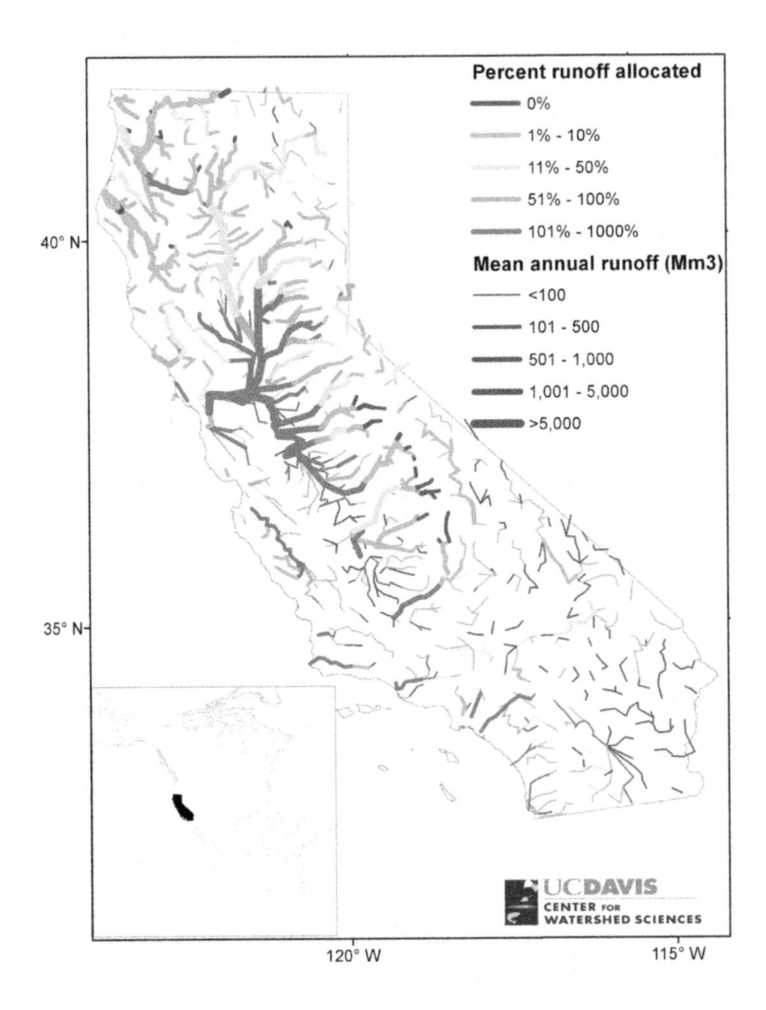


**FIGURE 13.5**

Cumulative water-right allocations relative to mean annual runoff, excluding water rights for hydropower generation. (Source: Grantham and Viers, 2014).

instead of maintaining ecosystem function. Under these legal systems, protecting inland waters become protracted legal battles.

Developing flexible tactics to negotiate competing water needs has become an area of increasing interest. Alternative approaches can provide flexibility to access water supplies independent of seniority-based frameworks. For example, cooperation between private landowners and regulatory agencies outside the formal requirements of water law can alleviate negative impacts of drought.

A wide range of methods exist to establish environmental flows to improve ecological health. But to establish effective water management, it is critical that these methods consider water supply and quality simultaneously.

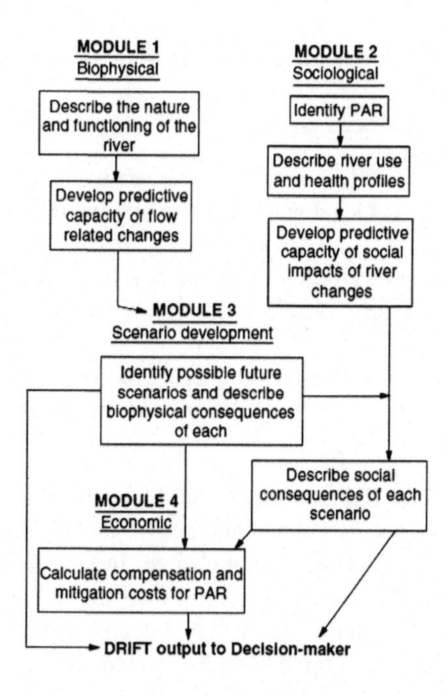


**FIGURE 13.6**

Decision support model DRIFT (Downstream Response to Imposed Flow Transformation) is used to evaluate how regulated rivers meet various socio-ecological goals (Source: King et al., 2003).

Although environmental flow experiments are developed to address specific questions, in particular reaches, critical information is provided but rarely within short time frame when a drought occurs. Thus several modeling approaches have been proposed.

For example, one approach relies on a coupled biophysical-socioeconomic model called DRIFT (Downstream Response to Imposed Flow Transformation) (Fig. 13.6). DRIFT is designed to develop scenarios to establish defined flow regimes that explicitly includes the socioeconomic factors.

DRIFT requires experienced scientists from a wide range of biophysical disciplines, e.g., Hydrology, Geomorphology, Chemistry, Botany, and Zoology. In the context where there are subsistence users of the river (People at Risk, PAR), the following socioeconomic disciplines are also employed: Sociology, Anthropology, Water Supply, Public Health, Livestock Health and Resource Economics. The DRIFT model has been applied in numerous locations, but most of the experience with it has been in South Africa and Lesotho. Thus to augment the model to address powerful stakeholders with water rights when no PAR exists is a scenario worth pursuing and testing.

## RIGHTS OF NATURE

As a consequence of human development, pollution, and climate change, freshwater systems are among the most endangered habitats in the world. Less than half of the longest are free of dams, a majority of the world's wetlands have disappeared since 1900, all of the large lakes have become over-fished. Most lakes show signs of eutrophication. Are there any intrinsic rights for these systems that can be used to protect them?

In some cases, inland waters have been granted rights. For example,

***India*** The highly polluted Ganges and Yamuna rivers were given the same status as a human being on March 20, 2017. This means legal guardians can now represent the waterways in court over any violation.

***New Zealand*** A river in the country's North Island became a legal person on March 15, 2017. A local Maori tribe has fought for nearly 150 years for the Whanganui River to be recognized as an ancestor.

***Equador*** In 2008 the South American country set a legal precedent by giving nature rights like those of humans in its constitution. This means entire ecosystems have the "right to exist, persist, maintain and regenerate."

The standard to have recognized rights in the United States is to have legal standing. People and corporations in the United States have legal standing, environmental resources do not, i.e., forests, rivers, etc. But that does not mean that this cannot change. In fact, in 1972, the Sierra Club argued that it was able to represent trees and seek restitution for their injuries. The case went to the Supreme Court and the court ruled against the Sierra Club in a 4–3 decision, but the opinion written by Justice Douglass describes his frustration that nature did not have standing in the courts:

> *Contemporary public concern for protecting nature's ecological equilibrium should lead to the conferral of standing upon environmental objects to sue for their own preservation. This suit would therefore be more properly labeled as Mineral King v. Morton.*
>
> *Inanimate objects are sometimes parties in litigation. A ship has a legal personality, a fiction found useful for maritime purposes. The corporation sole—a creature of ecclesiastical law—is an acceptable adversary and large fortunes ride on its cases. The ordinary corporation is a "person" for purposes of the adjudicatory processes, whether it represents proprietary, spiritual, aesthetic, or charitable causes.*
>
> *So it should be as respects valleys, alpine meadows, rivers, lakes, estuaries, beaches, ridges, groves of trees, swampland, or even air that feels the destructive pressures of modern technology and modern life. The river, for example, is the living symbol of all the life it sustains or nourishes—fish, aquatic insects, water ouzels, otter, fisher, deer, elk, bear, and all other animals, including man, who are dependent on it or who enjoy it for its sight, its sound, or its life. The river as plaintiff speaks for the ecological unit of life that is part of it. Those people who have a meaningful relation to that body of water—whether it be a fisherman, a canoeist, a zoologist, or a logger—must be able to speak for the values which the river represents and which are threatened with destruction ...*

The concept is captured quite well in Dr. Suess's book, "The Lorax," but further described in an article by Christopher Stone, "Should Trees have Standing?" written in 1972 (Fig. 13.7). Although the beautiful prose has a compelling argument, the practicality of who would speak for trees or rivers in a court of law is far from obvious. However, as in many cases about legal precedence changes, a range of administrative and case law could address these concerns. Thus there is no *a priori* reason that might disallow the extension of rights to nonhumans, and practicalities of how to manage standing could be developed over time.

**FIGURE 13.7**

Christopher Stone and Dr. Suess both describe how the issue of standing can play an important role in conservation.

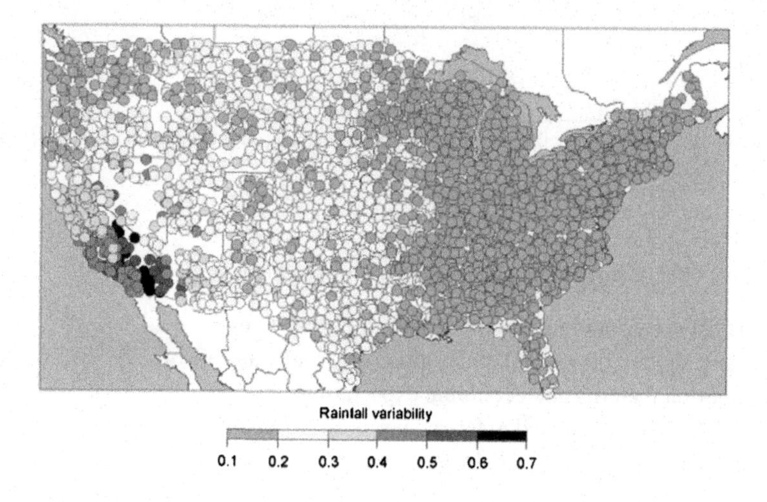


**FIGURE 13.8**

Variation in rainfall on the conterminous US (Source: Dettinger, 2011).

## POLITICS OF SCARCITY: MANAGING DROUGHT

Freshwater is becoming more precious every year. As demand from agriculture, manufacturing, energy production, and people increases, conflict is also likely to increase. In fact, water scarcity is already a frequent contributor to political conflicts, and even war. Even a multiyear drought has been linked to specific drivers of the civil unrest in Syria during the second decade of the 2000s. Given the political conflicts already, many believe future conflicts will be increasingly driven by water scarcity. Thus better management and sustainable use is key to maintain, use, and minimize conflict.

Variation in rainfall requires effective planning with respect to drought. Without a doubt, variation in rainfall is a fundamental constraint of the West, and in California, in particular (Fig. 13.8). The

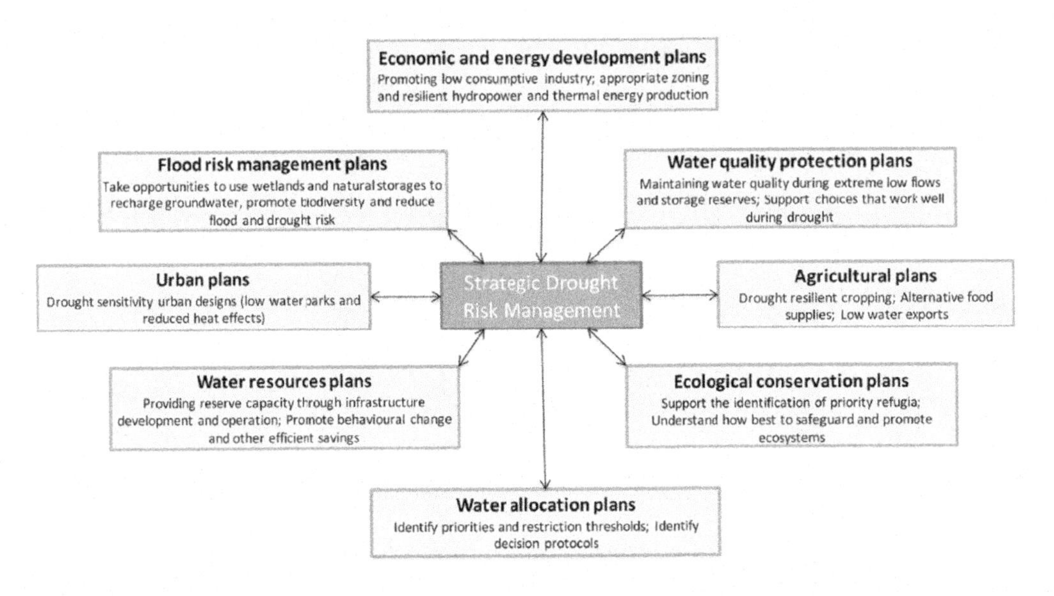


**FIGURE 13.9**

Risk management—the diverse set of ancillary concerns that influence drought risk management (Source: Sayers et al., 2017).

following areas have been identified as challenges in drought management:

1. Disconnections between population growth and water resources;
2. Geographic disconnection between population centers and water resources;
3. Shifting consumption patterns and economic expectations;
4. Increasing variability in climate and extended periods of dryness;
5. Increasing tension due to over-allocated water resources.

These challenges describe many parts of the world, including the western United States, well. But what are the steps to address these challenges. Some have identified specific strategies to manage drought and reduce the risk of conflict. For example, Strategic Drought Risk Management (SDRM) is a process designed improve environmental and political outcomes with the inevitable drought occurrences. Much of the process relies on establishing clear goals and objectives, appropriately assessing the system of interest and its uncertainty, and establishing effective communication and participation in the process.

However, to effectively manage drought risks, a host of other issues need to be considered (Fig. 13.9). For example, Drought Management is linked to Flood Management. In fact, flooding and flood control are not opposites of drought, but go hand-in-hand in the planning process. Whereas flood control and water storage has been done using dams, the integrity of inland waters have suffered. Is it possible to reverse the impacts on inland waters while reducing the risks of drought and impacts of flooding?

| Advantages | Disadvantages |
| --- | --- |
| Reduces consumption of fossil fuels for electricity production | Dirt can build up at dams, decreasing their effectiveness |
| Reduces production of greenhouse gases, such as $CO_2$ | Large-scale wildlife habitat destruction due to river valley flooding |
| Reduces production of pollution, such as particulate matter | Interferes with natural wildlife migration patterns, such as salmon |
| Can prevent uncontrolled flooding | Dam construction forces people to leave their homes if they live in or near the flooded river valley |
| Provides water for irrigation | Very expensive to build |
| Creates areas for certain types of recreation, such as boating and fishing | Reduces areas for certain types of recreation, such as fishing, camping, hunting, hiking |
| Is a renewable energy source! | Interferes with natural flow of water through environment |
| | If natural fisheries are affected, harms the livelihoods of people who rely on those fisheries to make a living |
| | Requires maintenance |
| | Can fail catastrophically! |

**FIGURE 13.10**

Risk Management should recognize the link between resilient freshwater and human systems (Source: Sayers et al., 2017).

## MANAGING RISKS OF THE "NONNATURAL DISASTERS"

Over the decades, Ecologists have defined the capacity to recover from disturbances using a range of metrics. In general, the concept of resilience continues to have value, especially in relation to ecosystem services. Degraded inland waters are less resilient to perturbation, and importantly for us, less able to perform ecosystem services, such as flood control or drought mitigation (Fig. 13.10).

For over a century, the United States has conducted a mostly failed experiment in flood control. We have channelized hundreds of rivers with thousands of miles levees to reimage the floodplain without floods. Beginning in the 1920s, dramatic images of floods and stories of lost lives and livelihoods, compelled the US government to fix the wrong problem. Instead of admitting the power of rivers was something to adapt to, the United States started building structures to hold back floods, straighten and smooth the bottom and sides of channels to hurry floodwater out of cities. Behind engineering confidence, a staggering number of homes, offices, and factories were built in the floodplain. And when structures failed, not catastrophically, the federal government measured risks like actuary tables (10-, 20-, 50-, and 100-year floods) and manage losses with insurance settlements. And with subsidized insurance premiums, the federal government has subsidized risky behavior, i.e., construction in the floodplain. And to add insult to injury, the maps used to estimate flooding and insurance rates are often wrong—in the wrong direction, flood-prone areas are not always mapped, and many do not have sufficient insurance needed to recover.

As one step, the United States Federal Emergency Management Agency has created new tools to identify properties at risk of flooding. Coupled with changes in insurance premiums that better align with risk, the federal government will not be encouraging as much building as before.

But the trend to change landuse is going to take much longer. Whereas some zoning and building codes are starting to change, these will not address the huge settlement issues, where riverine resilience has been castrated.

## POLICY PROCESSES AND THE CLEAN WATER ACT

### GETTING ON THE AGENDA

Given the importance of clean water to humans and the threats facing inland waters, it may seem odd that these issues are rarely on top of the political agenda. Some argue that issues compete in the public sphere in the context of a political resource limitation—the capacity of the public or legislature to address a finite number of issues. In this zero-sum view, to put an issue higher on the agenda, means a subsequent lowering of another issue. Interest groups battle for media exposure and promote their issues, at the expense (intentionally and unintentionally) of other issues.

Another aspect of the agenda is how public officials try to use issues as political office stepping stones. As policy entrepreneurs, political actors will use emerging issues to promote their own candidacy. Politicians in the late 1960s and early 1970s used the public's recognition of environmental issues to build their platforms. In this light, we should then wonder how issues reach the status of being in the public eye.

As it turns out, issues might remain an undercurrent for years or even decades before they become public-debated issues. In fact, issues have a cycle: of issue recognition, public outcry, and then the grinding public policy process aimed at addressing the issue. Environmental issues seem to have a particularly difficult time staying in the public arena. When the public becomes alarmed as a result of discovery and enthusiasm, environmental concerns are on the agenda for short periods of time as the public realizes the true costs, complexity of the issues, and hence moves on to a new issue, creating an attention cycle.

## WATER QUALITY AND THE FEDERAL POLLUTION WATER CONTROL ACT

Beginning with the Rivers and Harbors Act (1899), the United States federal government began regulating navigable waters. Although it was not designed to address water quality, the Supreme Court allowed Section 13 (Refuse Act) to regulate wastewater discharges from a steel mill in 1959. But even before this case, congress acknowledged that the growing problem of sewage and industrial discharge required more direct regulation on pollution.

In 1948 the Federal Water Pollution Control Act was passed and directed states to develop water quality standards and created funding mechanisms to construct treatment facilities. Even after several amendments, water quality improvements were minimal. For example, enforcement was based on the development of standards within states and territories—but by 1971 only half of the jurisdictions had been fully approved standards. Without standards, the federal government could not enforce the Water Pollution Act.

In 1969 two very important events happened: (1) the Santa Barbara Oil Spill (Fig. 13.11 and Fig. 13.12), the largest spill in US history until Exxon Valdez crashed and the Deep Water Horizon blow out and (2) the Cuyahoga River fire in Cleveland, Ohio. Although separated by several months, these events included televised images of sea birds covered in oil and images of fireboats trying to put out a fire burning the oil distillates and debris in a river. As images of these events were being channeled into homes, usually quiet committee hearings became media events. Thus water quality became part of the national political agenda.

Since Earth Day 1970 environmental issues have remained on the national agenda and the United States Congress passes numerous environmental laws: National Environmental Polity Act (1969), Lead-Based Paint Poison Act (1970); Clean Air Act (1970); Environmental Improvement Act (1970);

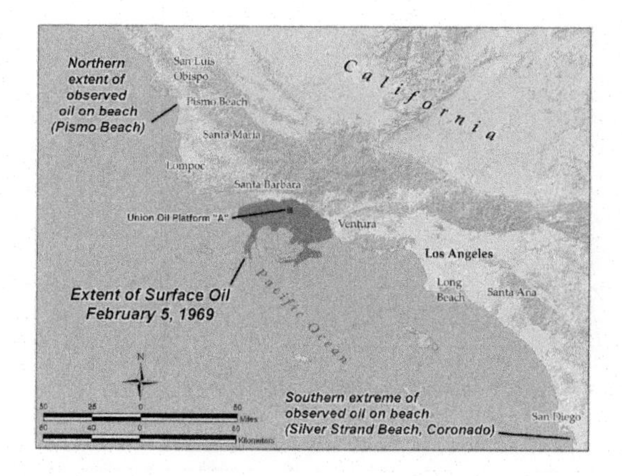


**FIGURE 13.11**

Map of oil spill extent off Santa Barbara coast, 1969 (Source: Wikipedia).

**FIGURE 13.12**

A duck covered in a thick coating of crude oil, picked up when it lighted on waters off Carpinteria State Beach in Santa Barbara County, Calif., after the oil spill in January 1969 (Source: Bettmann/Getty Images).

Federal Water Pollution Control Amendments (1972); Federal Insecticide, Fungicide, and Rodenticide Act (1972); Marine Protection, Research, and Sanctuaries Act (1972); Endangered Species Act (1973), and the Safe Drinking Water Act (1974).

Although the Federal Water Pollution Control Act was dramatically rewritten and given the name the Clean Water Act, it was not implemented until another set of amendments in 1978, but we will refer to the new name as we describe the changes made in 1972. First, there were aspirational goals, e.g., elimination of all discharges of pollutants into waters by the year 1985. These have not been met, but they have also never been removed from the act. Therefore the policy of the federal governments is to continue to make progress toward that goal. Furthermore, the CWA shifts away from 'tolerable limits,' to a permit system—dischargers must have permits to pollute water.

**FIGURE 13.13**

Storm drain by pass many environmental process and discharge runoff directly to surface waters, i.e.streams, wetlands, and lakes (Source: Resource Management Associates, Toms River, NJ).

On a more practical level, the law requires technology-based water treatment facilities. Thus in addition to relying on standards, discharges are to meet effluent limits based on the "best available technology" designed to improve water quality. Coupled with research and development funding and cost-sharing grants to build treatment facilities, the Clean Water Act drives technology development and its use in infrastructure to treat water pollution. For the most part, the CWA has resulted in substantial improvement of water quality.

Based on designated beneficial uses, e.g., drinking water, fishing, recreation, habitat, etc., each territory submits a list of water bodies that fail to meet their designated beneficial uses, or 'impaired'. In 1997, the EPA reported that 60% of all water bodies were impaired. The causes include uncontrolled urban runoff and nonpoint sources, such as agricultural runoff.

To address the stormwater problem, Congress defined industrial stormwater dischargers and municipal separate storm sewer systems (often called "MS4") as point sources and began requiring them to obtain NPDES permits for things like storm drains and such (Fig. 13.13). Although the permit exemption for agricultural discharges continued, Congress created several programs and grants, including a demonstration grant program at the EPA to expand the research and development of non-point controls and management practices. But since the late 2000s, another section of the Act gained attention to another program of the law—Section 303 and the Total Daily Maximum Load. In this case, the state or Environmental Protection Agency is required to estimate the amount of a pollutant that enters a water body by source and allocate how much each source is allowed to discharge, so the water body can meet its designated uses. Once this happens, the water body can be removed from the list of impaired water bodies.

## NORTH YUBA RIVER WATERSHED PROJECT

After experience, a number of catastrophic fires and their potential impacts on water resources, partnerships have begun addressing forest health issues.

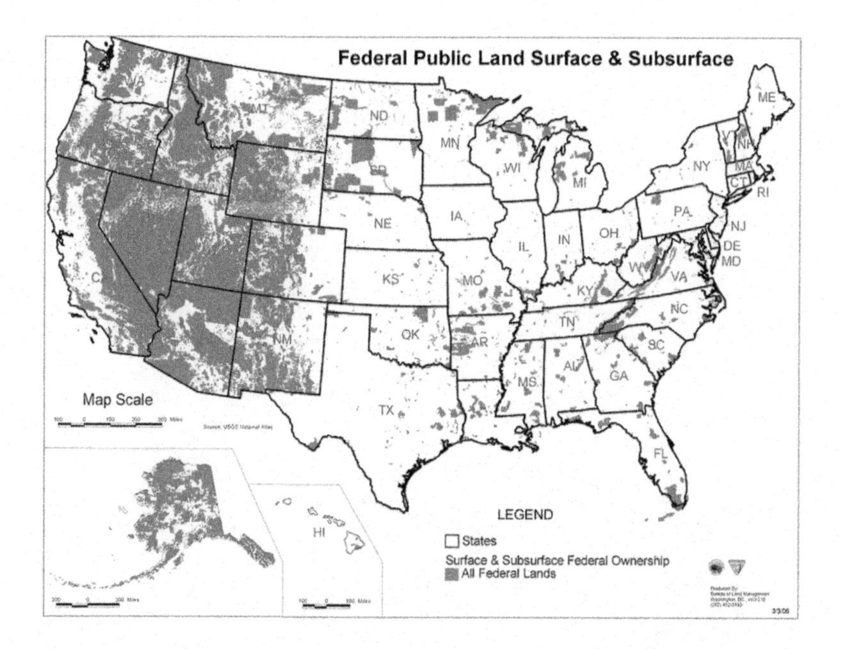


**FIGURE 13.14**

Ownership of Federal lands in the 50 states, including subsurface rights. This map includes as "federal lands," land held "in trust" for Native Americans, which are not generally considered "federal lands" in other contexts. The map highlights the limitations Native Americans have regarding their resources, in this case how the Department of Interior makes their land "invisible" (Source: Wikipedia).

In the West, the federal government owns most of the land (Fig. 13.14). In California, most of the remote areas are owned by the federal government and most of the Sierra Nevada Mountains is owned by the United States Forest Service. The management in the United States has changed dramatically with settlement. From a sophisticated use of controlled burning by native peoples, to a policy of fire exclusion, forest structure has changed dramatically in the West. In addition, forestry practices designed to obtain maximum yields planted trees at high densities. But after 1960, national forests have been managed under multiple-use policy, which calls for balancing timber yield with other values, such as wildlife, recreation, soil and water conservation, aesthetics, grazing, and wilderness protection. And after a number of drivers, imported lumber, over capitalization, and environmental concerns, the silvaculture industry has been dramatically reduced.

However, headwater forests in California are in poor condition, subject to catastrophic fires, thus the Yuba River Restoration Project has begun tree thinning, meadow restoration, prescribed burning, and invasive species management—all specifically designed to reduce the risk of severe fire, improve watershed health, and protect water resources. Using an adaptive management framework, researchers from several universities will monitor the impacts on water supply and other ecosystem services, providing data to quantify the benefits of restoration activities undertaken. It is hoped that by documenting the success of this project (reduce fire risks, water quality protections) future investment in forest restoration will improve overall forest health.

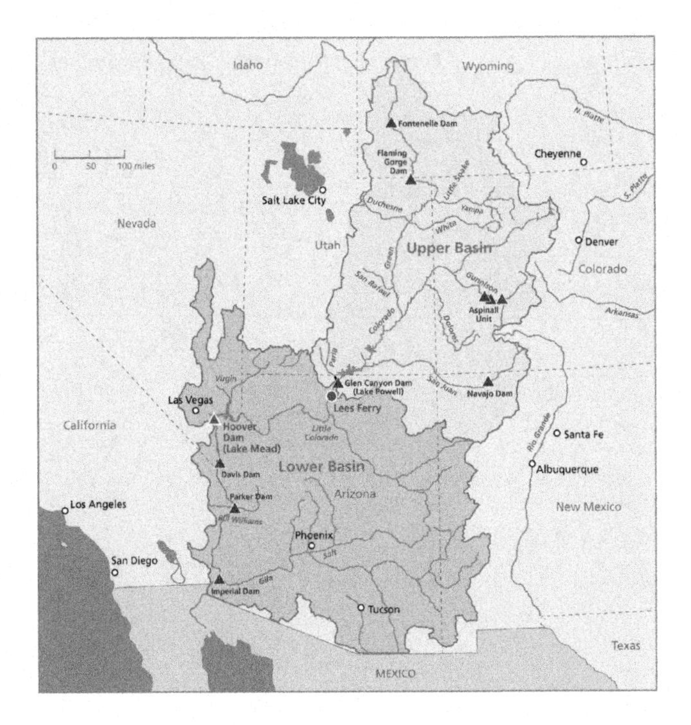


**FIGURE 13.15**

Map of the Colorado River basin and watershed.

## POLITICAL AND WATERSHED BOUNDARIES

### TRANSBOUNDARY WATERS: COLORADO RIVER

Water politics are a function of the availability of water and water resources, human development, and its necessity for all life forms. As such, the conflict and cooperation between jurisdiction of water resources transcend governance structures.

In 1922 seven states agreed on how the Colorado River would be allocated in what is known as the Colorado Compact. The compact allowed dividing the river into an upper (Colorado, New Mexico, Utah, and Wyoming) and lower basin (Nevada, Arizona and California); the population of each part would have about equal withdrawal shares (Fig. 13.15). More specifically, state level allotments were made in 1928 (for the lower portion) and 1948 (for the upper basin). These agreements also include some allocation to Mexico, but it was far from adequate to sustain the delta habitat in the Sea of Cortez. To partially address this problem, the United States and Mexico signed an agreement, "Minute 319" in 2012, which defined how surpluses or drought might influence allocations, based on the surface elevation of Lake Mead.

The compact enabled the widespread irrigation of the Southwest, as well as the subsequent development of state and federal water-works projects under the United States Bureau of Reclamation. Such projects included the Hoover Dam and Lake Powell.

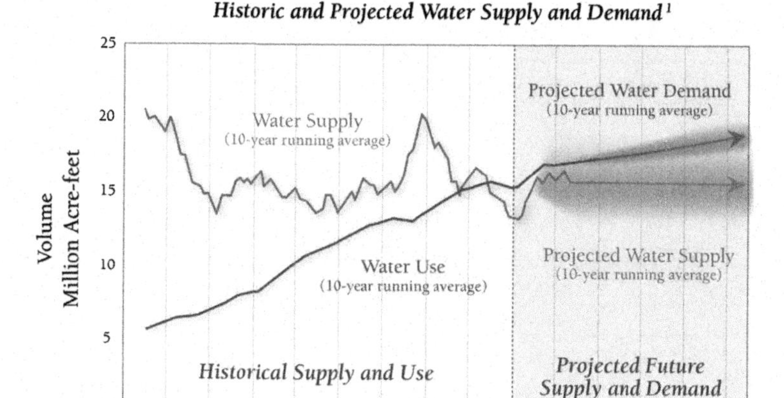


**FIGURE 13.16**

Historic and projected water supply and demand in the Colorado River (U.S. Department of the Interior, 2012).

The compact has received strong criticism, in the wake of a protracted decrease in precipitation in the region. The compact allocations were established based on average flow between 1905 and 1922. But based on Tree Ring data, this period is associated with abnormally high precipitation (Fig. 13.16). With more typical rainfall, the river has been "over allocated," which may be further exacerbated with drought and climate change. An agreement was reached by seven Western states in 2019 on a plan to manage the Colorado River amid a 19-year drought, but it is likely that this will require continual negotiation.

## KLAMATH RIVER: FISHERIES, FARMING, AND POLITICS OF CONFLICT

Meanwhile in the far north of the state, the Klamath River experienced the largest salmon kill in the history of the Western United States in 2002. According to the official estimate of mortality, about 34,000 fish died. Some estimated over 70,000 adult chinook salmon (*Oncorhynchus tshawytscha*). With a relatively high number of migrants, warm water temperatures, low waterflow because of a drought and irrigation withdrawals from the river, the chances for a successful spawning severely constrained.

With fish mortality in the background, hostilities between stakeholders, such as farmers, Native Americans and fishermen, and the federal government has been a reoccurring problem. At one point, in 2001, a group of farmers revolted and stormed the irrigation canals after the federal government cut off their water deliveries.

The Klamath River originates from the Upper Klamath Lake, a large but shallow lake (Fig. 13.17). But the lake regularly experiences algal blooms and low inputs.

**FIGURE 13.17**

Image of Klamath River wWhatershed, which crosses several jurisdictions.

In this relatively remote area of the country, the precursor to the Bureau of Reclamation constructed a system of dams, canals, diversions, and tunnels that ultimately provided irrigation water for 210,000 acres. With a series of hydroelectric dams, the Klamath Project provides cheap electricity to homesteaders in one of the last major areas of federally sponsored settlements.

Four dams in the Klamath River have become a source of conflict: Copco 1 (1918), Copco 2 (1925), J.C. Boyle (1958) and Iron Gate (1962). According to their owner, PacificCorp, the dams are outdated and hardly produce any electricity. They do not store or divert water and provide no flood control. But they have had a dramatic impact on the fisheries. After the dams were completed, spawning declined from 100,000 fish to 4000 individuals.

The Karuk, a native American tribe in the region has traditionally relied on the fishery as a major food source. However, the number of fish cannot sustain the native peoples. Because the Karuk suffer from high rates of food insecurity, they have launched an ecocultural restoration project to improve health outcomes. Removal of the dams could provide an important component of this project.

PacifiCorp supports the goal to remove them. To maintain the dams, PacifiCorp's relicensing would require the construction of a fish passage around the dams, a cost of $300 million. Along with another $150 million to retrofit the dam, the cost to tear the dam down would be cheaper.

**FIGURE 13.18**

Image of trash trapped behind a "trash dam" at Goat Canyon (Source: Surfrider).

Meanwhile, significant political opposition still remain. Tribal rights to water are senior to the farmers. In years of drought, tribes began exercising these rights and farmers worried that they might lose access to water allocations. But through a set of mediations, farmers agreed to compromise some volume if tribes would allow yearly deliveries. The agreement also set up funding for ecosystem restoration and fish recovery.

A second agreement was made to remove the dams, where the cost would be split between PacifiCorp and California. In addition, PacifiCorp would not be held accountable for the sediment that would be released with the dam removal.

More than 40 parties, including the Interior Department, California, Oregon, PacifiCorp, three tribes, environmental groups, fishermen and associations, representing 94% of the Klamath Project irrigators, signed the agreement. But federal support was required, and the House blocked any progress. Without federal approval, the agreement was split into two parts, and dam removal was moving forward until endorsement was rescinded by the United States Secretary of the Interior, David Bernhardt in 2019.

Currently, the dams are being run at reduced capacity to limit their impacts on the Lost River sucker (Deltistes luxatus) and shortnose sucker (Chasmistes brevirostris), and the water allocations agreements are in dispute. Meanwhile, some political leaders continue to oppose the project, citing improper government oversight, failure to solve outstanding water issues, and the impacts on electrical rate payers. Without a doubt these concerns sound compelling, but strike using a different set of values. Managing inland waters to build ecological and social resilience requires that a diverse set of values are carefully addressed.

## TIAJUANA RIVER AND INTERNATIONAL RELATIONS

The Tijuana River watershed extends about 150 km inland of the Pacific Ocean and most of it is in Mexico. However, the river is known for its pollution. Although the river had low toxicity at baseflow, the toxicity increased dramatically with the initial periods of precipitation-generated runoff. As runoff continued, toxicity declined, suggesting that chemicals that account for storm toxicity are depleted from

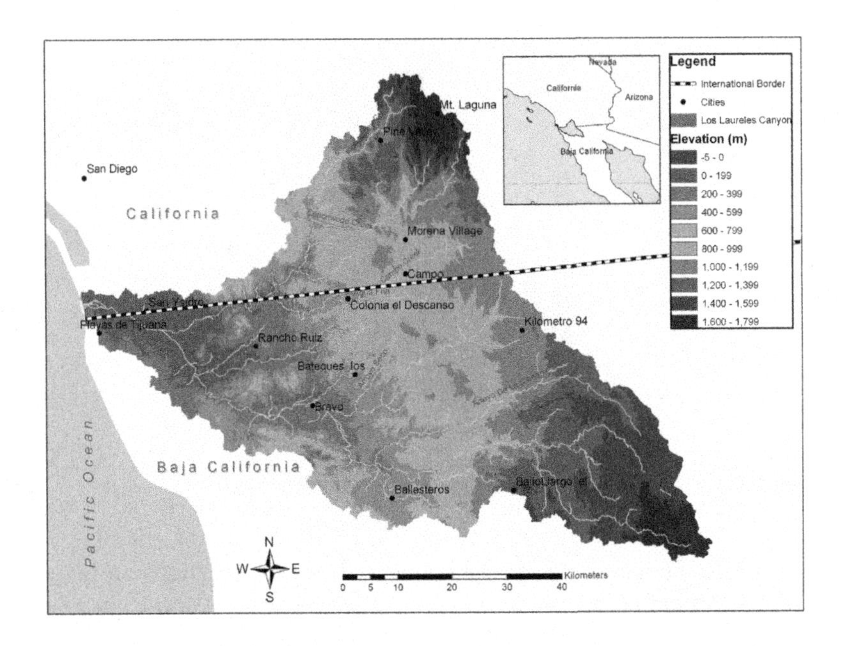


**FIGURE 13.19**

Image of Tiajuana River Watershed (Source: Gamble and Wilken-Robertson, 2008).

the landscape. Based on these results and concerns about untreated sewage and trash (Fig. 13.18) in the river, the various municipalities and resource agencies have sued, not Mexico, but the International Boundary Water Commission. The commission has negotiated a wide range of agreements between the two countries, including agreements about water quality. With this lawsuit stakeholders hope that prior agreements will be enforced or the commission will negotiate new agreements that improve the waters that enter the United States (Fig. 13.19). Unfortunately, given the complexity of immigration, poverty, industry supplying U.S. demand, and drug trafficking, this area of Mexico has limited capacity to address these issues. That being said, we find that the link to water quality is directly related to the well-being of the residents. With this in mind, state and federal agencies might approach water quality issues on the border using a more sophisticated approach.

## BENEFITS AND COSTS

### BENEFIT AND COST ANALYSIS

As part of a critique of the command-and-control approach to environmental regulation, Economists advocated for subjecting regulations to benefit cost analyses (BCA). At one end of the spectrum, these BCAs could be used to assess the efficiency of government regulation to maximize the benefits relative to costs, at the other extreme BCAs could be used as a political tool to promote or discredit any project by selecting assumptions that justify forgone conclusions.

**Table 13.1 Components of a benefits and costs analysis (BCA) for dams.**

| Advantages | Relative Complexity of Assigning a Cost |
|---|---|
| Reduces consumption of fossil fuels for electricity production | Low |
| Reduction production of greenhouse gases, such as $CO_2$ | Low |
| Can prevent uncontrolled flooding | Medium |
| Provides water for irrigation | Low |
| Creates areas for certain types of recreation, such as boating and fishing | Low |
| Is a renewable energy source | Low |
| **Disadvantages** | |
| Emits fossil fuels during construction and cement curing process | Low |
| Increase in $CH_4$ emissions in reservoir | Low |
| Sediment accumulates over time, decreasing effectiveness and useful life | Low |
| Large-scale wildlife habitat destruction due to river valley flooding | Medium |
| Dam construction forces people to leave their homes if they live in or near the flooded valley | High |
| Very expensive to build | Low |
| Reduces areas for certain types of recreation, such as fishing, camping, hunting, and hiking | Medium |
| Emits fossil fuels with new recreation types | Medium |
| Interferes with natural flow of water through environment | High |
| Reduces carbon sequestration potential | Medium |
| Natural fisheries are affected, harms the livelihood of people who rely on those fisheries to make a living | High |
| Requires maintenance | Low |
| Can fail catastrophically | High |

The United States Army Corp of Engineers is famous for their poor track record of developing BCAs. In general, their BCAs have been regarded over the years as "without merit in decision making," and so poorly done that not initiating the undertaking would be beneficial.

Even without an explanation of how BCAs are done, we should appreciate how calculating the complexity of BCAs lands in a biased fashion on the disadvantage side (Table 13.1). Based on this, any agency doing a BCA can easily skew the results of an analysis.

## ASSESSING NATURE'S CONTRIBUTION

What is the value of nature? How do we measure that? Would this change the relative merits between development choices? These are the questions posed by Economists as they struggle to "internalize" the environment in economics models. For some, sound policy relies on determining the "production functions" that describe how ecosystems generate services. The production functions could document service flows occurring locally (e.g., recreation, subsistence fishing), across regions (e.g., water purification and flood control), and globally (e.g., climate change mitigation).

**Table 13.2 Estimated value (USD) of wetland ecosystem services.**

| | Mangrove | Non-Vegetated Sediment | Salt/Brackish Marsh | Freshwater Marsh | Freshwater Woodland | TOTAL |
|---|---|---|---|---|---|---|
| N America | 30,014 | 550,980 | 29,810 | 1728 | 64,315 | 676,846 |
| Latin America | 8445 | 104,782 | 3129 | 531 | 6125 | 123,012 |
| Europe | 0 | 268,333 | 12,051 | 253 | 19,503 | 300,141 |
| Asia | 27,519 | 1,617,518 | 23,806 | 29 | 149,597 | 1,818,534 |
| Africa | 84,994 | 159,118 | 2466 | 334 | 9775 | 256,687 |
| Australasia | 34,696 | 147,779 | 2120 | 960 | 83,907 | 269,462 |
| TOTAL | 185,667 | 2,848,575 | 73,382 | 3836 | 333,223 | 3,444,682 |

However, collective decision-making usually treat ecosystem services as "free." But as ecosystem capital becomes scarcer, the value of these services has become an important part of our decision-making. By quantifying these values, we can begin to appreciate their value relative to environmental options. These data have been used to document the value of many types of ecosystems (Table 13.2).

For many, these calculations provide compelling evidence that the Earth's ecosystems have economic worth that is poorly captured in typical business models. However, for others the act of putting a price tag on nature is risky. By monetizing nature, the value of nature becomes finite, therefore, "buyable." Furthermore, the values associated with ecosystems are culturally defined. For example, the value of Lake Victoria for tourism versus subsistence farming are radically different. Or the value of a river, where ancestors were born and died may have deep cultural meaning that may be poorly captured in monetary value. As many will recognize the noninstrumental value of nature, e.g., nature has intrinsic value.

As an alternative, the Nature's Contribution to People is a framework that attempts to better capture local and indigenous-knowledge holders with a pluralistic approach (Fig. 13.20).

In the context of environmental conflicts around scarcity, decision-making legitimacy may be associated with positive environmental outcomes. Thus the use of Nature's Contribution to People may improve assessments because it often acts as a voice of under-represented voices and can increase the effectiveness and legitimacy of the policy decision-making and governance itself.

## WOTUS AND THE WRONG RIGHTS CLAIMS

Since the passage of the Clean Water Act, the jurisdiction of the law has been a source of ambiguity. As conceptualized the act applies to all "waters of the United States" or WOTUS. But what defines these waters? In the first incarnations, waters were associated with navigation, but with the passage of the federal Water Pollution Control Act the focus expanded to include waters that crossed state boundaries. The federal government mediated state conflicts with respect to water, perhaps a downstream user wanted to compel an upstream polluter to stop polluting. The role of the federal government as a way to resolve state-state conflicts was easy to justify in the federalist system of the United States. But as we learned, states did not do much on their own to protect water quality within or across state boundaries. But using a general concept that navigable waters were covered by the federal government through a different set of case laws, the federal government associated WOTUS with navigable waters.

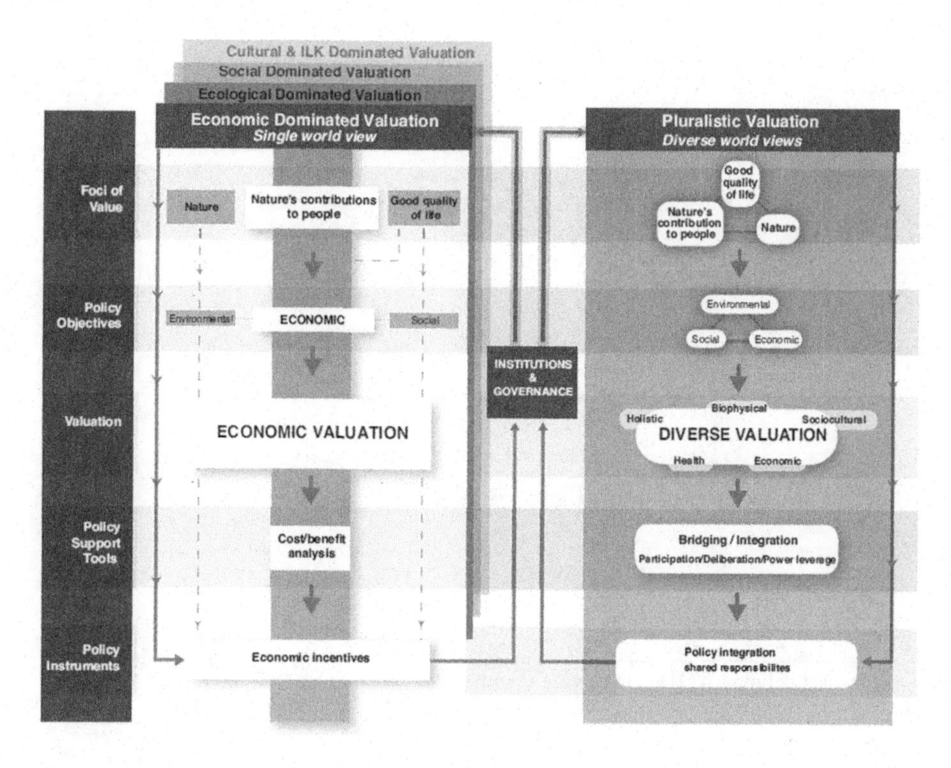


**FIGURE 13.20**

A stylized illustrative framework of contrasting approaches to the process of valuation (Source: Pascual et al., 2017). The right side panel emphasizes the importance of a pluralistic valuation approach, compared with value monism or unidimensional valuation approaches to human–nature relationships represented in the left side panel.

The USEPA and USACOE began asserting their jurisdiction to wetlands beyond a traditional definition of navigation, based on Supreme Court cases, and the Corp developed a migratory bird rule to define wetlands. But in successive decisions, the definition has been contracting again (Fig. 13.21), but often leaving the Corp with unclear directions. To develop consistent guidelines, the Obama administration attempted to clarify the rules and expanded WOTUS to what they considered to be the intent of the CWA.

Anyone who has worked on water or wetland issues would appreciate how these guidelines would ignite a firestorm of protests and legal challenges. First Congress tried to change the rule through legislation, something that has been tried for two decades by both Democratic and Republican party members. By eclipsing Congress's attempt to change the rules directly by threatened a veto, stakeholders mounted a successful campaign to have the rules 'suspected' in the courts.

Unfortunately framed as government overreach into waters that are on private properties, e.g., unnamed ditches, stakeholders have missed an important aspect of the goals of the CWA and the intent in identifying WOTUS. Waters are not owned by anyone, although rights to water are clearly granted. But along with these rights are responsibilities. The CWA articulates responsibilities that the states are

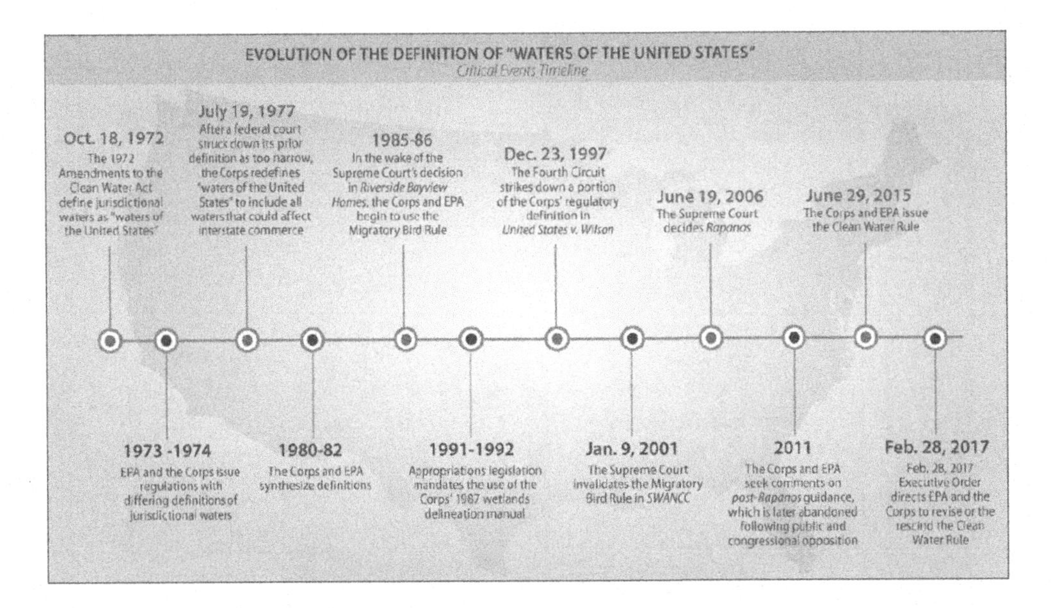


**FIGURE 13.21**

Timeline of major cases defining WOTUS (Source: Congressional Research Service).

required to address, e.g., prohibition of toxic discharges. If the waters are going to remain outside the jurisdiction of the CWA, then private landowners may actually have greater liability with respect to their discharges.

Finally, states can claim their waters too. For example, in California, the waters of the state are broadly construed to include all waters within the state's boundaries, whether private or public, including in both natural and artificial channels. As such, the state's authority to regulate water quality is not limited to the vagaries of the federal definitions. Whereas other states, such as Arizona, rely on federal rules to protect their waters, if WOTUS remains limited, it is likely that differences in water quality among states will increase.

## NATURAL RESOURCE RESTORATION AND HUMAN SERVICES

Developing positive views and political support to improve the health of inland waters requires a better understanding of human motivations and interests. For example, research has shown that restoration projects are more likely to gain public support if they simultaneously increase important human services that natural resources provide to people. Thus it is important to consider how societal functions (e.g., flood control, water quality) depend on functional wetlands and floodplains.

By explicitly and scientifically melding societal and ecosystem perspectives, a project along the Sacramento River and floodplain included restoration actions that simultaneously improved both ecosystem health and the services (e.g., flood protection and recreation) that ecosystem provides to people. In addition, by directly engaging local stakeholders to formulate, implement, and interpret the studies, the project leaders promoted a high level of trust that ultimately translated into better support for the project.

This example requires careful planning, attention to process, personal relationships with stakeholders, and skilled leadership. The success of maintaining and improving inland waters may rely more on our capacity to understand and communicate the biophysical processes and thoughtful engagement with stakeholders at every level.

## INTRINSIC VALUE AND ECOSYSTEMS SERVICES: A FALSE DICHOTOMY?

### LINKING SPECIES AND ECOSYSTEM FUNCTIONS

Although biodiversity is one characteristic of ecosystems and may have intrinsic value, biodiversity will also have functional value in ecosystems. After describing some of the complex diversity at the microscopic scale, any given taxa's ecological roles may seem far from unique: there are many detritivores, bacteria predators, and autotrophs. In other words, evolutionary divergences seem to create ecological redundancies. Does each species have unique functional value in the ecosystem?

Within regulatory frameworks (e.g., Endangered Species Act), Environmental Scientists attempt to discern the ecological functions of species diversity as a management issue. In the tension between endemic and introduced species, we find compelling evidence where introductions do not function as ecological redundancies, but alter the ecosystem in dramatic ways. As we learn more about ecosystems, we find that subtle differences within functional groups can have important community and ecosystem-wide processes and impact. Even diversity within a population can translate to differences in community or ecosystem processes.

Thus the question about the ethical value of redundancy is moot. Redundancy is a matter of scale and precision. As we learn more, redundancy is an imagined category, useful for research, but not for ethical decisions about the value of a taxon.

### RELATIONAL PLURALISM

The distinction between instrumental value and intrinsic value has become a simple trope to highlight differences in motivations. In some circles, one of these values is rated as valid and in other circles the value is seen as problematic. After years of entertaining scholarly arguments, the pragmatic view usually settles on both having value, in different ways for different people, and these vary within individuals based on the issue of concerns and will change over time. In fact, some people inconsistently apply these values.

In addition, some might argue that another category exists based on relational views of the world. We consider it environmentally worthy to protect as respect to our ancestors or our descendents, or people in the next village. For example, in a survey asking why residents care about the water in Colombia, the response was "downstream people drink water, and here the river is born. We much take care of the water in benefit of the downstream people."

To end our chapter, we should think to embrace all motivations folks have to protect inland waters. Farmers manage their fields every day, then know why water supply and quality are important. Mothers and fathers think carefully about the source of water their children drink. Fish and snails and algae help

**Table 13.3 Modified "Golden rules" in SDRM as proposed by Sayers et al. (2017).**

| "Golden Rule" | Status |
|---|---|
| 1. Define explicit long-term goals | |
| 2. Assess system behavior and uncertainties | |
| 3. Develop effective participation | |
| 4. Communicate risk | |
| 5. Acknowledge inherent controversies and trade-offs | |
| 6. Use limited resources effectively | |
| 7. Implement a portfolio of measures | |
| 8. Continuous review and adaptive | |

maintain the waters for farmers, parents, and children. Our job is to make all of these activities as easy as possible.

# NEXT STEPS

## CHAPTER STUDY QUESTIONS

1. Describe political conflicts and legal context for water supply allocations.
2. Explain how endangered species protections influence the ecology of inland waters.
3. Describe the policy tools used to resolve conflicts and protect water supplies, water quality, and aquatic ecosystems.

## LITERATURE RESEARCH ACTIVITIES

According to Sayers et al. (2017), the rules shown in Table 13.3 should be used to guide Strategic Drought Risk Management (SDRM). Create a table similar to Table 13.3, and evaluate the progress for drought management in a country, state, region, or jurisdiction of your interest.

CHAPTER

# UNFINISHED BUSINESS: FUTURES AND INLAND WATERS

14

*Water is taken for granted, only when there is less than too much and more than not enough.*
**Marc Los Huertos**

## CONTENTS



The aspirations to protect inland waters as a resource and as habitat must be reconciled with the socioeconomic realities on every corner of the planet. The efforts to accomplish these goals will require a robust acknowledgment of the issues that face inland waters and the political will to address them. Neither of these exists today, but there are numerous components that can be used as a foundation for better environmental outcomes for inland waters.

This chapter suggests that a robust approach can be built on three components.

After reading this chapter you should be able to

1. Describe three types of activities to protect and restore inland waters.

## THREE SUGGESTIONS TO BUILD ON

The following suggestions provide general components that when combined provide a path to improve the management of inland waters:

***Know your waters & watershed*** Learning about local and regional water sources and discharges and the land use patterns in a watershed can be valuable to residents, empowering them to protect the resources, e.g., Science in Your Watershed (USGS).

In addition, citizen science activities produce data and engage communities to evaluate the health of their Aquatic Systems. Examples of these include Secchi Dip-In, First Flush Water Quality Monitoring, Migratory Dragonfly Partnership, and various stream, wetland, lake clean up days. More sustained activities include adopting a water body. For example, "adopting a creek" programs are widespread and popular.

Ecology and Management of Inland Waters. https://doi.org/10.1016/B978-0-12-814266-0.00028-3

***Political engagement*** Monitoring the activities of government at local, regional, state, and federal levels is key to build the political will to protect inland waters.

***Education*** Passing our knowledge and environmental values to the next generation is key to sustaining momentum to protect and restore inland waters. These include formal and informal experiences, and are applicable to a wide range of peoples.

These suggestions are purposely vague, but provide structure for those concerned about the sustainability of our water use to support human and nonhuman activity.

And it is up to each of us to find the balance between these strategies—but when we do, we become part of a growing movement aimed towards better ecological outcomes for our inland waters.

# References


Atkinson, S.D., Bartholomew, J.L., 2010. Disparate infection patterns of Ceratomyxa shasta (Myxozoa) in rainbow trout (Oncorhynchus mykiss) and Chinook salmon (Oncorhynchus tshawytscha) correlate with internal transcribed spacer-1 sequence variation in the parasite. International Journal for Parasitology 40, 599–604. http://www.sciencedirect.com/science/article/pii/S0020751909004019.

Baker, B.J., Banfield, J.F., 2003. Microbial communities in acid mine drainage. FEMS Microbiology, Ecology 44, 139–152. https://doi.org/10.1016/S0168-6496(03)00028-X.

Baumgartner, A., Reichel, E., 1975. The World Water Balance: Mean Annual Global, Continental and Maritime Precipitation, Evaporation and Run-Off. Elsevier Scientific Pub. Co., Amsterdam. https://ccl.on.worldcat.org/oclc/2417333.

Bayley, P.B., 1995. Understanding large river: floodplain ecosystems. BioScience 45, 153–158.

Bilby, R.E., Bisson, P.A., 1992. Allochthonous versus autochthonous organic matter contributions to the trophic support of fish populations in clear-cut and old-growth forested streams. Canadian Journal of Fisheries and Aquatic Sciences 49, 540–551. https://doi.org/10.1139/f92-064.

Bisson, P.A., Montgomery, D.R., Buffington, J.M., 2017. Valley segments, stream reaches, and channel units, third edition. In: Methods in Stream Ecology, Volume 1. Elsevier, pp. 21–47.

Boddey, J.A., 2017. Plasmepsins on the antimalarial hit list. Science 358, 445–446.

Brett, M.T., Lubnow, F.S., Villar-Argaiz, M., Müller-Solger, A., Goldman, C.R., 1999. Nutrient control of bacterioplankton and phytoplankton dynamics. Aquatic Ecology 33, 135–145.

Brooker, M., 1985. The ecological effects of channelization. The Geographical Journal 151, 63–69.

Brooks, J.L., Dodson, S.I., 1965. Predation, body size, and composition of plankton. Science 150, 28–35. http://www.jstor.org/stable/1717947.

Burns, D.A., Fenn, M.E., Baron, J.S., Lynch, J.A., Cosby, B.J., 2011. National Acid Precipitation Assessment Program Report to Congress: an Integrated Assessment. Technical report. Council National Science Technology, Washington, D.C. http://pubs.er.usgs.gov/publication/70007175.

Carlton, W., 1999. Assessment of Radionuclides in the Savannah River Site Environment Summary. Technical report. Savannah River Site, US.

Carpenter, S.R., Kitchell, J.F., Hodgson, J.R., 1985. Cascading trophic interactions and lake productivity. Bioscience 35, 634–639. http://www.jstor.org/stable/1309989.

Ceballos, G., Ehrlich, P.R., Barnosky, A.D., García, A., Pringle, R.M., Palmer, T.M., 2015. Accelerated modern human-induced species losses: entering the sixth mass extinction. Science Advances 1, e1400253.

Colmer, T., Voesenek, L., 2009. Flooding tolerance: suites of plant traits in variable environments. Functional Plant Biology 36, 665–681.

Committee F.G.D., 2013. Classification of Wetlands and Deepwater Habitats of United States. FGDC-STD-004-2013, second edition. Wetlands Subcommittee, Federal Geographic Data Committee and U.S. Fish and Wildlife Service, Washington, DC.

Cowardin, L.M., Carter, V., Golet, F.C., LaRoe, E.T., 1979. Classification of Wetlands and Deepwater Habitats of the United States. US Department of the Interior, US Fish and Wildlife Service.

Dettinger, M., 2011. Climate change, atmospheric rivers, and floods in California – a multimodel analysis of storm frequency and magnitude changes. JAWRA Journal of the American Water Resources Association 47, 514–523.

Dillon, P.J., Rigler, F.H., 1974. The phosphorus–chlorophyll relationship in lakes. Limnology and Oceanography 19, 767–773. http://www.jstor.org/stable/2834379.

Dodds, W.K., 2002. Freshwater Ecology: Concepts and Environmental Applications. Academic Press.

Driscoll, C., Lawrence, G., Bulger, A.J., Butler, T., Cronan, C.S., Eagar, C., Lambert, K., Likens, G.E., Stoddard, J., Weathers, K., 2001. Acidic deposition in the northeastern United States: sources and inputs, ecosystem effects, and management strategies. BioScience 51, 180–198.

Duffy, M.A., Ochs, J.H., Penczykowski, R.M., Civitello, D.J., Klausmeier, C.A., Hall, S.R., 2012. Ecological context influences epidemic size and parasite-driven evolution. Science 335, 1636–1638. http://www.sciencemag.org/content/335/6076/1636.abstract.

Dvořák, P., Poulíčková, A., Hašler, P., Belli, M., Casamatta, D.A., Papini, A., 2015. Species concepts and speciation factors in cyanobacteria, with connection to the problems of diversity and classification. Biodiversity and Conservation 24 (4), 739–757. https://doi.org/10.1007/s10531-015-0888-6.

Eagles-Smith, C.A., Suchanek, T.H., Colwell, A.E., Anderson, N.L., 2008. Mercury trophic transfer in a eutrophic lake: the importance of habitat-specific foraging. Ecological Applications 18, A196–A212. http://www.esajournals.org/doi/abs/10.1890/06-1476.1.

Edmondson, W.T., Lehman, J.T., 1981. The effect of changes in the nutrient income on the condition of Lake Washington. Limnology and Oceanography 26, 1–29. http://www.jstor.org/stable/2835803.

Eiting, T.P., Smith, G.R., 2007. Miocene salmon (Oncorhynchus) from Western North America: Gill Raker evolution correlated with plankton productivity in the Eastern Pacific. Palaeogeography, Palaeoclimatology, Palaeoecology 249, 412–424. http://www.sciencedirect.com/science/article/pii/S0031018207000910.

Evermann, B.W., Goldsborough, E.L., 1907. The Fishes of Alaska. US Government Printing Office. 624.

Ferguson, B.K., 2016. Toward an alignment of stormwater flow and urban space. JAWRA Journal of the American Water Resources Association.

Findlay, D.L., Kasian, S.E.M., 1987. Phytoplankton community responses to nutrient addition in Lake 226, Experimental Lakes Area, northwestern Ontario. Canadian Journal of Fisheries and Aquatic Sciences 44, 35–46.

Firestone, M., Davidson, E., 1989. Microbiological basis of NO and N2O production and consumption in soil. In: Exchange of Trace Gases between terrestrial Ecosystems and the Atmosphere, vol. 47.

Fisher, S.G., Gray, L.J., Grimm, N.B., Busch, D.E., 1982. Temporal succession in a desert stream ecosystem following flash flooding. Ecological Monographs 52, 93–110. http://www.jstor.org/stable/2937346.

Flynn, T.M., O'Loughlin, E.J., Mishra, B., DiChristina, T.J., Kemner, K.M., 2014. Sulfur-mediated electron shuttling during bacterial iron reduction. Science 344, 1039–1042.

Fong, P., Foin, T.C., Zedler, J.B., 1994. A simulation model of lagoon algae based on nitrogen competition and internal storage. Ecological Monographs 64, 225–247. http://www.esajournals.org/doi/abs/10.2307/2937042.

Friedman, M., Keck, B.P., Dornburg, A., Eytan, R.I., Martin, C.H., Hulsey, C.D., Wainwright, P.C., Near, T.J., 2013. Molecular and fossil evidence place the origin of cichlid fishes long after gondwanan rifting. Proceedings of the Royal Society B 280, 20131733.

Gamble, Lynn H., Wilken-Robertson, Michael, 2008. Kumeyaay cultural landscapes of Baja California's Tijuana River Watershed. Journal of California and Great Basin Anthropology, 127–152.

Garcia, X.-F., Schnauder, I., Pusch, M., 2012. Complex hydromorphology of meanders can support benthic invertebrate diversity in rivers. Hydrobiologia 685, 49–68.

Glazier, D.S., Deptola, T.J., 2011. The amphipod gammarus minus has larger eyes in freshwater springs with numerous fish predators. Invertebrate Biology 130, 60–67.

Go, Y.Y., Balasuriya, U.B., Lee, C.-k., 2014. Zoonotic encephalitides caused by arboviruses: transmission and epidemiology of alphaviruses and flaviviruses. Clinical and Experimental Vaccine Research 3, 58–77.

Grantham, T.E., Viers, J.H., 2014. 100 years of California water rights system: patterns, trends and uncertainty. Environmental Research Letters 9, 084012.

Griffith, M., Perry, S., Perry, W., 1994. Secondary production of macroinvertebrate shredders in headwater streams with different baseflow alkalinity. Journal of the North American Benthological Society 13, 345–356.

Grimaldi, D.A., 2010. 400 million years on six legs: on the origin and early evolution of Hexapoda. Arthropod Structure & Development 39, 191–203. http://www.sciencedirect.com/science/article/pii/S1467803909000747.

Grindler, N.M., Allsworth, J.E., Macones, G.A., Kannan, K., Roehl, K.A., Cooper, A.R., 2015. Persistent organic pollutants and early menopause in us women. PLoS ONE 10, e0116057.

Grossart, H.-P., Rojas-Jimenez, K., 2016. Aquatic fungi: targeting the forgotten in microbial ecology. Current Opinion in Microbiology 31, 140–145.

Gunderson, L.H., Light, S.S., Holling, C., 1995. Lessons from the everglades. BioScience 45, S66–S73.

Hall, C.J., 2002. Nearshore marine paleoclimatic regions, increasing zoogeographic provinciality, molluscan extinctions, and paleoshores, California: Late Oligocene (27 Ma) to late Pliocene (2.5 Ma). Geological Society of America, Special Paper 357. pp. v–489.

Harada, K.-i., Oshikata, M., Uchida, H., Suzuki, M., Kondo, F., Sato, K., Ueno, Y., Yu, S.-Z., Chen, G., Chen, G.-C., 1996. Detection and identification of microcystins in the drinking water of Haimen City, China. Natural Toxins 4, 277–283. https://doi.org/10.1002/(SICI)(1996)4:6<277::AID-NT5>3.0.CO;2-1.

He, D., Fiz-Palacios, O., Fu, C.-J., Fehling, J., Tsai, C.-C., Baldauf, S.L., 2014. An alternative root for the eukaryote tree of life. Current Biology 24, 465–470.

Heimann, K., Cirés, S., 2015. Chapter 33 – n2-fixing cyanobacteria: ecology and biotechnological applications. In: Kim, S.-K. (Ed.), Handbook of Marine Microalgae. Academic Press, Boston, pp. 501–515. http://www.sciencedirect.com/science/article/pii/B9780128007761000339.

Hohmann-Marriott, M.F., Blankenship, R.E., 2011. Evolution of photosynthesis. Annual Review of Plant Biology 62, 515–548. http://www.annualreviews.org/doi/abs/10.1146/annurev-arplant-042110-103811.

Holland, H.D., 2006. The oxygenation of the atmosphere and oceans. Philosophical Transactions of the Royal Society B: Biological Sciences 361, 903–915.

Howard, J.K., Cuffey, K.M., 2003. Freshwater mussels in a California North Coast Range river: occurrence, distribution, and controls. Journal of the North American Benthological Society 22, 63–77. http://www.jstor.org/stable/1467978.

Huffaker, C., et al., 1958. Experimental studies on predation: dispersion factors and predator–prey oscillations. California Agriculture 27, 343–383.

Hughes, K.A., Houde, A.E., Price, A.C., Rodd, F.H., 2013. Mating advantage for rare males in wild guppy populations. Nature 503, 108.

Humphries, P., Keckeis, H., Finlayson, B., 2014. The river wave concept: integrating river ecosystem models. BioScience 64 (10), 870–882.

Hutchinson, G.E., 1965. The Ecological Theater and the Evolutionary Play. Yale University Press.

Ishikawa, A., Kabeya, N., Ikeya, K., Kakioka, R., Cech, J.N., Osada, N., Leal, M.C., Inoue, J., Kume, M., Toyoda, A., Tezuka, A., Nagano, A.J., Yamasaki, Y.Y., Suzuki, Y., Kokita, T., Takahashi, H., Lucek, K., Marques, D., Takehana, Y., Naruse, K., Mori, S., Monroig, O., Ladd, N., Schubert, C.J., Matthews, B., Peichel, C.L., Seehausen, O., Yoshizaki, G., Kitano, J., 2019. A key metabolic gene for recurrent freshwater colonization and radiation in fishes. Science 364, 886–889. https://science.sciencemag.org/content/364/6443/886.

Izaguirre, G., Jungblut, A.D., Neilan, B.A., 2007. Benthic cyanobacteria (Oscillatoriaceae) that produce microcystin-LR, isolated from four reservoirs in southern California. Water Research 41, 492–498. URL <Go to ISI>://000244108500026.

Johnson, S.W., Neff, A.D., Lindeberg, M.R., 2015. A handy field guide to the nearshore marine fishes of Alaska.

Jørgensen, B.B., 1990. A thiosulfate shunt in the sulfur cycle of marine sediments. Science 249, 152–154.

King, J., Brown, C., Sabet, H., 2003. A scenario-based holistic approach to environmental flow assessments for rivers. River Research and Applications 19 (5–6), 619–639.

Kocher, T.D., Stepien, C.A., 1997. Molecular Systematics of Fishes. Academic Press.

Kupferschmidt, K., 2017. Genomes rewrite cholera's global story. Science 358, 706–707.

Larson, G.L., Hoffman, R.L., McIntire, D.C., Buktenica, M.W., Girdner, S.F., 2007. Thermal, chemical, and optical properties of Crater Lake, Oregon. Hydrobiologia 574 (1), 69–84. https://doi.org/10.1007/s10750-006-0346-2.

Lekshmy, S., Jha, S.K., Sairam, R.K., 2015. Physiological and molecular mechanisms of flooding tolerance in plants. In: Elucidation of Abiotic Stress Signaling in Plants. Springer, pp. 227–242.

Levick, L.R., Goodrich, D.C., Hernandez, M., Fonseca, J., Semmens, D.J., Stromberg, J.C., Tluczek, M., Leidy, R.A., Scianni, M., Guertin, D.P., et al., 2008. The Ecological and Hydrological Significance of Ephemeral and Intermittent Streams in the Arid and Semi-Arid American Southwest. US Environmental Protection Agency, Office of Research and Development.

Lindeman, R.L., 1942. The trophic-dynamic aspect of ecology. Ecology 23, 399–417. http://www.jstor.org/stable/1930126.

Lytle, D.A., Merritt, D.M., Tonkin, J.D., Olden, J.D., Reynolds, L.V., 2017. Linking river flow regimes to riparian plant guilds: a community-wide modeling approach. Ecological Applications 27, 1338–1350.

Mailloux, R.J., Bériault, R., Lemire, J., Singh, R., Chénier, D.R., Hamel, R.D., Appanna, V.D., 2007. The tricarboxylic acid cycle, an ancient metabolic network with a novel twist. PLoS ONE 2, e690.

Martín-Durán, J.M., Wolff, G.H., Strausfeld, N.J., Hejnol, A., 2016. The larval nervous system of the penis worm *Priapulus caudatus* (ecdysozoa). Philosophical Transactions of the Royal Society B 371, 20150050.

McDonald, R.I., Weber, K., Padowski, J., Flörke, M., Schneider, C., Green, P.A., Gleeson, T., Eckman, S., Lehner, B., Balk, D., et al., 2014. Water on an urban planet: urbanization and the reach of urban water infrastructure. Global Environmental Change 27, 96–105.

McMenamin, M.A., 1982. Precambrian conical stromatolites from California and Sonora. Bulletin of the Southern California Paleontological Society 14.

Mooi, R.D., Gill, A.C., 2010. Phylogenies without synapomorphies—a crisis in fish systematics: time to show some character. Zootaxa 2450, 26–40.

Moss, B., 2010. Ecology of Freshwaters: A View for the Twenty-First Century, 4th edition. J. Wiley and Sons, Chichester, West Sussex, Hoboken, NJ.

Moyle, P.B., 2002. Inland Fish of California. The Regents of the University of California, Berkeley, CA.

Moyle, P.B., Thompson, L.C., Engilis Jr., A., Truan, M., Mosser, C.M., Purkey, D.R., Escobar, D.M., 2011. Innovative management options to prevent loss of ecosystem services provided by Chinook salmon in California: overcoming the effects of climate change. Final Report. RD 83301701-0. University of California, Davis.

Nelson, W.A., McCauley, E., Wrona, F.J., 2006. Ecology: mechanisms for consumer diversity (reply). Nature 439. https://doi.org/10.1038/nature04527, E2.

Noh, S.J., Lee, J.-H., Lee, S., Seo, D.-J., 2019. Retrospective dynamic inundation mapping of hurricane Harvey flooding in the Houston metropolitan area using high-resolution modeling and high-performance computing. Water 11, 597.

Nunoura, T., Chikaraishi, Y., Izaki, R., Suwa, T., Sato, T., Harada, T., Mori, K., Kato, Y., Miyazaki, M., Shimamura, S., Yanagawa, K., Shuto, A., Ohkouchi, N., Fujita, N., Takaki, Y., Atomi, H., Takai, K., 2018. A primordial and reversible tca cycle in a facultatively chemolithoautotrophic thermophile. Science 359, 559–563. https://science.sciencemag.org/content/359/6375/559.

O'leary, M.A., Bloch, J.I., Flynn, J.J., Gaudin, T.J., Giallombardo, A., Giannini, N.P., Goldberg, S.L., Kraatz, B.P., Luo, Z.-X., Meng, J., et al., 2013. The placental mammal ancestor and the post-k-pg radiation of placentals. Science 339, 662–667.

Pack, D.H., 1980. Precipitation chemistry patterns: a two-network data set. Science 208, 1143–1145. http://www.sciencemag.org/content/208/4448/1143.abstract.

Park, L.E., Downing, K.F., 2001. Paleoecology of an exceptionally preserved arthropod fauna from lake deposits of the miocene barstow formation, Southern California, U.S.A.. Palaios 16, 175–184. http://palaios.sepmonline.org/content/16/2/175.abstract.

Pascual, U., Balvanera, P., Díaz, S., Pataki, G., Roth, E., Stenseke, M., Watson, R.T., Dessane, E.B., Islar, M., Kelemen, E., et al., 2017. Valuing nature's contributions to people: the IPBES approach. Current Opinion in Environmental Sustainability 26, 7–16.

Pearson, L.A., Neilan, B.A., 2008. The molecular genetics of cyanobacterial toxicity as a basis for monitoring water quality and public health risk. Current Opinion in Biotechnology 19, 281–288. URL <Go to ISI>://000257457800014.

Persson, L., 1999. Trophic cascades: abiding heterogeneity and the trophic level concept at the end of the road. Oikos 85, 385–397. http://www.jstor.org/stable/3546688.

Persson, L., Byström, P., Wahlström, E., Westman, E., 2004. Trophic dynamics in a whole lake experiment: size-structured interactions and recruitment variation. Oikos 106, 263–274. https://doi.org/10.1111/j.0030-1299.2004.12767.x.

Pfeiffer, J.M., Breinholt, J.W., Page, L.M., 2019. Unioverse: a phylogenomic resource for reconstructing the evolution of freshwater mussels (Bivalvia, Unionoida). Molecular Phylogenetics and Evolution 137, 114–126. https://doi.org/10.1016/j.ympev.2019.02.016.

Regier, J.C., Shultz, J.W., Zwick, A., Hussey, A., Ball, B., Wetzer, R., Martin, J.W., Cunningham, C.W., 2010. Arthropod relationships revealed by phylogenomic analysis of nuclear protein-coding sequences. Nature 463, 1079–1083.

Richet, P., Bottinga, Y., Javoy, M., 1977. A review of hydrogen, carbon, nitrogen, oxygen, sulphur, and chlorine stable isotope fractionation among gaseous molecules. Annual Review of Earth and Planetary Sciences 5 (1), 65–110.

Rincón-Tomás, B., Khonsari, B., Mühlen, D., Wickbold, C., Schäfer, N., Hause-Reitner, D., Hoppert, M., Reitner, J., 2016. Manganese carbonates as possible biogenic relics in archean settings. International Journal of Astrobiology 15, 219–229.

Sallan, L., Friedman, M., Sansom, R.S., Bird, C.M., Sansom, I.J., 2018. The nearshore cradle of early vertebrate diversification. Science 362, 460–464.

Sayers, P.B., Yuanyuan, L., Moncrieff, C., Jianqiang, L., Tickner, D., Gang, L., Speed, R., 2017. Strategic drought risk management: eight 'golden rules' to guide a sound approach. International Journal of River Basin Management 15, 239–255.

Schierwater, B., Eitel, M., Jakob, W., Osigus, H.-J., Hadrys, H., Dellaporta, S.L., 2009, Kolokotronis, S.-O., DeSalle, R. Concatenated analysis sheds light on early metazoan evolution and fuels a modern "urmetazoon" hypothesis. PLOS Biology 7 (1), 1–9. https://doi.org/10.1371/journal.pbio.1000020.

Schindler, D.W., 1978. Factors regulating phytoplankton production and standing crop in the world's freshwaters. Limnology and Oceanography 23, 478–486. http://www.jstor.org/stable/2835456.

Schlesinger, W.H., 1997. Biogeochemistry: An Analysis of Global Change, second edition. Academic Press, San Diego, CA.

Schultheß, R., Wilke, T., Jørgensen, A., Albrecht, C., 2010. The birth of an endemic species flock: demographic history of the bellamya group (Gastropoda, Viviparidae) in Lake Malawi. Biological Journal of the Linnean Society 102, 130–143.

Schulze, S., Westhoff, P., Gowik, U., 2016. Glycine decarboxylase in C3, C4 and C3–C4 intermediate species. In: SI: 31: Physiology and Metabolism 2016. Current Opinion in Plant Biology 31, 29–35. https://doi.org/10.1016/j.pbi.2016.03.011.

Slobodkin, L.B., Bossert, P.E., 2010. Chapter 5 – Cnidaria. In: Thorp, J.H., Covich, A.P. (Eds.), Ecology and Classification of North American Freshwater Invertebrates, third edition. Academic Press, San Diego. ISBN 978-0-12-374855-3, pp. 125–142.

Sodemann, H., 2006. Tropospheric Transport of Water Vapour: Lagrangian and Eulerian Perspectives. ETH Zurich.

Song, Y., Liew, J.H., Sim, D.Z., Mowe, M.A., Mitrovic, S.M., Tan, H.T., Yeo, D.C., 2019. Effects of macrophytes on lake-water quality across latitudes: a meta-analysis. Oikos 128, 468–481.

Striker, G.G., 2012. Flooding stress on plants: anatomical, morphological and physiological responses. In: Botany, Volume 1. InTech, pp. 3–28.

Sumner, D.Y., Hawes, I., Mackey, T.J., Jungblut, A.D., Doran, P.T., 2015. Antarctic microbial mats: a modern analog for archean lacustrine oxygen oases. Geology 43, 887–890.

Timofeeva, Y.O., Karabtsov, A.A., Semal, V.A., Burdukovskii, M.L., Bondarchuk, N., 2014. Iron–manganese nodules in udepts: the dependence of the accumulation of trace elements on nodule size. Soil Science Society of America Journal 78, 767–778.

Torsvik, T.H., Cocks, L.R.M., 2016. Earth History and Palaeogeography. Cambridge University Press.

U.S. Geological Survey, Dieter, C.A., Maupin, M.A., Caldwell, R.R., Harris, M.A., Ivahnenko, T.I., Lovelace, J.K., Barber, N.L., Linsey, K.S., 2018. Estimated use of water in the United States in 2015. Circular. pp. 76. http://pubs.er.usgs.gov/publication/cir1441.

van der Valk, A.G., 1981. Succession in wetlands: a Gleasonian approach. Ecology 62, 688–696. http://www.jstor.org/stable/1937737.

Vanni, M.J., Layne, C.D., 1997. Nutrient recycling and herbivory as mechanisms in the "top-down" effect of fish on algae in lakes. Ecology 78, 21–40. http://www.jstor.org/stable/2265976.

Vidal, T., Irwin, B.J., Madenjian, C.P., Wenger, S.J., 2019. Age truncation of alewife in Lake Michigan. Journal of Great Lakes Research 45 (5), 958–968. https://doi.org/10.1016/j.jglr.2019.06.006.

Vilhena, D.A., Antonelli, A., 2015. A network approach for identifying and delimiting biogeographical regions. Nature Communications 6.

Vogel, S., 1994. Life in Moving Fluids: The Physical Biology of Flow, 2nd edition. Princeton University Press, Princeton, N.J. http://www.loc.gov/catdir/description/prin031/93046149.html.

Voosen, P., 2018. Sticky glaciers slowed tempo of ice ages. Science (New York, NY) 361, 739.

Wain, D.J., Rehmann, C.R., 2010. Transport by an intrusion generated by boundary mixing in a lake. Water Resources Research 46.

Walter, R.C., Merritts, D.J., 2008. Natural streams and the legacy of water-powered mills. Science 319, 299–304.

Wantzen, K.M., Junk, W.J., Rothhaupt, K.-O., 2008. An extension of the floodpulse concept (FPC) for lakes. Hydrobiologia 613 (1), 151–170. https://doi.org/10.1007/s10750-008-9480-3.

Wedemeyer, G.A., Saunders, R.L., Clarke, W.C., 1980. Environmental Factors Affecting Smoltification and Early Marine Survival of Anadromous Salmonids. Department of Fisheries and Oceans, Biological Station.

Wetzel, R.G., 2001. Limnology: Lake and River Ecosystems, 3rd edition. Academic Press, San Diego.

Wisheu, I.C., Keddy, P.A., 1992. Competition and centrifugal organization of plant communities: theory and tests. Journal of Vegetation Science 3, 147–156. http://www.jstor.org/stable/3235675.

World Health Organization, et al., 2015. World Malaria Report 2014. WHO, Geneva. Fecha de consulta 23, 247.

# Glossary

**Aerobic** Oxic, with oxygen. 8

**Allochthonous** Carbon fixed from outside the water column, e.g. riparian vegetation leaves. 171, 209, 348, 431, 432

**Alluvial** Made up of or found in the materials that are left by the water of rivers, floods, etc. 14

**Ambient** Relating to the immediate surroundings of something. 331

**Amictic** lakes are "perennially sealed off by ice from most of the annual seasonal variations in temperature." Amictic lakes exhibit inverse cold water stratification whereby water temperature increases with depth below the ice surface 0°C (less-dense) up to a theoretical maximum of 4°C (at which the density of water is highest). 248

**Amniote** A group of tetrapods (four-limbed animals with backbones or spinal columns) that have an egg equipped with an amnios, an adaptation to lay eggs on land rather than in water as anamniotes do. They include synapsids (mammals along with their extinct kin) and sauropsids (reptiles and birds), as well as their fossil ancestors. 17

**Anadromous** Species that migrate to the ocean to become sexually mature and return to inland waters to reproduce. 157, 210, 260, 299, 427

**Anaerobic** Anoxic, without oxygen. 8

**Aphotic zone** The portion of a lake or ocean where there is little or no sunlight. 246

**Archea** A domain that constitutes single-celled organisms, these microorganism are prokaryotes and have no cell nucleus. 8

**Autochthonous** Carbon fixed within the water column, e.g. submerge vegetation or algae. 209, 240, 432

**Bathymetry** Study of underwater depth of lake or ocean floors. 246

**Benthic** The ecological region at the lowest level of a body of water such as an ocean or a lake, including the sediment surface and some sub-surface layers. Organisms living in this zone are called benthos, e.g. the benthic invertebrate community, including crustaceans and polychaetes. 189

**Bilaterians** are animals with bilateral symmetry, i.e. they have a head ("anterior") and a tail ("posterior") as well as a back ("dorsal") and a belly ("ventral"); therefore they also have a left side and a right side. 14

**Calvin Cycle** The cycle spends ATP as an energy source and consumes NADPH2 as reducing power for adding high energy electrons to make the sugar. There are three phases of the cycle. 170

**Conservative behavior** Conservative behavior indicates that the concentration of a constituent or absolute magnitude of a property varies only due to mixing processes. 209, 219

**Cyanobacteria** A phylum of bacteria that obtain their energy through photosynthesis and are the only photosynthetic prokaryotes able to produce oxygen. 19, 131, 372

**Dark reaction** is a euphemism for a complex set of reactions. 170

**DDD** Dichlorodiphenyldichloroethane was synthesized in 1874 by the Austrian chemist Othmar Zeidler, who did not have an immediate application the compound. It is colorless and crystalline; it is closely related chemically and is similar in properties to DDT, but it is considered to be less toxic to animals than DDT. The molecular formula for DDD is $(ClC_6H_4)_2CHCHCl_2$ or $C_{14}H_{10}Cl_4$. 316

**DIC** Dissolved inorganic carbon. 217

**Dimictic** Lakes that mix twice a year. 248

**DOC** Dissolved organic carbon. Usually it's a measure of organic matter that passes through a 5 μm or 0.45 μm glass fiber filter, and then digested with a carbon analyzer. 203, 205

**DOM** Dissolved organic matter. Usually it's a measure of organic matter that passes through a 5 μm or 0.45 μm glass fiber filter, and then combusted in a muffle furnace. 203

**Ecdysozoa** A group of protostome animals,supported by morphological characters and includes all animals that grow by ecdysis, moulting their exoskeleton. 138

**Ectotherm** Animal that regulates their body temperature depending on external sources. 331

**Emergent** Aquatic plant with leaves and flowers that appear above the water surface. 189

**Endocrine disruptors** Substances that may interfere with the function of hormones in the body. 326

**Endolithic** Growing within cavities of rock. 189

**Endopelic** Growing within mud (sediment). 189

**Endorheic** A closed drainage basin that retains water and allows no outflow to other external bodies of water, such as rivers or oceans, but converges instead into lakes or swamps, permanent or seasonal, that equilibrate through evaporation. xl, 41, 47, 179, 240, 393

**Endosammic** Growing within sand. 189

**Epigean** Generally used for animals that neither burrow nor swim nor fly; it also indicates that the germination of a plant takes place above ground. 189

**Epilimnion** Upper portion of a lake. 206–208, 219, 247

**Epilithic** Growing attached to rock surfaces. 189

**Epipelic** Growing on mud (sediment). 189

**Epiphytic** Growing attached to other plants. 189

**Epipsammic** Growing on sand. 189

**Epizoic** Growing attached to animals. 189

**Eukaroyote** Change in populations of living organisms on Planet Earth through time; "descent with modification" (Darwin); "species of organisms originate as modified descendants of other species" (Hurry 1993). 8

**Euphotic zone** The layer of water in a lake or ocean closer to the surface that receives enough light for photosynthesis to occur. 201, 207

**Euryhaline** Wide salinity range, able to adapt to a wide range of salinities. 15

**Eutrophic** An ecosystem response to the addition of artificial or natural substances, such as nitrates and phosphates, through fertilizers or sewage, to an aquatic system. One example is the "bloom" or great increase of phytoplankton in a water body as a response to increased levels of nutrients. 248

**Exorheic** Basins whose surface waters can drain to the ocean. 41, 47, 180, 240, 393

**Facies** A body of rock with specified characteristics. Ideally, a facies is a distinctive rock unit that forms under certain conditions of sedimentation, reflecting a particular process or environment. 15

**Fecundity** In demography and biology, fecundity is the actual reproductive rate of an organism or population, measured by the number of gametes (eggs), seed set, or asexual propagules. 158

**Floc** Floc is a small, loosely aggregated mass of flocculent material suspended in or precipitated from a liquid. It consists of finely divided suspended particles in a larger, usually gelatinous mass, the result of physical attraction or adhesion to a coagulant compound. 323

**Flood stage** The level at which a body of water's surface has risen to a sufficient level to cause sufficient inundation of areas that are not normally covered by water. 287

**Fluvial** Fluvial is a term used in geography and Earth science to refer to the processes associated with rivers and streams and the deposits and landforms created by them. When the stream or rivers are associated with glaciers, ice sheets, or ice caps, the term glaciofluvial or fluvioglacial is used. 14

**Fumarole** An opening in a planet's crust which emits steam and gases such as carbon dioxide, sulfur dioxide, hydrogen chloride, and hydrogen sulfide. The steam forms when superheated water condenses as its pressure drops when it emerges from the ground. 215

**Fusiform** A body shape common to many aquatic animals, characterized by being tapered at both the head and the tail. Or in botany, a cell or other plant structure is spindle-like and tapers at both ends. 118

**Gill rakers** Plankton straining structures in the pharynx of fishes. 22

**Glochidia** A microscopic larval stage of some freshwater mussels, aquatic bivalve mollusks in the families Unionidae and Margaritiferidae. 114, 115

**Gondwana** A supercontinent that existed from the Neoproterozoic until the Jurassic. 18, 20

**Hetertrophic** An organism that cannot produce its own food, relying instead on the intake of nutrition from other sources of organic carbon, mainly plant or animal matter. 248

**Hox** Genes, from an abbreviation of homeobox, are a group of related genes that control the body plan of the embryo along the anterior-posterior (head-tail) axis. After the embryonic segments have formed, the Hox proteins determine the type of segment structures (e.g. legs, antennae, and wings in fruit flies or the different vertebrate ribs in humans) that will form on a given segment. Hox proteins thus confer segmental identity, but do not form the actual segments themselves. 13

**Hypertonic** refer a solution with higher osmotic pressure than another solution. 211

**Hypolimnion** Lower portion of a stratified lake. 206–208, 219, 247

**Hypoosmotic** Solution with a lesser concentration of solute (fresh water compared to salt water). If a cell from a fresh water fish is placed into a beaker of salt water, the cell is said to be hypoosmotic to the water. Isosmotic: solutions of equal solute concentrations are said to be isosmotic. 211

**Hyporheic** Region beneath and lateral to a stream bed, where there is mixing of shallow groundwater and surface water. 191, 331

**Hyporheic flow** Region of water exchange beneath and lateral to a stream bed. 300

**Hyporheic zone** Region beneath and lateral to a stream bed, where there is mixing of shallow groundwater and surface water. 300

**Inertia** The resistance of any physical object to any change in its state of motion (including a change in direction). In other words, it is the tendency of objects to keep moving in a straight line at constant linear velocity. 195

**Isopicnals** Lines of constant density, usually in the context of lakes and oceans. 189

**Lacustrine** "Of a lake" or "relating to a lake". 15

**Laminar** A fluid flowing in parallel layers with no disruption between the layers. 195

**Laurasia** The more northern of two supercontinents that formed part of the Pangaea supercontinent around 335 to 175 million years ago. The other being Gondwana. 18

**Littoral** The part of a sea, lake, or river that is close to the shore. 246

**Meromictic** lakes have layers of water that do not intermix. In meromictic lakes, however, the layers of the lake water can remain unmixed for years, decades, or centuries. Meromictic lakes can usually be divided into three sections or layers. The bottom layer is known as the monimolimnion; the waters in this portion of the lake circulate little, and are generally hypoxic and saltier than the rest of the lake. The top layer is called the mixolimnion, and essentially behaves like a holomictic lake. The area in between is referred to as the chemocline. 248

**Mesozoic** Dinosaurs, large marine reptiles (ichthyosaurs, mosasaurs), pterosaurs. 72

**Metalimnion** A thin but distinct layer in a large body of fluid (e.g.water, such as an ocean or lake, or air, such as an atmosphere) in which temperature changes more rapidly with depth than it does in the layers above or below. 206, 207, 247

**Monomictic** Lakes that mix once a year. 248

**Monophyletic** group is a taxon (group of organisms) which forms a clade, meaning that it consists of an ancestral species and all its descendants. The term is synonymous with the uncommon term holophyly. Monophyletic groups are typically characterized by shared derived characteristics (synapomorphies). 71

**Neuston** The region on or just below the surface of a body of water. 106

**Neustonic** The term neuston refers to the assemblage of organisms associated with the surface film of lakes, oceans, and slow-moving portions of streams. 189

**Non-conservative behavior** Non-conservative behavior indicates that the concentration of a constituent may vary as a result of biological or chemical process. 209, 219

**Oligotrophic** Environment that offers very low levels of nutrients. 248

**Ozone** Oxygen molecule of three atoms. Ozone is formed from dioxygen by the action of ultraviolet light and also atmospheric electrical discharges, and is present in low concentrations throughout the Earth's atmosphere. In total, ozone makes up only 0.6 ppm of the atmosphere. 8

**Paraphyletic** Groups that include all the descendants of a common ancestor are said to be monophyletic. A paraphyletic group is a monophyletic group from which one or more subsidiary clades (monophyletic groups) is excluded to form a separate group. 82

**Pelagic** Any water in a sea or lake that is neither close to the bottom nor near the shore can be said to be in the pelagic zone. 246

**Periphytic** Inhabit the surface of submerged plants and other underwater objects. 189

**Periphyton** Organisms that inhabit the surface of submerged plants and other underwater objects. 266

**Photic zone** The uppermost layer of water in a lake or ocean that is exposed to intense sunlight. 246

**Photosynthetically active radiation** or PAR designates the spectral range (wave band) of solar radiation from 400 to 700 nanometers that photosynthetic organisms are able to use in the process of photosynthesis. This spectral region corresponds more or less with the range of light visible to the human eye. 204

**Plesiomorphic** An ancestral character shared by two or more taxa – but also with other taxa linked earlier in the clade. 86

**Polymictic** Lakes that mix many times a year. 248

**Polyphyletic** group is one characterized by one or more homoplasies: Character states which have converged or reverted so as to appear to be the same but which have not been inherited from common ancestors. 71, 72

**Proxy** Preserved physical characteristics of the environment that can stand in for direct measurements, is commonly used to reconstruct past climate conditions. 245

**Rheotaxis** A form of taxis seen in many aquatic organisms, e.g.fish, whereby they will (generally) turn to face into an oncoming current which is referred to as positive rheotaxis. In a flowing stream, this behavior allows fish to hold their position instead of being swept downstream by the current. 260

**RuBisCO** Ribulose-1,5-bisphosphate carboxylase/oxygenase is an enzyme involved in the first major step of carbon fixation, a process by which atmospheric $CO_2$ is converted by autotrophs to energy-rich molecules in the Calvin Cycle. 170

**Schistosomiasis** Also known as snail fever and bilharzia, is a disease caused by parasitic flatworms called schistosomes in the genus Schistosoma is a genus of trematodes as Platyhelmithes. The urinary tract or the intestines may be infected. Symptoms include abdominal pain, diarrhea, bloody stool, or blood in the urine. Those who have been infected for a long time may experience liver damage, kidney failure, infertility, or bladder cancer. In children, it may cause poor growth and learning difficulty. 48, 115

**Secchi depth** Created in 1865 by Angelo Secchi, is a circular disk used to measure water transparency in oceans and lakes. The disk is mounted on a pole or line, and lowered slowly down in the water. The depth at which the pattern on the disk is no longer visible is taken as a measure of the transparency of the water. This measure is known as the Secchi depth and is related to water turbidity. 202, 203

**Secchi disk** Created in 1865 by Angelo Secchi, is a circular disk used to measure water transparency in oceans and lakes. The disc is mounted on a pole or line, and lowered slowly down in the water. The depth at which the pattern on the disk is no longer visible is taken as a measure of the transparency of the water. This measure is known as the Secchi depth and is related to water turbidity. 202

**Seston** The organisms (bioseston) and non-living matter (abioseston or tripton) swimming or floating in a water body. Bioseston can be often regarded as plankton, although it includes nekton as well. Abioseston comprises detritus as well. 199, 200

**Stygobitic** Inhabit groundwater systems or aquifers, such as caves, fissures, and cavities. 189

**Stygophilic** Inhabit both surface and subterranean aquatic environments. 189

**Thermocline** Temperature decreases rapidly from the mixed upper layer of a lake (epilimnion) to a colder deep water or hypolimnion. There is very little mixing between these two layers, thus forming a type of interface between the two layers. 206, 207, 247

**Thermotaxis** is a behavior in which an organism directs its locomotion up or down a gradient of temperature. 363

**Troglobites** An animal species, or population of a species, strictly bound to underground habitats, such as caves. 255

**Troglobitic** Small creatures that have adapted to a permanent life in a cave. 102

**Turbulent** flow includes complex flow non-parallel flow currents to the main flow direction that include vortices. 195

**Viscosity** A fluid is a measure of its resistance to gradual deformation by shear stress or tensile stress. For liquids, it corresponds to the informal notion of "thickness". For example, honey has a higher viscosity than water. 195

# List of Taxa

**Centroptilum elsa** a species of ephemeroptera. 153
**Chaoborus astictopus** Clear Lake knat. 316
**Chasmistes brevirostris** shortnose sucker. 457
**Cichla ocellaris** butterfly peacock bass, but may also refer to other related species; a very large species of cichlid from South America. 68
**Coccyzus americanus** Yellow-billed Cuckoo. 125
**Compsopogon coeruleus** Compsopogon coeruleus. xix, 408, 409
**copepods** copepods, crestaeau. 171
**Cottus confusus** shorthead sculpin. 432
**Craspedacusta sowerbyi** Craspedacusta sowerbii is a freshwater jellyfish in the phylum Cnidaria. 91
**Cryptosporidium parvum** one of several species that cause cryptosporidiosis, the SAR supergroup parasitic infects the mammalian intestinal tract. 84
**Culex quinquefasciatus** southern house mosquito. 318
**Culex tarsalis** Western Encephalitis Mosquito. 318
**Cyclops bicolor** Cryptocyclops bicolor is the only species of this genus in Norway. It is smaller than any other cyclopoids. 100
**Cyclops bicuspidatus** a planktonic species of copepod found throughout the world, except Australia. 100
**Cyclops leuckarti** Cyclops leuckarti Claus has a world-wide distribution and is found both in the plankton of lakes and in small ponds. 100
**Cyprinodon salinus salinus** Death Valley pupfish. xl, xlii
**Cyprinus carpio** Common Carp. 333

**Daphnia** Daphnia genus, crustacean. 101, 171, 210, 268
**Daphnia dentifera** Daphnia dentifera. 159
**Deltistes luxatus** Lost River sucker. 457
**Dorosoma petenense** threadfin shad. 392
**Dreissena polymorpha** zebra mussel. 72, 114

**Eichhornia crassipes** water hyacinth. 262, 409, 410
**Electrophorus electricus** Electric eel. 152
**Empidonax traillii** Willow Flycatcher. 125
**Entosphenus tridentatus** Pacific Lamprey. 156
**Epeorus** a genus belonging to the order Ephemeroptera. 150
**Erymystax dissimilis** streamlined chub. 73
**Erymystax insignis** blotched chub. 73
**Eudiaptomus gracilis** a genus of crustacean in the family Diaptomidae. xiv, 137
**Euglena gracilis** a freshwater species of single-celled alga (Excavata) in the genus Euglena. 86

**Fusarium euwallacea** fusarium wilt. 410

**Gambusia affinis** western mosquito fish. 320, 321
**Gammarus minus** a species of Amphipoda in the family Gammaridae. 171, 173
**Gasterosteus aculeatus** threespine stickleback. 258, 260, 265

**Giardia intestinalis** also known as Giardia duodenalis and Giardia lamblia, is a flagellated parasitic microorganism, that colonizes and reproduces in the small intestine, causing giardiasis. 87
**Giardia lamblia** also known as Giardia duodenalis and Giardia intestinalis, is a flagellated parasitic microorganism, that colonizes and reproduces in the small intestine, causing giardiasis. 36, 87, 325
**Gila elegansus** bonytail chub. 194

**Hexagenia limnbata** Okanagan Mayfly. 138, 139
**Hexaprotodon Liberiensis** pigmy hippopotamus. 406
**Hippopotamus amphibius** Common Hippopotamus. 409
**Hypomesus transpacificus** delta smelt. 407

**Juga Silicula** Shasta Juga, a medium-sized, aquatic snail. 154

**Lepomis macrochirus** bluegill sunfish. 263
**Leuctra** genus in the Leuctridae are a family of stoneflies, commonly known as rolled-winged stoneflies and needleflies. 171
**Limnephilus flavicornis** a case building caddisfly that uses silken thread, spun by glands placed on the labium to construct portable shelters. 109
**Lipotes vexillifer** Yangtze River Dolphin. 405

**Metschnikowia bicuspidata** single-celled fungal parasites of crustaceans, including Daphnia and fish. 159
**Microcystis** a cyanobaterium genus. 157
**Microcystis aeruginosa** Microcystis aeruginosa. 149
**Micropterus salmoides** largemouth bass. 274, 360, 392
**Mysis diluviana** opossum shrimp. 407, 408

**New Zealand mudsnail** *Potamopyrgus antipodarum.* 158
**Nitrosomonas** a genus of Gram-negative rod-shaped chemoautotrophic bacteria. This organism oxidizes ammonia into nitrite as a metabolic process, known as nitritation (a step of Nitrification). 372

**Oncorhynchus clarkii seleniris** Paiute cutthroat trout. xl
**Oncorhynchus kisutch** coho salmon. 432
**Oncorhynchus mykiss** steelhead trout. 71, 413
**Oncorhynchus nerka** Kokanee salmon. 407
**Oncorhynchus tshawytscha** Chinook salmon. 167, 309, 455
**Oryza sativa** Rice. 306
**Oscillatoria rubescens** Oscillatoria rubescens. 341, 344

**Paracapnia angulata** angulate snowfly, is a species of small winter stonefly in the family Capniidae. 171
**Peltoperla arcuata** Appalachian Roachfly. 171
**Perca fluviatilis** European perch. 278

**Percina tanasi** snail darter. 404
**Phaeodactylum tricornuntam** Phaeodactylum tricornuntam. 77
**Phoxinus eos** redbelly dace. 274
**Phoxinus neogaeus** finescale dace. 274
**Picea mariana** black spruce. 164
**Planktothrix** a genus of filamentous cyanobacteria. 157
**Plasmodium** a genus of unicellular eukaryotes that are obligate parasites of vertebrates and insects. 84, 320, 322
**Plasmodium falciparumspecies** a unicellular protozoan parasite of humans, and the deadliest species of Plasmodium that causes malaria in humans. 320
**Platyhelminthes** flat worms. 115, 158
**Podura aquatica** water springtail. 106
**Poecilia reticulata** rainbow fish. 160
**Populus fremontii** Fremont cottonwood. 411
**Potamopyrgus antipodarum** New Zealand mudsnail. 112, 159, 171, 412
**Psephurus gladius** Chinese Paddlefish. 405
**Pseudacris regilla** Pacific treefrog. 413
**Pseudemys alabamensis** Alabama Red-Bellied Turtle. 122
**Pycnopsyche** genus of northern caddisflies in the family Limnephilidae. 171

**Quercus agrifolia** coast live oak. 411

**Rallus obsoletus obsoletus** California Ridgway's Rail. 415
**Rana muscosa** yellow-legged frog. 413
**Rhizobia** diazotrophic bacteria that fix nitrogen after becoming established inside the root nodules of legumes or as free-living bacteria. 372
**Rutilus rutilus** Common roach. 278

**Salix** Willow spp.. 150
**Salix laevigata** red willow. 411
**Salmo** Salmon genus. 157
**Salmo salar** Atlantic Salmon. 414
**Salvelinus fontinalis** Brook Trout. 338
**Spartina alterniflora** smooth cordgrass. 254, 415, 433
**Spartina foliosa** rough cordgrass. 415
**Speleonectes atlantida** discovered in August 2009 in the Canary Islands, this eyeless crustacean species is in the Nectiopoda. 104
**Sphagnuma** sphagnum moss or peat moss. 252
**Spongilla lacustris** a species of sponge of the freshwater sponge family Spongillidae that lives in fresh water lakes. xiii, 91, 92

**Tachycineta albilinea** mangrove swallow. 263
**Tamarix** Salt cedar. 409
**Theliderma intermedia** Cumberland monkeyface pearly mussel. 73, 407
**Tipula abdominalis** giant crane fly. 171

**Toxoplasma gondii** Toxoplasma gondii. 84
**Typha** cattail. 144

**Umbra limit** central mudminnows. 274

**Vibrio cholerae** a Gram-negative, comma-shaped bacterium found in brackish or saltwater and attach themselves easily to the chitin-containing shells of crabs, shrimps, and other shellfish. Some strains cause the disease cholera, which can be derived from the consumption of undercooked or raw marine life species. When ingested, V. cholerae can cause diarrhea and vomiting in a host within several hours to 2–3 days of ingestion. 318
**Vireo belli** Least Bells Vireo. 411

# Index

**T**

Printed in the United States
By Bookmasters